FORBIDDEN ARCHEOLOGY'S

IMPACT

How A Controversial New Book Shocked The Scientific Community And Became An Underground Classic

FORBIDDEN ARCHEOLOGY'S

IMPACT

MICHAEL A. CREMO

BHAKTIVEDANTA BOOK PUBLISHING, INC.
Los Angeles • Sydney • Stockholm • Bombay

Readers interested in the subject matter of this book are invited to correspond with the author at:

Bhaktivedanta Book Publishing, Inc.
3764 Watseka Avenue
Los Angeles, CA 90034

First edition 1998

Design by Stewart Cannon, Logo Loco Graphics
Cover art by Charles Payne

Published by Bhaktivedanta Book Publishing, Inc. Science Books Division (BBT Science Books) for the Bhaktivedanta Institute.

Distributed by Torchlight Publishing, Inc.
3046 Oakhurst Avenue, Los Angeles, CA 90034
World Wide Web: www.torchlight.com

Library of Congress Cataloging-in-Publishing Data

Cremo, Michael A., 1948-

Forbidden archeology's impact : how a controversial new book shocked the scientific community and became an underground classic / by Michael A. Cremo

p. cm.

Includes bibliographical references and index

ISBN 0-89213-283-3 (hc)

1. Cremo, Michael A., 1948- Forbidden archeology.

2. Anthropology, Prehistoric. 3. Human evolution 4. Cremo, Michael A., 1948---Correspondence. I. Title

GN741.C745 1998 97-42314

569.9- -dc21 CIP

Dedicated to

His Divine Grace
A.C. Bhaktivedanta Swami Prabhupada

oṁ ajñāna-timirāndhasya jñānāñjana-śalākayā
cakṣur unmīlitaṁ yena tasmai śrī-gurave namaḥ

Contents

Foreword

by Colin Wilson

It was in the autumn of 1994 that the paranormal researcher Alexander Imich recommended to me a book called *Forbidden Archeology*. It sounded to me just the kind of thing I had been hoping to find. At the time, I was researching a book arguing that the Sphinx might be thousands of years older than historians believe. This was an argument that had first been put forward in the 1930s by a maverick Egyptologist named René Schwaller de Lubicz, who had noted that the Sphinx showed signs of water weathering, rather than erosion by wind-blowing sand. And since there has obviously been very little rain in the Sahara for thousands of years, this would seem to argue that the Sphinx was built long before the reign of the Pharaoh Cheops, about 2,500 BC—the date usually accepted by Egyptologists. This in turn suggested that there may have been a technically accomplished civilization at a time when, according to the history books, men were only just beginning to build the first cities of mud brick So a book suggesting that man might be far older than is usually believed sounded exactly what I was looking for. I lost no time in ordering it from my American bookshop.

For some reason, it took a long time to arrive—perhaps they sent it by surface mail instead of airmail—and I forgot all about it. The following March, 1995, I happened to be in a delightful little town called Marion, in Massachusetts, where I was one of a panel of speakers on the subject of the evolution of human consciousness, organized by the Marion Foundation. This proved to have an excellent library, and the first volume I saw was the huge work called *Forbidden Archeology* by Michael Cremo and Richard Thompson. And since I had a few hours before dinner, I lost no time in finding the librarian and asking if I could borrow it.

As soon as I settled down in an armchair with a glass of wine—in a house provided for us by Michael Baldwin, the conference organizer—I realized that this would be one of those happy pieces of serendipity that befall authors who are obsessively in search of material. (I often feel as if I have a staff of invisible helpers who drop books into my lap.) I had no idea what to expect—for all I knew, Cremo and Thompson might belong to the lunatic fringe. But the first few pages made me aware that this was not only a work of serious scholarship, but one whose implications were revolutionary.

Michael Cremo begins his introduction by mentioning the discovery by Mary Leakey of footprints at Laetoli, in Tanzania, which were indistinguishable from those of modern humans. But they were about 3.6 million years old.

Now I had been among the earliest readers of Robert Ardrey's *African Genesis* (1961), whose thesis is that our earliest human ancestor, then posited as *Australopithecus africanus* (southern ape man), appeared in Africa about two million years ago, and that he differed from other apes in that he had learned to kill with weapons—such as an antelope's foreleg bone, which makes a natural club. Ardrey goes on to argue that man is a descendant of this killer ape, and that explains why he is unable to resist making war on his neighbors.

Bob Ardrey subsequently became a close friend, and I dedicated a book to him. I accepted Bob's thesis about the origin of human violence, and made extensive use of it in a book called *A Criminal History of Mankind.* But if a 'human' footprint dating from 3.6 million years ago had appeared in Tanzania, then it certainly looked as if Ardrey must have been wrong. Or could the Laetoli footprint have been a freak one-off, due to some accident of stratification?

There was another problem. In 1974, Donald Johanson found the bone fragments of a hominid (human ancestor) which he called Lucy, at Hadar, in Ethiopia, and potassium-argon dating revealed that Lucy was about 3.5 million years old; so were remains of thirteen other hominids found in the following year. It was generally accepted that Johanson and his team had pushed back the antiquity of man another one and a half million years. But Lucy and her friends, designated members of the new species *Australopithecus afarensis,* were certainly not in the least like modern humans.

The very first fossil of *Australopithecus*, the Taung skull, had been discovered early in this century by Raymond Dart in South Africa. When anthropologist Sir Arthur Keith saw the Taung skull—upon which Dart based his belief that *Australopithecus* is our ancestor—he objected that it hardly seemed old enough. The Taung skull was then believed to be about a million years old, and Keith felt there was simply not enough time for *Australopithecus* to turn into our Cro-Magnon ancestor in a mere 900,000 years. But even a two million year dating raises questions. Is a mere two million years, or even 3.5 million years, really enough for man to evolve from an ape to a human being?

Cremo and Thompson did not think so. And they had some fascinating evidence to back it up.

It seemed that Richard Thompson, a founding member of the Bhaktivedanta Institute, was encouraged by its founder, Swami Prabhupada, to examine accounts of human origins, since Vedic literature holds that the human race is far, far older than modern science allows. In 1984, he began collecting material about human origins, and two years later asked Michael Cremo to take a look at it. Cremo was one of those children of the Sixties who had gone toward India on a spiritual quest, and had become a follower of Prabhupada in 1974, when he was 26 years old.

In looking through current anthropological textbooks, Cremo had been struck by how few reports of fossil evidence on human origins he could find from 1859, when *The Origin of Species* appeared, until the mid-1890s, when Java man was discovered. This inspired him to look into works of anthropology published in the nineteenth century, and to his surprise he found that a great deal of material had appeared during that time, only to vanish from the record in later years. Fortunately, there were enough footnotes to enable Cremo, with the help of researcher Stephen Bernath, to track down many of the original reports. And it became clear that they had been ignored in later years becaŭse they did not fit in with the emerging Darwinian orthodoxy. But it was not because their authors opposed Darwin for ideological reasons. In fact, most of them were Darwinists. Their reports simply presented evidence of human bones, or even artifacts, that seemed to date to periods when man was not supposed to be on earth.

For example, in the 1860s, Portugeuse geologist Carlos Ribeiro, head of the Geological Survey of Portugal, found flint tools in limestone beds of Tertiary origin (a period extending from 5 to 65 million years ago). At this time, the influential anthropologist Ernst Haeckel was suggesting that the "missing link" might be found in the Pliocene (5 million years ago), or even the late Miocene, just before that. But the layers in which Ribeiro found the flints were middle and early Miocene, dating back as much as twenty million years. He doubted the layers could really be so old. So in his geological maps of Portugal, Ribeiro showed the flint-bearing beds as belonging to the Pleistocene, which dates back a mere two million years. Criticised by the French geologist Edouard de Verneuil, and encouraged by worked flints found by the Abbe Bourgeois in Miocene beds in France, he took a deep breath and began to refer to the beds as Tertiary. And that is why Carlos Ribeiro and his flints disappeared from the academic record.

It's easy enough to explain, of course. Perhaps some Pleistocene flints had gotten into Miocene beds by some freak of nature, deposited there by the Tagus River, which flows through them. But as Cremo went on searching, such cases began to multiply to an extent where the repetition of this kind of explanation became absurd. There were just too many anomalies—enough to make up a 900-page book.

Now there are plenty of books around, written by born-again Christians, which argue that Darwin was entirely wrong, and that God created man in the Garden of Eden, just as the Bible says. But *Forbidden Archeology* was obviously in a completely different category. I could see, as I read its meticulously researched arguments, that the original reasons for the research were irrelevant. These reasons were, quite simply, that one of the historical commentaries on the Vedic hymns, the *Bhagavata Purana*, states that man has been around on earth for a vast period of time called the day of Brahma, which is composed of a thousand yuga cycles. Each yuga cycle last 12,000 "years of the gods." Since each year of the gods is 360 solar years long, one yuga cycle amounts to 4,320,000 years and a thousand of them yield a day of Brahma—4,320,000,000 years.

According to the cosmological calendars of ancient India, we are now about 2 billion years into the current day of Brahma. This is roughly the age assigned by paleontologists to the first fossil signs of life on earth—the blue-green algae and other single-celled creatures. But according to the *Bhagavata Purana*, there should also have been more advanced life forms present, including, perhaps, humans.

Some of the evidence reported in *Forbidden Archeology* does not far exceed the conventional limits for human antiquity. The Laetoli footprints, for example, are about four million years old. This date is not so incredible. Haeckel's nineteenth-century contemporaries would not have raised an eyebrow. It is such dates as the worked flint found in a Miocene bed, perhaps twenty million years old, that make us gasp.

Just (I suspect) to tease his readers with the more extreme possibilities suggested by the *Bhagavata Purana*, Cremo ends *Forbidden Archeology* with some really incredible dates: a nail found in Devonian sandstone, about 400 million years old, and a gold thread found in Carboniferous rock, about 320 million years old. A report that must have delighted Charles Fort speaks of a petrified shoe sole found in Triassic rock from Nevada, over 200 million years old.

Now, as readers of *Forbidden Archeology's Impact* will see, *Forbidden Archeology* caused a great deal of outrage. Jonathan Marks's review is typical: the book is dismissed as "fundamentalist" and as "Hinduoid creationist drivel." This is absurd, since the book contains no religious arguments of any sort, Hindu-oid or otherwise.

No, *Forbidden Archeology* is simply an extremely erudite and extremely amusing account of what might be called "the other side of the post-Darwinist story." And any fair-minded scientist would surely agree that Cremo and Thompson have presented a case that deserves an answer. We would like to know what is the scientific answer to the flints found in the Tagus River beds, and the engraved bone found by Frank Calvert in Miocene beds, or the human skeleton found in Pliocene strata at Castenedolo, Italy—not a humanoid skeleton like Lucy, but a normal one. Most absurd of all, what about the metallic grooved sphere (Cremo and Thompson include a photograph on p. 813 of *Forbidden Archeology*) found in South Africa in beds around 2.8 *billion* years old?

Erich von Däniken would, of course, have no problem with this—he would merely take it as evidence of a visit from ancient astronauts in the pre-Cambrian era. And, for all I know, that may be what it is. Cremo, perhaps with his tongue in his cheek, offers it simply as one more rather Fortean anomaly to demonstrate that the dates suggested by the *Bhagavata Purana* are not so extreme after all.

The nonspecialist reader will raise an obvious objection to all this. If man was around on earth—looking more or less like us—five million years ago or more, why have we not found more evidence of his presence? The answer, as Cremo points out, can be found in Richard Leakey's remark, in *People of the Lake*: "If someone went to the trouble of collecting together in one room all of the fossil remains discovered so far of our ancestors (and their biological relatives) . . . he would need only a couple of large trestle tables on which to spread them out." There must be a vast amount of evidence still buried beneath the earth. And some of that may indicate, like the Laetoli footprints, that man has been around for a very long time indeed.

And what relevance has all this to my own arguments in *From Atlantis to the Sphinx* (a terrible title chosen by my publisher, who felt that my own *Before the Sphinx* was too dull)? I have simply argued that civilization has probably been around at least ten thousand years longer than we assume. In *Maps of the Ancient Sea Kings*, Charles Hapgood, a professor of the history of science, took a careful look at old maps known as *portolans,* used by medieval mariners, and concluded that they had to have been based on originals dating back thousands of years—one shows the coast of Antarctica as it was before the ice, perhaps 5,000 BC. Hapgood concluded that there was a worldwide maritime civilization as long ago as 7,000 BC—or more than 3,000 years before "civilization" is supposed to have started in Sumeria. But since we now know that Jericho sprang up about 8,500 BC, this is not too incredible.

By "Atlantis" I mean what Schwaller de Lubicz meant by it—some unknown civilization of perhaps 12,000 years ago. The argument of *Forbidden Archeology* supports that argument by suggesting that an intelligent, upright creature called Man did not burst onto the world scene as recently as we assume. Cremo and Thompson have thrown down a perfectly reasonable challenge. If "respectable" scientists choose to ignore it, the rest of us will feel justified in concluding it is one more argument for the formidable case against scientific orthodoxy presented in *Forbidden Archeology.*

Colin Wilson
October 31, 1997
Gorran Haven, England

Introduction and Acknowledgments

My book *Forbidden Archeology: The Hidden History of the Human Race*, coauthored with Richard Thompson, inspired considerable controversy when it appeared in 1993. It quickly became what geologist Dr. Virginia Steen-McIntyre called "an underground classic."

After spending six reclusive years in researching and writing the book, I suddenly found myself, as the principal author, drawn into an international round of academic conferences, radio and television interviews, correspondence, and email exchanges. Interest in *Forbidden Archeology* spread all the way across the intellectual spectrum, ranging from professional archeologists and anthropologists to advocates of extraterrestrial contacts with earth.

The variety and intensity of the responses was surprising to me, indicating that the subject of the book—evidence for the extreme antiquity of the human race—was striking a sensitive spot at some deep level of the modern psyche. Some found the book threatening, others found it a welcome revelation.

My involvement with a spiritual tradition with roots in India also drew considerable attention. I did not explicitly evoke elements of my spiritual practice and belief in *Forbidden Archeology*. But my exposure to accounts of extreme human antiquity in the ancient Sanskrit writings of India did in fact, as I acknowledged in the Introduction to my book, inspire my research and writing.

In this book, I document the reactions to *Forbidden Archeology* from different knowledge communities. I have compiled in this book (1) journal articles by me and papers I have delivered on *Forbidden Archeology* at scientific conferences and other gatherings; (2) reviews of *Forbidden Archeology* printed in academic journals, with related correspondence; (3) reviews of *Forbidden Archeology,* and articles about it, printed in nonacademic publications; (4) selected correspondence with scientists, philosophers, and others about the book; (5) transcripts of selected radio and television interviews, along with selected internet exchanges about *Forbidden Archeology*.

I envision that this volume will provide source material for scholars in the fields of history of science, philosophy of science, sociology of scientific knowledge, archeology, anthropology, and religion. And I hope it will also be of interest to other readers, who have often asked me questions about reactions to *Forbidden Archeology*.

For their help in compiling this book and bringing it into print, I thank Lori Erbs, Alister Taylor, Greg Stein, Chris Glenn, Stewart Cannon, Roy Richard, Yvonne Collings and Charles Payne.

Michael A. Cremo
June 17, 1997
Los Angeles

1

Conference Papers and Journal Articles

1.1. Michael A. Cremo (1994) "Puranic Time and the Archeological Record." World Archeological Congress 3, New Delhi, India.

I presented this paper at the World Archaeological Congress 3, which was held December 4–11, 1994, in New Delhi, India. I gave it in the section on Cultural Concepts of Time, chaired by Dr. Tim Murray of Latrobe University in Australia and Dr. D. P. Agarwal of the Physical Research Laboratory in Amedhabad, India. It was originally published in a collection of papers circulated to members of the section in advance of the Congress. It is also scheduled to appear in a conference proceedings volume, to be published by Routledge. An edited version of this paper appears in Back to Godhead, *the magazine of the Hare Krishna movement (May/June 1995, pp. 36-40), and the original version appears in* ISKCON Communications Journal *(No. 4, July-December 1994, pp. 34-43). I read this same paper, with some minor changes, at the conference Revisiting Indus-Sarasvati Age & Ancient India, sponsored by the Greater Atlanta Vedic Temple Society in cooperation with the Hindu University of America and other organizations. The conference was held in Atlanta, Georgia, USA, October 4-6, 1996.*

Abstract

The time concept of modern archeology, and modern anthropology in general, resembles the general cosmological-historical time concept of Europe's Judeo-Christian culture. Differing from the cyclical cosmological-historical time concepts of the early Greeks in Europe, and the Indians and others in Asia, the Judeo-Christian cosmological-historical time concept is linear and progressive. Modern archeology also shares with Judeo-Christian theology the idea that humans appear after the other major species. The author subjectively positions himself within the Vaishnava

Hindu world view, and from this perspective offers a radical critique of modern generalizations about human origins and antiquity. Hindu historical literatures, particularly the *Puranas* and *Itihasas*, place human existence in the context of repeating time cycles called *yugas* and *kalpas,* lasting hundreds of millions of years. During this entire time, according to Puranic accounts, humans coexisted with creatures who in some ways resemble the earlier tool-making hominids of modern evolutionary accounts. If one were to take the Puranic record as objectively true, and also take into account the generally admitted imperfection and complexity of the archeological and anthropological record, one could make the following prediction. The strata of the earth, extending back hundreds of millions of years, should yield a bewildering mixture of hominid bones, some anatomically modern human and some not, as well as a similarly bewildering variety of artifacts, some displaying a high level of artistry and others not. Given the linear progressivist preconceptions of generations of archeologists and anthropologists, one could also predict that this mixture of bones and artifacts would be edited to conform to their deeply rooted linear-progressive time concepts. A careful study of the archeological record, and the history of archeology itself, broadly confirms these two predictions. Linear-progressivist time concepts thus pose a substantial barrier to truly objective evaluation of the archeological record and to rational theory-building in the area of human origins and antiquity.

Main Text

The practically employed time concept of the modern historical scientist, including the archeologist, strikingly resembles the traditional Judeo-Christian time concept. And it strikingly differs from that of the ancient Greeks and Indians.

This observation is, of course, an extreme generalization. In any culture, the common people may make use of various time concepts, linear and cyclical. And among the great thinkers of any given period, there may be many competing views of both cyclical and linear time. This was certainly true of the ancient Greeks. It can nevertheless be safely said that the cosmological concepts of several of the most prominent Greek thinkers involved a cyclic or episodic time similar to that found in the Puranic literatures of India. For example, we find in Hesiod's *Works and Days* a series of ages (gold, silver, bronze, heroic and iron) similar to the Indian *yugas.* In both systems, the quality of human life gets progressively worse with each passing age. In *On Nature* (fragment 17) Empedocles speaks of cosmic time cycles. In Plato's dialogues there are descriptions of revolving time (*Timaeus* 38a) and recurring catastrophes that destroy or nearly destroy human civilization (*Politicus* 268d, *ff*). Aristotle said in many places in his works that the arts and sciences had been discovered many times in the past (*Metaphysics* 1074b 10, *Politics* 1329b 25). In the teachings of Pythagoras, Plato and Empedocles regarding transmigration of souls, this cyclical pattern is extended to individual psychophysical existence.

When Judeo-Christian civilization arose in Europe, another kind of time became prominent. This time has been characterized as linear and vectorial. Broadly speaking, this time concept involves a unique act of cosmic creation, a unique appearance of the human kind and a unique history of salvation, culminating in a unique denouement in the form of a last judgment. The drama occurs only once; individually, human life mirrors this process. With some notable exceptions, orthodox Christian theologians did not accept transmigration of the soul.

Modern historical sciences share the basic Judeo-Christian assumptions about time. The universe we inhabit is a unique occurrence. Humans have arisen once on this planet. The history of our ancestors is regarded as a unique though unpredestined evolutionary pathway. The future pathway of our species is also unique. Although this pathway is officially unpredictable, the myths of science project a possible overcoming of death by biomedical science and mastery over the entire universe by evolving, space-traveling humans. One group, the Santa Fe Institute, sponsor of several conferences on "artificial life," predicts the transferal of human intelligence into machines and computers displaying the complex symptoms of living things (Langton 1991, p. *xv*). "Artificial life" thus becomes the ultimate transfiguring salvation of our species.

One is tempted to propose that the modern human evolutionary account is a Judeo-Christian heterodoxy, which covertly retains fundamental structures of Judeo-Christian cosmology, salvation history and eschatology, while overtly dispensing with the scriptural account of divine intervention in the origin of species, including our own. This is similar to the case of Buddhism as Hindu heterodoxy. Dispensing with the Hindu scriptures and God concepts, Buddhism nevertheless retained basic Hindu cosmological assumptions, such as cyclical time, transmigration and *karma.*

Another thing the modern human evolutionary account has in common with the earlier Christian account is that humans appear after the other life forms. In *Genesis*, God creates the plants, animals and birds before human beings. For strict literalists, the time interval is short—humans are created on the last of six of our present solar days. Others have taken the *Genesis* days as ages. For example, around the time of Darwin, European scientists with strong Christian leanings proposed that God had gradually brought into existence various species throughout the ages of geological time until the perfected earth was ready to receive human beings (Grayson 1983). In modern evolutionary accounts, anatomically modern humans retain their position as the most recent major species to occur on this planet, having evolved from preceding hominids within the past 100,000 or so years. And despite the attempts of prominent evolutionary theorists and spokespersons to counteract the tendency—even among themselves—of expressing this appearance in teleological fashion (Gould 1977, p. 14), the idea that humans are the crowning glory of the evolutionary process still has a strong hold on the public and scientific minds. Although anatomically modern humans are given an age of about 100,000 years, modern archeologists and anthropologists, in common with Judeo-Christian accounts, give civilization an age of a few thousand years and, again in common with Judeo-Christian accounts, place its earliest occurrence in the Middle East.

I do not here categorically assert a direct causal link between earlier Judeo-Christian ideas and those of the modern historical sciences. Demonstrating that, as Edward B. Davis (1994) points out in his review of recent works on this subject, needs much more careful documentation than has yet been provided. But the many common features of the time concepts of the two knowledge systems suggest these causal links do exist, and that it would be fruitful to trace connections in sufficient detail to satisfactorily demonstrate this.

I do, however, propose that the tacitly accepted and hence critically unexamined time concepts of the modern human sciences, whether or not causally linked with Judeo-Christian concepts, pose a significant unrecognized influence on interpretation of the archeological and anthropological record. To demonstrate how this might be true, I shall introduce my own experience in evaluating this record from the alien standpoint of the cyclical time concepts and accounts of human origins found in the *Puranas* and *Itihasas* of India.

My subjective path of learning has led me to take the Vaishnava tradition of India as my primary guide to life and to the study of the visible universe and what may lie beyond. For the past century or so, it has been considered quite unreasonable to bring concepts from religious texts directly into the realm of the scientific study of nature. Indeed, many introductory anthropology and archeology texts make a clear distinction between "scientific" and "religious" ways of knowing, relegating the latter to the status of unsupported belief, with little or no utility in the objective study of nature (see, for example, Stein and Rowe 1993, chapter 2). Some texts even go so far as to boast that this view has been upheld by the United States Supreme Court (Stein and Rowe 1993, p. 37), as if the state were the best and final arbiter of intellectual controversy. But I propose that total hostility to religious views of nature in science is unreasonable, especially for the modern historical sciences. Despite their pretensions to areligious objectivity, practitioners unconsciously retain or incorporate into their workings many Judeo-Christian cosmological concepts, especially concerning time, and implicitly employ them in their day-to-day work of observation and theory-building. In this sense, modern evolutionists share some intellectual territory with their Fundamentalist Christian antagonists.

But there are other ways to comprehend historical processes in nature. How this is so can be graphically sensed if one performs the mental experiment of looking at the world from a radically different time perspective—the Puranic time concept of India. I am not alone in suggesting this. Gene Sager, a professor of philosophy and religious studies at Palomar College in California, wrote in an unpublished review of my book *Forbidden Archeology* (Cremo and Thompson 1993): "As a scholar in the field of comparative religion, I have sometimes challenged scientists by offering a cyclical or spiral model for studying human history, based on the Vedic concept of the *kalpa.* Few Western scientists are open to the possibility of sorting out the data in terms of such a model. I am not proposing that the Vedic model is true….However, the question remains, does the relatively short, linear model prove to be adequate? I believe *Forbidden Archeology* offers a well-researched challenge. If we are to meet this challenge, we need to practice open-mindedness and proceed in a cross-cultural, interdisciplinary fashion" (personal communication, 1993). The World Archeological Congress provides a suitable forum for such cross-cultural, interdisciplinary dialogue.

This cyclical time of the *Puranas* operates only within the material cosmos. Beyond the material cosmos lies the spiritual sky, or *brahmajyoti.* Innumerable spiritual planets float in this spiritual sky, where material time, in the form of *yuga* cycles, does not act.

Each *yuga* cycle is composed of four *yugas.* The first, the *Satya-yuga*, lasts 4800 years of the demigods. The second, the *Treta-yuga*, lasts 3600 years of the demigods. The third, the *Dvapara-yuga*, lasts 2400 years of the demigods. And the fourth, *Kali-yuga*, lasts 1200 years of the demigods (*Bhagavata Purana* 3.11.19). Since the demigod year is equivalent to 360 earth years (Bhaktivedanta Swami 1973, p. 102), the lengths of the *yugas* in earth years are, according to standard Vaishnava commentaries, 432,000 years for the *Kali-yuga,* 864,000 years for the *Dvapara-yuga*, 1,296,000 years for the *Treta-yuga*, and 1,728,000 years for the *Satya-yuga.* This gives a total of 4,320,000 years for the entire *yuga* cycle. One thousand of such cycles, lasting 4,320,000,000 years, comprises one day of Brahma, the demigod who governs this universe. A day of Brahma is also called a *kalpa.* Each of Brahma's nights lasts a similar period of time. Life is only manifest on earth during the day of Brahma. With the onset of Brahma's night, the entire universe is devastated and plunged into darkness. When another day of Brahma begins, life again becomes manifest.

Each day of Brahma is divided into 14 *manvantara* periods, each one lasting 71 *yuga* cycles. Preceding the first and following each *manvantara* period is a juncture (sandhya) the length of a *Satya-yuga* (1,728,000) years. Typically, each *manvantara* period ends with a partial devastation. According to Puranic accounts, we are now in the twenty-eighth *yuga* cycle of the seventh *manvantara* period of the present day of Brahma. This would give the inhabited earth an age of about 2 billion years. Interestingly enough, the oldest undisputed organisms recognized by paleontologists—algae fossils like those from the Gunflint formation in Canada—are just about that old (Stewart 1983, p. 30). Altogether, 453 *yuga* cycles have elapsed since this day of Brahma began. Each *yuga* cycle involves a progression from a golden age of peace and spiritual progress to a final age of violence and spiritual degradation. At the end of each *Kali-yuga,* the earth is practically depopulated.

During the *yuga* cycles, human species coexist with other humanlike species. For example, in the *Bhagavata Purana* (9.10.20), we find the divine *avatara* Ramacandra conquering Ravana's kingdom, Lanka, with the aid of intelligent forest-dwelling monkey men, who fight Ravana's well-equipped soldiers with trees and stones. This occurred in the *Treta-yuga,* about one million years ago.

Given the cycle of *yugas,* the periodic devastations at the end of each *manvantara,* and the coexistence of civilized human beings with creatures who in some ways resemble the human ancestors of modern evolutionary accounts—what predictions might the Puranic account give regarding the archeological record? Before answering this question, we must also consider the general imperfection of the fossil record (Raup and Stanley 1971). Hominid fossils in particular are extremely rare. Furthermore, only a small fraction of the sedimentary layers deposited during the course of the earth's history have survived erosion and other destructive geological processes (Van Andel 1981).

Taking the above into account, I propose the Puranic view of time and history predicts a sparse but bewildering mixture of hominid fossils, some anatomically modern and some not, going back tens and even hundreds of millions of years and occurring at locations all over the world. It also predicts a more numerous but similarly bewildering mixture of stone tools and other artifacts, some showing a high level of technical ability and others not. And, given the cognitive biases of the majority of workers in the fields of archeology and anthropology over the past 150 years, we might also predict that this bewildering mixture of fossils and artifacts would be edited to conform with a linear, progressive view of human origins. A careful investigation of published reports by myself and Richard Thompson (1993) offers confirmation of these two predictions. What follows is only a sample of the total body of evidence catalogued in our lengthy book. The citations given are for the single reports that best identify particular finds. Detailed analyses and additional reports cited elsewhere (Cremo and Thompson 1993) offer strong confirmation of the authenticity and antiquity of these discoveries.

Incised and carved mammal bones are reported from the Pliocene (Desnoyers 1863, Laussedat 1868, Capellini 1877) and Miocene (Garrigou and Filhol 1868, von Dücker 1873). Additional reports of incised bones from the Pliocene and Miocene may be found in an extensive review by the overly skeptical de Mortillet (1883). Scientists have also reported pierced shark teeth from the Pliocene (Charlesworth 1873), artistically carved bone from the Miocene (Calvert 1874) and artistically carved shell from the Pliocene (Stopes 1881). Carved mammal bones reported by Moir (1917) could be as old as the Eocene.

Very crude stone tools occur in the Middle Pliocene (Prestwich 1892) and from perhaps as far back as the Eocene (Moir 1927, Breuil 1910, especially p. 402). One will note that most of these discoveries are from the nineteenth century. But such artifacts are still being found. Crude stone tools have recently be reported from the Pliocene of Pakistan (Bunney 1987), Siberia (Daniloff and Kopf 1986), and India (Sankhyan 1981). Given the current view that toolmaking hominids did not leave their African center of origin until about 1 million years ago, these artifacts are somewhat anomalous, what to speak of a pebble tool from the Miocene of India (Prasad 1982).

More advanced stone tools occur in the Oligocene of Europe (Rutot 1907), the Miocene of Europe (Ribeiro 1873, Bourgeois 1873, Verworn 1905), the Miocene of Asia (Noetling 1894), and the Pliocene of South America (F. Ameghino 1908, C. Ameghino 1915). In North America, advanced stone tools occur in California deposits ranging from Pliocene to Miocene in age (Whitney 1880). An interesting slingstone, at least Pliocene and perhaps Eocene in age, comes from England (Moir 1929, p. 63).

More advanced artifacts have also been reported in scientific and nonscientific publications. These include an iron nail in Devonian Sandstone (Brewster 1844), a gold thread in Carboniferous stone (*Times* of London, June 22, 1844), a metallic vase in Precambrian stone (*Scientific American*, June 5, 1852), a chalk ball from the Eocene (Melleville 1862), a Pliocene clay statue (Wright 1912, pp. 266-69), metallic tubes in Cretaceous chalk (Corliss 1978, pp. 652-53), and a grooved metallic sphere from the Precambrian (Jimison 1982). The following objects have been reported from Carboniferous coal: a gold chain (*The Morrisonville Times*, of Illinois, U.S.A., June 11, 1891), artistically carved stone (*Daily News* of Omaha, U.S.A., April 2, 1897), an iron cup (Rusch 1971), and stone block walls (Steiger 1979, p. 27).

Human skeletal remains described as anatomically modern occur in the Middle Pleistocene of Europe (Newton 1895, Bertrand 1868, de Mortillet 1883). These cases are favorably reviewed by Keith (1928). Other anatomically modern human skeletal remains occur in the Early and Middle Pleistocene of Africa (Reck 1914, L. Leakey 1960d, Zuckerman 1954, p. 310; Patterson and Howells 1967, Senut 1981, R. Leakey 1973), the Early Middle Pleistocene of Java (Day and Molleson 1973), the Early Pleistocene of South America (Hrdlicka 1912, pp. 319-44), the Pliocene of South America (Hrdlicka 1912, p. 346; Boman 1921, pp. 341-42), the Pliocene of England (Osborn 1921, pp. 567-69), the Pliocene of Italy (Ragazzoni 1880, Issel 1868), the Miocene of France and the Eocene of Switzerland (de Mortillet 1883, p. 72), and even the Carboniferous of North America (*The Geologist* 1862). Several discoveries from California gold mines range from Pliocene to Eocene (Whitney 1880). Some of these bones have been subjected to chemical and radiometric tests that have yielded ages younger than suggested by their stratigraphical position. But when the unreliabilities and weaknesses of the testing procedures are measured against the very compelling stratigraphic observations of the discoverers, it is not at all clear that the original age attributions should be discarded (Cremo and Thompson 1993, 753- 794).

Humanlike footprints have been found in the Carboniferous of North America (Burroughs 1938), the Jurassic of Central Asia (*Moscow News* 1983, no.4, p. 10), and the Pliocene of Africa (M. Leakey 1979). Shoe prints have been reported from the Cambrian (Meister 1968) and the Triassic (Ballou 1922).

In the course of negotiating a fashionable consensus that anatomically modern humans evolved from less advanced hominids in the Late Pleistocene, scientists gradually rendered unfashionable the considerable body of compelling contradictory evidence summarized above. It thus became unworthy of discussion in knowing circles. Richard Thompson and I have concluded (1993) that the muting of this evidence was accomplished by application of a double standard, whereby favored evidence was exempted from the severely skeptical scrutiny to which unfavored evidence was subjected.

One example from the many that could be cited to demonstrate the operation of linear progressive preconceptions in the editing of the archeological record is the case of the auriferous gravel finds in California. During the days of the California Gold Rush, starting in the 1850s, miners discovered many anatomically modern human bones and advanced stone implements in mine shafts sunk deeply into deposits of gold-bearing gravels capped by thick lava flows (Whitney 1880). The gravels beneath the lava were from 9 to 55 million years old, according to modern geological reports

(Slemmons 1966). These discoveries were reported to the world of science by J. D. Whitney, state geologist of California, in a monograph published by the Peabody Museum of Natural History at Harvard University. From the evidence he compiled, Whitney came to a nonprogressivist view of human origins—the fossil evidence he reported indicated that the humans of the distant past were like those of the present.

To this W. H. Holmes (1899, p. 424) of the Smithsonian Institution replied: "Perhaps if Professor Whitney had fully appreciated the story of human evolution as it is understood today, he would have hesitated to announce the conclusions formulated, notwithstanding the imposing array of testimony with which he was confronted." This attitude is still prominent today. In their college textbook, Stein and Rowe assert that "scientific statements are never considered absolute" (1993, p. 41). But they also make this very absolute statement: "Some people have assumed that humans have always been the way they are today. Anthropologists are convinced that human beings...have changed over time in response to changing conditions. So one aim of the anthropologist is to find evidence for evolution and to generate theories about it." Apparently, an anthropologist, by definition, can have no other view or purpose. Keep in mind, however, that this absolute commitment to a linear progressive model of human origins, ostensibly areligious, may have deep roots in Judeo-Christian cosmology.

One of the things Holmes found especially hard to accept was the similarity of the purportedly very ancient stone implements to those of the modern Indians. He wondered (1899, pp. 451-52) how anyone could take seriously the idea that "the implements of a Tertiary race should have been left in the bed of a Tertiary torrent to be brought out as good as new, after the lapse of vast periods of time, into the camp of a modern community using identical forms?" The similarity could be explained in several ways, but one possible explanation is the repeated appearance in the same geographical region of humans with particular cultural attributes in the course of cyclical time. The suggestion that such a thing could happen is bound to strike those who see humans as the recent result of a long and unique series of evolutionary changes in the hominid line as absurd—so absurd as to prevent them from considering any evidence as potentially supporting a cyclical interpretation of human history.

It is noteworthy, however, that a fairly open-minded modern archeologist, when confronted with the evidence catalogued in my book, himself brought up, in a somewhat doubting manner, the possibility of a cyclical interpretation of human history to explain its occurrence. George F. Carter, noted for his controversial views on early man in North America, wrote to me on January 26, 1994, concerning the stone tools from Table Mountain, California, which would have to be at least 9 million years old. He asked if

I thought that there had been separate human populations that manufactured similar stone tool industries, separated by 9 million years.

That is exactly what I would propose—that in the course of cyclic time, humans with a culture resembling that of modern North American Indians did in fact appear in California millions of years ago, perhaps several times. Carter found it difficult to accept that kind of reasoning. But that difficulty, which encumbers the minds of most archeologists and anthropologists, may be the result of a rarely recognized and even more rarely questioned commitment to a culturally acquired linear progressive time sense.

It would, therefore, be worthwhile to inspect the archeological record through other time lenses, such as the Puranic lens. Many will take my proposal as a perfect example of what can happen when someone brings their subjective religious ideas into the objective study of nature. Jonathan Marks (1994) reacted in typical fashion in his review of *Forbidden Archeology*: "Generally, attempts to reconcile the natural world to religious views end up compromising the natural world."

But until modern anthropology conducts a conscious examination of the effects of its own covert, and arguably religiously derived, assumptions about time and progress, it should put aside its pretensions to universal objectivity and not be so quick to accuse others of bending facts to fit religious dogma. *Om Tat Sat.*

References Cited

Ameghino, C. (1915) El femur de Miramar. *Anales de Museo nacional de historia natural de Buenos Aires*, 26: 433—450.

Ameghino, F. (1908) Notas preliminares sobre el *Tetraprothhomo argentinus,* un precursor de hombre del Mioceno superior de Monte Hermoso. *Anales de Museo nacional de historia natural de Buenos Aires*, 16: 105—242.

Ballou, W. H. (1922) Mystery of the petrified "shoe-sole" 5,000,000 years old. *American Weekly* section of the *New York Sunday American*, October 8, p. 2.

Bertrand, P. M. E. (1868) Crane et ossements trouves dans un carriere de l'avenue de Clichy. *Bulletin de la Societe d'Anthropologie de Paris (Series 2), 3*: 329—335.

Bhaktivedanta Swami, A. C. (1973) *Shrimad-Bhagavatam (Bhagavata Purana)*, Canto Three, Part Two. Los Angeles, Bhaktivedanta Book Trust.

Boman, E. (1921) Los vestigios de industria humana encontrados en Miramar (Republica Argentina) y atribuidos a la época terciaria. *Revista Chilena de Historia y Geografia, 49(43)*: 330—352.

Bourgeois, L. (1873) Sur les silex considérés comme portant les margues d'un travail humain et découverts dans le terrain miocène de Thenay. *Congrès International d'Anthropologie et d'Archéologie Préhistoriques, Bruxelles 1872, Compte Rendu,* pp. 81—92.

Breuil, H. (1910) Sur la présence d'éolithes a la base de l'Éocene Parisien. *L'Anthropologie,* 21: 385—408.

Brewster, D. (1844) Queries and statements concerning a nail found imbedded in a block of sandstone obtained from Kingoodie (Mylnfield) Quarry, North Britain. *Report of the British Association for the Advancement of Science, Notices and Abstracts of Communications,* p. 51.

Bunney, S. (1987) First migrants will travel back in time. *New Scientist, 114(1565):* 36.

Burroughs, W. G. (1938) Human-like footprints, 250 million years old. *The Berea Alumnus.* Berea College, Kentucky, November, pp. 46—47.

Calvert, F. (1874) On the probable existence of man during the Miocene period. *Journal of the Royal Anthropological Institute of Great Britain and Ireland,* 3: 127.

Capellini, G. (1877) Les traces de l'homme pliocène en Toscane. *Congrès International d'Anthropologie et d'Archéologie Préhistoriques, Budapest 1876, Compte Rendu.* Vol. 1, pp. 46—62.

Charlesworth, E. (1873) Objects in the Red Crag of Suffolk. *Journal of the Royal Anthropological Institute of Great Britain and Ireland,* 2: 91—94.

Corliss, W. R. (1978) *Ancient Man: A Handbook of Puzzling Artifacts.* Glen Arm, Sourcebook Project.

Cremo, M. A., and Thompson, R. L. (1993) *Forbidden Archeology: The Hidden History of the Human Race.* San Diego, Bhaktivedanta Institute.

Daniloff, R., and Kopf, C. (1986) Digging up new theories of early man. *U. S. News & World Report,* September 1, pp. 62—63.

Davis, Edward B. (1994) Review of Cameron Wybrow (Editor): *Creation, Nature, and Political Order in the Philosophy of Michael Foster (1903—1959); The Classic Mind Articles and Others, with Modern Critical Essays,* and Cameron Wybrow: *The Bible, Baconism, and Mastery over Nature: The Old Testament and Its Modern Misreading. Isis 53(1)*: 127—129.

Day, M. H., and Molleson, T. I. (1973) The Trinil femora. *Symposia of the Society for the Study of Human Biology, 2:* 127— 154.

De Mortillet, G. (1883) *Le Préhistorique.* Paris, C. Reinwald.

Desnoyers, J. (1863) Response à des objections faites au sujet d'incisions constatées sur des ossements de Mammiferes fossiles des environs de Chartres. *Compte Rendus de l'Académie des Sciences, 56*: 1199—1204.

Garrigou, F., and Filhol, H. (1868) M. Garrigou prie l'Académie de vouloir bien ouvrir un pli cacheté, déposé au nom de M. Filhol fils et au sien, le

16 mai 1864. *Compte Rendus de l'Académie des Sciences, 66*: 819—820.
The Geologist, London, 1862 Fossil man, 5: 470.
Gould, S. J. (1977) *Ever Since Darwin*. New York, W. W. Norton.
Grayson, Donald K. (1983) *The Establishment of Human Antiquity*. New York, Academic Press.
Holmes, W. H. (1899) Review of the evidence relating to auriferous gravel man in California. *Smithsonian Institution Annual Report 1898—1899*, pp. 419—472.
Hrdlicka, A. (1912) *Early Man in South America*. Washington, D.C., Smithsonian Institution.
Issel, A. (1868) Résumé des recherches concernant l'ancienneté de l'homme en Ligurie. *Congrès International d'Anthropologie et d'Archéologie Préhistoriques, Paris 1867, Compte Rendu*, pp. 75—89.
Jimison, S. (1982) Scientists baffled by space spheres. *Weekly World News*, July 27. (Key details of the report in the *Weekly World News*, a sensationalistic tabloid, were confirmed by correspondence with Dr. Roelf Marks in South Africa.)
Keith, A. (1928) *The Antiquity of Man*. Vol. 1. Philadelphia, J. B. Lippincott.
Langton, C. G. (1991) Preface. *In* Langton, C. G., *et al.*, eds. *Artificial Life II: Proceedings of the Workshop on Artificial Life Held February, 1990 in Santa Fe, New Mexico*. Santa Fe Institute Studies in the Sciences of Complexity, Proceedings Volume X. Redwood City, Addison-Wesley, pp. xiii—xv.
Laussedat, A. (1868) Sur une mâchoire de Rhinoceros portant des entailles profondes trouvée à Billy (Allier), dans les formations calcaires d'eau douce de la Limagne. *Compte Rendus de l'Académie des Sciences, 66*: 752—754.
Leakey, L. S. B. (1960) *Adam's Ancestors*, 4th edition. New York, Harper & Row.
Leakey, M. D. (1979) Footprints in the ashes of time. *National Geographic*, 155: 446—457.
Leakey, R. E. (1973) Evidence for an advanced Plio-Pleistocene hominid from East Rudolf, Kenya. *Nature*, 242: 447—450.
Marks, J. (1994) Review of *Forbidden Archeology: The Hidden History of the Human Race*, by Michael A. Cremo and Richard L. Thompson. 1993. San Diego: Bhaktivedanta Institute. *American Journal of Physical Anthropology, 93*: 140—141.
Meister, W. J. (1968) Discovery of trilobite fossils in shod footprint of human in "Trilobite Bed"—a Cambrian formation, Antelope Springs, Utah. *Creation Research Society Quarterly*, 5(3): 97—102.
Melleville, M. (1862) Note sur un objet travaillé de main d'homme trouve dans les lignites du Laonnais. *Revue Archéologique, 5*: 181—186.

Moir, J. R. (1917) A series of mineralised bone implements of a primitive type from below the base of the Red and Coralline Crags of Suffolk. *Proceedings of the Prehistoric Society of East Anglia, 2*: 116—131.

Moir, J. R. (1927) *The Antiquity of Man in East Anglia.* Cambridge, Cambridge University Press.

Moir, J. R. (1929) A remarkable object from beneath the Red Crag. *Man,* 29: 62—65.

Newton, E. T. (1895) On a human skull and limb-bones found in the Paleolithic terrace-gravel at Galley Hill, Kent. *Quarterly Journal of the Geological Society of London, 51*: 505—526.

Noetling, F. (1894) On the occurrence of chipped flints in the Upper Miocene of Burma. *Records of the Geological Survey of India, 27*: 101—103.

Osborn, H. F. (1921) The Pliocene man of Foxhall in East Anglia. *Natural History, 21*: 565—576.

Patterson, B., and Howells, W. W. (1967) Hominid humeral fragment from Early Pleistocene of northwestern Kenya. *Science, 156*: 64—66.

Prasad, K. N. (1982) Was *Ramapithecus* a tool-user? *Journal of Human Evolution, 11:* 101—104.

Prestwich, J. (1892) On the primitive character of the flint implements of the Chalk Plateau of Kent, with reference to the question of their glacial or pre-glacial age. *Journal of the Royal Anthropological Institute of Great Britain and Ireland, 21(3)*: 246—262.

Ragazzoni, G. (1880) La collina di Castenedolo, solto il rapporto antropologico, geologico ed agronomico. *Commentari dell' Ateneo di Brescia,* April 4, pp. 120—128.

Raup, D., and Stanley, S. (1971) *Principles of Paleontology.* San Francisco, W. H. Freeman.

Reck, H. (1914) Erste vorläufige Mitteilungen über den Fund eines fossilen Menschenskeletts aus Zentral-afrika. *Sitzungsbericht der Gesellschaft der naturforschender Freunde Berlins, 3: 81*—95.

Ribeiro, C. (1873) Sur des silex taillés, découverts dans les terrains miocène du Portugal. *Congrès International d'Anthropologie et d'Archéologie Préhistoriques, Bruxelles 1872, Compte Rendu*, pp. 95—100.

Rusch, Sr., W. H. (1971) Human footprints in rocks. *Creation Research Society Quarterly, 7*: 201—202.

Rutot, A. (1907) Un grave problem: une industrie humaine datant de l'époque oligocène. Comparison des outils avec ceux des Tasmaniens actuels. *Bulletin de la Société Belge de Géologie de Paléontologie et d'Hydrologie, 21*: 439—482.

Sankhyan, A. R. (1981) First evidence of early man from Haritalyangar area, Himalchal Pradesh. *Science and Culture*, 47: 358—359.

Senut, B. (1981) Humeral outlines in some hominoid primates and in Plio-pleistocene hominids. *American Journal of Physical Anthropology, 56*: 275—283.

Slemmons, D. B. (1966) Cenozoic volcanism of the central Sierra Nevada, California. *Bulletin of the California Division of Mines and Geology, 190*: 199—208.

Steiger, B. (1979) *Worlds Before Our Own*. New York, Berkley.

Stein, Philip L. and Rowe, Bruce M. (1993) *Physical Anthropology*. Fifth Edition. New York, McGraw-Hill.

Stewart, Wilson N. (1983) *Paleobotany and the Evolution of Plants*. Cambridge, Cambridge University Press.

Stopes, H. (1881) Traces of man in the Crag. *British Association for the Advancement of Science, Report of the Fifty-first Meeting*, p. 700.

Van Andel, T. H. (1981) Consider the incompleteness of the geological record. *Nature, 294*: 397—398.

Verworn, M. (1905) Die archaeolithische Cultur in den Hipparionschichten von Aurillac (Cantal). *Abhandlungen der königlichen Gesellschaft der Wissenschaften zu Göttingen, Mathematisch-Physikalische Klasse, Neue Folge, 4(4)*:3—60.

Von Dücker, Baron (1873) Sur la cassure artificelle d'ossements recueillis dans le terrain miocène de Pikermi. *Congrès International d'Anthropologie et d'Archéologie Préhistoriques. Bruxelles 1872, Compte Rendu*, pp. 104—107.

Whitney, J. D. (1880) The auriferous gravels of the Sierra Nevada of California. *Harvard University, Museum of Comparative Zoology Memoir 6(1)*.

Wright, G. F. (1912) *Origin and Antiquity of Man*. Oberlin, Bibliotheca Sacra.

Zuckerman, S. (1954) Correlation of change in the evolution of higher primates. *In* Huxley, J., Hardy, A. C., and Ford, E. B., eds. *Evolution as a Process*. London, Allen and Unwin, pp. 300—352.

1.2. Michael. A. Cremo (1995) "The Reception of *Forbidden Archeology*: An Encounter Between Western Science and a Non-Western Perspective on Human Antiquity." Kentucky State University Institute for Liberal Studies Sixth Annual Interdisciplinary Conference: Science and Culture.

I presented this paper at the Kentucky State University Institute of Liberal Studies, Sixth Annual Interdisciplinary Conference: Science and Culture, held at Frankfort, Kentucky, March 30 - April 1, 1995. A short abstract is published in the conference proceedings. A short excerpt from the paper appears in Back to Godhead *(March/April 1996, p. 26). The entire text of the paper appears in* ISKCON Communications Journal *(Vol. 5, No. 1, 1997).*

Abstract

Forbidden Archeology, by Michael Cremo and Richard Thompson of the Bhaktivedanta Institute, documents voluminous scientifically reported evidence contradicting current ideas about human antiquity. This suppressed evidence supports accounts of extreme human antiquity encountered in ancient India's Puranic literature. Responses to *Forbidden Archeology* from mainstream and nonmainstream knowledge communities illuminate Western science's descent from self-proclaimed epistemic superiority into a diverse multipolar global intellectual constellation from which may emerge a new consensus on human origins.

Main Text

In 1993 I published the book *Forbidden Archeology: The Hidden History of the Human Race*, coauthored with Richard L. Thompson (Cremo & Thompson, 1993). In his foreword, ethnomethodological sociologist Pierce J. Flynn, of California State University at San Marcos, noted:

> *Forbidden Archeology* does not conceal its own positioning on a relativist spectrum of knowledge production. The authors admit to their own sense of place in a knowledge universe with contours derived from personal experience with Vedic philosophy, religious perception, and Indian cosmology. Their intriguing discourse on the 'Evidence for Advanced Culture in Distant Ages' is light years from 'normal' Western science, and yet provokes a cohesion of probative thought. In my view, it is just this openness of subjective positioning that makes *Forbidden Archeology* an original and important contribution to postmodern scholarly studies now being done in sociology, anthropology, archeology, and the history of science and ideas. The authors' unique perspective provides postmodern scholars with an invaluable parallax view of historical scientific praxis, debate, and development.

In my own introduction to *Forbidden Archeology*, I noted (Cremo & Thompson, 1993):

> Richard Thompson and I are members of the Bhaktivedanta Institute, a branch of the International Society for Krishna Consciousness [ISKCON] that studies the relationship between modern science and the world view expressed in the Vedic literature. This institute was founded by our spiritual master, His Divine Grace A. C. Bhaktivedanta Swami Prabhupada, who encouraged us to critically examine the prevailing account of human origins and the methods by which it was established. From the Vedic literature, we derive the idea that the human race is of great antiquity. To conduct systematic research into the existing scientific literature on human antiquity, we expressed the Vedic idea in the form of a theory that various humanlike and apelike beings have coexisted for a long time.

To further contextually position myself, I offer that in 1976 I was initiated by Bhaktivedanta Swami Prabhupada (1896-1977) into the Brahma-Madhva-Gaudiya branch of Vaishnavism. The lineage of Gaudiya Vaishnavism, of which ISKCON is a modern institutional expression, extends back thousands of years, but the most recent representatives, of the nineteenth and twentieth centuries, are most important for this paper (see Goswami, S. D., 1980).

In the nineteenth century, India's British rulers offered Western education to Indian intellectuals. Their goal was to create a cadre of English-speaking and English-thinking Indians to assist them in their program of military, political, economic, religious and cultural domination. This educational program successfully induced many Indian intellectuals, including Gaudiya Vaisnavas, to abandon their traditional culture and wisdom for Western modes of science and theology.

But in the middle of the nineteenth century, Kedarnatha Dutta (1838-1914), an English-speaking magistrate in the colonial administration, became interested in Gaudiya Vaishnavism. After his initiation by a Gaudiya Vaishnava *guru,* he inaugurated a revival of Gaudiya Vaishnavism among the intelligent classes, in Bengal and throughout India.

The central goal of Gaudiya Vaishnavism is cultivation of *bhakti,* or devotion to the Supreme Personality of Godhead, known by the name Krishna, "the all-attractive one." The *bhakti* school also incorporates a strong philosophical tradition, grounded in a literal, yet by no means naive, reading of the Vedic and Puranic texts, including their accounts of history and cosmogony.

Kedarnatha Dutta, later known by the title Bhaktivinoda Thakura, communicated Gaudiya Vaishnava teachings not only to his Indian contemporaries but also to the worldwide community of intellectuals. He reached the latter by publishing several works in English, among them *Shri Chaitanya Mahaprabhu: His Life and Precepts,* which appeared in 1896.

In the early twentieth century, Bhaktivinoda Thakura's son Bimala Prasada Dutta, later known as Bhaktisiddhanta Sarasvati Thakura (1874-1936), carried on the work of his father, expanding Gaudiya Vaishnavism in India and sending a few disciples to England and Germany. The European expeditions did not, however, yield any permanent results, and the missionaries returned home.

In 1922, my own spiritual master, then known as Abhay Charan De, met Bhaktisiddhanta Sarasvati Thakura in Calcutta, India. A recent graduate of Scottish Churches College in Calcutta and follower of Gandhi, De was somewhat skeptical of this very traditional *guru.* But he found himself won over by Bhaktisiddhanta Sarasvati's sharp intelligence and spiritual purity. At this first meeting, Bhaktisiddhanta Sarasvati requested De to spread the Gaudiya Vaishnava teachings throughout the world, especially in English. In 1933 De became the formal disciple of Bhaktisiddhanta Sarasvati, and in 1936, the year of Bhaktisiddhanta's death, he received a letter from him renewing his request that De teach in the West. In 1965, at the age of 69, De, now known as Bhaktivedanta Swami, came to New York City, where a year later he started ISKCON, the institutional vehicle through which the teachings of Gaudiya Vaishnavism were to spread quickly around the world.

Among these teachings were those connected with the origin of life and the universe. To scientifically establish these teachings, Bhaktivedanta Swami in 1975 organized the Bhaktivedanta Institute. Bhaktivedanta Swami envisioned the process of introducing Gaudiya Vaishnava teachings on the origin of life and the universe as one of direct confrontation with prevailing Western scientific ideas, such as Darwinian evolution.

My own involvement in the Bhaktivedanta Institute, as a Western convert to Gaudiya Vaishnavism, can thus be seen in the historical context of the larger cultural interaction between Western science and an Asian Indian knowledge tradition with vastly different views on natural history.

The Vedic and Puranic texts speak of a divine origin and spiritual purpose to life. According to the *Puranas,* humans have existed on this planet for hundreds of millions of years and did not evolve from more apelike ancestors. The *Puranas* do, however, tell of intelligent races of apelike beings who coexisted with humans over vast periods of time.

In the 900 pages of *Forbidden Archeology*, my coauthor and I documented a great deal of scientifically reported evidence that, consistent with Puranic texts, extends the antiquity of our species millions of years into the past. This evidence was accumulated during eight years of research into the history of archeology and anthropology since the time of Darwin. We also documented how this evidence was systematically suppressed, through a social process of "knowledge filtration," by the adherents of an emerging consensus among Western scientists that humans were a fairly recent production of an evolutionary process.

Using the word *archeology* in Foucault's sense (Foucault, 1972), *Forbidden Archeology* is an archeology of archeology. It investigates the formation of archeological discourse over time, illuminating the subjects, objects, situations, themes and practices of this discourse, including its practices of exclusion and suppression.

The primary goal of this paper is not to convince readers of *Forbidden Archeology*'s picture of extreme human antiquity but to analyze the reception of *Forbidden Archeology* in various knowledge and discourse communities. Nevertheless, just to give an idea of the kinds of evidence advanced in *Forbidden Archeology*, I shall provide two examples (much abbreviated from Cremo & Thompson, 1993).

In 1880, Harvard University's Peabody Museum published a massive work by J. D. Whitney, State Geologist of California, on the geology of the gold mining regions of California (Whitney 1880). In this book, Whitney catalogued hundreds of artifacts and human skeletal remains found by miners, mining engineers and mine supervisors deep inside gold mines at dozens of locations. All of the evidence gathered by Whitney indicated that the objects could not have entered from other levels. The gold-bearing gravels from which the objects were taken are, according to modern geological reports, anywhere from 10 to 50 million years old (Slemmons, 1966). Given current doctrine that anatomically modern humans came into existence about 100,000 years ago, the evidence reported by Whitney is quite extraordinary.

Whitney's evidence was dismissed, however, by William H. Holmes of the Smithsonian Institution, who said (Holmes, 1899), "Perhaps if Professor Whitney had fully appreciated the story of human evolution as it is understood today, he would have hesitated to announce the conclusions formulated, notwithstanding the imposing array of testimony with which he was confronted." So here we find a credible report of evidence for extreme human antiquity dismissed principally because it contradicted the emerging scientific consensus that humans evolved fairly recently.

Such "knowledge filtration," with theoretical preconceptions governing the acceptance and rejection of evidence, continues to the present. In 1979, Mary Leakey discovered at Laetoli, Tanzania, a set of footprints indistinguishable from anatomically modern human footprints (Leakey, 1979). These were found in solidified volcanic ash deposits about 3.6 million years old. Fossils of the foot bones of the early hominids of that time do not fit the Laetoli prints. At present, human beings like ourselves are the only creatures known to science that can make prints like those found at Laetoli. Nevertheless, most scientists, because of their theoretical preconceptions, are not prepared to consider that a human like ourselves may have made the Laetoli prints 3.6 million years ago.

Multiply the above two examples a hundred times, and one will get some idea of the quantity of evidence for extreme human antiquity contained in *Forbidden Archeology*.

Having set the stage for the appearance of *Forbidden Archeology*, let us place ourselves in the position of its principal author and imagine his feelings as the book began to make its way out into the world of reactive language. How would it be received, especially given his "openness of subjective positioning"? I was, of course, hopeful. In particular, I hoped my book would have some impact on the community of scholars practicing the sociology of scientific knowledge (SSK).

With this in mind, I sent a copy of *Forbidden Archeology* to SSK scholar Michael Mulkay, who replied in a handwritten letter dated May 18, 1993. Mulkay said he did not have time to comment on the book and offered, by way of consolation, that two of his own books, which he regarded as his best, were met with silence. *[The original draft of this paper included a direct quotation from Mulkay's letter, but he later indicated to me that he did not wish such quotations to be included in this book. I have therefore altered the text of the paper.]*

Mainstream Archeology and Anthropology

So, were the 900 pages of *Forbidden Archeology* to be met with the profoundest of academic silences? I contemplated the prospect of no comment on my text in dark and dreary interior monologues worthy of a narrator of a tale by Edgar Allen Poe. Fortunately, the silence was soon broken, not by a tapping at my door but by a review (Marks, 1994) appearing in the January 1994 issue of *American Journal for Physical Anthropology (AJPA)*. Apparently, *Forbidden Archeology* posed a challenge that could not be ignored.

In an earlier, perceptive essay, A. J. Greimas had offered a semiotic exploration of a challenge's narrative dimension (Greimas, 1990). "A challenge," he said, "is a confrontation that is perceived as an affront." *Forbidden Archeology* was certainly perceived as such by Jonathan Marks, book review editor for *AJPA*.

Arrogating to himself the reviewing of *Forbidden Archeology* (instead of assigning it to an outside reviewer), Marks (1994) adopted a combative and derisive stance, characterizing the book as "Hindu-oid creationist drivel" and "a veritable cornucopia of dreck."

Why did Marks respond at all? According to Greimas (1990), a challenge consists of a challenging subject inviting a challenged subject to carry out a particular narrative program, while at the same time warning the challenged subject "as to his modal insufficiency (his 'not-being-able-to-do') for the carrying out of that program." The sending of *Forbidden Archeology* to the book review editor of *AJPA* was consciously intended as just such an invitation for physical anthropologists to carry out their narrative program (of establishing their truth of human evolution)—with the implication they would not be able to do so in the face of the evidence documented in *Forbidden Archeology*.

A challenge may further be classed (Greimas, 1990) as a "constraining communication." In other words, "When faced with an affirmation of his incompetence, the challenged subject cannot avoid answering because silence would inevitably be interpreted as an admission of that incompetence" (Greimas, 1990). Marks obviously felt constrained to respond to *Forbidden Archeology*, thus accepting its challenge contract and thereby placing the challenged subject (Marks, *AJPA,* physical anthropology, Western science) and the challenging subject (Cremo, *Forbidden Archeology*, Bhaktivedanta Institute, Gaudiya Vaishnavism) on an equal subjective footing. As Greimas (1990) noted, "If the challenge is to work properly there must be an *objective complicity* between the manipulator [i.e. the challenging subject] and the manipulated [i.e. the challenged subject]....It is unthinkable for a knight to challenge a peasant, and the converse is unthinkable also."

Marks's response to the challenge of *Forbidden Archeology* contained elements of bravado, showing him an able defender of physical anthropology, but also elements of unconscious fear, showing him as a threatened member of an unstable discipline in danger of dismemberment by dark forces within and without.

Regarding the latter, Marks (1994) first alluded to "the Fundamentalist push to get 'creation science' into the classroom." The Christian fundamentalist enemy is apparently alive and kicking, still "pushing" to get into territory physical anthropology regards as its own ("the classroom").

Marks (1994) then admitted that "the rich and varied origins myths of all cultures are alternatives to contemporary evolution." His use of the present tense ("are alternatives") instead of the past tense ("were alternatives") is a reflection that physical anthropology, in the postcolonial era, feels once more threatened by living alternative cosmologies that not long ago were securely categorized as cognitively dead myths. In other words, it is not only the Christian fundamentalists who are enemies, but all alternative cosmologies and (Marks 1994) "all religious-based science, like the present volume" (i.e. *Forbidden Archeology*).

Marks (1994) also alluded to "goofy popular anthropology" and its literature. These pose a secular, populist threat to modern physical anthropology and archeology (as in the case of the ongoing reports of Bigfoot and transoceanic diffusionist contacts between North America and the ancient civilizations of Asia, Africa and Europe).

Of greatest concern to Marks, however, were traitors in the ranks of the academic community itself. Marks (1994) described sociologist Pierce Flynn, who contributed a foreword to *Forbidden Archeology*, as "a curious personage." He went on to castigate Flynn for "placing this work within postmodern scholarship."

In short, Marks identified *Forbidden Archeology*, quite correctly in my view, with an array of perceived enemies at the boundaries of his discipline, and within the walls of the disciplinary sanctuary itself. These enemies included fundamentalists, creationists, cultural revivalists, religion-based sciences (especially Hindu-based), populist critiques of science, purveyors of anomalies and, finally, the post-modern academic critics of science in the fields of sociology, history and philosophy, what to speak of anthropology itself. Later, I shall suggest how we might see archeology and anthropology as disciplinary partners of their perceived enemies, sharing a common discursive domain.

After Marks's review appeared, I exchanged some letters with Matt Cartmill, editor of the *AJPA,* regarding a rejoinder. (I asserted that Marks had misrepresented the substance of the book.) But Cartmill declined to allow a rejoinder, saying finally [in a letter dated August 30, 1994] that he believed Marks's review was accurate.

All in all, I was pleased with Marks's response to *Forbidden Archeology*. The provocation had been designed to evoke just such a response, which I had anticipated would be tactically useful. And it was. First, Marks's refusal to come to grips with the substance of the book, namely the factual evidence, lent support to the book's theme of knowledge suppression. Second, his derisive name-calling helped get media attention for the popular edition of *Forbidden Archeology*, titled *The Hidden History of the Human Race*. Excerpts from Marks's review were prominently featured in the book, and when seen alongside positive reviews, gave the impression of a serious work that was stirring considerable controversy. Third, I envisioned that Marks's remarks would provide material for scholarly papers such as this one. And the main news for this paper is that Marks's review objectifies a cognitive clash between a science informed by Gaudiya Vaishnava teachings and traditional Western science, with the clash manifested in the privileged textual space of Western science itself.

If the pages of a discipline's journals are one locus of privileged discourse, the conferences of a discipline are another. A few months after the *AJPA* review, the publishing branch of the Bhaktivedanta Institute took a book display table at the annual meeting of the American Association of Physical Anthropologists. Eight physical anthropologists purchased copies of *Forbidden Archeology* on the spot, and I assume others ordered later from sales materials they took with them.

Also in 1994, Kenneth L. Feder reviewed *Forbidden Archeology* in *Geoarchaeology: An International Journal* (Feder 1994). Feder's tone was one of amazement rather than derision.

> *Forbidden Archeology: The Hidden History of the Human Race* is not the usual sort of publication reviewed in this or, for that matter, any other archeology or anthropology journal. Neither author is an archeologist or paleoanthropologist; one is a mathematician, the other a writer. So far as I can tell neither has any personal experience with the process of archaeological field work or laboratory analysis.

Nevertheless, *Forbidden Archeology* rated a four-page review in *Geoarchaeology,* one of the prominent archeology journals.

What was going on here? I submit that Feder and the editors of *Geoarchaeology* were reacting to *Forbidden Archeology* in much the same way as some art critics of the 1960s reacted to the Brillo box sculptures of Andy Warhol. The Brillo boxes were not art, it appeared to some critics, but these critics could not help commenting upon them as if they were art (Yau, 1993). And thus the boxes were, after all, art, or as good as art. I suppose it is true, in some sense, that *Forbidden Archeology* is not real archeology, and that I am not a real archeologist (or historian of archeology). And, for that matter, neither is this paper a "real" paper, and neither am I a "real" scholar. I am an agent of Gaudiya Vaishnavism, with an assigned project of deconstructing a paradigm, and this paper and *Forbidden Archeology* are part of that project. And yet *Forbidden Archeology* is reviewed in *AJPA* and *Geoarchaeology,* and this paper is read by me at an academic conference on science and culture, just like Warhol's Brillo boxes were displayed in galleries for purchase by collectors rather than stacked in supermarkets for throwing out later.

Perhaps what was so disconcerting about Warhol's Brillo boxes was that their idiosyncratic artificial reality somehow called into question the hitherto naively accepted "natural" supermarket reality of the ubiquitous everyday Brillo boxes, and hence of all everyday public culture "things." The same with *Forbidden Archeology*. Its intended artificialness, its not quite seamless mimicry of a "genuine" text, called fascinated attention to itself as it simultaneously undermined the natural artifactitious impression of the so-called "real" archeology texts. But there is a difference between *Forbidden Archeology* and Warhol's Brillo boxes. If you open up a Brillo box, you won't find any Brillo pads; but if you open up *Forbidden Archeology*, you will find archeology, although of an unusual sort.

After all, I am not simply a literary pop artist who delights in producing artificial archeology text surfaces; there is some substantiality to the project. I am a representative of Gaudiya Vaishnavism, and my purpose is to challenge some fundamental concepts of Western science. This did not escape Feder, who wrote: "The book itself represents something perhaps not seen before; we can fairly call it 'Krishna creationism' with no disrespect intended."

Feder (1994), sustaining his mode of amazement, then wrote about this new invasion, this "something perhaps not seen before":

> The basic premises of the authors are breathtaking and can be summarized rather briefly:
> • The prevailing paradigm of human evolution…is wholly untenable.
> • There is what amounts to a passive conspiracy (the authors call it a "knowledge filter") to suppress a huge body of data that contradicts our prevailing paradigm.
> • These suppressed data include archaeological evidence in the form of incised bones, lithics, and anatomically modern human skeletal remains that date to well before the commonly accepted appearance of *Australopithecus.*
> • This purported evidence indicates that "beings quite like ourselves have been around as far back as we care to look—in the Pliocene, Miocene, Oligocene, Eocene and beyond (p. 525). The authors cite "humanlike footprints" in Kentucky dating to about 300,000,000 (not a misprint) years ago (p. 456).
> • All evidence of human evolution from an apelike ancestor is suspect at best and much of it can be explained as the fossil remains of nonancestral hominids or even extinct apes (some *Homo erectus* specimens, it is proposed, might represent an extinct species of giant gibbon [p. 465]).
> • Some of these nonancestral hominids have survived into the present as indicated by reports of Bigfoot, Yeti, and the like.
> • There is evidence of anomalously advanced civilizations extending back millions of years into the past.

I very much appreciated this accurate representation of the "breathtaking" substance of *Forbidden Archeology* (as compared with Marks's calculated misrepresentation). Feder (1994) further noted:

> While decidedly antievolutionary in perspective, this work is not the ordinary variety of antievolutionism in form, content, or style. In distinction to the usual brand of such writing, the authors use original sources and the book is well written. Further, the overall tone of the work is far superior to that exhibited in ordinary creationist literature. Nonetheless, I suspect that creationism is at the root of the authors' argument, albeit of a sort not commonly seen before.

In the above passage, the Brillo box phenomenon again displays itself. The technically convincing imitation of a well-written archeology text somewhat disarmed Feder. But like the critics of Warhol, he showed he was not to be fooled by surface appearances and could see what the artist/author was really up to. He knew what was meant. However, I propose the "meaning" lies not so much beneath the surface of the *Forbidden Archeology* text as in the temporal and spatial continuities and discontinuities of this textual surface with other textual surfaces, not excluding Feder's review and this paper.

Feder (1994) then addressed the Gaudiya Vaishnava element of the *Forbidden Archeology* text.

> When you attempt to deconstruct a well-accepted paradigm, it is reasonable to expect that a new paradigm be suggested in its place. The authors of *Forbidden Archeology* do not do this, and I would like to suggest a reason for their neglect here. Wishing to appear entirely scientific, the authors hoped to avoid a detailed discussion of their own beliefs (if not through evolution, how? If not within the last four million years, when?) since, I would contend, these are based on a creationist view, but not the kind we are all familiar with.

Here Feder is being somewhat unfair. The authors were not avoiding anything. The book as conceived would have introduced an alternative paradigm based on Gaudiya Vaishnava texts. But as I said in the Introduction to *Forbidden Archeology* (Cremo & Thompson, 1993):

> Our research program led to results we did not anticipate, and hence a book much larger than originally envisioned. Because of this, we have not been able to develop in this volume our ideas about an alternative to current theories of human origins. We are therefore planning a second volume relating our extensive research results in this area to our Vedic source material.

This book is still in the research and writing stage. But in a paper presented at the World Archaeological Congress 3 (Cremo, 1994), I have outlined in some detail a summary presentation of the Gaudiya Vaishnava account of human origins and antiquity. Furthermore, the relevant source materials are easily available to most readers, many of whom will already know something of Indian cosmology. Feder continued:

> The authors are open about their membership in the Bhaktivedanta Institute, which is a branch of the International Society for Krishna Consciousness, and the book is dedicated to their "spiritual master," the group's founder. They make a reasonable request regarding their affiliation with this organization: "That our theoretical outlook is derived from the Vedic literature should not disqualify it" (p. xxxvi). Fair enough, but what is their "theoretical outlook?" Like fundamentalist Christians they avoid talking about the religious content of their perspective, so we can only guess at it.

Again, somewhat unfair. The specific religious content of our perspective was openly acknowledged in the Introduction to *Forbidden Archeology* (Cremo & Thompson, 1993). Although the detailed development of this perspective was postponed to a forthcoming volume, pointers to the perspective were clear enough to remove it from the realm of guesswork. Feder (1994) himself did not have to look very far to find out about it:

> What does Hindu literature say humanity originated from and when? According to Hindu cosmology, the cosmos passes through cycles called *kalpas,* each of which corresponds to 4.32 billion earth years (Basham, 1959). During each *kalpa,* the universe is created and then absorbed. Each *kalpa* is divided into 14 *manvantaras,* each lasting 300,000,000 million years (the age of the Kentucky footprints), and separated by lengthy periods. Within each *manvantara* the world is created with human beings more or less fully formed, and then destroyed, only to be created once again in the next *manvantara.*

Yes. And I said essentially the same thing in my above-mentioned World Archeological Congress paper (Cremo 1994). Feder (1994) then began to bring his review to a close: "We all know what happens when we mix a literal interpretation of the Judeo-Christian creation myth with human paleontology: we get scientific creationism." We also get scientific evolutionism.

The two are more closely related than the partisans of either would care to admit. As I noted in my World Archaeological Congress-3 paper (Cremo 1994), Judeo-Christian cosmology, based on linear vectorial time, "involves a unique act of creation, a unique appearance of the human kind and a unique history of salvation, culminating in a unique denouement in the form of a last judgment. The drama occurs only once." I went on to say (Cremo 1994):

> Modern historical sciences share the basic Judeo-Christian assumptions about time and humanity. The universe we inhabit is a unique occurrence. Humans have arisen once on this planet. The history of our species is regarded as a unique though unpredestined evolutionary pathway. The future pathway of our species is also unique. Although this pathway is officially unpredictable, the myths of science project a possible overcoming of death by evolving, space-traveling humans ...One is tempted to propose that the modern human evolutionary account is a Judeo-Christian heterodoxy, which covertly retains fundamental structures of Judeo-Christian cosmology, salvation history and eschatology, while overtly dispensing with the Biblical account of divine intervention in the origin of species, including our own.

But let us get back to Feder's final words:

> It seems we now know what happens when we mix a literal interpretation of the Hindu myth of creation with human paleontology; we get the antievolutionary Krishna creationism of *Forbidden Archeology*, where human beings do not evolve and where the fossil evidence for anatomically modern humans dates as far back as the beginning of the current *manvantara* (Feder, 1994).

Of course, I did not invent the fossil evidence showing that anatomically modern humans existed "as far back as the beginning of the current *manvantara*." Abundant examples are present in the archeological literature of the past 150 years. *Forbidden Archeology* merely displayed that evidence and demonstrated how it was unfairly set aside by misapplication of evidential standards.

If I were a scholar trying to make a career in the modern university system, a review like Feder's would be disheartening. But I am not trying to advance an academic career for myself. I stand outside that system. Indeed, I am part of another system. And my goal is to engage the system to which Feder belongs in a textual exchange with the system to which I belong. And in that sense *Forbidden Archeology* can be called successful. In the privileged textual space of Western science we see the intrusion of alien texts, not as passive objects of study but as vital resisting and aggressing entities. There has been a change in the discursive field. A new vortex has formed. The texts of Gaudiya Vaishnava cosmology, mediated by an array of convincingly contoured archeological textuality (*Forbidden Archeology*), have become displayed in a new textual space, a space that must now configure itself differently. *Forbidden Archeology* is not a random event, a ripple that will soon fade, but the foreshock of a tectonic movement of cultures. Transnationalism and multiculturalism are not merely concepts entertained by departmental chairs and university administrators; they are brute objective realities.

These realities are reflected in the responses to *Forbidden Archeology* in *AJPA* and *Geoarchaeology.* Reviews and brief notices of *Forbidden Archeology* have also appeared in *L'Homme* (35, 173-174), *Journal of Field Archeology* (21, 112), *Antiquity* (67, 904), and *Ethology, Ecology, and Evolution* (6, 461). Another full review is forthcoming in *L'Anthropologie.*

Responses from individual scholars are also illuminating. In a letter to me dated November 26, 1993, K. N. Prasad, former president of the Archaeological Society of India, praised *Forbidden Archeology* as "an excellent reference book, which will act as a catalyst for further research on a subject of immense interest." An important audience for *Forbidden Archeology*, which is an open defense of the reality of India's Puranic literature, is the English-educated elite represented by Prasad. It may be recalled that *Forbidden Archeology* traces its own lineage to Bhaktivinoda Thakura, the English-educated magistrate, who in the late nineteenth century initiated an effort to reclaim the Indian intelligentsia from its immersion in nascent Western modernity.

Bhaktivinoda Thakura, as we have seen, also initiated an approach on behalf of Gaudiya Vaishnavism to the Western intelligentsia itself. *Forbidden Archeology* is part of that ongoing approach. William W. Howells, one of the major architects of the current paradigm of human evolution, wrote to me on August 10, 1993: "Thank you for sending me a copy of *Forbidden Archeology*, which represents much careful effort in critically assembling published materials. I have given it a good examination ....To have modern human beings ...appearing...at a time when even simple primates did not exist as possible ancestors...would be devastating to the whole theory of evolution, which has been pretty robust up to now....The suggested hypothesis would demand a kind of process which could not possibly be accommodated to the evolutionary theory as we know it, and I should think it requires an explanation of that aspect. It also would give the Scientific Creationists some problems as well! Thank you again for letting me see the book. I look forward to viewing its impact." I sent *Forbidden Archeology* to Howells, conscious of its relation to the project begun in the last century by Bhaktivinoda Thakura.

And finally a few words from Richard Leakey's letter to me of November 6, 1993: "Your book is pure humbug and does not deserve to be taken seriously by anyone but a fool. Sadly there are some, but that's part of selection and there is nothing that can be done." These words, like those of Jonathan Marks (1994), were reproduced on the cover and in the front matter of the popular edition of *Forbidden Archeology*, inspiring sales and media coverage.

All in all, *Forbidden Archeology*, inspired by Gaudiya Vaishnava teachings on human origins and antiquity, seems to have created a minor sensation within mainstream archeology and anthropology. But that is only part of the story. Indeed, we have only begun to trace the impact of *Forbidden Archeology*.

History, Sociology, and Philosophy

Let us now turn from mainstream science to mainstream science studies—the history, sociology and philosophy of science.

On November 12, 1993, David Oldroyd of the School of Science and Technology Studies at the University of New South Wales in Australia wrote to the Bhaktivedanta Institute. He stated that he had been asked to write an essay on *Forbidden Archeology for Social Studies of Science*, one of the main journals in the field of science studies and asked for a review copy. Although Oldroyd expressed no opinion about the book, he said he did think it worthy of attention.

The promised review has not yet come out. But I did receive a letter (November 12, 1993) from one of Oldroyd's graduate students, Jo Wodak, who was about to begin her honors thesis on *Forbidden Archeology*. She expressed some surprise that she had not encountered in her studies the evidence documented in the book and noted that she had herself come across some examples of suppression of evidence. *[In the original draft of this paper, I included quotations from letters to me by Oldroyd and Wodak, but they have since indicated that they do not wish to have their correspondence with me directly quoted in this book. I have therefore altered the paper to conform to their wishes. Whatever references I have made to their correspondence with me should not be taken as contradictory to their published opinion of* Forbidden Archeology, *included in Chapter Two of this book.]*

Wodak's readiness to accept the possibility of suppression of controversial evidence and unorthodox views is welcome. From the standpoint of Gaudiya Vaishnavism, *Forbidden Archeology*'s deconstruction of modern science involves two issues. The first is the origin and antiquity of the human species. The second is the process by which knowledge of the first issue may best be obtained. In this regard, *Forbidden Archeology* is essentially an argument for the epistemic superiority of received transcendental knowledge to empirically manufactured knowledge. Exposing the shortcomings of the latter increases the viability of the former (the sacred texts of Gaudiya Vaishnavism).

In December of 1994, I attended the World Archaeological Congress-3 in New Delhi, India, where I presented my paper (Cremo, 1994) titled "Puranic Time and the Archeological Record." Several Indian archeologists and anthropologists congratulated me on the paper and requested copies. They found the image of a Western convert to Gaudiya Vaishnavism presenting such a paper at a major scientific gathering intriguing. While standing in one of the lobbies of the Taj Palace Hotel between sessions, I was approached by Tim Murray, an archeologist from La Trobe University in Australia. "Oh, so you're Cremo," he said. He had recognized my name on my badge and announced to me that he had recently written a review of *Forbidden Archeology* for *British Journal for the History of Science (BJHS)*. Murray told me that he teaches history of archeology and that he had recommended *Forbidden Archeology* to his graduate students. He told them that if one was going to make a case for extreme human antiquity, *Forbidden Archeology* was the way to do it. I have not yet seen the *BJHS* review.

At the 1994 annual meeting of the History of Science Society (HSS), A. Bowdoin Van Riper, an authority on history of human antiquity investigations, requested from a Bhaktivedanta Institute publishing representative a review copy of *Forbidden Archeology*. Van Riper said he sometimes reviewed books for *Isis,* the journal of the HSS. The representative told Van Riper that he was not authorized to give away a free copy. When informed of this later, I personally sent a copy of *Forbidden Archeology* to Van Riper, who replied on January 5, 1995: "The premise is audacious, to say the least, and intriguing to me if only for that reason." I am hopeful he will review the book for *Isis.*

There is also a chance for a review of *Forbidden Archeology* in *Bulletin of the History of Archaeology.*

On September 28, 1994, Henry H. Bauer, book review editor for *Journal of Scientific Exploration*, wrote to our publishing branch requesting a copy of *Forbidden Archeology*. Bauer explained that one of the goals of the journal was to advance the study of anomalous phenomena in various areas of scientific investigation.

I am eagerly looking forward to the reviews of *Forbidden Archeology* in *Social Studies of Science, British Journal for the History of Science*, and

Journal of Scientific Exploration. Until they come out, it is difficult to gauge the impact the book is having in the science studies community. It is interesting that *Forbidden Archeology* is being treated as both a science text and a science studies text.

Religion Studies

From the beginning, I thought the acknowledged Gaudiya Vaishnava foundation of *Forbidden Archeology* would draw the attention of religion scholars. A review is forthcoming in *Science & Religion News*, published by the Institute on Religion in an Age of Science.

Early in 1994, Mikael Rothstein of the Institute for the History of Religion at the University of Copenhagen wrote a major article about *Forbidden Archeology* for publication in *Politiken*, Denmark's largest and most influential newspaper (Rothstein, 1994). In a letter to me dated February 2, 1994, Rothstein said:

> The text refers to your points through examples and compares your message to that of the evolutionists at Darwin's time in order to demonstrate how the positions have changed. Today the creationists deliver the provoking news. Previously this was the function of the evolutionists. The article acknowledges your solid argumentation, which is often more than hard to refute, but I do not present any judgment as to whether you are right in your conclusions....However, I find the book amazing in many ways and hopefully I have made my modest contribution to get it sold....For the record: The article is placed in the specific science-section of the paper, and it is entitled (as you may understand after all) "Forbidden Archeology." The subtitle reads: "Religious scientists provoke the theory of evolution." "Religious" because I mention your affiliation with ISKCON and state that you have a religious interest in your otherwise scholarly enterprise.

The *Politiken* article also mentions my affiliation with the Bhaktivedanta Institute, and identifies my spiritual commitment to Vaishnavism. The article resulted in inquiries from one of Denmark's largest publishers about translation rights for *Forbidden Archeology*.

Out of the Mainstream

Up to now, we have been looking at reactions to *Forbidden Archeology* from mainstream scholars and journals. Now let us move out of the mainstream. As we do, you will notice a change in climate, as we encounter some unqualified endorsements of *Forbidden Archeology*.

In the fall 1994 issue of *Journal of Unconventional History, Forbidden Archeology* was one of several books discussed in a review essay by Hillel Schwarz:

> *Forbidden Archeology* takes the current conventions of decoding to their extreme. The authors find modern *homo sapiens* to be continuous contemporaries of the apelike creatures from whom evolutionary biologists usually trace human descent or bifurcation, thus confirming those Vedic sources that presume the nearly illimitable antiquity of the human race—all toward the implicit end of preparing us for that impending transformation of global consciousness at which Bhaktivedanta brochures regularly hint.... Despite its unhidden religious partisanship, the book deserves a reckoning in this review for its embrace of a global humanity distinct from other primates...Meditating upon our uniqueness (I am here supplying the missing links of the thesis) we may come to realize that what can change (awaken) humanity is no mere biochemical exfoliation but a work of the spirit, in touch with (and devoted to) the ancient, perfect, perfectly sufficient, unchanging wisdom of the Vedic masters (Schwarz, 1994).

William Corliss is the publisher of several "sourcebooks" of well-documented anomalous evidence in different fields of science. Most university libraries have copies. Corliss also sells books by other authors, which he lists in a supplement to his newsletter. In *Science Frontiers Book Supplement* number 89 (September-October 1993), Corliss prominently featured *Forbidden Archeology*:

> *Forbidden Archeology* has so much to offer anomalists that it is difficult to know where to start. One's first impression is that of a massive volume bearing a high price tag. Believe me, *Forbidden Archeology* is a great bargain, not only on a cents-per-page basis but in its systematic collection of data challenging the currently accepted and passionately defended scenario of human evolution.... Here are fat chapters on incised bones, eoliths, crude tools, and skeletal remains—all properly documented and detailed, but directly contradicting the textbooks and museum exhibits.... The salient theme of this huge book is that human culture is much older than claimed (Corliss, 1993).

I liked having *Forbidden Archeology* in Corliss's catalog. I knew it might generate some sales to university libraries. But more importantly, Corliss was a pipeline to thousands of readers who were deeply interested in the whole subject of anomalous evidence. This audience (the serious scientific anomaly community) was quite prepared to accept that whole areas of science were completely wrong and that evidence was being unfairly suppressed. But this audience was also interested in good documentation for such claims.

Forbidden Archeology has also been reviewed in *Fortean Times* (Moore, 1993), a journal dedicated to the study of "Fortean phenomena" (extreme scientific anomalies), in *FATE* (Swann, 1994), a popular magazine featuring accounts of the paranormal, and in the journals and newsletters of societies focusing on anomalous archeological discoveries and evidence for pre-Columbian contacts between the Americas and the Old World (e.g. Hunt, 1993). Needless to say, the reviews are positive.

In *Forbidden Archeology*, I documented the work of scientists who held positions in mainstream science institutions but who had reported on anomalous archeological discoveries. Some of these scientists, as expected, were very pleased with *Forbidden Archeology*. Virginia Steen-McIntyre, a geologist, had reported a date of over 250,000 years for the Hueyatlaco site near Puebla, Mexico. Thereafter, her career trajectory took a sharp turn downwards. About *Forbidden Archeology*, she said in a letter dated October 30, 1993:

> What an eye-opener! I didn't realize how many sites and how much data are out there that don't fit modern concepts of human evolution. Somewhere down the line the god of the Vedas and the God of the Bible will clash .... But until then the servants of both can agree on one thing—human evolution is for the birds! ...I'm doing my bit getting the publicity out for your book. Have ordered a copy for the local library.... I'm also sending the book review that appeared in Sept./Oct. *Science Frontiers Book Supplement* to various friends and colleagues (almost 50 so far). I predict the book will become an underground classic. Whether it will break into the mainstream media is questionable—the Illuminati are tightly in control there.

Forbidden Archeology, like a robot surveyor on Mars, was sending back signals to me as it mapped a complex cognitive domain. I am reporting in this paper primarily the preliminary basic mappings of that terrain, but this snippet from Virginia Steen-McIntyre provides a higher resolution look at the mapping process in a confined space. Observe the connections—the copy to the public library, the link with the Corliss newsletter book supplement, the mailing to friends and colleagues (working in geology, archeology and anthropology), the pointer to other discourse communities (the conspiracy theorists) and reference to the metarelationships among Vedas, Bible and Science. I shall return to this theme later—*Forbidden Archeology* as robot mapper of alien discourse terrains, prober of new channels and portals of complex border crossing connectivity. For those concerned about the future of scholarly life on this planet, take note—this is where we are heading.

New Age

In *International Journal of Alternative and Complementary Medicine (IJACM)*, John Davidson (1994) reviewed *Forbidden Archeology*, saying:

> Michael Cremo and Richard Thompson are ...to be congratulated on spending eight years producing the only definitive, precise, exhaustive and complete record of practically all the fossil finds of man, regardless of whether they fit the established scientific theories or not. To say that research is painstaking is a wild understatement. No other book of this magnitude and caliber exists. It should be compulsory reading for every first year biology, archaeology and anthropology student—and many others, too!

IJACM can fairly be placed within or on the borders of the New Age scientific discourse community. The same is true of the *Adventures Unlimited 1994 Catalog*, which featured *Forbidden Archeology* in its new book section, with a blurb describing *Forbidden Archeology* as "a thick (nearly 1000-page) scholarly work that confronts traditional science and archaeology with overwhelming evidence of advanced and ancient civilizations." The catalog cover has these additional subtitles: "Inside...Ancient Wisdom, Lost Cities, Anti-Gravity, Tesla Technology, Secret Societies, Free Energy Science, Exotic Travel...and more!" and "Frontiers in Travel. Archaeology, Science & History." As an author with pretensions to academic respectability, am I embarrassed to find my book in such company? No. I simply notice that *Forbidden Archeology* has mapped both *American Journal of Physical Anthropology* and *Adventures Unlimited 1994 Catalog* as part of its discursive domain. And in addition to simply noting the mapping, I will offer a suggestion that the easy mobility of the exploratory text called *Forbidden Archeology* through the different regions of its domain points to the disintegration of what one might call the Enlightenment Consensus. The Enlightenment Consensus was marked by orogenic episodes that cut an existing domain of discourse (in which Newton could write both his *Principia* and his *Alchemy*) into noncommunicating domains of science and "pseudoscience." All that is now changing, perhaps faster than we can accurately measure.

Barbara and Dennis Tedlock, editors of *American Anthropologist*, have noted (Tedlock & Tedlock, 1995):

> New Age titles in bookstores outnumber anthropological ones, and the kinds of titles we once disliked seeing in the same section with anthropology or archaeology now occupy whole sections of their own—the shamanism, goddess worship, and New Age sections. These shifts reflect social and cultural developments that are well under way not only in this country but all over the world—north or south, east or west. The urban participants in the New Age and related movements have lines of communication that reach into the remotest deserts, jungles, and mountains of our own traditional field research. In some ways these developments look like a privatization of the educational tasks we once saw as our own.

Nonmainstream Religion

In the realm of nonmainstream religion, *Forbidden Archeology* has found its way into many unusual spaces. In a letter to me dated June 28, 1993, Duane Gish of the Institute for Creation Research (ICR), a Christian fundamentalist organization, noted that he found *Forbidden Archeology* "quite interesting and perhaps useful to us." In 1994, I visited the ICR in Santee, California, and spoke with Gish, who purchased copies of *Forbidden Archeology* for the ICR library and research staff.

During 1994, I appeared on fundamentalist Christian radio and television programs, which were, of course, favorable to the antievolution message of *Forbidden Archeology*. This was true despite the displays of sectarian feeling Christian fundamentalists sometimes manifest in relation to Asian religions, especially those popularly labeled as "new religions."

Siegfried Scherer, a microbiologist at a German university and also a young-earth Christian creationist, contributed a jacket blurb for *Forbidden Archeology* even though aware of the Gaudiya Vaishnava backgrounds of the authors.

The Vishwa Hindu Parishad (VHP) is a worldwide Hindu religious and cultural organization. It is generally seen as conservative, even fundamentalist. On August 2, 1993, Kishor Ruperelia, general secretary of the VHP in the United Kingdom, faxed this message to *Back to Godhead*, the bimonthly magazine of the International Society for Krishna Consciousness (ISKCON):

> I have just received the May/June 1993 issue (Vol. 27, No. 3) of the magazine *Back to Godhead*, and I am writing with reference to the condensed form article on the book *Forbidden Archeology*, written by ISKCON researchers Michael Cremo and Richard Thompson.... Having read the article, I consider it very important that a meeting be held between the authors of the book and some of the Indian scholars who are in the USA at present to participate in a world conference organized by the VHP of America under style of "Global Vision 2000" to take place in Washington DC on Aug. 6th 7th & 8th.... Inspired by the Vedic writings and encouraged by His Divine Grace A. C. Bhaktivedanta Swami Prabhupada, the scholarly authors have made a tremendous and painstaking effort to compile and compare umpteen evidences to make archeological scholars rethink about the predominant paradigm on human origin and antiquity.

Of course, this is just the sort of reaction one might expect from a conservative, traditionalist Hindu cultural and religious organization such as the VHP. Less expected would be the extremely favorable review of *Forbidden Archeology* by Islamic scholar Salim-ur-rahman that appeared in a Pakistani newspaper. Here is an excerpt:

> *Forbidden Archeology* is a serious and thought-provoking book, reminding us that the history of the human race may be far more older than we are led to imagine.... In a way, this all ties up with the remarks attributed to the Holy Prophet (PBUH), Hadhrat Ali and Imam Iafar Sadiq in which they said that the Adam we are descended from was preceded by numerous Adams and their progeny.

Forbidden Archeology possesses a remarkable capacity for border crossing—here we find literalist followers of Christianity, Hinduism and Islam (not always the best of friends in some situations) according a respectful welcome to a text with Gaudiya Vaishnava foundations. And it does not stop there.

Libraries and Media

By having a librarian do an online computer search, I have learned that dozens of university libraries have acquired *Forbidden Archeology*, even though the Bhaktivedanta Institute publishing branch has not yet made a systematic approach to them (we have been waiting for some of the forthcoming reviews in academic journals to actually come forth). There have also been some spontaneous requests for inspection copies by university teaching professors, and a search of sales reports from our book trade distributor shows a good number of orders from university bookstores.

Internationally, several publishers have been expressing interest in *Forbidden Archeology* and its popular version *The Hidden History of the Human Race*. A German edition of *Forbidden Archeology* is already in print and selling well. A Mexican publisher has recently acquired Spanish translations rights to *Hidden History*. Inquiries have also been received from Indian, Russian, Slovenian, Indonesian, Dutch, Japanese, French, Belgian, Danish and Swedish publishers. For example, Monique Oosterhof of the Dutch firm Arena wrote on June 14, 1994:

> We are very interested in the book *The Hidden History of the Human Race* by Michael A. Cremo and Richard L. Thompson. If the Dutch rights are still free, could you please send us a copy of the book? Thank you in advance. Arena is one of the leading Dutch publishers. We publish international literature of high level: Benoit Groult, Viktor Jerofejev, Charles Johnson, Shere Hite, Eduardo Mendoza, Laura Esquivel, Helen Zahavi, Bernice Rubens, Meir Shalev, Klaus Mann, Carmen Martin Gaite, Harold Brodkey, Joan Didion, Kaye Gibbons etc.

During the fall and winter of 1994, I went on an author's tour to promote *Hidden History* in the United States. I was guest on over 60 radio and television shows, ranging from the sensationalistic *Sightings* television show produced by Paramount to the high-brow *Thinking Allowed*, which airs on 80 PBS television stations nationwide. Surprisingly, I found the hosts receptive to the basic message that Western science was not telling the truth about human origins. This was also true of the people who called in on the talk radio shows. Even more surprisingly, I found a great deal of interest in the Gaudiya Vaishnava alternative to the current theory of human evolution. In terms of domain mapping, I found the nationally syndicated radio talk shows of Laura Lee, Art Bell and Bob Hieronimus to be quite significant. Each host focuses exclusively on scientific anomalies, ranging from UFOs and crop circles to archeological mysteries of the kind found in *Forbidden Archeology*.

As another illustration of the connectivity manifested by *Forbidden Archeology* in its exploratory domain mapping, I offer the following. *Forbidden Archeology* was very positively reviewed in the *Hazelton Standard-Speaker*, a Pennsylvania newspaper (Conrad, 1993). The title of the article, which featured a large blow-up of the cover of *Forbidden*

Archeology, was "New book claims man existed on earth long before the apes." On December 18, 1994, Laura Cortner, executive producer of Hieronimus & Co.: 21st Century Media Source, wrote:

> We are very interested in reviewing a copy of your book *Forbidden Archeology* which we read about in an article by Ed Conrad in the Hazelton (PA) *Standard-Speaker*, 11/17/93. We have long been interested in doing a special program on the subject of archeological finds that challenge the earliest recorded history of humans for quite some time, and we are encouraged to learn of an academic book with authors we could interview on the radio. Our programs are designed to educate our listeners on a wide variety of subjects that are usually not covered in the major media, and we know your book will be of interest to them.

Shortly thereafter, I was guest for two hours on the Bob Hieronimus radio show. I returned to the show for another appearance later in 1994. At the request of the producers, copies of the abridged version of *Forbidden Archeology* were provided to them for distribution to other guests appearing on the show. On February 7, 1995, Laura Cortner wrote to me:

> Enclosed is the latest letter of praise we have received from one of our recent guests on 21st Century Radio to whom we have presented a copy [an English zoologist who specializes in "living fossils"]. Thank you very much for supplying us with extra copies to continue this type of promotion. In the last month we have also sent copies to two of the creators of Howdy Doody, and to Peter Occhiogrosso, author of *The Joy of Sects: A Spiritual Guide to the World's Religions* (Doubleday).

Tracking *Forbidden Archeology* can be quite dizzying, as it moves from the pages of *American Journal of Physical Anthropology* to the *Hazelton Standard-Speaker* to the airwaves of 21st Century Radio and then into the hands of an English zoologist and the creators of Howdy Doody.

Cyberspace

Forbidden Archeology has also invaded cyberspace. Not long ago, a friend told me that the introduction to *Forbidden Archeology*, complete with a color image of the book cover, had appeared on somebody's home page on the World Wide Web (WWW). Net surfers can check it out at:

http://zeta.cs.adfa.oz.au/Spirit/Veda/Forbidden-Archeology/forbidden-arch.html

A search through the WWW detected the presence of *Forbidden Archeology* in several online bookstores. *Forbidden Archeology* has also been responsible for some searing flame wars in discussion groups such as talk.origins. I participated under my Gaudiya Vaishnava initiation name, Druta Karma. Anomalous human skeletal remains found in deposits over 2 million years old at Castenedolo, Italy, in the late nineteenth century were one of the hotter topics. At one point, my chief opponent posted this text (name deleted):

> Subject: Re: Castenedolo (Help!)
> Organization: HAC — Johns Hopkins University, Baltimore
> I have spent the last two days trying to find out as much as I could about the Castenedolo finds. The result is disappointing. I found only one reference that even mentioned Castenedolo, and that was a reference given by Mr. Karma....Mr. Karma's post seems impressive, and I am (I admit) not easy to impress....Basically, I've reached a dead-end. I can't comment on the finds much, but I do have some questions. It seems that Mr. Karma has effectively dealt with my objections. I currently consider Mr. Karma's Castenedolo post unchallenged on talk.origins. Anyone else willing to give it a try?

Conclusion

It's time to interrupt the transmissions from *Forbidden Archeology* as it continues to map a new terrain of discourse. The preliminary mapping illuminates an ongoing process of global cultural realignment and transition, wherein Western science finds itself retreating, somewhat unwillingly, from its previous position of self-proclaimed epistemic superiority and coming into an intellectual world-space where it finds itself just one of many knowledge traditions.

The responses to *Forbidden Archeology* from within the network of modern science show a degree of resistance to this developing reality and a hope that the system as it is, perhaps with some adjustments, will survive intact.

But, as one can see, the expanding topology of the terrain mapped by *Forbidden Archeology* reaches far beyond artificial interdisciplinary realignments within the modern university system. The familiar unities are dissolving. As foreseen by Foucault (1972, p. 39): "One is forced to advance beyond familiar territory, far from the certainties to which one is accustomed, towards a yet uncharted land and unforeseeable conclusion."

Specifically, *Forbidden Archeology* charts the domain of the discourse of human origins and antiquity. One can no longer hope that this will remain the inalienable property of a certain discipline, such as archeology or anthropology. Neither can salvation be found in forging new interdisciplinary links with other fragmenting disciplines. One cannot even be certain that disciplines such as anthropology can avoid marginalization, a possibility much discussed among anthropologists themselves.

The map generated by *Forbidden Archeology* points to the emergence of a diverse multipolar global intellectual constellation from which may emerge a new academic consensus on human origins. Those participating most effectively in this process will be those who have mastered the techniques of complex boundary crossings, able to move freely with open minds, through scientific disciplines such as anthropology and archeology, the science studies branches of history, philosophy and sociology, the academic study of religion, the populist purveyors of scientific anomalies, the range of New Age interests and the whole world of religion-based sciences and cosmologies of traditional spiritual cultures, such as Gaudiya Vaishnavism.

References Cited

Basham, A. L. (1959). *The wonder that was India.* New York, NY: Grove Press.

Conrad, E. (1993). New book claims man existed on earth long before the apes. *Hazelton [PA]Standard-Speaker*, November 17.

Corliss, W. (1993) Forbidden archeology. *Science Frontiers Book Supplement, 89*, 1.

Cremo, M. A. (1994). Puranic Time and the Archeological Record. *Theme Papers: Concepts of Time. World Archaeological Congress-3, New Delhi, December 4-11, 1994.* Bound volume of precirculated papers, issued on behalf of the Academic Committee of WAC-3.

Cremo, M. A., & Thompson, R. L. (1993). *Forbidden archeology: The hidden history of the human race.* San Diego, CA: Bhaktivedanta Institute.

Davidson, J. (1994) Fascination over fossil finds. *International Journal of Alternative and Complementary Medicine*, August, 28.

Feder, K. L. (1994). Review of *Forbidden archeology: The hidden history of the human race.* Michael A. Cremo and Richard L. Thompson, 1993, Govardhan Hill Pub., San Diego. *Geoarchaeology: An International Journal, 9,* 337-340.

Foucault, M. (1972). *The archaeology of knowledge.* New York, NY: Pantheon.

Goswami, S. D. (1980) *Shrila Prabhupada-lilamrta, A Biography of His Divine Grace A. C. Bhaktivedanta Swami Prabhupada, Founder-Acharya of the International Society for Krishna Consciousness. Volume 1. A Lifetime in Preparation, India 1896-1965.* Los Angeles, CA: Bhaktivedanta Book Trust.

Greimas, A. J. (1990). *Narrative semiotics and cognitive discourses.* London, UK: Pinter.

Holmes, W. H. (1899). Review of the evidence relating to auriferous gravel man in California. *Smithsonian Institution Annual Report 1898-1899*, 419-472.

Hunt, J. (1993). Antiquity of modern humans: Re-evaluation. *Louisiana Mounds Society Newsletter, 64*, 2-3.

Leakey, M. (1979) Footprints in the ashes of time. *National Geographic, 155*, 446-457.

Marks, J. (1994). *Review of Forbidden archeology: The hidden history of the human race*, by Michael A. Cremo and Richard L. Thompson. 1993. San Diego: Bhaktivedanta Institute. *American Journal of Physical Anthropology, 93*, 140-141.

Moore, S. (1993). *Forbidden archeology: The hidden history of the human race*, by Michael A. Cremo and Richard L. *Thompson. Fortean Times, 72*, 59.

Rothstein, M. (1994). Forbudt arkaeologi. *Politiken,* January 31, section 3, page 1.

Schwarz, H. (1994). Earth born, sky driven: Book review. *Journal of Unconventional History, 6*, 68-76.

Slemmons, D. B. (1966). Cenozoic volcanism of the central Sierra Nevada, California. *Bulletin of the California Division of Mines and Geology, 190,* 199-208.

Salim-ur-rahman (1994) Spanner in the works. *The Friday Times,* April 21.

Swann, I. (1994). *Forbidden archeology: The hidden history of the human race.* Michael A. Cremo and Richard L. Thompson. Fate, January, 106-107.

Tedlock, B., & Tedlock, D. (1995). From the editors. *American Anthropologist, 97,* 8-9.

Whitney, J. D. (1880). The auriferous gravels of the Sierra Nevada of California. *Harvard University, Museum of Comparative Zoology Memoir 6*(1).

Yau, J. (1993). *In the realm of appearances: The art of Andy Warhol.* Hopewell, NJ: Ecco.

1.3. Michael A. Cremo (1995) "Forbidden Archeology: Evidence for Extreme Human Antiquity and the Ancient Astronaut Hypothesis." Ancient Astronaut Society World Conference, Bern, Switzerland.

I presented this paper at the Ancient Astronaut Society World Conference, held in Bern, Switzerland, August 17-19, 1995. A slightly edited version of the paper appears in Ancient Skies *(Vol. 22, No. 4, September-October 1995, pp. 1-4).* Ancient Skies *is the newsletter of the Ancient Astronaut Society (1921 St. Johns Avenue, Highland Park, IL 60035, U.S.A.). A German translation of the complete text is scheduled to appear in a collection of papers edited by Ulrich Dopatka and published by ECON Publishers of Germany.*

Proponents of the ancient astronaut hypothesis say that many features of human civilization can be explained by past extraterrestrial contacts. Some researchers explain the very origin of the human species through such contacts. Generally, the proposed scenarios involve accepting the standard evolutionary account of hominid evolution, with a relatively late extraterrestrial intervention giving rise to anatomically modern human beings. In this paper, I shall not directly discuss specific theories of human origins involving extraterrestrial intervention. Instead, I will offer a general caution that such speculations regarding human origins should take into account the actual physical evidence for human antiquity. Careful investigation shows that the full range of this evidence is not reported in current literature. When all relevant evidence is taken into account, it becomes apparent that one should not link extraterrestrial theories of human origins too closely to the currently accepted scientific ideas of human evolution.

According to standard ideas, the human line branched off from the ancestors of the modern chimpanzees about 5 million years ago. The first hominids, or humanlike primates, were the australopithecines. Further developments led to *Homo habilis* and *Homo erectus*. Finally, at about 100,000 years ago, anatomically modern humans like ourselves appeared. Scientists say the factual evidence supports this view and no other. But in 1984, at the request of my colleague Dr. Richard Thompson of the Bhaktivedanta Institute, I launched an 8-year investigation into the entire history of archeology and anthropology, with some astonishing results. These results were reported in the massive 900-page book *Forbidden Archeology* (Cremo and Thompson 1993), which provoked shockwaves of outrage from mainstream archeologists and anthropologists. This book has since been released in abridged popular form as *The Hidden History of the Human Race* (Cremo and Thompson 1994). The basic message of these books is this—the complete archeological record shows that humans have been present on this planet for hundreds of millions of years.

Our Ancient Astronaut Society colleague Walter J. Langbein put it very nicely in a review of *Forbidden Archeology* for the Austrian journal *PARA:* "If we imagine the history of humanity as a giant museum, containing all knowledge on this topic, then we shall find that several of the rooms of this museum have been locked. Scientists have locked away the facts that contradict the generally accepted picture of history. Michael A. Cremo and Richard L. Thompson have, however, opened many of the locked doors and allowed lay persons as well as scientists to see inside" (Langbein 1994).

So what kind of evidence do we find in the locked rooms of the museum of the history of humanity? Let us begin with one example, which is typical of many others documented in *Forbidden Archeology*. In the year 1852, the journal *Scientific American* (June 5) carried an intriguing report of a metallic vase blasted out of solid rock at Dorchester, Massachusetts, near the city of Boston. Describing the vase, which was about 4.5 inches high, the report said: "The body of this vessel resembles zinc in color, or a composition metal in which there is a considerable portion of silver. On the side there are six figures of a flower or bouquet beautifully inlaid with silver, and around the lower part of the vessel, a vine, or wreath, also inlaid with silver. The chasing, carving, and enlaying are exquisitely done by the art of some cunning workman. This curious and unknown vessel was blown out of the solid pudding stone, fifteen feet below the surface." The "pudding stone" is known to geologists as the Roxbury Conglomerate, and according to the United States Geological Survey, this rock is of Precambrian age, over 600 million years old. Given that the vase is of human manufacture, this is quite extraordinary. According to standard views, there would have been no life on land at this time, and only simple forms of marine life in the earth's oceans.

Before proceeding on to other examples of evidence for extreme human antiquity, let me pause to explain why I undertook this research effort. As a member of the Bhaktivedanta Institute, which is the science studies branch of the International Society for Krishna Consciousness, I have deeply studied the ancient Sanskrit writings of India. Among these writings are the *Puranas,* or histories. The *Puranas* contain accounts of human civilizations existing on this planet for hundreds of millions of years. During this time, human beings like ourselves coexisted with intelligent races of apelike creatures, such as the Vanaras. For example, in the Indian epic called the *Ramayana,* which deals with events that took place over one million years ago, we find the humanlike *avatara* Ramacandra leading an army of Vanaras. This suggests that in the distant past humans coexisted with more apelike humanoid creatures and did not evolve from them.

Furthermore, the Puranic time is cyclical, rather than linear. In other words, there is a recurrent patterning of time. This time concept is very similar to the time concept of the classical Greek thinkers, such as Plato and Aristotle. Aristotle, for example, said that the great achievements of human civilization had been invented several times in the past, in the course of cyclical time (*Metaphysics* 1074 b 10, *Politics* 1329 b 25).

As I mentioned in a paper that I presented at the World Archeological Congress in New Delhi last December (Cremo, in press), one can from the Puranic model of time make two predictions about the archeological record. The first prediction is that one should expect to find a bewildering mixture of human fossils and artifacts, some appearing quite advanced and others appearing quite primitive, going back hundreds of millions of years. And in truth, one does find this. And the second prediction is that this archeological evidence for coexistence of humans and more apelike creatures will be edited by establishment scientists to conform to their linear progressive concept of time and evolution, with simple forms existing earlier and more complex ones later. And, in fact, one does find this. Scientists have indeed selectively suppressed the abundant evidence for the extreme antiquity of anatomically "modern" humans. This suppression has taken place by a process of knowledge filtration, whereby reports conforming to certain preconceived notions are preserved in scientific discourse and reports not conforming to these preconceived notions are dropped from scientific discourse. I am not here talking of the normal process of sifting good reports from bad, by impartial application of reasonable standards. Instead, I am referring to the unfair application of a double standard in the treatment of evidence.

Supporters of the ancient astronaut hypothesis are well aware of how this knowledge filtration process operates in suppressing evidence for extraterrestrial contacts with human civilizations. So we should not be surprised to see the same process of suppression operating in other critical areas of scientific inquiry. Certainly, we should not be complacent and assume without question that what we are told by spokespersons for the scientific establishment about human origins and antiquity is necessarily correct.

The process of knowledge filtration in the study of human origins has been going on systematically for about 150 years. Let us now consider some specific cases of how the process of knowledge filtration operates. One such case is from the nineteenth century, when the active suppression of evidence for extreme human antiquity was beginning. In 1849, gold was discovered in California. Miners rushed there to extract it. In the beginning they simply panned gold from the streams, but later they began digging mineshafts into the sides of mountains. Many such mines were opened at Table Mountain in Tuolumne County. The deposits at Table Mountain are covered by hundreds of feet of solid basalt, preventing objects from entering from above. The gold-bearing gravels near the bedrock are said by modern geologists to be from 33 to 55 million years old (Slemmons 1966). Miners recovered human artifacts and human skeletal remains from mineshafts at this level. Such artifacts and fossils were also found in gravels from the higher sub-basaltic gravels, which are at least 9 million years old. These objects were found not only at Table Mountain but at many other locations in the same region. Among the artifacts discovered in the mines were obsidian spear points and stone mortars and pestles, such as the stone mortar and pestle recovered by J. H. Neale 1,500 feet inside the Montezuma Tunnel mineshaft at Table Mountain. This specimen, found in the gravels near the bedrock, would thus be between 33 and 55 million years old. All of these discoveries were collected and reported to the scientific world by Dr. J. D. Whitney (1880), the state geologist of California, in a book published by Harvard University's Peabody Museum of Natural History.

One might argue that the discoveries were made by miners rather than professional scientists. But one must recognize that most of the Java *Homo erectus* discoveries and many of the *Australopithecus* discoveries in Africa were made by paid native collectors who then turned them over to professional scientists. Furthermore, a professional scientist, Clarence D. King, a geologist of the United States Geological Survey, made an important discovery of a stone grinding implement at Table Mountain (Becker 1891, pp. 193-194). He personally took it out of the rock, in which it was tightly embedded. The rock at that level was at least 9 million years old.

So, what happened to these very well-documented discoveries, placing human beings in California up to 55 million years ago? William H. Holmes, a very powerful anthropologist working at the Smithsonian Institution, in Washington, D. C., used his prestige and influence to discredit Dr. Whitney and his discoveries. Why? Because they contradicted the emerging theory that human beings had evolved fairly recently from more apelike ancestors.

Holmes (1899, p. 424) wrote: "Perhaps if Professor Whitney had fully appreciated the story of human evolution as it is understood today, he would have hesitated to announce the conclusions formulated, notwithstanding the imposing array of testimony with which he was confronted." In other words, if the facts do not conform to the favored theory, then the facts, even an imposing array of them, must be set aside.

Holmes found it troubling that the implements found in the mines resembled those of the California Indians of recent history. Striking a similar note, archeologist George F. Carter, noted for his controversial views on early man in North America, wrote to me on January 26, 1994. He asked if I thought that there had been several appearances of toolmaking humans in California over the course of millions of years. Yes, and why not? That is exactly what the evidence suggests—an extreme antiquity for the human race with recurring cultural manifestations in the course of cyclical time.

So here we have some very good evidence for the existence of human beings as much as 55 million years ago in California. Some might be more happy with remains of metal machines than stone tools, but that is simply a matter of taste. Stone tools of the kind found in California are just as good evidence for a human presence as metal machinery. Furthermore, our present kind of industrial civilization may be a rather unique occurrence in cyclical time. Even in terms of classical history, our current level and mode of industry is a product of the past couple of centuries. Preceding it were many highly developed urban civilizations that were not based on widespread modern manufacturing technologies. And in the not too distant future humanity might return to that more traditional pattern. And while this does not rule out the existence of high technology artifacts in the distant past, it may mean that they were somewhat rarer than we might expect, given our current level of industrial civilization. In other words, we should not *necessarily* expect to find lots of computers and automobiles in ancient deposits.

The California discoveries also provide very good evidence for suppression of evidence by Holmes and others. Someone might argue that such things may have happened in the nineteenth century but not today. But as we shall see in our next case, suppression of evidence that contradicts the idea of a recent human origin is still going on. In the 1970s, Virginia Steen-McIntyre was a young geologist working for the United States Geological Survey. She took part in the dating of an archeological site in Mexico—at a place called Hueyatlaco near the city of Puebla. Anthropologist Cynthia Irwin-Williams had recovered advanced stone tools from this site. Tools of this level of sophistication are normally attributed to anatomically modern humans, not to apemen such as *Homo erectus*. Using four different methods (uranium series, tephra hydration, fission track and stratigraphy), Virginia Steen-McIntyre and her colleagues obtained dates of over 250,000 years for the Hueyatlaco site (Steen-McIntyre *et al.* 1981).

This was unexpected for two reasons. According to standard views, human beings did not enter North America until about 25,000 years ago, at most, with conservative scientists favoring an entry time of 12,000 years. Furthermore, human beings capable of making such tools did not come into existence until about 100,000 years ago. The principal anthropologist at the site was quite unhappy with the dates obtained by Virginia-Steen McIntyre and her colleagues. She wanted a date of 25,000 or less, not 250,000 years! Virginia Steen-McIntyre, however, remained firm in her conviction that the age of 250,000 years was correct. But she paid a considerable price for her conviction. She found it difficult to get her report published, she was labeled a publicity seeker within her profession, she lost a teaching position she held at an American university and found that her career as a geologist was blocked.

Describing her experience with the knowledge filtration process, Virginia Steen-McIntyre wrote in a letter (March 30, 1981) to Estella Leopold, associate editor of *Quaternary Research*: "The problem as I see it is much bigger than Hueyatlaco. It concerns the manipulation of scientific thought through the suppression of 'Enigmatic Data,' data that challenges the prevailing mode of thinking. Hueyatlaco certainly does that! Not being an anthropologist, I didn't realize the full significance of our dates back in 1973, nor how deeply woven into our thought the current theory of human evolution has become. Our work at Hueyatlaco has been rejected by most archaeologists because it contradicts that theory, period."

The case of Virginia Steen-McIntyre shows the suppression of evidence by very direct means. In other cases, the process of knowledge filtration is more subtle. It can take the form of an inability to properly evaluate evidence because of strongly held theoretical preconceptions. For example, in 1979 researchers in Tanzania found sets of footprints in volcanic ash deposits about 3.6 million years old. According to Mary Leakey (1979) and other scientists, these footprints are indistinguishable from modern human footprints. The usual explanation is that the footprints were made by *Australopithecus,* the apeman of that period. But careful study shows that none of the fossil foot bones of *Australopithecus* fit the Laetoli prints (Tuttle 1985). Among other things, *Australopithecus* had toes much longer than those of modern human beings. Others have suggested that an as yet unknown apeman with humanlike feet made the prints. This is possible, but as of today, the only creatures known to science that could make the prints are human beings like ourselves. But most scientists, including those involved in the Laetoli discoveries, would not even dream of considering this possibility. They are absolutely convinced that anatomically modern human beings evolved about 100,000 years ago and could not possibly have been present 3.6 million years ago in Africa.

One might ask if there are any human skeletal remains of that age, and the answer is yes. For example, such fossils occur at Castenedolo, in northern Italy, near Brescia. There the Italian geologist Giuseppe Ragazzoni (1880) collected bones of 4 individuals from a blue clay formation of Middle Pliocene age—about 3 or 4 million years old. The skeletal remains show the Castenedolo individuals were anatomically modern. Some have suggested that the skeletons arrived in their positions by burial in fairly recent times, but as a professional geologist Ragazzoni was well aware of this possibility. He carefully inspected the overlying layers of sediment and found them undisturbed. A skeleton of similar age was found by other researchers at Savona, Italy, and details of its discovery were reported to the scientific world by Arthur Issel (1868).

But many influential scientists were committed to a fairly recent appearance of the modern human type by evolution from primitive apelike creatures, and they opposed such discoveries on theoretical grounds. Some scientists thought this unfair. Speaking of the finds of Pliocene age at Castenedolo, Savona and elsewhere, the Italian anatomist Giuseppe Sergi (1884, p. 310) protested: "By means of a despotic scientific prejudice, call it what you will, every discovery of human remains in the Pliocene has been discredited."

British archeologist R. A. S. Macalister provides a good example of such scientific prejudice. In 1921 (p. 183), he wrote about the Castenedolo finds: "There must be something wrong somewhere." But why? Had not the bones been discovered by a professional geologist in a layer of undisturbed Pliocene clay? That was not good enough for Macalister (1921, p. 184), who said of the Castenedolo bones: "If they really belonged to the stratum in which they were found, this would imply an extraordinarily long standstill for evolution. It is much more likely that there is something amiss with the observations." Quite simply, Macalister was inclined to reject the Castenedolo fossils because they violated his evolutionary preconceptions. He further stated (Macalister 1921, p. 185): "The acceptance of a Pliocene date for the Castenedolo skeletons would create so many insoluble problems that we can hardly hesitate in choosing between the alternatives of adopting or rejecting their authenticity." Macalister, of course, rejected their authenticity, and given his prominent position, this rejecting carried tremendous authority. Here we see the process of knowledge filtration operating. Good evidence is set aside, simply because it violates the concept of a recent evolutionary origin of the modern human type.

Keep in mind that the Castenedolo fossils show that there were anatomically modern human beings present on earth at the same time the Laetoli footprints were made in East Africa, about 4 million years ago. There is also some fragmentary fossil evidence from Africa itself. In 1965, anthropologists Bryan Patterson and William W. Howells found at Kanapoi, Kenya, a fragment of a humerus (upper arm bone). Upon examining it, they found it to be almost exactly like a modern human humerus (Patterson and Howells 1967). Other researchers have found it to be different from those of the australopithecines (McHenry and Corruccini 1975). The Kanapoi humerus is 4-5 million years old. Considered alone, it is merely suggestive of a human presence, but when taken in the context of the Laetoli footprints and the Castenedolo skeletons, which are of the same age, the case for a human presence 4 million years ago becomes quite strong.

And we can go much further back in time. The French anthropologist Gabriel de Mortillet (1883, p. 72), in his book *Le Préhistorique,* tells of a complete anatomically modern human skeleton found in a Miocene formation at Midi de France (at least 5 million years old and perhaps as much as 25 million years old) and another such skeleton found in an Eocene formation at Delémont, Switzerland (at least 38 million years old). There are human artifacts of similar antiquity, among them the stone tools found by Carlos Ribeiro (1867), head of the Geological Survey of Portugal, in Miocene formations near Lisbon; stone tools found by Louis Bourgeois (1873) in a Miocene formation at Thenay, France; and stone tools found by Fritz Noetling (1894), of the Geological Survey of India, in a Miocene formation in Burma. These discoveries, made in the late nineteenth century, were published in scientific journals and discussed in scientific conferences. The only reason for their absence from current textbooks is that they contradict the idea of a recent human origin.

From the United States comes a human skeleton found 90 feet deep in coal in Macoupin County, Illinois. Immediately above the skeleton were 2 feet of unbroken slate rock. The coal in which the skeleton was found is from the Carboniferous period, making the fossil about 300 million years old. The report of this discovery was printed in the December 1862 edition of a scientific journal called *The Geologist.*

Here again, there are human artifacts of similar antiquity. The June 11, 1891 edition of the *Morrisonville Times* newspaper, of Morrisonville, Illinois, in the United States, carried a report of a gold chain discovered inside a solid piece of coal. The chain was found by Mrs. S. W. Culp, wife of the newspaper's publisher, when she was breaking a lump of coal. According to the Illinois State Geological Survey, the coal containing the chain is of Carboniferous age, about 300 million years old. In 1897, a coal miner working in a mine near Webster, Iowa, in the United States, found an unusual carving on a piece of stone. The *Daily News* of Omaha, Nebraska (April 2, 1897) said: "The stone is of dark grey color and about two feet long, one foot wide and four inches in thickness. Over the surface of the stone, which is very hard, lines are drawn at angles forming perfect diamonds. The center of each diamond is a fairly good face of an old man..." Also of interest is an iron pot found in a block of coal in the year 1912 by Frank J. Kenwood. The discovery occurred at the Municipal Electric Plant in Thomas, Oklahoma, where the coal was burned to generate power. Kenwood wrote in a notarized affidavit: "I came upon a solid chunk of coal which was too large to use. I broke it with a sledge hammer. This iron pot fell from the center, leaving the impression of mould of the pot in the piece of coal" (Rusch 1971, p. 201). The coal was traced by Kenwood to the Wilburton Mine. According to the Oklahoma Geological Survey, the coal in that mine is about 312 million years old.

Such intriguing discoveries continue to be made. Over the past several decades, miners in South Africa have found hundreds of metallic spheres with grooves running around their equators. One specimen has three such parallel grooves encircling it. The spheres are composed of a very hard substance. Roelf Marx, curator of the museum of Klerksdorp, South Africa, wrote about the spheres in a letter dated September 12, 1984: "They are found in pyrophyllite, which is mined near the little town of Ottosdal in the Western Transvaal. This pyrophyllite is a quite soft secondary mineral...and was formed by sedimentation about 2.8 billion years ago. On the other hand the globes...are very hard and cannot be scratched, even by steel" (Cremo and Thompson 1993, p. 813). No one has yet come up with a satisfactory natural explanation for the spheres, which appear to be human artifacts.

In this paper, I have mentioned only a few of the hundreds of well-documented cases showing the extreme antiquity of the human race on this planet. Undoubtedly there is an extraterrestrial dimension to the origin of the human species. But extraterrestrial contacts and interventions should be considered within the framework of a terrestrial human presence extending back hundreds of millions of years, in the course of a time flow that may manifest repetitive patterning.

References Cited

Becker, G. F. (1891) Antiquities from under Tuolumne Table Mountain in California. *Bulletin of the Geological Society of America, 2:* 189-200.

Bourgeois, L. (1873) Sur les silex considérés comme portant les margues d'un travail humain et découverts dans le terrain miocène de Thenay. *Congrès International d'Anthropologie et d'Archéologie Préhistoriques, Bruxelles 1872, Compte Rendu*, pp. 81—92.

Cremo, M. A (in press) Puranic time and the archeological record. *In* Murray, T., ed., *Proceedings of the World Archaeological Congress 3, New Delhi, India, 1994.* Volume on Concepts of Time. London, Routledge.

Cremo, M. A., and Thompson, R. L. (1993) *Forbidden Archeology: The Hidden History of the Human Race.* San Diego, Bhaktivedanta Institute.

Cremo, M. A., and Thompson, R. L. (1994) *The Hidden History of the Human Race.* San Diego, Govardhan Hill.

De Mortillet, G. (1883) *Le Préhistorique.* Paris, C. Reinwald.

Holmes, W. H. (1899) Review of the evidence relating to auriferous gravel man in California. *Smithsonian Institution Annual Report 1898—1899*, pp. 419—472.

Issel, A. (1868) Résumé des recherches concernant l'ancienneté de l'homme en Ligurie. *Congrès International d'Anthropologie et d'Archéologie Préhistoriques, Paris 1867, Compte Rendu*, pp. 75— 89.

Langbein, W. J. (1994) *Forbidden Archeology. PARA*, January.
Leakey, M. D. (1979) Footprints in the ashes of time. *National Geographic, 155*: 446—457.
Macalister, R. A. S. (1921) *Textbook of European Archaeology*. Vol. 1. *Paleolithic Period*. Cambridge, Cambridge University Press.
McHenry, H. M. and Corruccini, R. S. (1975) Distal humerus in hominoid evolution. *Folia Primatologica 23*: 227-244.
Noetling, F. (1894) On the occurrence of chipped flints in the Upper Miocene of Burma. *Records of the Geological Survey of India, 27*: 101—103.
Patterson, B., and Howells, W. W. (1967) Hominid humeral fragment from Early Pleistocene of northwestern Kenya. *Science, 156*: 64—66.
Ragazzoni, G. (1880) La collina di Castenedolo, solto il rapporto antropologico, geologico ed agronomico. *Commentari dell' Ateneo di Brescia*, April 4, pp. 120—128.
Ribeiro, C. (1873) Sur des silex taillés, découverts dans les terrains miocène du Portugal. *Congrès International d'Anthropologie et d'Archéologie Préhistoriques, Bruxelles 1872, Compte Rendu*, pp. 95—100.
Rusch, Sr., W. H. (1971) Human footprints in rocks. *Creation Research Society Quarterly, 7*: 201—202.
Sergi, G. (1884) L'uomo terziario in Lombardia. *Archivio per L'Antropologia e la Etnologia, 14*: 304-318.
Slemmons, D. B. (1966) Cenozoic volcanism of the central Sierra Nevada, California. *Bulletin of the California Division of Mines and Geology, 190*:199-208.
Steen McIntyre, V., Fryxell R., and Malde, H. E. (1981) Geologic evidence for age of deposits at Hueyatlaco archaeological site, Valsequillo, Mexico. *Quaternary Research 16:* 1-17.
Tuttle, R. H. (1985) Ape footprints and Laetoli impressions: a response to the SUNY claims. *In* Tobias, P. V. ed. *Hominid Evolution: Past, Present, and Future*. New York, Alan R. Liss, pp. 129-133.
Whitney, J. D. (1880) The auriferous gravels of the Sierra Nevada of California. *Harvard University, Museum of Comparative Zoology Memoir 6*(1).

1.4. Michael A. Cremo (1996) "Screams from the Stream: Mainstream Science Reacts to *Forbidden Archeology*." *The Anomalist*, vol. 4, pp. 94-103.

This article appeared in The Anomalist (No. 4, Autumn 1996, pp. 94-103), *a journal edited by Patrick Huyghe. The version below differs slightly from the version published in* The Anomalist. *The differences result from the elimination of direct quotations from personal correspondence from certain individuals, at their request. In such cases I have substituted brief paraphrases.*

Imagine a typical Western showdown. One gunfighter (let's call him Stranger) confronts another (let's call him Boss) in a saloon. Stranger, his hand poised above the pistol on his hip, challenges, "Okay, Boss, draw!" What will Boss do? Will he scornfully turn his back on Stranger and return to his drink, signaling that Stranger's challenge is beneath his notice? Or will he let loose a blast of invective, trying to bluff Stranger into withdrawing his challenge? Or will he actually accept the challenge and draw his gun?

When my book *Forbidden Archeology*, coauthored with Richard L. Thompson, was published, I felt like Stranger. The book, which catalogued extensive scientific evidence for extreme human antiquity, constituted a direct challenge to prevailing scientific accounts of human origins.

Although members of the anomalist and alternative science communities praised *Forbidden Archeology*, this was not altogether unexpected. I was more eager to see the reactions from members of the mainstream scientific community. After all, our book was more directly aimed at them than anyone else. And so, having probed the body of mainstream science with a book of archeological anomalies, I thought it would be worthwhile to monitor their responses.

Initially, I wondered if there would be any reaction at all. When I sent a manuscript of *Forbidden Archeology* to Michael Mulkay, a leading sociologist of scientific knowledge, he replied in a letter dated May 18, 1993. He said he did not have time to read it and, by way of consolation, told me that his two best books had met with silence from reviewers.

When I was researching and writing *Forbidden Archeology*, I discovered that orthodox scientists have often employed silence as the most effective way of responding to evidence that challenges an established doctrine. Our book was definitely challenging an established doctrine. I suspected it would be ignored, but the silence was eventually broken by Jonathan Marks in a review appearing in the January 1994 issue of *American Journal of Physical Anthropology (AJPA)*.

Marks characterized *Forbidden Archeology* as "Hindu-oid creationist drivel" and "a veritable cornucopia of dreck." In my introduction to the book, I openly acknowledged our affiliation with the Bhaktivedanta Institute, the science studies branch of the International Society for Krishna Consciousness, and stated that our historical critique of archeology was motivated by our readings in India's Puranic literature, which place the origin of human civilization in the very remote past. But the text of the book itself contains nothing more than reports and analyses of archeological and anthropological discoveries, mostly derived from scientific journals of the past 150 years. The probable ages of these discoveries are, interestingly enough, consistent with the accounts of human antiquity found in the Puranas and other ancient Sanskrit writings.

One might wonder why Marks responded at all to *Forbidden Archeology*. Let's return to the scene of our Western showdown. Stranger has challenged Boss to draw his gun and shoot. Boss's first option is to look contemptuously at Stranger and instead of drawing his gun just turn silently to the bar and resume drinking his whiskey. If Stranger really is not worth fighting, and everyone in the saloon accepts Boss's silent refusal to acknowledge Stranger's challenge as proper, then Boss wins. But if Boss senses that the onlookers will interpret his refusal to answer the challenge as a sign of weakness, then he might feel constrained to respond to Stranger. In this case, Boss might first try to bluff Stranger with some name-calling and insults. That is what Marks chose to do. *Forbidden Archeology* was too substantial a challenge to be ignored.

But Marks's review, although filled with contemptuous bravado, showed him a member of an embattled establishment threatened on many fronts. He first mentioned "the Fundamentalist push to get 'creation science' into the classroom." Christian creationists are, of course, the classic adversaries of Darwinian evolutionists. Marks then conceded that "the rich and varied origins myths of all cultures are alternatives to contemporary evolution." But this politically correct admission did not disguise his conviction that none of these alternative cosmologies, springing once more to life in the postcolonial world, are to be taken seriously. Marks went on to disparage "all religious-based science, like the present volume" (i.e. *Forbidden Archeology*).

Marks also sneered at "goofy popular anthropology" and its literature. A flood of well-written books give friendly treatment to shamanism, goddess worship, Bigfoot, alien abductions, and pre-Columbian contacts between the Americas and the ancient civilizations of Asia, Africa, and Europe. These pose a significant secular threat to his discipline.

The reality of this threat was noted by Barbara and Dennis Tedlock, editors of *American Anthropologist*: "New Age titles in bookstores outnumber anthropological ones, and the kinds of titles we once disliked seeing in the same section with anthropology or archaeology now occupy whole sections of their own—the shamanism, goddess worship, and New Age sections. These shifts reflect social and cultural developments that are well under way not only in this country but all over the world—north or south, east or west. The urban participants in the New Age and related movements have lines of communication that reach into the remotest deserts, jungles, and mountains of our own traditional field research. In some ways these developments look like a privatization of the educational tasks we once saw as our own."

Of special concern to Marks were subversives within the academic community itself. Marks called California State University (San Marcos) sociologist Pierce Flynn, who contributed a favorable foreword to *Forbidden Archeology*, "a curious personage." He faulted Flynn for "placing this work within postmodern scholarship," or any realm of scholarship whatsoever. In short, Marks grouped our book, correctly I would say, with a multitude of mainstream anthropology's perceived enemies—creationists, cultural revivalists, religion-based sciences (especially Hindu-based), populist critiques of science, anomalists and, finally, the postmodern academic critics of science in the fields of sociology, history and philosophy.

The theme of an embattled orthodoxy was repeated in a review of *Forbidden Archeology* that appeared in *L'Homme,* one of the foremost anthropology journals in the world. Wiktor Stoczkowski, an anthropologist with the French National Center for Scientific Research in Paris, wrote: "Historians of science repeat tirelessly that the Biblical version of origins was replaced in the nineteenth century by the evolution theory. This simple story is substituted, in our imaginations, for the more complex reality that we are today confronted with a remarkable variety of origins accounts. Those of official science are far from being uniform. Prehistory told by scientists committed to Marxist theory is not the same as that presented by feminist scholars. Furthermore, the version of prehistory found in children's school books is different from that found in professional scientific publications. And this in turn is very different from that of the Jehovah's Witnesses, the American Creationist, the Catholic Church, or those who seek to explain our origin by extraterrestrial intervention. I have skipped over many other versions. And *Forbidden Archeology* gives us one more ...inspired by the Vedic philosophy."

The anthropologist Richard Leakey, like Marks, reacted to *Forbidden Archeology* by name-calling rather than discussing the evidence reported in the book on its merits. In a letter to me dated November 8, 1993, Leakey said: "Your book is pure humbug and does not deserve to be taken seriously by anyone but a fool. Sadly there are some, but that's part of selection and there is nothing that can be done."

Responses to the book from mainstream professional scientists were not, however, totally negative. In my introduction, I had written: "We anticipate that many workers will take *Forbidden Archeology* as an invitation to productive discourse on (1) the nature and treatment of evidence in the field of human origins and (2) the conclusions that can most reasonably be drawn from this evidence." To some extent, this has occurred. Mainstream scholars have grudgingly acknowledged the quality of the research that went into the book. In a letter to me dated August 10, 1993, William W. Howells, a highly respected Harvard physical anthropologist, said that our book "represents much careful effort in critically assembling published materials."

In a major review of *Forbidden Archeology* published in *Geoarchaeology,* Ken Feder, an anthropologist from Central Connecticut State University, wrote: "While decidedly antievolutionary in perspective, this work is not the ordinary variety of antievolutionism in form, content, or style. In distinction to the usual brand of such writing, the authors use original sources and the book is well written. Further, the overall tone of the work is far superior to that exhibited in ordinary creationist literature."

In another major review in the *British Journal for the History of Science*, archeologist and historian of archeology Tim Murray, of Latrobe University in Australia, said: "I have no doubt that there will be some who will read this book and profit from it. Certainly it provides historians of archaeology with a useful compendium of case studies in the history and sociology of scientific knowledge, which can be used to foster debate within archaeology about how to describe the epistemology of one's discipline." Murray also guardedly admitted that the religious perspective of the authors of *Forbidden Archeology* might have some utility: "The 'dominant paradigm' has changed and is changing, and practitioners openly debate issues which go right to the conceptual core of the discipline. Whether the Vedas have a role to play in this is up to the individual scientists concerned."

Some of the reviewers just mentioned have even gone so far as to address the factual evidence presented in *Forbidden Archeology*. Returning to our gunfighter analogy, this is the equivalent of Boss responding to Stranger's challenge by drawing his gun and shooting. Assuming that Stranger stands his ground and fires back, all that remains is to see who's left standing when the shooting stops. Or perhaps both will stop shooting and come to some mutually acceptable agreement. But, at the moment, conflict over the evidence seems likely to continue for some time. As an example, let's consider Ken Feder's critique of the evidence presented in *Forbidden Archeology*. Although it deserves a more complete response, I shall here simply call attention to only some of his many misstatements.

One topic discussed in our book is eoliths, very primitive stone tools that look like broken rock but are actually objects of human manufacture. Many mainstream scientists accept eoliths that fit within expected time frames of human evolution and reject those that are too old. In his *Geoarchaeology* review, Feder said, concerning the anomalously old eoliths, that my coauthor and I "do not address the fact that putative Eolithic tools are invariably the same raw material as their surrounding geological deposits, a fact that supports natural rather than cultural origin for the materials." Not true. Eoliths are not always picked out of large masses of the same kind of broken rock. For example, we noted in *Forbidden Archeology* (pp. 142-143) that at Foxhall, England, the flint eoliths are found in layers of black sediment. Pieces of flint are rare in these deposits. Archeologist J. M. Coles took this to mean that the flints arrived there by human agency.

Feder claimed *Forbidden Archeology* ignored "wear pattern analysis." Again, not true. For example, in my introduction, I mentioned signs of use on a working edge as one factor in determining if an eolith was really a human tool. And in the chapter on eoliths (p. 134), there is a report by two geologists who mentioned seeing signs of use on eoliths from the Red Crag of East Anglia in England. Other discussions of use wear patterns can be found throughout the book. Concerning the stone tool industry found at Thenay, France, we quoted (p. 235) S. Laing, who said: "The inference [that an stone object is cultural] is strengthened...if the microscope discloses parallel striae and other signs of use on the chipped edge, such as would be made by scraping bones or skins, while nothing of the sort is seen on the other natural edges." In connection with a stone tool industry found at Aurillac, France, we quoted (p. 252) Max Verworn, who discusses wear patterns on the edges of the stone tools.

Many of the scientific reports of anomalously old human artifacts and skeletal remains found in *Forbidden Archeology* date back to the late nineteenth and early twentieth centuries. Feder asked why such discoveries, if genuine, are not still being made today. In a letter to Feder dated August 16, 1994, I noted: "Anomalously old human remains are in fact being discovered today, but are interpreted so as to fit the current evolutionary paradigm. Some examples are the Laetoli footprints, 3 million years old and anatomically modern in all details, interpreted as australopithecine, even though the known foot fossils of this genus do not match the prints; the 1481 femur, described by Richard Leakey as indistinguishable from an anatomically modern human femur, but attributed to *Homo habilis,* although the new OH-62 discovery makes this attribution questionable; and the original Trinil *Pithecanthropus* femur, found in the same 800,000 year old layer as the famous *Pithecanthropus* skull and now characterized by many professionals as modern human without discussion of its long accepted age and provenance."

These cases are discussed in detail in *Forbidden Archeology*. Feder also complained: "When a nineteenth century researcher using a standard microscope of that time claims that striations found on bones dating back tens of millions of years are butchering marks, this is the equivalent, in the authors' view, of a modern researcher identifying cut marks using a scanning electron microscope. I doubt that many working in the field would agree." Feder did not, however, cite any field workers to back up his claim.

In *Forbidden Archeology* (p. 38) we noted that archeologist John Gowlett of Oxford University says: "Under a microscope [*the optical kind*], marks made by man are distinguishable in various ways from those made by carnivores. Dr. Henry Bunn (University of California) observed through an optical microscope at low magnification that stone tools leave v-shaped cuts, which are much narrower than rodent gnawing marks."

Regarding use marks on stone tools, John J. Shea, a field worker in archeology, said in a recent article in *Evolutionary Anthropology* that "most microwear analysts" consider optical microscopes capable of magnifying 160 times "sufficient to identify the presence or absence of most kinds of wear and to provide some information about tool motions and worked materials."

What to make of all this? It would appear that mainstream anthropology, of which archeology is a subdiscipline, is somewhat factionalized. On the right, we have fundamentalist Darwinians ("evolution is a fact, not a theory") who react to any antievolutionary challenge with silence or contemptuous name-calling. Some of them will, however, discuss anomalous evidence, but with a view to discrediting it by any means possible. On the left, there are postmodern scholars who are more willing to acknowledge alternative worldviews. Both right and left appear to recognize that mainstream science is no longer quite so mainstream.

We used to hear a lot about science and fringe science. Today the situation is somewhat changed. The so-called fringe sciences have turned into real competitors. As we have seen, mainstream anthropologists and archeologists appear quite perplexed and threatened by surprisingly strong competition from religion-based science, New Age science, anomalists, and others. The same is true in other disciplines.

The reductionist, materialist consensus that has for several centuries dominated science appears to be breaking down. After a period of flux, I expect a new consensus to emerge, one that embodies principles and phenomena familiar to the readers of this journal. To speed the process along, I urge anomalists to keep mainstream scientists in mind as an important audience for their work. They are going to have to be part of the new consensus, and it's time they get accustomed to negotiating with us.

References Cited

Cremo, Michael A. and Thompson, Richard L. (1993) *Forbidden Archeology*. San Diego, Bhaktivedanta Institute.

Feder, K. L. (1994). Review of *Forbidden archeology. Geoarchaeology: An International Journal, 9*, 337-340.
Marks, J. (1994). Review of *Forbidden archeology. American Journal of Physical Anthropology, 93*, 140-141.
Murray, Tim (1995) Review of *Forbidden archeology. British Journal for the History of Science, 28*, 377-379.
Shea, John J. (1992) "Lithic Microwear Analysis in Archeology." *Evolutionary Anthropology, 1*, 143-150.
Stoczkowski, Wiktor (1995) Review of *Forbidden archeology. L'Homme, 35*, 173-174.
Tedlock, B., & Tedlock, D. (1995). "From the editors." *American Anthropologist, 97*, 8-9.

1.5. Michael A. Cremo (1997) "*Forbidden Archeology*: A Three-Body Interaction Among Science, Hinduism and Christianity." Unpublished. Submitted by invitation to *Hindu-Christian Studies Bulletin.*

In 1996 Dr. Klaus Klostermaier, a professor of religion at the University of Manitoba in Winnipeg, Canada, asked me to contribute an article to the Hindu Christian Studies Bulletin. *The article was submitted for publication, but it was not included in the issue of the* Bulletin *for which it was commissioned. It is currently being held by the editors of the* Bulletin *for possible inclusion in a future issue.*

The interactions among science, Hinduism and Christianity are as complex as those in the three-body problem of astrophysics. In practice, astrophysicists select a central body, say the Earth, with a second body, the Moon orbiting it, and then try to determine the perturbations induced in the motion of the Moon by the attraction of the third body, the Sun. There is no general solution for this problem. This means that independent of observations one cannot calculate very far in advance (or very far into the past) the exact position of the Moon relative to the other two bodies. The perturbations of the Moon's orbit, induced by the attractions of the Earth and Sun, are incalculably complex, as are the movements of Hinduism in relation to the twin influences of science and Christianity. The reactions provoked by my book *Forbidden Archeology*,[1] from scientists, scholars and religionists,[2] provide useful data for examination of a three-body problem in the study of Hinduism, Christianity and science.

I introduced myself as an American citizen, of Italian Catholic heritage and educated in secular schools, who converted to Gaudiya Vaisnavism in

1973, at age twenty-five. In 1976, His Divine Grace A. C. Bhaktivedanta Swami Prabhupada, founder-*acharya* of the International Society for Krishna Consciousness(ISKCON), accepted me as his disciple. Bhaktivedanta Swami Prabhupada traced his lineage through nine generations of Gaudiya Vaishnava *gurus* to Chaitanya Mahaprabhu, who appeared in the latter part of the fifteenth century. Since 1973, I have strictly followed the ISKCON *bhakti* regimen, including rising before dawn, attending temple worship, hearing readings from *Bhagavata Purana*, practicing japa meditation for about two hours a day, and making occasional pilgrimages to sites sacred to Gaudiya Vaisnavas, such as Mayapur, West Bengal, appearance place of Chaitanya Mahaprabhu, and Vrndavana, Uttar Pradesh, appearance place of Krishna.

In 1984, I began working with the Bhaktivedanta Institute, the science studies branch of ISKCON. The Institute was founded in 1974 for the purpose of examining (and challenging) materialistic scientific ideas about the origin of life and the universe from the standpoint of Vedic knowledge. I use the term Vedic in its broad Vaishnava sense to include the *Vedas, Puranas* and *Itihasas.* The Bhaktivedanta Institute, following the teachings of Bhaktivedanta Swami Prabhupada, generally favors a literal reading of the Vedic texts. In the realm of scientific discourse, this means using Vedic texts as sources of hypotheses, which can then be employed to explain evidence. My book *Forbidden Archeology*, which presents abundant scientifically reported evidence consistent with Puranic accounts of extreme human antiquity and documents social factors underlying the exclusion of this evidence from contemporary scientific discourse, was published by the Bhaktivedanta Institute in 1993.

As can be seen from this brief autobiographical sketch, I was pulled from the orbit of modern liberal Christianity and secular science into the orbit of traditional Hinduism. I thus have considerable empathy for the Indian intellectuals of the nineteenth century, the *bhadralok,* who were pulled from the orbit of traditional Hinduism into the orbits of modern liberal Christianity and secular science.

As we explore the history of that time, we encounter, along with the Christian missionaries, the British Orientalists, the members of the Brahmo Samaj and the Arya Samaj, and the Theosophists, such personalities as Krishna Mohan Banerjea (1813-1885), the Bengali convert and disciple of Scottish missionary Alexander Duff. Born a *brahmana,* Banerjea became a Christian and underwent a liberal education at Calcutta's Hindu College. Afterwards he flaunted the rules of his caste, going so far as to get drunk with some friends and throw pieces of raw beef into the courtyard of a *brahmana's* house, causing considerable uproar in the neighborhood.[3] In writing *Forbidden Archeology* and directing to it the attention of evolutionary scientists, I have performed an act roughly equivalent to throwing beef into a *brahmana's* courtyard. The book, anti-Darwinian as well as religiously inspired, has provoked considerable reaction in orthodox scientific circles.

Among the more emotional respondents to *Forbidden Archeology* was Jonathan Marks, who, in his review, published in *American Journal of Physical Anthropology,* called it "Hindu-oid creationist drivel" and "a veritable cornucopia of dreck."[4] Here we find a tonal echo of some of the reactions of early European scholars and missionaries to Hinduism. We may recall, for example, the words of William Hastie, leader of a Scottish missionary organization in Calcutta, who as late as 1882 denigrated India as "the most stupendous fortress and citadel of ancient error and idolatry" and condemned Hinduism as "senseless mummeries, licentiousness, falsehood, injustice, cruelty, robbery, [*and*] murder."[5]

Not all reviewers were so dismissive of *Forbidden Archeology*'s scholarly worth. In a lengthy review article in *Social Studies of Science* (provocatively titled "Vedic Creationism: A Further Twist to the Evolution Debate"), Jo Wodak and David Oldroyd asked, "So has *Forbidden Archeology* made any contribution at all to the literature on palaeoanthropology?" They concluded, "Our answer is a guarded 'yes', for two reasons." First, "the historical material ...has not been scrutinized in such detail before," and, second, the book does "raise a central problematic regarding the lack of certainty in scientific 'truth' claims."[6]

In *L'Anthropologie*, Marylène Pathou-Mathis wrote: "M. Cremo and R. Thompson have willfully written a provocative work that raises the problem of the influence of the dominant ideas of a time period on scientific research. These ideas can compel the researchers to orient their analyses according to the conceptions that are permitted by the scientific community." She concluded, "The documentary richness of this work, more historical and sociological than scientific, is not to be ignored."[7]

And in *British Journal for the History of Science*, Tim Murray noted in his review of *Forbidden Archeology*: "I have no doubt that there will be some who will read this book and profit from it. Certainly it provides the historian of archaeology with a useful compendium of case studies in the history and sociology of scientific knowledge, which can be used to foster debate within archaeology about how to describe the epistemology of one's discipline."[8]

I will not dwell much further upon the academic integrity and utility of the archeological evidence presented in *Forbidden Archeology*. I want to focus instead on how the book fits into the larger history of interactions among science, Hinduism and Christianity. Although these interactions defy simplistic explanation, it is possible to trace a broad pattern of development.

In the late eighteenth century and early nineteenth century, some European scholars, such as John Playfair, were intrigued by the vast time scales of Vedic histories and attributed considerable antiquity to Hindu astronomical texts. Playfair, for example, put the composition of the *Surya-siddhanta* before the beginning of the *Kali-yuga*, or over 5,000 years ago.[9] And that implied an even longer history of refined astronomical observation. But other European scholars, deeply influenced by Christian chronology, were unhappy with such assertions. John Bentley, for example, put his knowledge of astronomical science to work in discrediting the proposals of Playfair and others.

About one of his opponents, Bentley wrote:

> By his attempt to uphold the antiquity of Hindu books against absolute facts [*Bentley's*], he thereby supports all those horrid abuses and impositions found in them, under the pretended sanction of antiquity, *viz.* the burning of widows, the destroying of infants, and even the immolation of men. Nay, his aim goes still deeper; for by the same means he endeavours to overturn the Mosaic account, and sap the very foundations of our religion: for if we are to believe in the antiquity of Hindu books, as he would wish us, then the Mosaic account is all a fable, or a fiction.[10]

Bentley regarded the vast time periods of Hindu cosmology as a recent imposition by the *brahmanas,* who desired "to arrogate to themselves that they were the most ancient people on the face of the earth."[11] Unable to tolerate a chronology that "threw back the creation [in the current *kalpa*] to the immense distance of 1,972,947,101 years before the Christian era,"[12] Bentley held that the Puranic histories should be compressed to fit within the few thousand years of the Mosaic account.[13] Sir William Jones also brought the expansive Hindu chronology into line with the Biblical time scale.[14]

At the same time Bentley and Jones were using science and textual criticism to dismantle the Hindu chronology, their contemporaries in Europe were using the same methods to dismantle the Mosaic chronology. The process accelerated with the advent of Darwinism, leaving only a minority of Christian intellectuals committed to a divine creation of Adam and Eve about six thousand years ago. In India, many Hindu intellectuals, influenced by science and liberal Christianity, similarly gave up the historical accounts of the *Puranas,* which place humans on earth millions of years ago. Today, the Darwinian evolutionary account of human origins remains dominant among intellectuals in India and throughout the world, although the postmodern tendency toward relativism has somewhat weakened its hold.

This is the background against which *Forbidden Archeology* appeared. It has been quite interesting for me to monitor academic reactions to the book, especially the attempts of reviewers to grapple with its Hindu inspiration and relationship with Christian and Darwinian accounts of human origins and antiquity.

In his review of *Forbidden Archeology* for *Geoarchaeology,* Kenneth L. Feder wrote: "The book itself represents something perhaps not seen before; we can fairly call it 'Krishna creationism' with no disrespect intended."[15] After describing the contents of the book, Feder added, "While decidedly antievolutionary in perspective, this work is not the ordinary variety of antievolutionism in form, content, or style. In distinction to the usual brand of such writing, the authors use original sources and the book is well written. Further, the overall tone of the work is superior to that exhibited in ordinary [*i.e., Christian*] creationist literature."[16]

Comparisons between *Forbidden Archeology* and Christian creationist literature are common in the academic reviews of the book. Murray wrote in *British Journal for the History of Science*, "This is a piece of 'Creation Science' which, while not based on the need to promote a Christian alternative, manifests many of the same types of argument."[17] He further characterized *Forbidden Archeology* as a book that "joins others from creation science and New Age philosophy as a body of works which seek to address members of a public alienated from science, either because it has become so arcane or because it has ceased to suit some in search of meaning for their lives."[18]

Some of the comparisons are less polite. Marks acrimoniously wrote in *American Journal of Physical Anthropology*, "The best that can be said is that more reading [*of the scientific literature*] went into this Hindu-oid creationist drivel than seems to go into the Christian-oid creationist drivel."[19] Paleoanthropologist Colin Groves wrote:

> A book like this, simply because it is superficially scholarly and not outright trash like all the Christian creationist works I have read, might indeed make a useful deconstructionist exercise for an archaeology or palaeoanthropology class. So it's not without value. You could do worse, too, than place it in front of a Gishite with the admonition "Look here: these guys show that human physical and cultural evolution doesn't work. Therefore it follows that the Hindu scriptures are true, doesn't it?"[20]

Stripping away the armor of defensive ridicule in such statements, we find a materialistic science not yet totally secure in its ongoing global three-body interaction with unreconstructed Christianity and traditional Hinduism.

Wiktor Stoczkowski, reviewing *Forbidden Archeology* in *L'Homme,* accurately noted, "Historians of science repeat tirelessly that the Biblical version of origins was replaced in the nineteenth century by the evolution theory. In our imaginations, we substitute this simple story for the more complex reality that we are today confronted with a remarkable variety of origins accounts."[21] Among those accounts Stoczkowski included that of the Biblical creationists. "*Forbidden Archeology,*" he added, "gives us one more, dedicated to 'His Divine Grace A. C. Bhaktivedanta Swami Prabhupada' and inspired by the Vedic philosophy that disciples study in the United States at the Bhaktivedanta Institute, a branch of the International Society for Krishna Consciousness."[22]

The main text of *Forbidden Archeology* is solely dedicated to documentation and analysis of evidence consistent with Vedic accounts of extreme human antiquity. The religious affiliation of the authors and their commitment to Vedic historical accounts are briefly mentioned in the Introduction. Reviewers have therefore taken upon themselves to expand upon these topics for the benefit of their readers. It is somewhat novel to find substantive discourse on *yugas* and *manvantaras,* the Bhaktivedanta Institute and ISKCON in the pages of mainstream journals of archeology, anthropology and science studies. Up to this time, such references have largely been confined to the pages of religious studies journals.

In the first few pages of their *Social Studies of Science* review article, Wodak and Oldroyd gave extensive background information on: ISKCON ("a modern variant of the Bhakti sects that have dominated Hindu religious life over the last one and a half millennia"); the teachings of the movement's founder ("for Prabhupada, science gives no adequate account of the origin of the universe or of life"); the Bhaktivedanta Institute (they comment on "the boldness of its intellectual programme"); and Vedic chronology ("partial dissolutions, called *pralaya,* supposedly take place every 4.32 billion years, bringing catastrophes which whole groups of living forms can disappear"). One also encounters many references to the *Rg Veda, Vedanta, the Puranas, the atma, yoga,* and *karma.*[23]

In common with other reviewers, Wodak and Oldroyd draw a connection between *Forbidden Archeology* and the work of Christian creationists. "As is well known," they note, "Creationists try to show that humans are of recent origin, and that empirical investigations accord with human history as recorded in the Old Testament. *Forbidden Archeology* (*FA*) offers a brand of Creationism based on something quite different, namely ancient Vedic beliefs. From this starting point, instead of claiming a human history of mere millennia, *FA* argues for the existence of *Homo sapiens* way back into the Tertiary, perhaps even earlier."[24]

Despite the considerable attention Wodak and Oldroyd devoted to *Forbidden Archeology*'s Vedic inspiration, the greater part of their review article focused on the book's substance, about which they commented:

> It must be acknowledged that *Forbidden Archeology* brings to attention many interesting issues that have not received much consideration from historians; and the authors' detailed examination of the early literature is certainly stimulating and raises questions of considerable interest, both historically and from the perspective of practitioners of SSK [sociology of scientific knowledge]. Indeed, they appear to have gone into some historical matters more deeply than any other writers of whom we have knowledge.[25]

Another example of extensive references to ISKCON and Vedic concepts can be found in Feder's *Geoarchaeology* review of *Forbidden Archeology*:

> The authors are open about their membership in the Bhaktivedanta Institute, which is a branch of the International Society for Krishna Consciousness, and the book is dedicated to their "spiritual master," the group's founder. They make a reasonable request regarding their affiliation with this organization: "That our theoretical outlook is derived from the Vedic literature should not disqualify it." (p. xxxvi). Fair enough, but what is their "theoretical outlook?"[26]

Feder, citing Basham's *The Wonder That Was India*, goes on to give a succinct account of Hindu cosmology's *kalpas,* each of which lasts 4.32 billion years and "is divided into 14 *manvantaras,* each lasting 300,000,000 years." Feder then explains how "within each *manvantara* the world is created with human beings more or less fully formed, and then destroyed, only to be created once again in the next *manvantara*."[27]

In the concluding paragraph of his review, Feder gives his own comments on our three-body problem:

> We all know what happens when we mix a literal interpretation of the Judeo-Christian creation myth with human paleontology; we get scientific creationism. It seems we now know what happens when we mix a literal interpretation of the Hindu myth of creation with human paleontology; we get the antievolutionary Krishna creationism of *Forbidden Archeology*, where human beings do not

> evolve and where the fossil evidence for anatomically modern humans dates as far back as the beginning of the current *manvantara*.[28]

A more favorable estimation of *Forbidden Archeology*'s Vedic roots was offered by Hillel Schwarz in *Journal of Unconventional History,* which, as the title suggests, is situated on the outer edges of respectable scholarship's domain. But it is at such edges that advances in understanding often occur. Schwarz observed: "*Forbidden Archeology* takes the current conventions of decoding to their extreme. The authors find modern *Homo sapiens* to be continuous contemporaries of the apelike creatures from whom evolutionary biologists usually trace human descent or bifurcation, thus confirming those Vedic sources that presume the nearly illimitable antiquity of the human race."[29]

Schwarz was not put off by the authors' underlying motives for writing *Forbidden Archeology*. "Despite its unhidden religious partisanship," said Schwarz, "the book deserves a reckoning in this review for its embrace of a global humanity permanently distinct from other primates." He accurately detected the book's implicit thesis, namely, that "humanity is no mere biochemical exfoliation but a work of the spirit, in touch with (and devoted to) the ancient, perfect, perfectly sufficient, unchanging wisdom of the Vedic masters."[30]

One might wonder what the Christian creationists think of Hindu-inspired *Forbidden Archeology*. Perhaps sensing an ally in their battle against Darwinism, they have reacted somewhat favorably. A reviewer of the abridged version of *Forbidden Archeology* stated in *Creation Research Society Quarterly*: "This book is must reading for anyone interested in human origins." After expressing his surprise at finding the book in a major U. S. chain store, the reviewer noted that its "theoretical outlook is derived from the Vedic literature in India, which supports the idea that the human race is of great antiquity." The reviewer made clear that he did not share this view: "As a recent earth creationist, I would not accept the evolutionary time scale that the authors appear to accept. However," he added, "the authors have shown that even if you accept the evolutionary view of a vast age for the earth, the theory of human evolution is not supported."[31]

Up to this point, mainstream religious studies scholars have not, to my knowledge, published any reviews of *Forbidden Archeology* in their professional journals (although many did receive copies for review). But they have not been totally silent. Historian of religion Mikael Rothstein of the University of Copenhagen wrote in a review article published in the science section of *Politiken,* Denmark's largest newspaper, that in the nineteenth century Darwinism challenged the creationist views of Christian religion. Today, he said, the roles have been reversed. Religion, not science, is the primary source of intellectual provocation. And *Forbidden Archeology* is "in principle just as provoking as *The Origin of Species*." Rothstein informs his readers that the authors of *Forbidden Archeology* belong to the Bhaktivedanta Institute, the "academic center" for ISKCON, which he correctly characterized as "part of the Vaishnava religion from India." Noting that the authors are Hindu "monks" as well as scholars, he stated, "Their otherwise thorough academic argumentation can thus find support in the Vaishnava mythology, which actually describes the history of man and the geological development of the earth in a way that is compatible with their results." According to Rothstein, people who have grown up with the idea of Darwinian evolution can by reading *Forbidden Archeology* "get a glimpse of the feeling the people of the Church experienced when Darwin's theory was presented."[32]

Gene Sager, a professor of religious studies at Palomar College in California, wrote about *Forbidden Archeology*:

> As a scholar in the field of comparative religion, I have sometimes challenged scientists by offering a cyclical or spiral model for studying human history, based on the Vedic concept of the *kalpa.* Few Western scientists are open to the possibility of sorting out the data in terms of such a model. I am not proposing that the Vedic model is true....However, the question remains, does the relatively short, linear model prove to be adequate? I believe *Forbidden Archeology* offers a well-researched challenge. If we are to meet this challenge, we need to practice open-mindedness and proceed in a cross-cultural, interdisciplinary fashion.[33]

I have not yet seen any reviews of *Forbidden Archeology* in academic journals published in India. But I have gotten responses from Indian scholars in other arenas. When I presented a paper based on *Forbidden Archeology* at the World Archeological Congress 3, held in New Delhi in 1994,[34] a number of Indian scholars approached me privately and expressed their appreciation of my efforts to uphold the Puranic chronology. My World Archeological Congress paper also drew me an invitation to speak at a conference on Vedic history in the United States, organized by several Hindu organizations.[35] I earlier received an invitation, which I was not able to accept, from Kishor Ruperalia, general secretary for the Vishwa Hiṇdu Parishad in the United Kingdom, to speak at a conference organized by the VHP. Ruperalia wrote about *Forbidden Archeology*, "Inspired by the Vedic writings and encouraged by His Divine Grace A. C. Bhaktivedanta Swami Prabhupada, the scholarly authors have made a tremendous and painstaking effort...to make archeological scholars rethink the predominant paradigm on human origins and antiquity."[36]

Where does all this leave us in terms of our three-body question? In astrophysics, there are some special cases of the three-body problem that do allow for reasonably accurate solutions. If one of the bodies (a man-made earth satellite, for example) can be assigned an infinitely small mass, this simplifies the matter somewhat. In terms of the global interactions among science, Christianity and Hinduism, as related to any substantive discussion of human origins and antiquity, the three-body problem has been solved, in the minds of many modern intellectuals, by assigning traditional Hindu concepts of human origins and antiquity an infinitely small mass. The problem is then reduced to establishing the relative positions of the accounts of human origins and antiquity offered by modern Darwinian evolutionists and their Christian fundamentalist opponents. And the result is a somewhat stable and predictable system. We find a Christian fundamentalist body revolving in a fixed orbit of perpetual subordination to the central body of a Darwinian consensus negotiated between modern science and liberal Christianity (and liberal Hinduism). But the substantial and widespread reactions to *Forbidden Archeology* suggest that traditional Hindu views of human origins and antiquity have again acquired sufficient mass to cause real perturbations in scientific and religious minds, thus introducing new elements of complexity into the relationships among Hinduism, Christianity and science.

What predictions might be made about future states of the three-body question I have posed? We seem to be entering an era when the boundaries between religion and science will, as in times past, no longer be so clear cut. This is especially true in the metaphysical areas of science, i.e., those dealing with phenomena beyond the range of normal experimentation and observation, such as Darwinian evolution. Indeed, Karl Popper, the philosopher of science who established falsifiability as a criterion for the validity of a scientific theory, said: "I have concluded that Darwinism is not a testable scientific theory, but a metaphysical research programme—a possible framework for testable scientific theories."[37] And there may be other such frameworks, perhaps even some derived from the Vedic texts.

In July of 1996, I took part in a roundtable discussion at the Institute for Oriental Studies of the Russian Academy of Sciences in Moscow. After I made a presentation about *Forbidden Archeology* and another work in progress [establishing the antiquity of the *Rg Veda* at five thousand years], Indologist Evgeniya Y. Vanina made these comments:

> I think that the statement you have made, and your paper, are very important because they touch upon the cooperation of

> science and religion—not just science and religion but how to look at the texts of the classical tradition as sources of information. There is a tendency among scholars to say whatever the *Vedas*—and the *Puranas,* the *Ramayana,* and the *Mahabharata*—are saying, it is all myth and concoction, and there is no positive information in it .... I think that such a negativist attitude toward the ancient and early medieval Indian texts as sources of information should definitely be discarded.[38]

Of course, the most likely persons to search for items of positive information in such texts are those who believe in them. For the past century or so, there has not been much room in the academic enterprise for believers, either in religious studies or the sciences. But this may be changing.

G. William Barnard of the religious studies department at Southern Methodist University suggested that "this all too-frequently found notion, that scholars who have no religious inclinations are somehow more objective and therefore are better scholars of religion than those who are pursuing a spiritual life, is fundamentally flawed."[39] Reflecting on the contribution that could be made by genuine spiritual traditions, Barnard advocated "a scholarship that is willing and able to affirm that the metaphysical models and normative visions of these different spiritual traditions are serious contenders for truth, a scholarship that realizes that these religious worlds are not dead corpses that we can dissect and analyze at a safe distance, but rather are living, vital bodies of knowledge and practice that have the potential to change our taken-for-granted notions."[40]

And in a perceptive article in *American Anthropologist*, Katherine P. Ewing observed: "While espousing cultural relativism, the anthropological community has maintained a firm barrier against belief." But this fear against "going native" has a detrimental effect on the search for truth. "To rule out the possibility of belief in another's reality," said Ewing, "is to encapsulate that reality and, thus, to impose implicitly the hegemony of one's own view of the world."[41] Ewing argued that belief may be a valid stance to take in fieldwork in cultural anthropology.

Even Jonathan Marks, one of *Forbidden Archeology*'s most strident critics, admitted that (in theory) "the rich and varied origins myths of all cultures are alternatives to contemporary evolution."[42] And Tim Murray wrote in his review that archeology is now in a state of flux, with practitioners debating "issues which go to the conceptual core of the discipline." Murray then proposed, "Whether the Vedas have a role to play in this is up to the individual scientists concerned."[43] This amounts to the smallest and most backhanded of concessions that the *Vedas* may have some utility in the conceptual reconstruction of modern scientific accounts of human origins and antiquity. But at this point in the three-body interaction among science, Hinduism and Christianity it must nonetheless be regarded as significant.

Some scholars, particularly those who identify themselves as postmodern, have already recognized the utility of the approach taken in *Forbidden Archeology*. Sociologist Pierce J. Flynn found positive value in the authors' status as believers.

> The authors admit to their own sense of place in a knowledge universe with contours derived from personal experience with Vedic philosophy, religious perception, and Indian cosmology.... In my view, it is just this openness of subjective positioning that makes *Forbidden Archeology* an original and important contribution to postmodern scholarly studies now being done in sociology, anthropology, archeology, and the history of science and ideas. The authors' unique perspective provides postmodern scholars with an invaluable parallax view of historical scientific praxis, debate, and development.[44]

I first met Pierce when I was living near the ISKCON temple in the Pacific Beach neighborhood of San Diego, California. He would sometimes bring his classes on field trips to the ISKCON temple, which provided an example of an alternative religious community for his sociology of religion students. I had volunteered to be their guide. During a conversation after one of the field trips, I mentioned to Pierce that I was working on a book that examined the question of human origins and antiquity from the viewpoint of the Vedic histories. He immediately grasped its significance and assured me that the book would be of interest to many scholars. When the book was finished, I therefore asked him to contribute a foreword. Pierce Flynn's remarks, written before the publication of *Forbidden Archeology* and, thus, before the many reviews in academic and scientific journals corroborated his estimation of the book's potential impact, were quite prescient.

And speaking of prescience, I predict we are moving into a period in which the Vedic texts, and scholars openly professing intellectual commitment to the Vedic texts, are going to be playing a larger role in the three-body interaction among science, Hinduism and Christianity. Although we are not going to immediately see a major realignment of the bodies under consideration, careful observers will note some significant perturbations in their orbits, which may eventually propagate into large scale shifts of the kind that have occurred so often in the history of ideas.

Learned societies, such as the Society for Hindu-Christian Studies, may prove to be the best observatories from which to track these changes in planetary configurations. A learned society, properly constituted, can be a place for catalysis—a place where professional scholars take leave of the shelter of their secular institutions, where independent scholars take leave of the shelter of their isolation, and religiously-oriented scholars take leave of the shelter of their faith communities—all joining in dialogue and a mutual search for truth, according to common principles of intellectual inquiry and integrity.

Notes

1. Michael A. Cremo and Richard L. Thompson, *Forbidden Archeology: The Hidden History of the Human Race,* San Diego: Bhaktivedanta Institute, 1993.

2. Scholarly reviews and notices of *Forbidden Archeology*, in addition to the ones cited in this article, appear in *Journal of Field Archeology 19*:112; *Antiquity 67*:904; *Ethology, Ecology, and Evolution* 6:461; *Creation/Evolution 14(1)*:13-25; and *Journal of Geological Education 43*:193. *Forbidden Archeology*, and its abridged popular edition *The Hidden History of the Human Race*, have also attracted considerable attention in New Age and alternative science circles. Furthermore, I have appeared on about one hundred radio and television programs, including an NBC television special *The Mysterious Origins of Man* (originally broadcast in February 1996). A collection of academic and popular reviews, along with my academic papers and publications related to *Forbidden Archeology*, selected correspondence, and selected transcripts of radio and television interviews, is scheduled to come out in late 1997 under the title *Forbidden Archeology's Impact.*

3. Martin Maw, *Visions of India: Fulfillment Theology, the Aryan Race Theory, and the Work of British Protestant Missionaries in Victorian India*, Frankfurt: Peter Lang, 1990, p. 45.

4. Jonathan Marks, "*Forbidden Archeology*" (book review), *American Journal of Physical Anthropology*, vol. 93(1), 1994, p. 141. Marks's statements were printed along with other favorable and unfavorable review excerpts just inside the cover of the abridged popular edition of *Forbidden Archeology*, helping to draw media attention to the book.

5. William Hastie, *Hindu Idolatry and English Enlightenment*, 1882. Quoted in Martin Maw, *Visions of India*, p. 8.

6. Jo Wodak and David Oldroyd, "Vedic Creationism: A Further Twist to the Evolution Debate," *Social Studies of Science*, vol. 26, 1996, p. 207. I regard this 22-page review article to be the most significant scholarly response to *Forbidden Archeology*.

7. Marylène Pathou-Mathis, "*Forbidden Archeology*" (book review), *L'Anthropologie*, vol. 99(1), 1995, p. 159. The cited passage is translated from the French original.

8. Tim Murray, "*Forbidden Archeology,*" *British Journal for the History of Science*, vol. 28, 1995, p. 379.

9. John Playfair, "Remarks on the Astronomy of the Brahmins," *Transactions of the Royal Society of Edinburgh*, vol. II, pt. 1, 1790, pp. 135-192.

10. John Bentley, *A Historical View of the Hindu Astronomy*, London: Smith, Elder, and Company, 1825, p. xxvii.

11. John Bentley, *A Historical View of the Hindu Astronomy*, p. 84.

12. John Bentley, *A Historical View of the Hindu Astronomy*, p. 84

13. John Bentley, *A Historical View of the Hindu Astronomy*, p. 84. Bentley preferred a date of 4225 B.C. for the start of the first *manvantara.*

14. Sir William Jones, *The Works of Sir William Jones*, Vol. I, London: Robinson and Evans, 1799, p. 313. In a table, Jones places the beginning of the first *manvantara* at 4006 B.C. and makes the first Manu a contemporary of Adam.

15. Kenneth L. Feder, "*Forbidden Archeology*" (book review), *Geoarchaeology,* vol. 9(4), 1994, p. 337. Feder's review contains inaccurate statements about the factual material presented in *Forbidden Archeology.* For example, he says (p. 339) that the authors "do not address the issue of use-wear on any of these unexpectedly ancient 'tools.'" But *Forbidden Archeology* cites S. Laing's 1894 report on the Thenay, France, implements: "The inference [that an object is cultural] is strengthened...if the microscope discloses parallel striae and other signs of use on the chipped edge, such as would be made by scraping bones or skins, while nothing of the sort is seen on the other natural edges" (*F. A.* p. 235). L. Bourgeois, the discoverer of the Thenay implements, reported such signs of use on them (*F. A.* p. 227). Max Verworn also gives quite detailed attention to wear pattern analysis in connection with the Aurillac implements (*F. A.* p. 252). A detailed response to Feder will be included in my forthcoming book, *Forbidden Archeology's Impact.*

16. Kenneth L. Feder, "*Forbidden Archeology*," p. 338.

17. Tim Murray, "*Forbidden Archeology*," p. 378.

18. Tim Murray, "*Forbidden Archeology*," p. 379.

19. Jonathan Marks, "*Forbidden Archeology*," p. 141.

20. Colin Groves, "Creationism: The Hindu View. A Review of *Forbidden Archeology*," *The Skeptic* (Australia), vol. 14(3), pp. 43-45. This review was forwarded to me electronically by a friend who saw it posted on an Internet discussion group. After *Forbidden Archeology* and its authors were featured on the NBC television special *The Mysterious Origins of Man* in February 1996, hundreds of messages about *Forbidden Archeology*, and its "Krishna creationism," were posted, as part of heated discussions, to Usenet groups such as sci.archeology, sci.anthropology, and alt.origins.

21. Wiktor Stoczkowski, "*Forbidden Archeology* (book review)," *L'Homme,* vol. 35, 1995, p. 173. Quoted passages are translated from the French original.

22. Wiktor Stoczkowski, "*Forbidden Archeology* (book review)," p. 173.

23. Jo Wodak and David Oldroyd, "Vedic Creationism," pp. 192-195.

24. Jo Wodak and David Oldroyd, "Vedic Creationism," p. 192.

25. Jo Wodak and David Oldroyd, "Vedic Creationism," p. 198.

26. Kenneth L. Feder, "*Forbidden Archeology*," pp. 339-340.

27. Kenneth L. Feder, "*Forbidden Archeology*," p. 340.

28. Kenneth L. Feder, "*Forbidden Archeology*," p. 340. Of course, one might also propose that we know what happens when we mix the Darwinian myth of transforming species with human paleontology. We get the modern account of human evolution, with ancient apes transforming into anatomically modern *Homo sapiens* in the most recent geological times. The real question, in the game of science played fairly, is which "myth," or theory, best fits all of the relevant evidence.

29. Hillel Schwarz, "Earth Born, Sky Driven" (collective book review), *Journal of Unconventional History,* vol. 6(1), 1994, p. 75.

30. Hillel Schwarz, "Earth Born, Sky Driven," p. 76.

31. Peter Line, "The Hidden History of the Human Race"(book review), *Creation Research Society Quarterly*, vol. 32, 1995, p. 46.

32. Mikael Rothstein, "Forbudt Arkaeologi," *Politiken,* 31 January, 1994, Section 3, p. 1. Quoted passages are translated from the original Danish.

33. The quoted passage is an excerpt from an unpublished review of *Forbidden Archeology* by Dr. Sager. Another passage from the review is printed with other endorsements in the abridged version of *Forbidden Archeology*.

34. Michael A. Cremo, "Puranic Time and the Archeological Record," in: *Theme Papers: Concepts of Time*. D. P. Agarwal, M. P. Leone, and T. Murray, (eds.), New Delhi, Academic Committee of the World Archeological Congress 3, 1994, pp. 23-35. This is a volume of machine-copied original drafts of papers intended for precirculation to conference participants. It is, however, available in some libraries. The official conference proceedings are to be published by Routledge.

35. International Conference on Revisiting Indus-Sarasvati Age and Ancient India, Atlanta, Georgia, U.S.A., October 4-6, 1996.

36. The quoted passage is from a faxed letter to the editors of *Back to Godhead*, dated August 3, 1993. Ruperalia was writing in response to an article about *Forbidden Archeology* that appeared in the May/June 1993 issue of *Back to Godhead*, the bimonthly magazine of the International Society for Krishna Consciousness.

37. Karl Popper, "Darwinism as a Metaphysical Research Programme," *Methodology and Science,* vol. 9, 1976, p. 104. The emphasis is Popper's. The

article is taken from Karl Popper's *An Intellectual Biography: Unended Quest.*

38. Dr. Vanina is the Chief of the Department of History and Culture of the Center of Indian Studies of the Institute for Oriental Studies of the Russian Academy of Sciences. Her quoted remarks are transcribed from a tape recording, made with the consent of the participants in the roundtable discussion.

39. G. William Barnard, "Transformations and Transformers: Spirituality and the Academic Study of Mysticism," *Journal of Consciousness Studies,* vol. 1(2), 1994, p. 256.

40. G. William Barnard, "Transformations and Transformers," pp. 257-258.

41. Katherine P. Ewing, "Dreams from a Saint: Anthropological Atheism and the Temptation to Believe," *American Anthropologist*, vol. 96(3), 1994, p. 572.

42. Jonathan Marks, "*Forbidden Archeology*," p. 140.

43. Tim Murray, "*Forbidden Archeology*," p. 379.

44. Pierce J. Flynn, Foreword to *Forbidden Archeology*, pp. xix-xx. Jonathan Marks, in his review of *Forbidden Archeology*, labeled Flynn a "curious personage." Concerning Flynn's opinion of *Forbidden Archeology*, Marks stated: "Dr. Pierce J. Flynn ...places this work within postmodern scholarship. I'd like to think postmodern scholars would distance themselves from it; even in the postmodern era, there has to be a difference between scholarship and non-scholarship."

1.6. Michael A. Cremo (1997) "The Later Discoveries of Boucher de Perthes at Moulin Quignon and Their Bearing on the Moulin Quignon Jaw Controversy." XXth International Congress of History of Science, Liège, Belgium.

This paper was accepted for presentation at the XXth International Congress of History of Science, at Liège, Belgium, July 20-26, 1997, in the section Earth Sciences in the Contemporary Period (Since 1800).

Abstract

When Jacques Boucher de Perthes reported stone tools in the Pleistocene gravels of northern France at Abbeville, he was ignored by the French scientific establishment. Later, he was vindicated by English scientists, who came to the Abbeville region and confirmed his discoveries. But some of these same English scientists later turned on him when he reported the discovery of the famous Moulin Quignon jaw. Eventually the discovery was proved a hoax. That is how the standard history goes. But when considered in detail, the hoax theory does not emerge with total clarity and certainty. Boucher de Perthes felt the English scientists who opposed him were influenced by political and religious pressures at home. In order to restore his reputation and establish the authenticity of the Moulin Quignon jaw, Boucher de Perthes conducted several additional excavations at Moulin Quignon, which yielded hundreds of human bones and teeth. But by this time, important minds had been made up, and no attention was paid to the later discoveries, which tended to authenticate the Moulin Quignon jaw. This lack of attention persists in many histories of archeology. This paper details the later discoveries of Boucher de Perthes at Moulin Quignon, addresses possible reasons for their scanty presence in (or complete omission from) many histories of the Moulin Quignon affair, and offers some suggestions about the role the historian of archeology might play in relation to the active work of that science.

Main Text

My book *Forbidden Archeology*, coauthored with Richard L. Thompson, examines the history of archeology and documents numerous discoveries suggesting that anatomically modern humans existed in times earlier than now thought likely. According to most current accounts, anatomically modern humans emerged within the past one or two hundred thousand years from more primitive ancestors. Much of the evidence for greater human antiquity, extending far back into the Tertiary, was discovered by scientists in the nineteenth and early twentieth centuries. Current workers are often unaware of this remarkable body of evidence. In their review article about *Forbidden Archeology*, historians of science Wodak and Oldroyd (1996, p. 197) suggest that "perhaps historians bear some responsibility" for this lack of attention.

> Certainly, some pre-FA histories of palaeoanthropology, such as Peter Bowler's, say little about the kind of evidence adduced by C&T, and the same may be said of some texts published since 1993, such as Ian Tattersall's recent book. So perhaps the rejection of Tertiary [and early Pleistocene] *Homo sapiens*, like other scientific determinations, is a social construction in which historians of science have participated. C&T claim that there has been a 'knowledge filtration operating within the scientific community', in which historians have presumably played their part.

I am also guilty of this knowledge filtering. In this paper, I give an example of my own failure to free myself from unwarranted prejudice.

The collective failure of scientists and historians to properly comprehend and record the history of investigations into human antiquity has substantial consequences on the present development of human antiquity studies. Current workers should have ready access to the complete data set, not just the portion marshaled in support of the current picture of the past and the history of this picture's elaboration. The value of historians' work in maintaining the complete archive of archeological data in accessible form can thus be significant for ongoing human antiquity studies. This approach does not, as some have suggested, entail uncritical acceptance of all past reporting. But it does entail suspension of naive faith in the progressive improvement in scientific reporting.

In February of 1997, I lectured on *Forbidden Archeology* to students and faculty of archeology and earth sciences at the University of Louvain, Belgium. Afterwards, one of the students, commenting on some of the nineteenth century reports I presented, asked how we could accept them, given that these reports had already been rejected long ago and that scientific understanding and methods had greatly improved since the nineteenth century. I answered, "If we suppose that in earlier times scientists accepted bad evidence because of their imperfect understanding and methods, then we might also suppose they rejected good evidence because of their imperfect understanding and methods. There is no alternative to actually looking critically at specific cases." This is not necessarily the task of the working archeologist. But the historian of archeology may here play a useful role.

The specific case I wish to consider is that of the discoveries of Jacques Boucher de Perthes at Moulin Quignon. This site is located at Abbeville, in the valley of the Somme in northeastern France. In *Forbidden Archeology*, I confined myself to the aspects of the case that are already well known to historians (Cremo and Thompson 1993, pp. 402-404). To summarize, in the 1840s Boucher de Perthes discovered stone tools in the Middle Pleistocene high-level gravels of the Somme, at Moulin Quignon and other sites. At first, the scientific community, particularly in France, was not inclined to accept his discoveries as genuine. Some believed that the tools were manufactured by forgers. Others believed them to be purely natural forms that happened to resemble stone tools. Later, leading British archeologists visited the sites of Boucher de Perthes's discoveries and pronounced them genuine. Boucher de Perthes thus became a hero of science. His discoveries pushed the antiquity of man deep into the Pleistocene, coeval with extinct mammals. But the exact nature of the maker of these tools remained unknown. Then in 1863, Boucher de Perthes discovered at Moulin Quignon additional stone tools and an anatomically modern human jaw. The jaw inspired much controversy and was the subject of a joint English-French commission. To do justice to the entire proceedings

(Falconer *et al.* 1863, Delesse 1863) would take a book, so I shall in this paper touch on only a few points of contention.

The English members of the commission thought the recently discovered stone tools were forgeries that had been artificially introduced into the Moulin Quignon strata. They thought the same of the jaw. To settle the matter, the commission paid a surprise visit to the site. Five flint implements were found in the presence of the scientists. The commission approved by majority vote a resolution in favor of the authenticity of the recently discovered stone tools. Sir John Prestwich remained in the end skeptical but nevertheless noted (1863, p. 505) that "the precautions we took seemed to render imposition on the part of the workmen impossible."

That authentic flint implements should be found at Moulin Quignon is not surprising, because flint implements of unquestioned authenticity had previously been found there and at many other sites in the same region. There was no dispute about this at the time, nor is there any dispute about this among scientists today. The strange insistence on forgery and planting of certain flint implements at Moulin Quignon seems directly tied to the discovery of the Moulin Quignon jaw, which was modern in form. If the jaw had not been found, I doubt there would have been any objections at all to the stone tools that were found in the gravel pit around the same time.

In addition to confirming the authenticity of the stone tools from Moulin Quignon, the commission also concluded that there was no evidence that the jaw had been fraudulently introduced into the Moulin Quignon gravel deposits (Falconer *et al.* 1863, p. 452). The presence of grey sand in the inner cavities of the jaw, which had been found in a blackish clay deposit, had caused the English members of the commission to suspect that the jaw had been taken from somewhere else. But when the commission visited the site, some members noted the presence of a layer of fine grey sand just above the layer of black deposits in which the jaw had been found. (Falconer *et al.* 1863, pp. 448-449). This offered an explanation for the presence of the grey sand in the Moulin Quignon jaw and favored its authenticity.

Trinkaus and Shipman (1992, p. 96) insinuate, incorrectly, that the commission's favorable resolution simply absolved Boucher de Perthes of any fraudulent introduction of the jaw (hinting that others may have planted it). But that is clearly not what the commission intended to say, as anyone can see from reading the report in its entirety.

Here are the exact words of Trinkaus and Shipman:

> In any case, the commission found itself deadlocked. There was only one point of agreement: "The jaw in question was not fraudulently introduced into the gravel pit of Moulin Quignon [by Boucher de Perthes]; it had existed previously in the spot where

> M. Boucher de Perthes found it on the 28th March 1863." This lukewarm assertion of his innocence, rather than his correctness, was hardly the type of scientific acclaim and vindication that Boucher de Perthes yearned for. [the interpolation in brackets is by Trinkaus and Shipman]

But the commission (Falconer *et al.* 1863, p. 452) also voted in favor of the following resolution: "All leads one to think that the deposition of this jaw was contemporary with that of the pebbles and other materials constituting the mass of clay and gravel designated as the black bed, which rests immediately above the chalk." This was exactly the conclusion desired by Boucher de Perthes. Only two members, Busk and Falconer, abstained. The committee as a whole was far from deadlocked.

Their scientific objections having been effectively countered, the English objectors, including John Evans, who was not able to join the commission in France, were left with finding further proof of fraudulent behavior among the workmen at Moulin Quignon as their best weapon against the jaw. Taking advantage of a suggestion by Boucher de Perthes himself, Evans sent his trusted assistant Henry Keeping, a working man with experience in archeological excavation, to France. There he supposedly obtained definite proof that the French workmen were introducing tools into the deposits at Moulin Quignon.

But careful study of Keeping's reports (Evans 1863) reveals little to support these allegations and suspicions. Seven implements, all supposedly fraudulent, turned up during Keeping's brief stay at Moulin Quignon. Five were found by Keeping himself and two were given to him by the two French workers who were assigned by Boucher de Perthes to assist him. Keeping's main accusation was that the implements appeared to have "fingerprints" on them. The same accusation had been leveled by the English members of the commission against the tools earlier found at Moulin Quignon. In his detailed discussion of Keeping, which is well worth reading, Boucher de Perthes (1864a, pp. 207-208) remarked that he and others had never been able to discern these fingerprints. Boucher de Perthes (1864a, p. 197, 204) also observed that Keeping was daily choosing his own spots to work and that it would have been quite difficult for the workers, if they were indeed planting flint implements, to anticipate where he would dig. I tend to agree with Boucher de Perthes (1864a, pp. 194-195) that Keeping, loyal to his master Evans, was well aware that he had been sent to France to find evidence of fraud and that he dared not return to England without it. Evans's report (1863), based on Keeping's account, was published in an English periodical and swayed many scientists to the opinion that Boucher des Perthes was, despite the favorable conclusions of the scientific commission, the victim of an archeological fraud.

Not everyone was negatively influenced by Keeping's report. In *Forbidden Archeology*, I cited Sir Arthur Keith (1928, p. 271), who stated, "French anthropologists continued to believe in the authenticity of the jaw until between 1880 and 1890, when they ceased to include it in the list of discoveries of ancient man."

I also was inclined to accept the jaw's authenticity but, given the intensity of the attacks by the English, in *Forbidden Archeology* I simply noted, "From the information we now have at our disposal, it is difficult to form a definite opinion about the authenticity of the Moulin Quignon jaw." I stated this as a mild antidote to the nearly universal current opinion that the Moulin Quignon jaw and accompanying tools were definitely fraudulent. But because Evans and his English accomplices had so thoroughly problematized the evidence, I could not bring myself to suggest more directly that the Moulin Quignon jaw was perhaps genuine.

Boucher des Perthes, however, entertained no doubts as to the authenticity of the jaw, which he had seen in place in the black layer toward the bottom of the Moulin Quignon pit. He believed it had been rejected because of political and religious prejudice in England. Stung by accusations of deception, he proceeded to carry out a new set of excavations, which resulted in the recovery of more human skeletal remains. These later discoveries are hardly mentioned in standard histories, which dwell upon the controversy surrounding the famous Moulin Quignon jaw.

For example, the later discoveries of Boucher des Perthes rate only a line or two in Grayson (1983, p. 217):

> Evans's demonstration of fraud and the strongly negative reaction of the British scientists ensured that the Moulin Quignon mandible would never be accepted as an undoubted human fossil. The same applied to additional human bones reported from Moulin Quignon in 1864.

Trinkaus and Shipman (1992, p. 96) are similarly dismissive:

> Desperately, Boucher de Perthes continued to excavate at Moulin Quignon. He took to calling in impromptu commissions (the mayor, stray geology professors, local doctors, lawyers, librarians, priests, and the like) to witness the event when he found something, or thought he was about to find something, significant.... Soon, the English and French scientists stopped coming to look at his material or paying any real attention to his claims.

Although aware of these later discoveries, I did not discuss them in *Forbidden Archeology*. I thus implicated myself in the process of inadvertent suppression of anomalous evidence posited in *Forbidden Archeology* (Cremo and Thompson 1993, p. 28):

> This evidence now tends to be extremely obscure, and it also tends to be surrounded by a neutralizing nimbus of negative reports, themselves obscure and dating from the time when the evidence was being actively rejected. Since these reports are generally quite derogatory, they may discourage those who read them from examining the rejected evidence further.

The cloud of negative reporting surrounding the Moulin Quignon jaw influenced not only my judgment of this controversial find but also discouraged me from looking into the later discoveries of Boucher des Perthes. So let us now look into these discoveries and see if they are really deserving of being totally ignored or summarily dismissed.

Boucher de Perthes (1864b), stung by the accusations he had been deceived, carried out his new investigations so as to effectively rule out the possibility of deception by workmen. First of all, they were carried out during a period when the quarry at Moulin Quignon was shut down and the usual workmen were not there (1864b, p. 219). Also, Boucher de Perthes made his investigations unannounced and started digging at random places. He would usually hire one or two workers, whom he closely supervised. Furthermore, he himself would enter into the excavation and break up the larger chunks of sediment with his own hands. In a few cases, he let selected workers, who were paid only for their labor, work under the supervision of a trusted assistant. In almost all cases, witnesses with scientific or medical training were present. In some cases, these witnesses organized their own careful excavations to independently confirm the discoveries of Boucher de Perthes.

Here follow excerpts from accounts by Boucher de Perthes and others of these later discoveries. They are taken from the proceedings of the local Société d'Émulation. Most French towns had such societies, composed of educated gentlemen, government officials, and businessmen.

On April 19, 1864, Boucher de Perthes took a worker to the gravel pit and on the exposed face of the excavation pointed out some places for a worker to dig. Boucher de Perthes (1864b, p. 219) "designated every spot where he should strike with his pick." In this manner, he discovered a hand axe, two other smaller worked flints, and several flint flakes. Then the worker's pick "struck an agglomeration of sand and gravel, which broke apart, as did the bone it contained." Boucher de Perthes (1864b, p. 219) stated, "I took from the bank the part that remained, and recognized the end of a human femur." This find occurred at a depth 2.3 meters, in the hard, compacted bed of yellowish-brown sand and gravel lying directly above the chalk. In this, as in all cases, Boucher de Perthes had checked very carefully to see that the deposit was undisturbed and that there were no cracks or fissures through which a bone could have slipped down from higher levels (p. 219). Digging further at the same spot, he encountered small fragments of bone, including an iliac bone, 40 centimeters from the femur and in the same plane (p. 219).

On April 22, Boucher de Perthes found a piece of human skull 4 centimeters long in the yellow-brown bed. This yellow-brown bed contains in its lower levels some seams of yellow-grey sand. In one of these seams. Boucher de Perthes found more skull fragments and a human tooth (1864b, p. 220).

On April 24, Boucher de Perthes was joined by Dr. J. Dubois, a physician at the Abbeville municipal hospital and a member of the Anatomical Society of Paris. They directed the digging of a worker in the yellow-brown bed. They uncovered some fragments too small to identify. But according to Dubois they displayed signs of incontestable antiquity. Boucher de Perthes and Dubois continued digging for some time, without finding anything more. "Finally," stated Boucher de Perthes (1864b, p. 221), "we saw in place, and Mr. Dubois detached himself from the bank, a bone that could be identified. It was 8 centimeters long. Having removed a portion of its matrix, Mr. Dubois recognized it as part of a human sacrum. Taking a measurement, we found it was lying 2.6 meters from the surface." About 40 centimeters away, they found more bones, including a phalange. They then moved to a spot close to where the jaw was discovered in 1863. They found parts of a cranium and a human tooth, the latter firmly embedded in a pebbly mass of clayey sand (p. 222). The tooth was found at a depth of 3.15 meters from the surface (p. 223).

On April 28, Boucher de Perthes began a deliberate search for the other half of the sacrum he had found on April 24. He was successful, locating the missing half of the sacrum bone about 1 meter from where the first half had been found. He also found a human tooth fragment in a seam of grey sand. Studying the edge of the break, Boucher de Perthes noted it was quite worn, indicating a degree of antiquity (1864b, p. 223).

On May 1, accompanied for most of the day by Dr. Dubois, Boucher de Perthes found three fragments of human skulls, a partial human tooth, and a complete human tooth (1864b, p. 223). On May 9, Boucher de Perthes (pp. 223-224) found two human skull fragments, one fairly large (9 centimeters by 8 centimeters).

On May 12, Boucher de Perthes carried out explorations in the company of Mr. Hersent-Duval, the owner of the Moulin Quignon gravel pit. They first recovered from the yellow bed, at a depth of about 2 meters, a large piece of a human cranium, 8 centimeters long and 7 centimeters wide. "An instant later," stated Boucher de Perthes (1864b, p. 224), "the pick having detached another piece of the bank, Mr. Hersent-Duval opened it and found a second fragment of human cranium, but much smaller. It was stuck so tightly in the mass of clay and stones that it took much trouble to separate it."

On May 15, Boucher de Perthes extracted from one of the seams of grey sand in the yellow-brown bed, at a depth of 3.2 meters, a human tooth firmly embedded in a chunk of sand and flint. The tooth was white. Boucher de Perthes (1864b, p. 225) noted: "It is a very valuable specimen, that replies very well to the ...objection that the whiteness of a tooth is incompatible with its being a fossil." He then found in the bed of yellow-brown sand "a human metatarsal, still attached in its matrix, with a base of flint" (p. 225). In the same bed he also found many shells, which also retained their white color. Boucher de Perthes (1864b, p. 226) observed: "Here the color of the bank, even the deepest, does not communicate itself to the rolled flints, nor to the shells, nor to the teeth, which all preserve their native whiteness." This answered an earlier objection to the antiquity of the original Moulin Quignon jaw and a detached tooth found along with it.

On June 6, Boucher de Perthes (1864b, p. 230) found in the yellow-brown bed, at a depth of 4 meters, the lower half of a human humerus, along with several less recognizable bone fragments. On June 7, he recovered part of a human iliac bone at the same place (p. 231). On June 8 and 9, he found many bone fragments mixed with flint tools, including many hand axes. Later on June 10, he returned with three workers to conduct bigger excavations. He found two fragments of tibia (one 14 centimeters long) and part of a humerus (p. 231). These bones had signs of wear and rolling. They came from a depth of 4 meters in the yellow-brown bed. Please note that I am just recording the discoveries of human bones. On many days, Boucher de Perthes also found fragments of bones and horns of large mammals. Boucher de Perthes (1864b, p. 232) noted that the human bones were covered with a matrix of the same substance as the bed in which they were found. When the bones were split, it was found that traces of the matrix were also present in their internal cavities (p. 232). Boucher de Perthes (1864b, p. 232) noted that these are not the kinds of specimens that could be attributed to "cunning workers." On this particular day, Boucher de Perthes left the quarry for some time during the middle of the day, leaving the workers under the supervision of an overseer. Boucher de Perthes (1864b, pp. 233-235) then reported:

> In the afternoon, I returned to the bank. My orders had been punctually executed. My representative had collected some fragments of bone and worked flints. But a much more excellent discovery had been made—this was a lower human jaw, complete except for the extremity of the right branch and the teeth.

My first concern was to verify its depth. I measured it at 4.4 meters, or 30 centimeters deeper than the spot where I had that morning discovered several human remains. The excavation, reaching the chalk at 5.1 meters, faced the road leading to the quarry. It was 20 meters from the point, near the mill, where I found the half-jaw on March 28, 1863.

The jaw's matrix was still moist and did not differ at all from that of all the other bones from that same bed. The matrix was very sticky, mixed with gravel and sometimes with pieces of bone, shells, and even teeth.

The teeth were missing from the jaw. They were worn or broken a little above their sockets, such that the matrix that covered them impaired their recognition. The deterioration was not recent, but dated to the origin of the bank.

Although I did not see that jaw *in situ,* after having minutely verified the circumstances of its discovery, I do not have the least doubt as to its authenticity. Its appearance alone suffices to support that conviction. Its matrix, as I have said, is absolutely identical to that of all the other bones and flints from the same bed. Because of its form and hardness, it would be impossible to imitate.

The worker in the trench, after having detached some of the bank, took it out with his shovel. But he did not see the jaw, nor could he have seen it, enveloped as it was in a mass of sand and flint that was not broken until the moment that the shovel threw it into the screen. It is then ...that it was seen by the overseer.

He recognized it as a bone, but not seeing the teeth, he did not suspect it was a jaw. Mr. Hersent-Duval, who happened to come by at that moment, was undeceived. He signaled the workers and told them to leave it as it was, in its matrix, until my arrival, which came shortly thereafter.

After a short examination, I confirmed what Mr. Hersent had said. It was not until then that the workers believed. Until that moment, the absence of teeth and the unusual form of the piece, half-covered with clay, had caused even my overseer himself to doubt.

I therefore repeat: here one cannot suspect anyone. Strangers to the quarry and the town, these diggers had no interest in deception. I paid them for their work, and not for what they found.....Dr. Dubois, to whom I was eager to show it, found it from the start to have a certain resemblance to the one found on March 28, 1863.

On June 17, Hersent-Duval had some workers dig a trench. They encountered some bones. Hersent-Duval ordered them to stop work, leaving the bones in place. He then sent a message for Boucher de Perthes to come. Boucher de Perthes arrived, accompanied by several learned gentlemen of Abbeville, including Mr. Martin, who was a professor of geology and also a parish priest. Boucher de Perthes (1864b, pp. 235-237) stated:

> Many fragments, covered in their matrix, lay at the bottom of the excavation, at a depth of 4 meters. At 3 meters, one could see two points, resembling two ends of ribs.
>
> Mr. Martin, who had descended with us into the trench, touched these points, and not being able to separate them, thought that they might belong to the same bone. I touched them in turn, as did Abbey Dergny, and we agreed with his opinion.
>
> Before extracting it, these gentlemen wanted to assure themselves about the state of the terrain. It was perfectly intact, without any kind of slippage, fissures, or channels, and it was certainly undisturbed. Having acquired this certainty, the extraction took place by means of our own hands, without the intermediary of a worker.
>
> Mr. Martin, having removed part of the envelope of the extracted bone, recognized it as a human cranium. And the two points at first taken as two ends of ribs, were the extremities of the brow ridges. This cranium, of which the frontal and the two parietals were almost complete, astonished us with a singular depression in its upper part.
>
> This operation accomplished, we occupied ourselves with the bones fallen to the bottom of the quarry. They were three in number, covered by a mass of clay so thick that one could not tell the kind of creature to which they belonged. Much later, they were identified by Dr. Dubois as a human iliac bone, a right rib, and two pieces of an upper jaw, perhaps from the same head as the partial cranium, because they came from the same bed.
>
> Having continued our excavation, we found yet another human bone, and we probably would have encountered others, if we had been able, without the danger of a landslide, to carry out the excavation still further.
>
> All of this was recorded by Abbey Dergny, in a report signed by him and professor Martin ...one of the most knowledgeable and respected men of our town.

On July 9th, a commission composed of the following individuals made an excavation at Moulin Quignon: Louis Trancart, mayor of Laviers; Pierre

Sauvage, assistant to the mayor of Abbeville, and member of the Société d'Émulation of that town; F. Marcotte, conservator of the museum of Abbeville, and member of the Société d'Émulation and the Academy of Amiens; A. de Caïeu, attorney, and member of the Société d'Émulation and the Society of Antiquaries of Picardy; and Jules Dubois, M.D., doctor at the municipal hospital of Abbeville, and member of many scientific societies (Dubois 1864a, p. 265).

At the quarry they carried out excavations at two sites. Marcotte, who had proclaimed his skepticism about the discoveries, was chosen to direct the digging of the workers. "He had the base of the excavation cleared away until it was possible to see the chalk, upon which directly rested the bed of yellow-brown sand," said Dubois (1864a, p. 266) in his report on the excavation of the first site in the quarry. "After we assured ourselves that the wall of the cut was clearly visible to us and that it was free of any disturbance, the work commenced under our direct inspection." After 15 minutes of digging, Marcotte recovered a bone that Dubois (1864a, p. 266) characterized as probably a piece of a human radius 8 centimeters long. The bone was worn and covered by a tightly adhering matrix of the same nature as the surrounding terrain. The excavation proceeded for a long time without anything else being found, until Mr. Trancart found part of a human femur or humerus (p. 267). Some minutes later, Trancart recovered a broken portion of a human tibia.

The commission then moved to the second site, about 11 meters away. It is movements like these that remove suspicions the bones were being planted. Dubois (1864a, p. 267) stated: "Here again we had to clear away the base of the section to reveal the actual wall of the quarry. The same precautions were taken to assure the homogeneity of the bed and the absence of any disturbance." At this site, Marcotte found a piece of a human femur, about 13 centimeters long (p. 268). It came from the bed of yellow-brown sand which lies directly on the chalk. Boucher de Perthes (1864b, p. 237) noted that two hand axes were also found on the same day.

On July 16, the members of the commission that carried out the July 9 excavation were joined at Moulin Quignon by Mr. Buteux and Mr. de Mercey, members of the Geological Society of France; Baron de Varicourt, chamberlain of His Majesty the King of Bavaria; Mr. de Villepoix, member of the Société d'Émulation; and Mr. Girot, professor of physics and natural history at the College of Abbeville. In addition to the members of the formal commission, a dozen other learned gentlemen, including Boucher de Perthes, were present for the new excavations.

Dubois noted in his report that the quarry wall at the chosen spot was undisturbed and without fissures. About the workers, Dubois (1864b, p. 270) stated, "Needless to say, during the entire duration of the work, they were the object of continuous surveillance by various members of the commission." In examining a large chunk of sediment detached by a pick, the commission members found a piece of a human cranium, comprising a large part of the frontal with a small part of the parietal (p. 270). It was found at a depth of 3.3 meters in the yellow-brown bed that lies just above the chalk (p. 271).

Dubois's report (1864b, p. 271) stated:

> Immediately afterwards, one of the workers was ordered to attack the same bank at the same height, but 3 meters further to the left. The other worker continued to dig at the extreme right. Is it necessary to repeat that all necessary precautions were taken to establish the integrity of the bed there and that the two workers each continued to be the object of scrupulous surveillance?
>
> We went a long time without finding anything resembling a bone. The excavation on the far right side yielded no results whatsoever. Finally, after about three and a half hours, there came to light the end of a bone, of medium size, situated horizontally in the bed. After its exact position was confirmed, Mr. Marcotte himself took from the sand a complete bone, about 13 centimeters long.... It was the right clavicle of an adult subject of small size.... Measurements showed it was lying 3 meters from the surface, and 2.3 meters horizontally from our starting point.

Further excavation caused a landslide. The debris was cleared away, however, and the excavation proceeded, yielding a human metatarsal. Several members of the commission, including the geologist Buteux, saw it in place. It was found at a depth of 3.3 meters just above the chalk in the yellow-brown bed. It was situated about 4 meters horizontally from the line where the excavation started (Dubois 1864b, p. 272). According to Boucher de Perthes (1864b, p. 238), the bones from this excavation, and apparently others, were deposited to the Abbeville museum.

I find the account of this excavation extraordinary for several reasons. First of all, it was conducted by qualified observers, including geologists capable of judging the undisturbed nature of the beds. Second, a skilled anatomist was present to identify the bones as human. Third, it is apparent that the workers were carefully supervised. Fourth, some of the human bone fragments were found at points 3 to 4 meters horizontally from the starting point of the excavation and depths over 3 meters from the surface. This appears to rule out fraudulent introduction. Fifth, the condition of the bones (fragmented, worn, impregnated with the matrix) is consistent with their being genuine fossils. I do not see how such discoveries can be easily dismissed.

Summarizing his discoveries, Boucher de Perthes (1864b) stated:

> The osseous remains collected in the diverse excavations I made in 1863 and 1864 at Moulin Quignon, over an area of about 40 meters of undisturbed terrain without any infiltration, fissure, or [p. 239] channel, have today reached two hundred in number. Among them are some animal bones, which are being examined (pp. 238-239).
>
> Among the human remains, one most frequently encounters pieces of femur, tibia, humerus, and especially crania, as well as teeth, some whole and some broken. The teeth represent all ages—they are from infants of two or three years, adolescents, adults, and the aged. I have collected, *in situ,* a dozen, some whole, some broken, and more in passing through a screen the sand and gravel take from the trenches (p. 240).
>
> Doubtlessly, a lot has been lost. I got some proof of this last month when I opened a mass of sand and gravel taken from a bank long ago and kept in reserve. I found fragments of bone and teeth, which still bear traces of their matrix and are therefore of an origin beyond doubt (p. 241).

Armand de Quatrefages, a prominent French anthropologist, made a report on Boucher de Perthes's later discoveries at Moulin Quignon to the French Academy of Sciences. Here are some extracts from the report (De Quatrefages 1864):

> In these new investigations, Boucher de Perthes has employed only a very few workers. In the majority of cases, he himself has descended into the excavation and with his own hands has broken apart and crumbled the large pieces of gravel or sand detached by the picks of the workers. In this manner, he has procured a great number of specimens, some of them very important. We can understand that this way of doing things guarantees the authenticity of the discoveries.
>
> On hearing the first results of this research, I encouraged Boucher de Perthes to persevere, and to personally take every necessary precaution to prevent any kind of fraud and remove any doubts about the stratigraphic position of the discoveries....
>
> As the discoveries continued, Boucher de Perthes sent to me, on June 8, 1864, a box containing several fragments of bones from human skeletons of different ages. I noted: 16-17 teeth from first and second dentitions; several cranial fragments, including a

> portion of an adult occipital and the squamous portion of a juvenile temporal; pieces of arm and leg bones, some retaining their articulator ends; pieces of vertebrae and of the sacrum. The specimens were accompanied by a detailed memoir reporting the circumstances of their discovery.
>
> I examined these bones with M. Lartet. We ascertained that most of them presented very nicely the particular characteristics that were so greatly insisted upon in denying the authenticity of the Moulin Quignon jaw. In accord with M. Lartet, I felt it advisable to persuade M. Boucher de Perthes to make further excavations, but this time in the presence of witnesses whose testimony could not in the least be doubted.... Among the more important specimens found in these latest excavations are an almost complete lower jaw and a cranium.
>
> All of these finds were made in the course of excavations that were mounted in an on-and-off fashion, without any definite pattern. That is to say, Boucher de Perthes would suddenly proceed to the sites, sometimes alone and sometimes with friends. Doing things like this very clearly renders any kind of fraud quite difficult. During the course of an entire year and more, the perpetrator of the fraud would have had to go and conceal each day the fragments of bone destined to be found by those he was attempting to deceive. It is hardly credible that anyone would adopt such means to attain such an unworthy goal or that his activities would have remained for so long undetected.
>
> Examination of the bones does not allow us to retain the least doubt as to their authenticity. The matrix encrusting the bones is of exactly the same material as the beds in which they were found, a circumstance that would pose a serious difficulty for the perpetrators of the daily frauds.... Because of the precautions taken by Boucher de Perthes and the testimony given by several gentlemen who were long disinclined to admit the reality of these discoveries, I believe it necessary to conclude that the new bones discovered at Moulin Quignon are authentic, as is the original jaw, and that all are contemporary with the beds where Boucher de Perthes and his honorable associates found them.

I am inclined to agree with De Quatrefages that the later discoveries of Boucher de Perthes tend to confirm the authenticity of the original Moulin Quignon jaw.

At this point, I wish to draw attention to a report by Dr. K. P. Oakley on the Moulin Quignon fossils. It is one of the few scientific reports from the twentieth century giving any attention at all to the later discoveries of Boucher de Perthes. Oakley gave the following results from fluorine content testing (Oakley 1980, p. 33). The original Moulin Quignon jaw had 0.12 percent fluorine, a second jaw (the one apparently found on June 10) had a fluorine content of 0.05 percent. By comparison, a tooth of *Paleoloxodon* (an extinct elephantlike mammal) from Moulin Quignon had a fluorine content of 1.7 percent, whereas a human skull from a Neolithic site at Champs-de-Mars had a fluorine content of 0.05 percent. Fluorine, present in ground water, accumulates in fossil bones over time. Superficially, it would thus appear that the Moulin Quignon jaw bones, with less fluorine than the *Paleoloxodon* tooth, are recent.

But such comparisons are problematic. We must take into consideration the possibility that much of a fossil bone's present fluorine content could have accumulated during the creature's lifetime. It is entirely to be expected that the tooth of an animal such as an elephant might acquire a considerable amount of fluorine from drinking water and constantly chewing vegetable matter—much more fluorine than the bone in a human jaw, not directly exposed to water and food. Also, the amount of fluorine in ground water can vary from site to site, and even at the same site bones can absorb varying amounts of fluorine according to the permeability of the surrounding matrix and other factors. Furthermore, fluorine content varies even in a single bone sample. In a typical case (Aitken 1990, p. 219), a measurement taken from the surface of a bone yielded a fluorine content of 0.6 percent whereas a measurement taken at 8 millimeters from the surface of the same bone yielded a fluorine content of just 0.1 percent. As such, Oakley's fluorine content test results cannot be taken as conclusive proof that the Moulin Quignon jaws were "intrusive in the deposits" (Oakley 1980, p. 33).

If the Moulin Quignon human fossils of Abbeville are genuine, how old are they? Abbeville is still considered important for the stone tool industries discovered by Boucher de Perthes. In a recent synoptic table of European Pleistocene sites, Carbonell and Rodriguez (1994, p. 306) put Abbeville at around 430,000 years, and I think we can take that as a current consensus.

Fossil evidence for the presence of anatomically modern humans at Abbeville is relevant to one of the latest archeological finds in Europe. Just this year, Thieme (1997, p. 807) reported finding advanced wooden throwing spears in German coal deposits at Schöningen, Germany. Thieme gave these spears an age of 400,000 years. The oldest throwing spear previously discovered was just 125,000 years old (Thieme 1997, p. 810).

The spears discovered by Thieme are therefore quite revolutionary. They are causing archeologists to upgrade the *cultural* level of the Middle

Pleistocene inhabitants of Europe, usually characterized as ancestors of anatomically modern humans, to a level previously associated exclusively with anatomically modern humans. Alternatively, we could upgrade the *anatomical* level of the Middle Pleistocene inhabitants of northern Europe to the level of modern humans. The skeletal remains from Moulin Quignon, at least some of which appear to be anatomically modern, would allow this. They are roughly contemporary with the Schöningen spears. Unfortunately, not many current workers in archeology are aware of the Moulin Quignon discoveries, and if they are aware of them, they are likely to know of them only from very brief (and misleading) negative evaluations.

Why have historians and scientists alike been so skeptical of the Moulin Quignon finds? I suspect it has a lot do to with preconceptions about the kind of hominid that should be existing in the European Middle Pleistocene. The following passage from Trinkaus and Shipman (1992, p. 97) is revealing:

> That any knowledgeable scientist should take the Moulin Quignon jaw seriously as a human fossil appears difficult to fathom in retrospect. Yet, despite the support for the Neander Tal fossils as an archaic, prehistoric human, few knew what to expect. Clearly, many...still expected human fossils to look just like modern humans; it was only a matter of finding the specimen in the appropriately prehistoric context.

It is clear that Trinkaus and Shipman would expect to find only ancestors of the modern human type in the European Middle Pleistocene. And today it would be hard to find a "knowledgeable scientist" who did not share this expectation. It is clear to me, however, that this fixed expectation may have obscured correct apprehension of the human fossil record in Europe and elsewhere. So perhaps it is good for researchers with different expectations to look over, from time to time, the history of archeology.

My own expectations are conditioned by my committed study of the Sanskrit historical writings of Vedic India (the Puranas), which contain accounts of extreme human antiquity. In his review of *Forbidden Archeology*, Murray (1995, p. 379) wrote:

> For the practising quaternary archaeologist current accounts of human evolution are, at root, simply that. The "dominant paradigm" has changed and is changing, and practitioners openly debate issues which go right to the conceptual core of the discipline. Whether the Vedas have a role to play in this is up to the

individual scientists concerned.

I am hopeful that some individual scientists will in fact decide that the Vedas do have a role to play in changing the conceptual core of studies in human origins and antiquity.

But let us return to the more limited question before us. As far as the finds of human bones at Moulin Quignon are concerned, I would be satisfied if a professor of archeology at a European university, perhaps in France or Belgium, would assign some graduate students to reopen the investigation.

Literature Cited

Aitken, M. J. (1990) *Science-based Dating in Archaeology.* London, Longman.

Boucher de Perthes, J. (1864a) Fossile de Moulin-Quignon: Vérification Supplémentaire. *In* Boucher de Perthes, J., *Antiquités Celtiques et Antédiluviennes. Memoire sur l'Industrie Primitive et les Arts à leur Origin* (Vol. 3). Paris, Jung-Treutel, pp. 194-214.

Boucher de Perthes, J. (1864b) Nouvelles Découvertes d'Os Humains dans le Diluvium, en 1863 et 1864, par M. Boucher de Perthes. Rapport a la Société Impériale d'Émulation. *In* Boucher de Perthes, J., *Antiquités Celtiques et Antédiluviennes. Memoire sur l'Industrie Primitive et les Arts à leur Origin* (Vol. 3). Paris, Jung-Treutel, pp. 215-250.

Carbonell, E. and Rodriguez, X. P. (1994) Early Middle Pleistocene deposits and artefacts in the Gran Dolina site (TD4) of the 'Sierra de Atapuerca' (Burgos, Spain). *Journal of Human Evolution, 26*: 291-311.

Cremo, M. A., and Thompson, R. L. (1993) *Forbidden Archeology: The Hidden History of the Human Race.* San Diego, Bhaktivedanta Institute.

Delesse, A. (1863) La mâchoire humaine de Moulin de Quignon. *Mémoires de la Société d'Anthropologie de Paris, 2*: 37-68.

De Quatrefages, A. (1864) Nouveaux ossements humains découverts par M. Boucher de Perthes à Moulin-Quignon. *Comptes Rendus Hebdomadaires de l'Académie des Sciences, 59*:107-111.

Dubois, J. (1864a) Untitled report of excavation at Moulin Quignon, on July 9, 1864. Société Impériale d'Émulation. Extrait du registre des procès-verbaux. Séance du 21 Juillet 1864. *In* Boucher de Perthes, J., *Antiquités Celtiques et Antédiluviennes. Memoire sur l'Industrie Primitive et les Arts à leur Origin* (Vol. 3). Paris, Jung-Treutel, pp. 265-268.

Dubois, J. (1864b) Untitled report of excavation made at Moulin Quignon on July 16, 1864. Société Impériale d'Émulation. Extrait du registre des procès-verbaux. Suit de la séance du 21 Juillet 1864. In Boucher de Perthes, J., *Antiquités Celtiques et Antédiluviennes. Memoire sur l'Industrie Primitive et les Arts à leur Origin* (Vol. 3). Paris, Jung-Treutel, pp. 269-272.

Evans, John (1863) The human remains at Abbeville. *The Athenaeum*, July 4, pp.19-20.

Falconer, H., Busk, George, and Carpenter, W. B. (1863) An account of the proceedings of the late conference held in France to inquire into the circumstances attending the asserted discovery of a human jaw in the gravel at Moulin-Quignon, near Abbeville; including the *procès verbaux* of the conference, with notes thereon. *The Natural History Review, 3* (new series): 423- 462.

Grayson, D. K. (1983) *The Establishment of Human Antiquity*. New York, Academic Press.

Keith, A. (1928) *The Antiquity of Man*. Philadelphia, J. B. Lippincott.

Murray, T. (1995) Review of *Forbidden Archeology. British Journal for the History of Science, 28*: 377-379.

Oakley, K. P. (1980) Relative dating of fossil hominids of Europe. *Bulletin of the British Museum of Natural History* (*Geology*), vol. 34.

Prestwich, J. (1863) On the section at Moulin Quignon, Abbeville, and on the peculiar character of some of the flint implements recently discovered there. *Quarterly Journal of the Geological Society of London*, vol. 19, first part, pp. 497-505.

Thieme, H. (1997) Lower Paleolithic hunting spears from Germany. *Nature,* February 27, 385: 807-810.

Trinkaus, E. and Shipman, P. (1992) *The Neandertals*. New York, Vintage.

Wodak, J. and Oldroyd, D. (1996) . "'Vedic Creationism': A Further Twist to the Evolution Debate." *Social Studies of Science*, vol. 26, pp. 192-213.

2

Mainstream Academic Reviews, Notices, and Related Correspondence

2.1. Archeology, Anthropology, and Biology

Traditionally, archeology is considered one of four subdisciplines of anthropology, the others being physical anthropology, cultural anthropology, and linguistics. Forbidden Archeology *received notices and reviews in most of the principal journals of anthropology and archeology. As could be expected, none of the reviewers accepted the book's conclusions, but some nevertheless offered appreciations for its thorough research and literary quality. If requested by the copyright holder, notices of permission are given in a specific form. All related correspondence, unless otherwise noted, is published by permission of the authors.*

2.1.1. Cyprian Broodbank (1993) Notice of *Forbidden Archeology.* New books section, *Antiquity,* vol. 67, p. 904. Reprinted by permission of *Antiquity.*

All the reasons and evidence why modern humans are not rather recent but most ancient, a very large, very odd compilation of every anomaly in a very pink jacket.

2.1.2. Anonymous (1994) Notice of *Forbidden Archeology.* "Publications," *Journal of Field Archeology,* vol. 21, p. 112.

Notices in the "Publications" section of Journal of Field Archeology *are brief summaries drawn from the flap copy of the book or other promotional material. They do not represent critical reviews.* Journal of Field Archeology *therefore does not give formal permission for reproduction of such notices.*

Michael Cremo, a research associate in history and philosophy of science, and Richard Thompson, a mathematician, challenge the dominant views of human origins and antiquity. This volume combines a vast amount of both accepted and controversial evidence from the archaeological record with sociological, philosophical, and historical critiques of the scientific method to challenge existing views and expose the suppression of information concerning history and human origins.

2.1.3. Jonathan Marks (1994) Review of *Forbidden Archeology*. *American Journal of Physical Anthropology,* vol. 93(1), pp. 140-141. © 1994 *American Journal of Physical Anthropology*. Reprinted by permission of Wiley-Liss, Inc., a division of John Wiley & Sons, Inc.

Jonathan Marks is professor of anthropology at Yale University. His was the first major academic review of Forbidden Archeology *to reach print. Excerpts from his extremely negative review were printed in the abridged popular version of* Forbidden Archeology, *which proved helpful in attracting the attention of radio and television. Marks and I were both guests on one of the radio shows during my fall 1994 media tour.*

The major fallacy of the Fundamentalist push to get "creation-science" into the classroom is the assumption that Genesis is the only alternative to evolution. Consequently, any knock at post-Darwinism is a blow to Genesis. Obviously, the rich and varied origin myths of all cultures are alternatives to contemporary evolution, though none can reasonably be considered science. Generally, attempts to reconcile the natural world to religious views end up compromising the nature world. The result is "science made stupid" which was actually the title of a very clever book a few years ago written by writer/illustrator Tom Weller. Weller's book, however, was intended to be humorous, while religion-based science, like the present volume, unfortunately is not.

The explicit aim of the authors is to reconcile paleoanthropology to the Vedic ideas that "the human race is of great antiquity" and that "various human and apelike beings have coexisted for a long time" (p. xxxvi). That does not sound particularly challenging; but, unsatisfied with the apparently easy harmony between normal science and their nebulous theology, the authors decided to redo anthropology.

The argument is simple: think of all the generalizations we can make about human evolution. Now think of all the exceptions, paradoxes, mistakes, and hoaxes. Now switch them. That is this book. As the Firesign Theatre once proclaimed: Everything you know is wrong! (But then, they were trying to be humorous too.)

For unclear reasons, given the looseness of their religious thesis, this book is antievolutionary. The authors are trying to argue that humans have always been on earth, even unto the Pre-Cambrian, when there was not much for them to eat or breathe. Nevertheless, a solid sphere with three grooves from that remote age indicates a human presence. At least according to the *Weekly World News*, their source. But, mixed in among references from *Cryptozoology,* the Bigfoot literature, and obscure nineteenth century books and newspaper articles are papers from this Journal and other more reputable sources.

There is no discussion of theory or process, only fossils. This is basically the old creationist gambit of juxtaposing different interpretations of the same remains and concluding that we know absolutely nothing. Indeed, it comes with a glowing jacket blurb by Philip Johnson, the creationist lawyer who wrote *Darwin on Trial.*

The other curious personage is Dr. Pierce J. Flynn, who has contributed both a jacket blurb and a Foreword [see section X of this chapter]. Flynn, whose affiliation is given as the Department of Arts and Sciences, Cal State-San Marcos, places this work within postmodern scholarship. I'd like to think postmodern scholars would distance themselves from it; even in the postmodern era, there has to be a difference between scholarship and non-scholarship. The prospect of variation in quality within the published literature, and criteria by which to establish it, is central to the endeavor of professional scholarship.

The best that can be said is that more reading went into this Hindu-oid creationist drivel than seems to go into the Christian-oid creationist drivel. At any rate, this is a must for anyone interested in keeping up with goofy popular anthropology; at well over 900 pages, it is a veritable cornucopia of dreck. And, according to the publicity blurb, "a condensed popular version is also in the works, scheduled for a spring 1994 release." Watch for it.

Related Correspondence

After seeing the review of Forbidden Archeology *in* AJPA, *I exchanged some letters with Dr. Matt Cartmill, the editor. Dr. Cartmill has asked that I not publish the letters he wrote to me. Repecting his wish, I shall simply give the dates of his letters along with the briefest indications of the contents.*

2.1.3.1. Letter to Dr. Matt Cartmill, Editor, *American Journal of Physical Anthropology,* April 16, 1994.

I would have submitted a more timely response to Jonathan Marks' unremittingly negative review of my book *Forbidden Archeology: The*

Hidden History of the Human Race (January 1994 issue), but I was out of the country when the review came out and did not see it until recently. My main concern is not that he labeled my interpretation of the evidence contained in the book as "Hindu-oid creationist drivel" but that he misrepresented the quality of the evidence itself, insinuating that it was mostly "goofy" stuff from tabloids, with a few other things from more respectable sources thrown in. I hope you will allow me some space in Notes and Comments to counter this misimpression, for the benefit of those of your readers who have not actually seen the book.

Included with the above letter was the following draft of a submission to the Notes and Comments *section of* American Journal of Physical Anthropology.

2.1.3.2. Michael A. Cremo. Beyond "Dreck" and "Drivel": A Response to Jonathan Marks's Review of *Forbidden Archeology*, Submitted to Dr. Matt Cartmill for Publication in *American Journal of Physical Anthropology*, April 16, 1994.

In his review of *Forbidden Archeology* (AJPA 93:140-141), Jonathan Marks alluded sarcastically to the case of a Precambrian grooved metallic sphere—insinuating this is typical of the material Richard Thompson and I included in our book.

In fact, the notorious grooved sphere is tucked away with a few similar reports in an appendix, which we introduced (p. 795) as follows: "The reports of this extraordinary evidence emanate, with some exceptions, from nonscientific sources....We ourselves are not sure how much importance should be given to this highly anomalous evidence. But we include it for the sake of completeness."

Marks implied that we uncritically accepted an account of the sphere contained in a tabloid newspaper, which he mockingly called our "source." He neglected to mention our correspondence with Roelf Marx (p. 813), the curator of the Klerksdorp Museum in South Africa, where the sphere and others like it are housed (they have turned up by the hundreds in nearby pyrophyllite deposits). He also neglected to mention our discussion of an unpublished natural explanation for the spheres given by South African geologist A. Bisschoff (pp. 813-14). Having found this explanation inadequate, we simply suggested that in the absence of a truly adequate natural explanation, one might have to consider the possibility, as remote as it might seem, that the extremely hard, perfectly round metallic spheres, especially the one with three groves around its equator, were produced artificially (p. 814). [*Not all of the spheres are perfectly round, nor are all of them of the same degree of hardness. The spheres are metallic in the sense of being composed of hematite, an ore of iron.*]

In any case, the main body of *Forbidden Archeology* contains not reports of Precambrian grooved spheres but reports of incised bones and shells, stone tools, and human skeletal remains—the usual stuff of anthropological and archeological discourse. Almost all of these reports were written by professional archeologists and anthropologists and appeared in standard academic journals over the past 150 years. Other than that these reports place anatomically modern humans, or other advanced hominids, at unexpectedly remote times—mostly Pliocene and Miocene—they are quite orthodox, in terms of their authors' qualifications, affiliations, and modes of presentation.

It is such well-documented material that takes *Forbidden Archeology* out of the "goofy popular anthropology" category, to which it was consigned by Marks. Physical anthropologist W. W. Howells has said (personal communication, 1993) that *Forbidden Archeology* "represents much careful effort in critically assembling published materials." And scholars from several other disciplines, such as history of science, are taking a serious interest in the book.

At the 1994 annual meeting of the AAPA in Denver, several members who purchased *Forbidden Archeology* told me they found its thorough compilation of anomalous evidence related to human origins and antiquity professionally useful to them.

I myself can offer this suggestion why physical anthropologists might want to have a look at the material contained in *Forbidden Archeology*. Recently, C. C. Swisher and his coworkers (1994) dated some of the Java *Homo erectus* finds at 1.9 million years. This suggests that *Homo erectus* migrated from Africa about 1 million years earlier than most workers have thought likely. In light of this, perhaps it would be useful to have another look at the crude stone implements found at Foxhall, and other sites in England, by J. Reid Moir (1927) in Late Pliocene contexts. The artificial nature of the implements and their Late Pliocene provenience were accepted by authorities ranging from Henry Fairfield Osborn (1921) to Louis Leakey (1960, pp. 66, 68). Moir's finds are not well known today, being absent from most current literature. But one will find in *Forbidden Archeology* a thorough discussion of them, with key excerpts from contemporary reports and illustrations of the artifacts.

One might also want to have a look at the many incised mammal bones from the Late Pliocene/Early Pleistocene St. Prest site in France. These were initially communicated by J. Desnoyers (1863) to the French Academy of Sciences and were accepted by leading French anthropologists such as Armand de Quatrefages (1884, p. 17). These incised bones, like Moir's tools, are not mentioned in most current literature, but one will find them in *Forbidden Archeology*.

As ideas change about the time of migration of hominids from Africa, and questions arise about the African origins hypothesis in general, concerned workers might find the above and much other material contained in *Forbidden Archeology* useful. Of course, Richard Thompson and I put our own Vedic spin on all this (humans and other hominids have coexisted for millions of years), but one can easily factor that out and simply give attention to the reported facts in light of one's own research interests. There is a lot of good, relevant stuff in the book that one would have a hard time finding elsewhere.

In addition to reports of anomalous evidence, *Forbidden Archeology* also contains an overview of the current controversies in physical anthropology—such as the number of species at Hadar, Ethiopia; the number of species represented in the *Homo habilis* materials; the degree of arboreality in the Hadar hominids; the nature of the *Homo erectus* to *Homo sapiens* transition; coexistence of *Homo erectus* and *Homo sapiens* in China; and the Neanderthal question. Also discussed are the implications of the Black Skull for australopithecine phylogeny, the debate on the Laetoli footprints, and the significance of OH-62 and KNM-WT 15000. Yes, there is in *Forbidden Archeology* a discussion of the Bigfoot literature, but if [*prominent British physical anthropologist*] John Napier (1973) could write a whole book about it, Richard Thompson and I felt it would be okay for us to include a chapter focusing on the extensive scientific reporting of this controversial topic.

Literature Cited

Cremo, M. A. and Thompson, R. L. (1993) *Forbidden Archeology: The Hidden History of the Human Race*. San Diego: Bhaktivedanta Institute.

De Quatrefages, A. (1884) *Hommes Fossiles et Hommes Sauvages.* Paris: B. Baillière.

Desnoyers, J. (1863) Response à des objections faites au sujet d'incisions constatées sur des ossements de Mammiferes fossiles des environs de Chartres. *Comptes Rendus de l'Académie des Sciences* (Paris), 56:1199-1204.

Leakey, L. S. B. (1960) *Adam's Ancestors*, 4th edition. New York: Harper & Row.

Marks, J. (1994) Review of *Forbidden Archeology: The Hidden History of the Human Race*, by Michael A. Cremo and Richard L. Thompson. *American Journal of Physical Anthropology,* 93:140-141.

Moir, J. R. (1927) *The Antiquity of Man in East Anglia.* Cambridge: Cambridge University Press.
Napier, J. R. (1973) *Bigfoot: The Yeti and Sasquatch in Myth and Reality.* New York: Dutton.
Osborn, H. F. (1921) *The Pliocene man of Foxhall in East Anglia.* Natural History, *21*:565-576.
Swisher III, C. C, Curtis, G. H., Jacob, T., Getty A. G., Suprijo, A., and Widiasmoro (1994) Age of the earliest known hominids in Java, Indonesia, *Science 263*:1118-1121.

2.1.3.3. Letter from Dr. Matt Cartmill, May 20, 1994.

Dr. Cartmill declined to publish my reply to Dr. Marks.

2.1.3.4. Letter to Dr. Matt Cartmill, June 27, 1994.

I take it from your reply concerning my response to Jonathan Marks's review of *Forbidden Archeology* that you have not seen the book. I am therefore sending you a copy, along with a copy of the recently released abridged version. You assumed I might have taken "offense" at Marks's "derisive language." Not true. I was absolutely delighted by it. He is demeaning himself, rather than the book, which he is actually helping to sell [*excerpts from Marks's review are printed along with excerpts from other reviews, favorable and unfavorable, in the abridged edition of Forbidden Archeology*]. But there is another reason why I was delighted with his derisive language and misrepresentation of the book. As you acknowledged in your luncheon address in Denver this year, there is currently an element within anthropology and the social sciences that views science as a kind of "cognitive colonialism" that enforces its views by essentially political methods. Marks's name-calling review of *Forbidden Archeology* fits this paradigm perfectly. So not only is Marks's review helping sell many copies of the abridged and unabridged editions of *Forbidden Archeology*, it is also going to be quite useful in presentations I shall be making to the audience of "postmodern" scholars [*see "The Reception of* Forbidden Archeology,*" presented at Kentucky State University Institute for Liberal Studies Sixth Annual Interdisciplinary Conference on Science and Culture, in Chapter 1 of this book*]. For that purpose, his review could not have been better if I had written it myself. But, because I do have some sense of fair play, I wanted to give you the choice of whether or not to leave Marks's review as your journal's final word on the subject. If you do, I predict it is going to prove embarrassing.

In any case, I do believe that Marks's review satisfies one of the criteria you employ for deciding whether or not to publish a rejoinder, namely, "that the substance of the book has been grossly misrepresented." To categorize the book as "goofy popular anthropology" is a rather gross misrepresentation. At least two other journals have indicated they will review the book. Like it or not, scholars in many fields are taking the book seriously. Of course, Marks was smart enough to cover himself. If you read the very fine print, you will find buried in his piece a fragment of a sentence that acknowledges the book contains some material from reputable sources. But let's just face the truth squarely—the overall effect of his review is to grossly misrepresent the substance of the book. Have a look and judge for yourself.

Since you did say you only "reluctantly" decided against publishing my reply, I thought I would give you a chance to reconsider your decision. I am therefore resubmitting the reply.

2.1.3.5. Letter to Dr. Matt Cartmill, August 16, 1994.

I am enclosing a copy of a review of my book *Forbidden Archeology* from *Geoarchaeology*. In my opinion, this review [*see section 2.1.4, in this chapter*], although unfavorable to the ideas set forth in *Forbidden Archeology*, does not misrepresent the substance of the book. If you compare it with Jonathan Marks's review, which does grossly misrepresent the substance of the book, I think you will see what I mean. "Drivel," "dreck," and "goofy popular anthropology" do not rate reviews in serious academic journals, and more such reviews are forthcoming. I haven't yet heard from you about my latest request for a rejoinder, but I take that as a sign you are still thinking about it, which is good.

2.1.3.6. Letter from Dr. Matt Cartmill, August 30, 1994.

Dr. Cartmill again refused to publish my reply to Dr. Marks.

2.1.3.7. Letter to Dr. Matt Cartmill, October 6, 1994.

Since you seem to have a great appreciation for Charles Fort, I am enclosing a review of *Forbidden Archeology* from the *Fortean Times* [*see Chapter 3*].

Concerning Fort's sense of humor, compared to which you find mine less richly developed, let's just say that perhaps my sense of humor is somewhat subtler. And as I don't think that you are an unsubtle person, you might, if you try, find some rarefied hints of humor in the literary production that has led to this unique exchange of correspondence. Think about it.

It appears that you still have not understood what *Forbidden Archeology* is really about.... The book is not simply a catalog of anomalies. It is an investigation into what "the pattern of evidence" really is. Until the evidence and its pattern is established, one cannot (can one?) proceed to explain the pattern.

Furthermore, why should I be the one who has to offer an alternative explanation? Let us suppose researcher A proposes a certain explanation X for a pattern of evidence Y. And researcher B points out that the pattern of evidence is actually Z, which invalidates the proposed explanation X. Why exactly is it that researcher B has to offer an alternative explanation? I don't see how that follows. If researcher B is correct, then I would say that there are many possibilities. Researcher A could offer a new explanation based on pattern Z. Researcher B could offer a new explanation. Both A and B could cooperate to offer a new explanation. Or both A and B could agree that at the moment no new explanation for evidence set Z is possible. From the strictly scientific point of view, any of these alternatives is valid, as far as I can see. What are your scientific reasons for proposing that B is obligated to provide an alternative explanation?

I also think your characterization of our treatment of the report from *The Weekly World News* is not honest. Yes, the offending newspaper report was submitted to me by someone. I did not accept it at face value. The actual source of the information that led to the inclusion of the grooved metallic sphere in a special appendix to *Forbidden Archeology* was not the *The Weekly World News* but correspondence with a scientist at the natural history museum where the object was displayed. He confirmed that this unusual object was in fact found in a Precambrian deposit, along with many similar objects. He also provided correspondence related to a rather unsatisfactory proposed natural explanation for the object advanced by local scientists. So what is your objection? Do you think the object does not exist? Do you think it was not found in the formation as reported? Do you have a credible natural explanation for the object? Do you think it is a hoax? Do you think that contemplating an artificial origin to the objects is impossible, given your view of the history of life on earth? Or is it simply your contention that any object or fossil mentioned in any tabloid at any time automatically ceases to be real?

In any case, the whole point of the appendix was that if one wants to consider the whole range of evidence related to human origins and antiquity, one also has to consider reports emanating from nonscientific sources. My coauthor and I did, however, distinguish this category of report from other kinds of published reports, and to suggest that we did not do so is simply incorrect. If you go through *Forbidden Archeology*, you will find that a great deal of attention is given to evaluating "variation in quality in published reports." As a matter of fact, the whole argument of the book is based on such evaluation. If you would care to discuss our analysis of the variation in quality of published reports regarding any of the particular cases, I would be happy to do so.

At the moment, I am in the midst of an author's tour for *Hidden History of the Human Race.* It is taking me to about 20 cities in the U.S., with many radio and TV engagements. We are getting a lot of mileage out of John Marks' review, an excerpt of which we have included in the front of the book. Media people love it. Actually, Marks was invited onto one of the many radio shows I have been doing. I wish he would come on all of them.

I am also working on my next book, which will offer the desired alternative paradigm for human origins. It involves some "paranormal" elements. It will surely appeal to some of your postmodernist colleagues among the cultural anthropologists.

I can understand why you are so concerned about those people. I have been going through the latest issue of *American Anthropologist*, where there is a quite interesting article by Katherine P. Ewing, one of your local Duke scholars, titled "Dreams from a Saint: Anthropological Atheism and the Temptation to Believe." In her abstract, she says, "The anthropological taboo against going native is examined in the context of the ethnographer's own dreams during fieldwork among Sufi saints in Pakistan. The essay demonstrates that accounts of these dreams shaped social relations in the field and argues that relativist neutrality is a cover for a refusal to believe that is impossible to hide. This refusal constitutes an implicit insistence that the relationship between ethnographer and subject be shaped by the parameters of a hegemonic Western discourse of rationality." Barbara and Dennis Tedlock are to be congratulated for their fairness as editors.

I've also found examples of the same openmindedness in the Journal of the Anthropology of Consciousness section of the AAA. I always look for the word "discourse" in article titles. That's a good marker for postmodern content. Anyway, it's good to see the "hegemonic Western discourse of rationality" breaking down. It opens up some intriguing possibilities.

I would be most pleased if you would carry out your intention to donate your copy of *Forbidden Archeology* to the medical center library and retain *Hidden History of the Human Race* in your own collection.

2.1.4. Kenneth L. Feder (1994) Review of *Forbidden Archeology*. *Geoarchaeology,* vol. 9, pp. 337-340. © 1994 *Geoarchaeology*. Reprinted by permission of John Wiley & Sons, Inc.

Kenneth L. Feder is an anthropologist at Central Connecticut State University. He writes on creation/evolution questions. Although Feder rejected the conclusions of Forbidden Archeology, *I regarded his review as positive for many reasons. He acknowledged the book was well written and researched, he accurately represented the contents of the book, he discussed the evidence and arguments in the book seriously, and he dealt respectfully with the book's underlying Vedic source material.*

The voluminous (900+ pages) *Forbidden Archaeology: The Hidden History of the Human Race* is not the usual sort of publication reviewed in this or, for that matter, any other archaeology or anthropology journal. Neither author is an archaeologist or paleoanthropologist; one is a mathematician, the other a writer. So far as I can tell, neither has any personal experience with the process of archaeological fieldwork or laboratory analysis.

The book itself represents something perhaps not seen before; we can fairly call it "Krishna creationism" with no disrespect intended. The basic premises of the authors are breathtaking and can be summarized rather briefly:

> The prevailing paradigm of human evolution (divergence from a common ancestor with the apes at 5-7 million years ago; a bipedal ancestor with an apelike brain by 4 million years ago; a bipedal, tool-making ancestor with a brain size beyond the range of apes by 2+ million years ago; creatures that can be subsumed within the genus *Homo* by 1.7 million years ago; archaic forms of our species by 0.4 million years ago; anatomically modern *Homo sapiens* by 0.1-0.2 million years ago) is wholly untenable.
>
> There is what amounts to a passive conspiracy (the authors call it a "knowledge filter") to suppress a huge body of data that contradicts our prevailing paradigm.
>
> These suppressed data include archaeological evidence in the form of incised bones, lithics, and anatomically modern human skeletal remains that date to well before the commonly accepted appearance of *Australopithecus.*
>
> This purported evidence indicates that "beings quite like ourselves have been around as far back as we care to look–in the Pliocene, Miocene, Oligocene, Eocene and beyond" (p. 525). The

authors cite "humanlike footprints" in Kentucky dating to about 300,000,000 (not a misprint) years ago (p. 456).

All evidence of human evolution from an apelike ancestor is suspect at best and much of it can be explained as the fossil remains of nonancestral hominids or even extinct apes (some *Homo erectus* specimens, it is proposed, might represent an extinct species of giant gibbon [p. 465]).

Some of these nonancestral hominids have survived into the present as indicated by reports of Bigfoot, Yeti, and the like.

There is evidence of anomalously advanced civilizations extending back millions of years into the past.

While decidedly antievolutionary in perspective, this work is not the ordinary variety of antievolutionism in form, content, or style. In distinction to the usual brand of such writing, the authors use original sources and the book is well written. Further, the overall tone of the work is far superior to that exhibited in ordinary creationist literature. Nonetheless, I suspect that creationism is at the root of the authors' argument, albeit of a sort not commonly seen before. It is impossible in the context of this short review to deal in an in-depth way with any of the myriad cases cited by the authors to buttress their claims. Instead, their general approach can be summarized.

The authors base virtually their entire book on a literature search and most (though not all) of that literature dates to the late nineteenth and early twentieth centuries. In so doing, the authors have resurrected nineteenth-century claims of "Tertiary Man" (Grayson, 1983) apparently superimposing on this a belief in the instantaneous appearance of anatomically modern *Homo sapiens* at some point in the very distant past, asserting that the evidence for this is at least as good, and usually better, than that cited for a much later and evolutionary origin for our species.

The authors maintain that the analytical techniques applied by nineteenth century scientists to incised bones and "eoliths" that led some to conclude that these very ancient items were the result of human activity are nearly the same techniques as those applied today to accepted evidence. Therefore, the authors assert, the conclusions reached by nineteenth and early twentieth century researchers that these very ancient objects were cultural in origin are of equal validity to the identification of more recent (late Pliocene) cultural objects by modern scientists. Thus, when a nineteenth century researcher using a standard microscope of the time claims that striations found on bones dating back tens of millions of years are butchering marks, this is the equivalent, in the authors' view, of a modern researcher identifying cut marks using a scanning electron microscope. I doubt that many working in the field would agree.

The authors also ignore trace element analysis of raw materials. They do not address the fact that putative Eolithic tools are invariably the same raw material as their surrounding geological deposits, a fact that supports a natural rather than cultural origin for the materials. They ignore wear pattern analysis and, therefore, do not address the issue of use-wear on any of these unexpectedly ancient "tools." All of their drawn figures of eoliths are taken from original sources, but not from the objects themselves. We have no way of knowing how accurate the drawings are, and none has a scale. The authors, I am certain, would complain that I am demanding a higher level of proof than I might for evidence that supports the current paradigm, and they would be right. But claiming that this is unfair, as they do throughout their book, reflects a lack of recognition of how science must work. Extraordinary claims have always required extraordinary levels of proof in whatever the field.

Why does the discovery of so much of the evidence cited by the authors date to the nineteenth century and so little to more recent times? In the view of the authors, until the discovery of Java Man, scientists were far more open to evidence of much older, anatomically modern human beings. With the discovery of that fossil, the modern evolutionary paradigm became established and contrary evidence was swept under the rug. So, the argument goes, paleoanthropologists no longer find stone tools or anatomically modern skeletons in Eocene deposits because they no longer look in deposits dating to this epoch. Maybe. But the authors must know that our current human evolutionary paradigm does not prevent paleontologists from investigating these older deposits, which is, in fact, what they do all of the time. Curiously (of course, not), these scientists do not find human fossils in deposits next to dinosaurs. Their biostratigraphic evidence invariably supports general evolutionary sequences.

The authors graciously acknowledge that their allegation of suppression of data refers to "an ongoing social process of knowledge filtration that appears quite innocuous" and not to a "satanic plot" being carried out by "scientific conspirators" (p. xxvi). But here the authors are being, at least in part, disingenuous. In several instances throughout the book, such plots (though not satanic) indeed are alleged by them: They claim that the fossils of anatomically modern human beings in anomalously old strata have been "perhaps intentionally discarded" (p. 419); they suggest that "perhaps the original investigators of Choukoutien purposefully held back from reporting stone tools and fire" (p. 547); they speculate that the Piltdown fraud might have been committed by "a select number of scientists at the British museum" who first perpetrated the hoax to prove their view of evolution, then orchestrated its unmasking in an expose that was itself a hoax (pp. 523-524); (the authors believe that the Piltdown cranium was a genuine hominid fossil that indicates the presence of a modern human in an unexpectedly old stratigraphic layer).

When you attempt to deconstruct a well-accepted paradigm, it is reasonable to expect that a new paradigm be suggested in its place. The authors of *Forbidden Archaeology* do not do this, and I would like to suggest a reason for their neglect here. Wishing to appear entirely scientific, the authors hoped to avoid a detailed discussion of their own beliefs (If not through evolution, how? If not within the last four million years, when?) since, I would contend, these are based on a creationist view, but not the kind we are all familiar with.

The authors are open about their membership in the Bhaktivedanta Institute, which is a branch of the International Society for Krishna Consciousness, and the book is dedicated to their "spiritual master," the group's founder. They make a reasonable request regarding their affiliation with this organization: "That our theoretical outlook is derived from the Vedic literature should not disqualify it" (p. xxxvi). Fair enough, but what is their "theoretical outlook?" Like fundamentalist Christian creationists, they avoid talking about the religious content of their perspective, so we can only guess at it.

Where does Hindu literature say humanity originated and when? According to Hindu cosmology, the cosmos passes through cycles called *kalpas*, each of which corresponds to 4.32 billion earth years (Basham, 1959). During each *kalpa*, the universe is created and then absorbed. Each *kalpa* is divided into 14 *manvantaras*, each lasting 300,000,000 years (the age of those Kentucky footprints), and separated by lengthy periods. Within each *manvantara* the world is created with human beings more or less fully formed, and then destroyed, only to be created once again in the next *manvantara*.

We all know what happens when we mix a literal interpretation of the Judeo-Christian creation myth with human paleontology; we get scientific creationism. It seems that we now know what happens when we mix a literal interpretation of the Hindu myth of creation with human paleontology; we get the antievolutionary Krishna creationism of *Forbidden Archaeology*, where human beings do not evolve and where the fossil evidence for anatomically modern humans dates as far back as the beginning of the current *manvantara*.

References

Basham, A.L. (1959). *The Wonder That Was India*. New York: Grove Press.

Grayson, D. (1983). *The Establishment of Human Antiquity*. New York: Academic Press.

Related Correspondence

2.1.4.1. Letter to Dr. Kenneth L. Feder, August 16, 1994.

Dr. Feder did not reply to this letter.

I very much appreciated your review of *Forbidden Archeology* in *Geoarchaeology* 9:4. I found it generally fair and reasonable, quite in contrast to the juvenile name-calling "review" given by Jonathan Marks in the *American Journal of Physical Anthropology* 93.1. *Journal of Field Archeology* gave a brief notice, and I am told there are additional reviews forthcoming in journals of anthropology and other disciplines. So you are not alone in commenting. I take your review as an invitation to archeologists and geologists to have a look at the book. What more could I ask? Thank you.

I do, however, have some critical comments for your consideration. These are offered in friendly fashion and do not at all change my overall appreciation of your thoughtful review.

You wrote that my coauthor and I "do not address the fact that putative Eolithic tools are invariably the same raw material as their surrounding geological deposits, a fact that supports natural rather than cultural origin for the materials." That is a very common "pat answer" to the eolith question, but it is rather misleading. It evokes a picture of huge masses of flint, with much natural fracturing, from which one picks out a few specimens that appear to be worked. But what does one make of this description by geologists J. Hamal-Nandrin and Charles Fraipont? "At Thorington Hall, the rarity of stones and their dispersal does not permit us to suppose that the flints have been naturally retouched by impact or pressure. One can observe that in the level where the flints are found one does not find any worn and fractured flints other than the ones appearing to be the result of intentional work" (*F.A.*, p. 134). They also say, "At Thorington Hall the detritus bed lacks many rocks. It contains coprolites, phosphate nodules, and only some small flint pebbles." Thus this does not appear to be a flint deposit, and the only fractured flints are the ones identified by contemporary experts (not the original discoverer) as the result of intentional work.

I could give other examples. At Foxhall, the worked flints are found in layers of black sediment and are quite rare (*F. A.* p. 142-43). There are horizontally stratified clean sand deposits above and below these layers. J. M Coles, a modern English archeologist, argued that the relative rarity of flints in the black layers, as well as the fact that it is hard to account for their presence by natural processes, indicates that they arrived in these layers by human agencies. Is there flint nearby? Of course there is. In southeastern

England, flint is quite a common rock. It also happens to be the material of choice for stone tools. But the indication is that flint does not occur naturally in the black layers in question.

I don't know how you can say that we totally ignored wear pattern analysis. Of course, the real question is not whether we ignored it but whether the original investigators ignored it. They certainly did not. As I noted on p. xxvi of the section of the introduction about eoliths, "Often, the working edge bore signs of use." I will admit that there is not much on this in Chapter 3, which is the chapter specifically about eoliths. But see p. 134, where Hamal and Nandrin noted that the edges of some Crag specimens showed signs of utilization as tools. And also see the enclosed page of illustrations from an article by J. Reid Moir, where examples of Eolithic flint implements with signs of use on working edges are given. This material was not included in the book, but I am supplying it to you for your information.

Some of the industries included in Chapter 4 are sometimes placed among the eoliths, and there we do have some more explicit examples of attention to wear use patterns. For example, S. Laing (1894) noted in regard to the Thenay specimens: "The inference [that an object is cultural] is strengthened...if the microscope discloses parallel striae and other signs of use on the chipped edge, such as would be made by scraping bones or skins, while nothing of the sort is seen on the other natural edges" (*F.A.* p. 235). Bourgeois reported such signs of use (*F. A.* p. 227), as did others apparently. Please also consider p. 252, where Max Verworn gives quite detailed attention to wear pattern analysis in connection with the Aurillac implements, which some would place among the eoliths. His method, which is explained in even greater detail in his original paper in German, included rigorous analysis of the length, depth, and direction of wear markings on edges. It is examples like these that convince me that adequate attention to use wear patterns was given by scientists of the nineteenth and early twentieth centuries.

You say paleontologists, if not anthropologists, are still looking in older deposits and are not finding human skeletons next to dinosaur bones. Of course, we ourselves did not make any claims about scientists finding human skeletons among dinosaur bones. Most of our scientific reporting of human skeletal remains comes from the Tertiary. If you followed our thesis closely, you would have noted that we had an explanation for this. Given the ideas current among working anthropologists and archeologists of the later nineteenth century, Tertiary humans were a distinct possibility, even among evolutionists. Therefore, if such things were found, they were reported scientifically because they fit within the paradigm of at least some scientists. But having humans at (or before) the time of the dinosaurs would have been outside even their expanded realm of tolerance. So if such things were found by scientists, they would not have been within their interpretative scope ("absolutely must be intrusive") and thus would not have been reported by them. Such reports (of artifacts as well as bones) did in fact, as we might predict, emanate from nonscientific sources (although there were some exceptional notices given by scientists, particularly in the early nineteenth century).

In any case, both you and I will admit that professional scientists were reporting anatomically modern human skeletal remains in the earlier Pleistocene and the rest of the Cenozoic during the nineteenth and early twentieth centuries. Ragazzoni, for example, was a geologist. When looking for shells in a Pliocene formation, he came upon some human bones, apparently *in situ* at Castenedolo. How do you explain the absence of such reporting today?

Here are some possible answers.

(1) The earlier discoveries were genuine, but hominid fossils are quite rare. It was therefore a statistical fluke that such things were discovered in the nineteenth century but are not being discovered today.

(2) Anomalously old human remains are in fact being discovered today but are interpreted so as to fit the current evolutionary paradigm. Some examples are the Laetoli footprints, 3 million years old and anatomically modern in all details, interpreted as australopithecine, even though the known foot fossils of this genus do not match the prints; the 1481 femur, described by Richard Leakey as indistinguishable from an anatomically modern human femur but attributed to *Homo habilis*, although the new OH-62 discovery makes this attribution questionable; and the original Trinil *Pithecanthropus* femur, found in the same 800,000 year old layer as the famous *Pithecanthropus* skull and now characterized by many professionals as modern human without discussion of its long accepted age and provenance.

(3) Such things are being found, but the discoverers assume they are obviously intrusive and thus do not report them.

(4) Such things are found, recognized for what they are, and not reported for reasons of academic self-preservation. We do have some anecdotal testimony from George Carter that young archeologists have found anomalously old stone tools at North American sites and not reported them out of fear of damage to their careers (*F. A.* p. 201).

Furthermore, when researching *Forbidden Archeology*, Richard Thompson and I visited Thomas Démére, a paleontologist at the San Diego Museum of Natural History, to discuss the nature of shark tooth marks on whale bone. After looking at some specimens, we asked this paleontologist if he had ever observed human incisions on fossil whale bone or other mammalian bone. His reply was, "I stay away from anything that has to do with man. Too controversial." This is an example of the kind of knowledge filtration process we talk about in *Forbidden Archeology*. There is nothing satanic about it. You just have a young professional, brought up on the standard ideas, doing his job in a professional way, and sticking to his own niche. If he sees some human marks on some bone, he just doesn't say anything about it professionally. He knows that North American anthropology is a hotly debated topic. His interest is paleontology, not anthropology. He doesn't think he is suppressing anything. He just wants to follow his own professional interests without getting tangled up in some other discipline's controversies. That's the message I got. It is comments like Démére's that lead me to believe that the knowledge filtration process described in *Forbidden Archeology* is in fact operating in paleontology as well as paleoanthropology.

You say "extraordinary claims demand extraordinary proof." But claims are judged extraordinary not in relation to some independent standard (which might be okay) but in relationship to the currently dominant consensus in a particular discipline. This means that the parties to the dominant consensus are in control of what claims should be deemed extraordinary, and hence almost impossible to prove (to them and those who accept their authority). That may, as you say, be the way things are in science, but it does not strike me as a particularly good way to treat evidence. It shades too far into ideology and partisan politics.

The real question is what are the standards of good evidence. We suggest that the comparative method provides a fairer and more objective way of evaluating evidence. Suppose that we have two sets of evidence, set (a) which is accepted by a mainstream scientific discipline, and set (b) which is not. If a careful comparison of (a) and (b) shows that they are equivalent, then both should be given equal legitimacy in scientific discussions. Either both (a) and (b) should be accepted, or they should both be rejected, or perhaps they should both be considered ambiguous. In our view, if (a) and (b) are of comparable quality, then (b) should not be consigned to oblivion, while (a) is prominently publicized. After all, the climate of scientific opinion may change later on, and (b) may no longer be regarded as constituting an extraordinary claim.

As for your assertion that electron microscopes might offer an improvement over optical microscopes employed by nineteenth and early twentieth century scientists studying incisions on bone, there might be some truth to that. But the degree of improvement in making a determination of human action is debatable. Are you suggesting that all determinations made with optical microscopes should be disregarded? More generally, are you suggesting that all scientific work carried out before (what year exactly?) should be set aside because the instrumentation was not the same as today's? If physicists today were to redo the Michelson-Morley experiment, they would surely use lasers and modern high-precision optics to get better measurements. But does that invalidate the experiment that established the absence of ether drift? The apparatus was good enough to get the result.

Similarly, optical microscopes and even the naked eye are sufficient for confirming human action and ruling out certain alternative explanations for marks on bone, such as animal gnawing or shark bites. As we note on page 38, John Gowlett says, "Under a microscope [the optical kind], marks made by man are distinguishable in various ways from those made by carnivores. Dr. Henry Bunn (University of California), observed through an optical microscope at low magnification that stone tools leave v-shaped cuts, which are much narrower than rodent gnawing marks." So as of 1984, we see that a prominent archeologist is saying that optical microscopes at low magnification are useful. I think this is especially true in cases where the marks are in places that would normally be cut in butchering operations.

Implicit in your criticism is that electron microscope work might overturn positive identifications made by optical microscopic analysis. It is more likely, it seems to me, that electron microscope work would tend to confirm the optical microscope work and even increase the number of positive identifications of human work by detecting patterns of deliberate incision not immediately apparent under lower magnification on bones lacking other signs of intentional work, such as proper placement of the incision for a specific butchering operation. See p. 49 of *F.A.*, where Blumenshine and Selvaggio maintain that human work on bone, particularly breaking bones for marrow, is underestimated by current field workers. They used the scanning electron microscope to confirm human work in many cases where others had not seen it.

Finally, your accusation that we neglected to bring out a new paradigm is somewhat unfair. First of all, I don't agree that a deconstructor of a paradigm is in any way obligated to bring out a replacement. Let's suppose a mathematician offers a proof of Fermatt's last theorem. Then another mathematician demonstrates it is not a proper proof. Is the second mathematician obligated to give his own proof before the other supposed proof can be set aside? No. The mathematicians will just admit they don't have a proof and wait for the next attempt.

However, in this case, we do have an alternative paradigm. In the Introduction to *Forbidden Archeology*, I wrote: "Our research program led to results we did not anticipate, and hence a book much larger than originally envisioned. Because of this we have not been able to develop in this volume our ideas about an alternative to current theories of human origins. We are therefore planning a second volume relating our extensive research results in this area to our Vedic source material." You give no hint of this in your review and state that we deliberately avoided discussing our alternative paradigm. That is not true. Originally we did plan to put the deconstruction and the alternative in the same volume. But this simply did not prove practical. I am now working on the follow-up book, another big one, and hope to see it in print in 1996.

By making the source of our forthcoming alternative paradigm very clear, we made the broad outlines of it obvious to anyone, such as yourself, who has even a passing acquaintance with the Vedic literature. Nevertheless, I think there will be some surprises in store. Just to give you a sample of what's coming, and to demonstrate to you that I am not trying to hide anything, I am sending you a draft of a paper that has been submitted to one of the sessions at the World Archeological Congress in New Delhi later this year [*the paper is included in Chapter 1 of this book*]. It's listed in their second announcement, but the session organizer tells me he is still making the final selections of papers to be presented. I am also enclosing a copy of the session description, which I find quite progressive. If you would care to make any comments on the paper, or upon anything I've said in this letter, I would be grateful.

2.1.4.2. Letter to Dr. Vance Holliday, Book Review Editor of *Geoarchaeology*, August 16, 1994.

Thank you for selecting my book *Forbidden Archeology* for review in *Geoarchaeology.* I was very pleased with the tone of Kenneth Feder's review. Of course, I don't agree with everything he said (please see the enclosed copy of my letter to him). But all things considered, the review was a fair one. The book was honestly directed toward an academic audience, and it is satisfying to see an academic response, even a negative one, that does not grossly distort the substance of the book. Such distortion unfortunately did occur in the case of *The American Journal of Physical Anthropology*, wherein Jonathan Marks dismissed *Forbidden Archeology* as "drivel," a "cornucopia of dreck," and "goofy popular anthropology." I have been going back and forth with Matt Cartmill about a rejoinder. I think it would be to his advantage, and his journal's, to allow one. In any case, the review that appeared in *Geoarchaeology* is a credit to you and your journal, and I am grateful.

2.1.4.3. Excerpt from Letter to Dr. Jack Donahue, Editor of *Geoarchaeology*, August 16, 1994.

I received no reply to this letter.

Although I was in general satisfied with Feder's review, there are some misstatements to which I would like to reply, if there is a forum in *Geoarchaeology* for doing so. The three matters I would like to address are: (1) eoliths being invariably of the same material as the surrounding deposit, (2) lack of attention to use markings, (3) the accusation of deliberately avoiding discussion of an alternative paradigm. If *Geoarchaeology* does have a forum for such replies, and you think the one I have outlined is warranted, please let me know and I will send it to you.

2.1.5. Anonymous (1994) Review of *Forbidden Archeology*. "Book News," *Ethology, Ecology & Evolution*, vol. 6, p. 461. Reprinted by permission of *Ethology, Ecology & Evolution*.

The reviewer says that the book "does not seem to produce appreciable result." But at this stage of the game, the wide notice received by Forbidden Archeology *in academic and scientific journals arguably is an appreciable result.*

The authors of this very ambitious book, neither of whom is a specialist in the field or a biologist, claim to reconsider completely the interpretation of the fossil record concerning the origin of Man. In fact, the theme of the book is that *Homo sapiens* "existed on earth millions of years ago" and that this fact has been suppressed or ignored by the scientific establishment because it contradicts the dominant views of human origins and antiquity. To prove this theory, the authors go over the history of the principal discoveries bearing on human evolution and they review much of the evidence which concerns human origins, particularly that which does not agree with the "dominant paradigm". The authors are members of the Bhaktivedanta Institute, "a branch of the International Society for Krishna Consciousness" and, in the Introduction, M. A. Cremo reveals that he and his co-author derived from Vedic literature "the idea that the human race is of great antiquity." Although this unconventional view is the starting point from which to rediscuss many points which are not clear, or presumed to be so by the authors, about the origin of Mankind, it does not seem to produce appreciable results.

2.1.6. Marylène Patou-Mathis (1995) Review of *Forbidden Archeology. L'Anthropologie*, vol. 99(1), p. 159. Translated from the original French. Reprinted by permission of Masson Editeur.

I found this review, in one of anthropology's oldest and most distinguished journals, noteworthy for its fair treatment of an extremely controversial book. Patou-Mathis seems to have been genuinely intrigued by Forbidden Archeology.

This enormous volume, 914 pages, dedicated to the hidden history of humanity, is surprising. What is, then, this forbidden archaeology? Moving past the first moment of surprise, one quickly opens this book and flips through it with interest. The first part presents the abnormal evidence (not accepted)—for example, cut and fractured bones, supposedly from man, discovered in tertiary caves; or the eoliths that have made such a lot of ink flow; or human remains found in California in the Pliocene or Eocene periods; and the footprints of humans observed in the Pennsylvanian (Upper Carboniferous) of Rockcastle in Kentucky. The authors tell us about the historical records of these discoveries and the polemics they gave rise to but don't give final judgments. The book's second part discusses conventional evidence: hominids from Java and China (Choukoutien among others) but also the fossil of Piltdown. Africa, with the most ancient discoveries of remains of *Australopithecus*, isn't forgotten. There again, for about a hundred pages, the authors describe the historical records and discussions relative to these fossils: the pros and cons of their relationship with the true hominids. Three appendices end the book. One concerns the chemical and radiometric analyses of human bones, the ages of which are disputed. A second concerns evidence for the existence of cultures in very ancient periods (Tertiary, Secondary). The third appendix summarizes the abnormal evidence for human antiquity: from the Precambrian (metallic spheres from the site of Ottosdalin in the Republic of South Africa) to the end of the Pleistocene.

M. Cremo and R. Thompson have willfully written a provocative work that raises the problem of the influence of the dominant ideas of a time period on scientific research. These ideas can compel the researchers to publish their analyses according to the conceptions that are permitted by the scientific community. If the evidence given isn't always convincing (far from it) regarding a very ancient origin of humanity, the documentary richness of this work, more historical and sociological than scientific, isn't to be overlooked.

2.1.7. Wiktor Stoczkowski (1995) Review of *Forbidden Archeology*. *L'Homme*, vol. 35, pp. 173-174.

The book by Michael Cremo and Richard Thompson[1] promises to lift the veil of silence that conceals disturbing ideas on the earliest antiquity of mankind. According to the authors, Darwinian orthodoxy tendentiously eliminates archeological indications showing that *Homo sapiens* is not a recent product of evolution and that for a long time he shared the Earth with numerous races of simian hominids from which he cannot be descended. The profession of belief is clear and laconic. There are 800 laborious pages of proofs, the academic appearance of which will, without a doubt, mislead more than one reader. In order to remove the possibility of the simian ancestry of man, the authors are occupied with demonstrating that man is older than the other kinds of hominids. After having given a new interpretation of classical fossils, they reveal to us the existence of human bones that were discovered in Illinois in rock from the Carboniferous period as well as human footprints from the same period in Kentucky and from the Jurassic period in Turkmenistan. Man was not only living in these remote periods, but also he had already an advanced civilization. As evidence they cite fossil anchors found in the depths of quarries, a mysterious inscription on a piece of marble extracted from its natural rock, a piece of money from the middle Pleistocene, a fossilized shoe sole from the Triassic, and even a metal vase from the Precambrian (600 million years ago). Official science, charge Cremo and Thompson, refuses to take into account these vestiges because they threaten the established conception of the origin of man.

"Our attitude regarding life and its future is influenced by our views on life's origin," declare Jehovah's Witnesses in another book, published by Watchtower Bible and Tract Society and dedicated to enlightening its readers about the weakness of the theory of evolution, on the grounds of, and in support of, the *Book of Genesis*.[2] This formula sums up very well what is at stake when the problem of origins is considered in our Western culture, no less than elsewhere. And since we are in the habit of thinking that the Western vision of the world is equivalent to that of science, it is not useless to remind ourselves that there are now 11 million copies in print of the book published by Watchtower (translated into 16 languages). We are mercifully silent about the sizes of the print runs of scholarly works of which we are proud. Historians of science repeat tirelessly that the Biblical version of origins was replaced in the nineteenth century by the evolution theory. This simple story is substituted, in our imaginations, for the more complex reality that we are today confronted with a remarkable variety of origins accounts. Those of official science are far from being uniform.

Prehistory told by scientists committed to Marxist theory is not the same as that presented by feminist scholars. Furthermore, the version of prehistory found in children's school books is different from that found in professional scientific publications. And this version in turn is very different from that of the Jehovah's Witnesses, the American Creationist, the Catholic Church, or those who seek to explain our origin by extraterrestrial intervention. I have skipped over many other versions. And *Forbidden Archeology* gives us one more, dedicated to "His Divine Grace Bhaktivedanta Swami Prabhupada" and inspired by the Vedic philosophy that disciples study at Bhaktivedanta Institute in the U.S.A., a branch of ISKCON, the International Society for Krishna Consciousness.

Forbidden Archeology isn't to be recommended to those who are trying to inform themselves about prehistory, but it would be useful for readers interested in modern beliefs. The man on the street "believes" in the theory of evolution like he believed (and continues to believe) in the Mosaic genesis, the Vedic genesis, and others. However, the peculiarity of modern religious belief doesn't lie in the act of believing, but in the manner in which it justifies itself. The 800 pages of meticulous archeological descriptions that accompany the Vedic creed in *Forbidden Archeology* tell us much about the role that faith attributes to empirical confirmation in our days. Modern irrationality is distinguished by unbridled research for scientific evidence to support every belief. And Western beliefs attempt today to not only be scientific and empirical but also systematic: hence, the complicity claimed by Cremo and Thompson with postmodern anthropology, which lifts relativity of interpretation to the level of an epistemological principle; hence the pretentious ramblings on the "human construction" of scientific facts, supported with names such as Paul Feyerabend, Thomas Kuhn, Steven Shapin, Steve Woolgar and Bruno Latour. We, the Westerners, want to be so learned that we cannot even abandon understanding for naiveté without covering our attempt with science. The same ambiguity is found moreover on the side of official science. Have we not seen recently an illustrious French paleoanthropologist support the investigation of Yeti? And what of the prehistories that certain scholars delight in telling in front of cameras, before the altar of popular audience response? It is significant that works like *Forbidden Archeology* are nourished not only by sacred texts, but equally by scientific publications, parts of which seem to evolve in an imaginary universe. This book gives us a curious collection of ideas, each of which has already had, at one time or another, a place in acknowledged scientific work. All of this indicates that science itself also contributes to the store of traditional ideas, from which dilettantes, learned or simple, can nourish their thinking and model on old mythical structures. What is new is our cult of the empirical method, which we worship without often understanding its true principles. "To believe is to believe in not believing."[3] Indeed, the modern act of believing consists of posing as an act of scientific knowledge.

Notes

1. Michael Cremo is an editor at the Bhaktivedanta Book Trust. Richard Thompson introduces himself as a researcher working in several areas (math, biology, geology, physics).
2. Anonymous, *Life: How did it appear?* New York: Watchtower, 1985.
3. J. Pouillon, *The Believed and the Known.* Paris: Le Seuil, 1993.

Related Correspondence

2.1.7.1. Letter to Wiktor Stoczkowski, May 20, 1997.

I enjoyed seeing your review of my book *Forbidden Archeology* in *L'Homme.* What I most appreciated was your recognition that the modern Western intellectual is indeed confronted with a variety of human origins accounts, including a new one represented by *Forbidden Archeology*.

You are essentially correct. *Forbidden Archeology* is belief "posing as an act of scientific knowledge." In this regard, I draw your attention to a passage from a paper I presented at the Sixth Annual Interdisciplinary Conference on Science and Culture, held by Kentucky State University's Institute for Liberal Studies in 1995:

> In 1994, Kenneth L. Feder reviewed *Forbidden Archeology* in *Geoarchaeology: An International Journal* (Feder 1994). Feder's tone was one of amazement rather than derision.
>
> > "*Forbidden Archeology: The Hidden History of the Human Race* is not the usual sort of publication reviewed in this or, for that matter, any other archeology or anthropology journal. Neither author is an archeologist or paleoanthropologist; one is a mathematician, the other a writer. So far as I can tell neither has any personal experience with the process of archaeological field work or laboratory analysis."
>
> Nevertheless, *Forbidden Archeology* rated a four-page review in *Geoarchaeology,* one of the prominent archeology journals.

> What was going on here? I submit that Feder and the editors of *Geoarchaeology* were reacting to *Forbidden Archeology* in much the same way as some art critics of the 1960s reacted to the Brillo box sculptures of Andy Warhol. The Brillo boxes were not art, it appeared to some critics, but these critics could not help commenting upon them as if they were art (Yau, 1993). And thus the boxes were, after all, art, or as good as art. I suppose it is true, in some sense, that *Forbidden Archeology* is not real archeology, and that I am not a real archeologist (or historian of archeology). And, for that matter, neither is this paper a "real" paper, and neither am I a "real" scholar. I am an agent of Gaudiya Vaishnavism, with an assigned project of deconstructing a paradigm, and this paper and *Forbidden Archeology* are part of that project. And yet *Forbidden Archeology* is reviewed in *American Journal of Physical Anthropology* and *Geoarchaeology,* and this paper is read by me at an academic conference on science and culture, just like Warhol's Brillo boxes were displayed in galleries for purchase by collectors rather than stacked in supermarkets for throwing out later.
>
> Perhaps what was so disconcerting about Warhol's Brillo boxes was that their idiosyncratic artificial reality somehow called into question the hitherto naively accepted natural supermarket reality of the ubiquitous everyday Brillo boxes, and hence of all everyday public culture things. The same with *Forbidden Archeology.* Its intended artificialness, its not quite seamless mimicry of a genuine text, called fascinated attention to itself as it simultaneously undermined the natural artifactitious impression of the so-called real archeology texts. But there is a difference between *Forbidden Archeology* and Warhol's Brillo boxes. If you open up a Brillo box, you won't find any Brillo pads; but if you open up *Forbidden Archeology*, you will find archeology, although of an unusual sort.
>
> After all, I am not simply a literary pop artist who delights in producing artificial archeology text surfaces; there is some substantiality to the project. I am a representative of Gaudiya Vaishnavism, and my purpose is to challenge some fundamental concepts of Western science.

In the most immediate sense, it is not so much that *Forbidden Archeology* is enlisting the empirical method to inspire support for beliefs arising from Vedic texts. The primary immediate motive was to attack the Darwinian paradigm, without simultaneously offering a fully developed alternative—be it Vedic or not (although a Vedic one is hinted at and will in fact be offered in a forthcoming book of mine). That the evidence reported in *Forbidden Archeology* is consistent with Vedic accounts of extreme human antiquity is, for the readers of *Forbidden Archeology*, intentionally of secondary importance. Ultimately, however, the goal of the book is in fact to lay the foundations for an effort to get the Western (and Western-influenced) intelligentsia to give some serious consideration to the Vedic accounts of human origins and antiquity. In that ultimate sense, your judgment is correct. *Forbidden Archeology* is belief "posing as an act of scientific knowledge."

But the same judgment must be rendered against current Darwinian ideas of human origins. Philosopher of science Karl Popper, who gave us the concept of falsifiability as a criterion for the validity of a scientific theory, wrote in his autobiography: "Darwinism is not a testable scientific theory, but a metaphysical research programme." I agree with Popper, and will add some additional thoughts.

The primary phenomenon alleged by Darwinism, the transformation of species, one into another, is unobserved. Various categories of evidence are said to be consistent with the unobserved primary phenomenon, but these in no way demonstrate or prove it. Transformation of species is not an observable empirical fact. To insist the opposite is an expression of belief that, to use your language, "poses as an act of scientific knowledge."

Regarding various beliefs about human origins and antiquity, including the Vedic and the Darwinian, the question for the empiricist becomes "Which belief is better at posing as an act of scientific knowledge?" The question is put to the empiricist because, as you have correctly observed, we today worship at the altar of empiricism, which we can translate as science (if by science we mean a mode of explanation dependent upon observable physical verification). The question for the empiricist is compounded when vast amounts of physical evidence contradict his Darwinian belief about human origins and antiquity, and when it can be demonstrated that this evidence is excluded from current normal scientific discourse primarily because it contradicts the Darwinian belief. And when this evidence turns out to be consistent with another belief—the Vedic belief, which normally justifies itself not through empiricism but through acceptance of knowledge from sources superior to normal human intelligence—that obviously is quite challenging, challenging enough to be commented upon by you in the pages of *L'Homme*.

Of course, you have not addressed the evidence reported in *Forbidden Archeology*, so there is not too much more for me to say, other than to thank you for taking notice of my book.

2.1.8. Langenheim Jr., R.L. (1995) Notice of *Forbidden Archeology*. "Books for Geoscientists," *Journal of Geological Education*, vol. 43(2), p. 193. Reprinted by permission of *Journal of Geological Education*.

Langenheim predictably asserts that most of the reports in Forbidden Archeology *"can be readily dismissed or are unverifiable." The same can be said of the reports used to support the current Darwinian view of human origins. It all depends upon what standard one employs to judge the reports, and upon how fairly one applies the standard to different sets of reports.* Forbidden Archeology *documents the application of a double standard in the evaluation of reports.*

This is a catalogue and discussion of numerous Precambrian to Pleistocene fossils and artifacts accepted by some as evidence of "anatomically modern humans." The authors tend to credit these reports at face value and argue that these "facts" are being ignored by the scientific establishment. Most of the reports, however, can readily be dismissed or are unverifiable. A few may have scientific merit. The descriptions and extensive bibliography of obscure references will be very useful to a "mainstream" scientist called upon to contend with creation scientists and others who may introduce these "facts" in paleontologic or anthropologic discourse.

2.2. History, History of Science, and Sociology of Scientific Knowledge

When writing Forbidden Archeology, *I anticipated that it would be of interest to scholars in science studies disciplines, such as history of science, philosophy of science, and sociology of scientific knowledge. I was not mistaken.*

2.2.1. Hillel Schwartz (1994) "Earth Born, Sky Driven: A review of *The Sky in Mayan Literature* (edited by Anthony F. Aveni), *Beyond 1492: Encounters in Colonial North America* (by James Axtell), *Forbidden Archeology: The Hidden History of the Human Race* (by Michael A. Cremo and Richard L. Thompson), *Love and Theft: Blackface Minstrelsy and the American Working Class* (by Eric Lott), *Children of the Earth: Literature, Politics and Nationhood* (by Marc Shell)." *Journal of Unconventional History*, vol. 6(1), pp. 68-76. Reprinted by permission of *Journal of Unconventional History* and Hillel Schwartz.

Hillel Schwartz is a historian. The following passages related to Forbidden Archeology *have been excerpted from Schwartz's article, unique for its overall favorable judgment on the book.*

Current jabber about the distortions of the Other, about figuration and disfiguration, about social and historical construct, about the "packing" and "unpacking" of symbols, about—in short—codes and decoding, must be put in the context of this new convention, which has been nurtured since World War II by cryptographers, cyberneticists, biochemists, linguists, French psychiatrists, structural anthropologists, literary analysts and, today, representational historians. For even if the most radical postmodern stance in any field is to subvert all stances, all "optics," it is a stance anchored in the belief that to decode is to make ready for change—a change which has to do penultimately with welcoming a multifocal world whose denizens need to cherish, and identify with, the global diversity of humanity, or animality, or biosphericality.

Whether those who engage in such critical tactics or "optics" would warm to a global embrace, or would instead resist globalism as another subtle form of tyranny and "hegemony," it is clear that the present urge to decode goes beyond intellectual curiosity, psychological intrigue with forms of disguise, or a desperate lust to solve unsolved mysteries. Our urge to decode has to do (as does each of the books I have taken in tow for this review) with revealing identities and, I would argue, with the need to reveal identities-in-common at this *fin de siecle* and *fin de millennium*, when all seems in flux…

Forbidden Archeology takes the current conventions of decoding to their extreme. The authors find modern *Homo sapiens* to be continuous contemporaries of the apelike creatures from whom evolutionary biologists usually trace human descent or bifurcation, thus confirming those Vedic sources that presume the nearly illimitable antiquity of the human race—all toward the implicit end of preparing us for that impending transformation of global consciousness at which Bhaktivedanta brochures regularly hint. Decoding certain chipped flints or "eoliths" as, many of them, very ancient stone tools, and recoding evidence others have rejected either as hoaxes or natural phenomena (metallic spheres, shoe prints, iron nails and gold threads in old stone, carvings, footprints), Cremo and Thompson discern the working presence of anatomically modern humans perhaps as far back as the Cambrian era, long before the age of dinosaurs, or at least in the early Pleistocene, tens of millions of years before the carbon-dates assigned to the Leakeys' [*Johanson's, actually*] celebrated hominid (?) skeleton, Lucy. *Forbidden Archeology* reads surprisingly well for what is basically an 828-page critical catalogue of two centuries of archeological evidence doubted or spurned by Western scientists. Despite its unhidden religious partisanship, the book deserves a reckoning in this review for its embrace of a global humanity permanently distinct from other primates. Other beings deserve our deepest respect, but humanity stands alone, its bloodlines unsullied by apes or *ramapithecus.* There are, as it were, originary castes of beings, each worthy, each separate. However, evolutionary processes may operate to refine life within a caste, they do not operate to translate a being from one caste to another. Meditating upon our uniqueness (I am here supplying the missing links of the thesis), we may come to realize that what can change (awaken) humanity is no mere biochemical exfoliation but a work of the spirit, in touch with (and devoted to) the ancient, perfect, perfectly sufficient, unchanging wisdom of the Vedic masters.

2.2.2. Tim Murray (1995) Review of *Forbidden Archeology. British Journal for the History of Science*, vol. 28, pp. 377-379. Reprinted by permission of the Council of the British Society for the History of Science and Tim Murray.

Tim Murray is head of the archeology department at Latrobe University in Melbourne, Australia. He is also a historian of archeology. In December of 1994, I met Murray at the World Archeological Congress 3 in New Delhi, India, where I presented my paper "Puranic Time and the Archeological Record" in a section he cochaired. During one of the breaks, he mentioned that he had been asked to review Forbidden Archeology *for* British Journal for the History of Science. *He said that although he disagreed with the book's conclusions and methods, he hoped I would find his review fair. He had already submitted the review, but it had not yet been published.*

Since the last eighteenth-century discussions of human antiquity and of the physical and cultural evolution of humanity have been marked by severe disputation and accusations of fraud, histories of palaeoanthropology and of quaternary geology (such as Grayson's *The Establishment of Human Antiquity,* New York, 1983, and more recently Van Riper's *Men among the Mammoths*, Chicago, 1993) have all canvassed the reasons for disputation and some (such as Spencer's *Piltdown: A Scientific Forgery*, Oxford, 1990) have delved deep into disciplinary psychology to establish the motivation for fraud. No one could deny that mainstream quaternary archaeology is unaware of its capacity to generate controversy. Furthermore, a knowledge of the discipline (and of its practitioners) clearly demonstrates that there is no single point of view about the meaning of the palaeoanthropological fossil record. Indeed it should be emphasized that practitioners have had altogether too much fun fighting amongst themselves to be much concerned with other possible combatants. Cremo and Thompson's massive work clearly demonstrates that others now want to play the game.

Whatever else *Forbidden Archeology* might be, it is a book with a point of view. Despite more than 900 pages of discussion, this can be fairly simply summarized. First, there is a contention that quaternary archaeologists have ignored what is described as being clear and unambiguous evidence (fossils and artefacts) of a much higher human antiquity than that accepted by 'the scientific community'. Note that Cremo and Thompson are not claiming that the scientists have rejected evidence of there being ancestral forms of fully modern human beings other than those currently recognized (i.e. members of the genus Australopithecus and earlier forms of the genus *Homo*). Instead, they are claiming that evidence of fully modern human beings has been found in the Tertiary geological record, and that knowledge of these radical data has been suppressed by practitioners for the last century or so. Secondly, the explanation for this 'Major Scientific Cover-Up' (their words, not mine) is to be found in the 'evolutionary prejudices' of 'powerful groups of scientists' who are members of the 'scientific establishment' who together act as a 'knowledge filter', reinforcing the dominance of 'evolutionary prejudices' by dispensing with anomalous and potentially destabilizing data. Thirdly, that Cremo and Thompson are not clear whether this filtration process is conscious (in the sense of cover-up or fraud) or simply the unconscious recommitment to normal science by research drones who have all power for original thought squashed out of them by the system.

Cremo and Thompson rest their case on two general assumptions. First, that the plausibility of conventional quaternary archaeology and paleoanthropology depends not on the actual evidence adduced by practitioners, but on the cognitive plausibility of evolutionary theory. Secondly, that scientists will move hell and high water to 'preserve the paradigm' and thus dispose of inconvenient evidence or 'freeze out' inconvenient practitioners. It is worth noting that in this, as in any good conspiracy theory, there are goodies and baddies, seekers after truth and representatives of the 'dominant paradigm'. At stake is the potential liberation of the human mind through deeper understanding of the meaning of human history. For Cremo and Thompson, if you do not accept the plausibility of evolutionary theory, then the flimsy edifice of quaternary archaeology that it supports crashes to the ground, leaving the way free to pursue another pathway towards enlightenment. For them, the vast store of anomalies (the documentation of which takes up the bulk of the volume), when taken together, provides compelling support for an attack on the paradigm of human evolution and on the data which have, up to this point, been seen to support it.

It should be noted that theirs is far from being a disinterested analysis, as *Forbidden Archeology* is designed to demolish the case for biological and cultural evolution and to advance the cause of a Vedic alternative. This is a piece of 'Creation Science' which, while not based on the need to promote a Christian alternative, manifests many of the same types of argument: first, an attempt to characterize the opposition as motivated by the need to preserve their view of the world rather than a desire to practice unfettered inquiry; secondly, to explain the currently marginal position of your alternative as being the result of prejudice, conspiracy and manipulation rather than of any fault of the theory itself; thirdly, to present the opposition (in this case mainstream palaeoanthropology and quaternary archaeology) as being united as a 'secret college' to manipulate the public mind and to exclude non-professionals from being able to control science for the benefit of all.

I have no doubt that there will be some who will read this book and profit from it. Certainly it provides the historian of archaeology with a useful compendium of case studies in the history and sociology of scientific knowledge, which can be used to foster debate within archaeology about how to describe the epistemology of one's discipline. On another level the book joins others from creation science and New Age philosophy as a body of works which seek to address members of a public alienated from science, either because it has become so arcane or because it has ceased to suit some in search of meaning in their lives.

Above all, this is a book about belief. Cremo and Thompson believe the Vedas provide a more accurate and internally consistent explanation for life of earth, but for all the talk about logic and consistency their system on the whole simply would not function without the existence of a 'supreme conscious being'. In an interesting example of projection Cremo and Thompson distinguish between their true and justified belief and the views of their evolutionary opponents, which are characterized by them as being 'unscientific'. For them, followers of evolutionary theory do so out of ignorance, fear, or blind faith, with the need to believe overcoming dispassionate assessment and objective argument.

What to do with this book and its claims? One path is to take each case raised by Cremo and Thompson and by a steady process of attrition to demolish their account. This can be (and is being) done. But this does not go to the heart of the volume or explain why the authors believe so strongly in the existence of Tertiary humanity. For that we have to go to the Vedas and in my view this can only be a personal journey. For the practising quaternary archaeologist, current accounts of human evolution are, at root, simply that. The 'dominant paradigm' has changed and is changing, and practitioners openly debate issues which go right to the conceptual core of the discipline. Whether the Vedas have a role to play in this is up to the individual scientists concerned. Although Cremo and Thompson might characterize archaeologists and palaeoanthropologists as being at the wrong end of a knowledge filter, it is fair comment that nothing in the 900 or so pages of *Forbidden Archeology* seems to undermine Cremo and Thompson's belief that Vedic literature got it right long before the advent of archaeological inquiry.

Related Correspondence

2.2.2.1. Letter to Dr. Tim Murray, June 25, 1995.

Murray was editing the volume of the World Archeological Congress 3 proceedings in which my paper was to appear.

In one of your faxes regarding the WAC3 proceedings, you mentioned that your review of *Forbidden Archeology* for *British Journal of the History of Science* will soon be out. I've noticed that the University of Florida library tends to get journals rather late, so if it's not too much trouble, could you please fax me a copy of the printed version when you receive it? Another review is just out. Stoczkowski in *L'Homme* (Jan.-Mar. issue), like you, does not share my "views or approach." But then I would not expect him (or you) to do that. I am, however, honored that you (and he—and, I suppose, the book review editors) considered *Forbidden Archeology* worthy of notice. What I found interesting in Stoczkowski's review was his accurate depiction of the metastate of archeology today. He said that the standard history of science tells us that empirical science triumphed over religious views in Europe in the nineteenth century. But the practical truth, he reflected, is that contemporary archeology and anthropology must contend with a variety of other voices, including the Jehovah's Witnesses (he noted their books sell in millions of copies around the world), the Christian Creationists in the United States, the advocates of extraterrestrial intervention in human origins, and now this—the Vedic view represented by *Forbidden Archeology*. And what to speak of the relativist, constructivist critique of science, which he also dwelled upon.

2.2.2.2. Letter to Dr. Tim Murray, January 12, 1996.

I recently read your review of *Forbidden Archeology* in *British Journal for the History of Science*. I liked it. At this point I am not expecting unqualified endorsements from mainstream archeologists and historians of archeology, but I do appreciate the book being taken somewhat seriously.

There are a few places where I thought you did not quite accurately represent *Forbidden Archeology*. For example, you took some pieces of ad copy from the jacket of the popular abridged edition of the book, which appeared under a different title, and represented them as if they were part of the text of *Forbidden Archeology*. That gives a somewhat misleading impression of what a reader will find in *Forbidden Archeology*. I don't object that you used the ad copy, just that you did not identify it as such and did not say it was not from the book actually under review.

Forbidden Archeology relies more on an epistemological critique of scientific methods as applied in archeology than on conspiracy theories. Some of your comments seem to ignore that, and give the mistaken impression that *Forbidden Archeology* should be grouped in the category of conspiracy literature.

For example, you said, "...there is a contention that quaternary archeologists have ignored what is described as being clear and unambiguous evidence (fossils and artefacts) of a much higher human antiquity than has been accepted by 'the scientific community.'"

The argument actually presented in the book is somewhat more subtle than that. We concluded that all evidence in archeology tends to be unclear and ambiguous. As we noted (p. 24): "All paleoanthropological evidence tends to be complex and uncertain. Practically any evidence in this field can be challenged, for if nothing else, one can always raise charges of fraud." In other words, one can always challenge the dating methods, the statistical analysis, the provenance, etc. But there appears to be a double standard. Evidence in harmony with the current consensus tends to be treated leniently whereas evidence that departs from the current consensus is subjected to extremely corrosive scrutiny.

The process of setting aside certain categories of evidence by application of a double standard we called "knowledge filtration." You said that the authors "are not clear whether this filtration process is conscious (in the sense of cover-up or fraud) or simply the unconscious recommitment to normal science." In my Introduction to *Forbidden Archeology* (p. xxvi) I noted: "When we speak of suppression of evidence, we are not referring to scientific conspirators carrying out a satanic plot to deceive the public. Instead, we are talking about an ongoing social process of knowledge filtration that appears quite innocuous but has a substantial cumulative effect."

Of course, in a few places we do identify instances of conscious fraud (as in the Piltdown case).

You wrote that *Forbidden Archeology* manifested "an attempt to characterize the opposition as motivated by the need to preserve their view of the world rather than a desire to practice unfettered inquiry," implying this was not true [*of "the opposition," i.e. modern archeological science*]. Yet you said elsewhere in your review (p. 379): "What to do with this book and its claims? One path is to take each case raised by Cremo and Thompson and by a steady process of attrition demolish their account. This can be (and is being) done." This sounds much like a stratagem described in *Forbidden Archeology* (pp. 25-26) whereby "prominent scientists will publish systematic attacks against...unwanted findings." We noted (p. 26) that "in the parlance of some scientists...these attacks are known as 'demolition jobs.'" Now it would seem to me that if someone has decided in advance to demolish *Forbidden Archeology* by discrediting every one of the hundreds of cases discussed therein, this provides a very good example of someone "motivated by the need to preserve their view of the world rather than a desire to practice unfettered inquiry." It would seem that a less prejudicial person might have proposed to study each case in the book to see whether or not it should be rejected or accepted, without announcing an a priori decision to destroy all of them. What could the motive of such a person be? By the way, who is doing the demolition job?

In any case, none of this detracts from my overall satisfaction with your review. With the exception of pushing the book too much in the direction of conspiracy literature, I found that you fairly accurately represented its substance and spirit, as well as its potential utility to the practitioners of a discipline in the process of fundamentally redefining itself. I especially appreciated the following comments: "I have no doubt that there will be some who will read this book and profit from it. Certainly it provides the historian of archaeology with a useful compendium of case studies in the history and sociology of scientific knowledge, which can be used to foster debate within archaeology about how to describe the epistemology of one's discipline." (p. 379)

I was thinking it might be interesting to put together a section for a conference, maybe the next WAC, that would bring together you, me, Dr. Grayson, Dr. Van Riper, and perhaps some others.

Finally, there appears to be an editing mistake in the review (p. 378). The review says: "No one could deny that mainstream quaternary archaeology is unaware of its capacity to generate controversy." I think the sentence was intended to read: "No one could deny that mainstream quaternary archaeology is aware of its capacity to generate controversy."

On another matter, how is the volume of proceedings for the WAC section on Time Concepts coming along?

2.2.3. Jo Wodak and David Oldroyd (1996). "'Vedic Creationism': A Further Twist to the Evolution Debate." *Social Studies of Science*, vol. 26, pp. 192-213. Reprinted by permission of Sage Publications and the authors.

David Oldroyd recently retired from his position as a historian of science at the School of Science and Technology Studies at the University of New South Wales, in Sydney, Australia. Jo Wodak was, at the time this article was written, his graduate student. I regard their review article as the most significant scholarly response to Forbidden Archeology.

A New/Old Brand of 'Creationism'

The Creation Science (CS) movement has attracted much attention in recent years, being sufficiently significant to warrant a massive historical study by Ronald Numbers.[1] As is well known, Creationists try to show that humans are of recent origin, and that empirical investigations accord with human history as recorded in the Old Testament. *Forbidden Archeology* (*FA*) offers a brand of Creationism based on something quite different, namely ancient Vedic beliefs. From this starting point, instead of claiming a human history of mere millennia, *FA* argues for the existence of *Homo sapiens* way back into the Tertiary, perhaps even earlier. Yet the argument is similar to that of CS in that it purports to be based on the results of scientific enquiry. In particular, *FA* draws on palaeoanthropology, supporting its theses by close examination and reappraisal of the early literature of this science—writings largely disregarded by current theorists in the field, but which Michael Cremo and Richard Thompson (C&T) invite the reader to reconsider.[2]

The Vedic Background

In their Preface, C&T acknowledge the Vedic inspiration of their research, but what this amounts to is not explained. They claim that such explanation is unnecessary, for their critique of current human evolutionary theory is so well supported by an 'abundance of facts' as not to need theoretical/philosophical justification. However, the authors promise that a forthcoming second volume will show how Vedic theory agrees with 'the facts' more completely than either 'Darwinian evolutionists' or 'Biblical creationists' (240-41). This, of course, is not much help to readers of *FA*, given that its conclusions are really only persuasive if the underlying Vedic assumptions are accepted. For this reason, a brief digress on the movement is in order, for while the influence of CS in the US and elsewhere is well known,[3] the presence and purpose of the Bhaktivedanta Institute are less familiar.

The Bhaktivedanta Institute (or Hare Krishna) movement is a devotional monotheist Hindu sect which offers salvation to its adherents irrespective of their social origins and without requiring them to renounce the social world. It is a modern variant of the Bhakti sects that have dominated Hindu religious life over the last one and a half millenia. The intellectual basis of the movement is 'Vedic Science', a monist philosophical system that reduces all phenomena to consciousness—conceived as the 'transcendental non-material energy of God'. Bhaktivedanta scholars, like Sankara[4] and other early Vedantin philosophers, locate the source of 'Vedic Science' in the *Rig Veda,* the most ancient Hindu scripture (and the earliest text in an Indo-European language) whose oral beginnings date from around 1000 BC. Whether these ideas are derived from this profound but poetically rather than philosophically focused text, or from somewhat later Upanishadic, Buddhist, Jain and Puranic philosophical speculation, is a question beyond the scope of this Review. What matters, for our purpose, is the nature of the philosophic system embodied in 'Vedic Science' and the way it surreptitiously informs C&T's argument. A significant feature of the system is the way it links mediaeval Puranic speculations on the number of species in the cosmos (supposedly some eight and a half million, with four hundred humanoid ones)[5] and on cosmic chronology with the idea—present in the *Rig Veda*—that the biological forms which emerged at the moment of creation were prearranged according to their different levels and degrees of conscious capacity.[6]

Vedanta began to be influential in the West in the mid-nineteenth century, but its convergence with the Bhaktis as the Bhaktivedanta movement did not take place until a century later, when the movement was taken across the Pacific by His Divine Grace A.C. Bhaktivedanta Swami Prabhupada (1896-1977). Swami Prabhupada—to whom *FA* is dedicated—moved from India to New York in 1965, where he established the International Society for Krishna Consciousness. In 1968, a community was founded in West Virginia, and the movement spread thereafter to many parts of the US. The present 'intellectual' centre is the Bhaktivedanta Institute (BI), located in San Diego since 1976 (as was the CS research centre until recently), first under the directorship of T.D. Singh and now of Richard Thompson. Housed in a ground-floor apartment in an unpretentious suburban building—when one of us (Oldroyd) visited the Institute in March 1994, only two people were 'in residence', for (he was informed) most of the group lead an agrarian life out of town, and C&T were busy promoting their book on the world circuit[7]—the Institute's modest setting belies the boldness of its intellectual programme. This is: to expose the 'insufficiencies and inaccuracies in modern science' and, 'utilizing Vedic insights into consciousness, the self and the origin of the universe', to offer an advance study into the nature, origin and purpose of life, at the same time establishing 'new scientific paradigms'.

Swami Prabhupada's writings, which include a record of ideas he propounded—guru-like—in 'morning walks' in Los Angeles in 1973, display a strong and dismissive distaste for Western science.[8] For Prabhupada, science gives no adequate account of the origin of the universe or of life; 'nothing is extinct'; and selection requires an intelligent 'chooser'. Thus, the design argument is at (or close to) the bottom of Prabhupada's rejection of Western science.[9] Similar ideas are developed in other BI publications.[10]

In the *Rig Veda* and the Puranas, life or the *atma* is an eternal, transcendental entity and consciousness is its most important property. More important than free will and desire, consciousness is the 'transcendental, nonmaterial energy of God'. The *atma* supposedly travels from one form of body to another—if all is going well, from less to more conscious physical form—under the laws of *karma* (action and reaction) and desire, so this is evolution of a wholly spiritual kind. In the human form, the *atma* reaches full consciousness, and it may now free itself from the law of *karma* through the practice of yoga. Yet misuse of free will makes devolution or reversion to a less conscious species possible, so there is an explicit moral and hierarchical dimension to Vedic evolutionary theory philosophy which strict Darwinism does not share.

A major difference between Vedic 'science' and both CS and orthodox Western science is the much longer and cyclic time-scale envisaged by the former. The life of our universe is thought to be 311,040 billion years, and its present age is a little over 150,000 billion years. A final annihilation at the end of its allotted span is foreseen; but then there will be a new recreation. The same holds for the many thousands of other co-existing universes. Partial dissolutions, called *pralaya,* supposedly take place every 4.32 billion years, bringing catastrophes in which whole groups of living forms can disappear. Because of this constant cycle of creation and destruction, the fossil record does not present a systematic system, globally or universally. Indeed, Vedic philosophy also holds that just as every species is divinely created, so no species becomes more than temporarily extinct. Presumably somewhere in other parallel universes, all the dinosaur species still exist!

Vedic Palaeoanthropology and SSK

Introducing *FA*, P.J. Flynn (PhD),[11] of California State University, San Marcos, suggests that C&T's ideas find support in the SSK literature. He writes (xix):

> Recent studies of the emergence of Western scientific knowledge accentuate that 'credible' knowledge is situated at an intersection between physical locales and social distinctions. Historical, sociological, and ethnomethodological studies of science by scholars such as Harry Collins, Michael Mulkay, Steven Shapin, Thomas Kuhn, Harold Garfinkel, Michael Lynch, Steve Woolgar, Andrew Pickering, Bruno Latour, Karin Knorr-Cetina, Donna Haraway, Allucquere Stone, and Malcolm Ashmore all point to the observation that scientific disciplines, be they paleoanthropology or astronomy, 'manufacture knowledge' through locally constructed representational systems and practical devices for making their discovered phenomenon visible, accountable, and consensual to a larger disciplinary body of tradition.

Well, here is a galaxy of talent (to which David Bloor could perhaps be added; see note 7), behind which one may construct an argument and a half! Indeed, C&T maintain (xxiv) that their approach 'resembles that taken by practitioners of the sociology of scientific knowledge (SSK)'. They also see themselves as adopting a 'critical approach' to science like that of Feyerabend, who, they say, argued that science had 'attained too privileged a position in the intellectual field'. Also, they see their work as resembling Martin J.S. Rudwick's, 'who has explored in detail the nature of scientific controversy'. However, although C&T certainly discuss social interactions between palaeoanthropologists, they make little explicit effort to relate their work to the technical SSK literature.

The argument of *FA* may be put in a nutshell as follows. In the early years of palaeoanthropology (late nineteenth and early twentieth century), much empirical information was interpreted as evidence favouring the existence of Tertiary humans. From about the 1930s onwards, this 'paradigm' began to change, and the palaeoanthropological community turned to the view that humans first evolved in the Pleistocene. But according to C&T, the old evidence was never satisfactorily refuted. It was simply reinterpreted, and the idea of Tertiary humans was dropped. It would, of course, be compatible with SSK views of science that such a paradigm shift (if so it may be called) would indeed involve complex social interactions and negotiations. And, we believe, it is certainly possible that the paradigm shift took place for reasons that had to do with interest, power or whatever, as part of the process of adding new empirical information and altering theoretical interpretation. Moreover, we would think it *possible* that once the new paradigm was well established it would be difficult to shift it; and we might now be unable to construe the facts correctly because of the all-pervasive and seemingly all-persuasive new paradigm. So what C&T have done is to re-examine the old literature; and on the basis of their detailed historical study they argue that the old arguments were never satisfactorily disproved and should be reconsidered with open minds. In fact, they are claiming that there has been a kind of conspiracy among palaeoanthropologists—indeed a stifling of the views of anyone who has been bold enough to argue for the extreme antiquity of human beings.

The Technique of Argument of *FA*

A science based upon Vedic philosophy represents a fascinating contrast to Western thinking—though as mentioned in note 9, some think that a happy union of the two can be achieved. Yet as already noted, overt Vedic philosophy in *FA* is conspicuous by its absence (and also in Cremo's interview with Laura Lee). Like Creation Scientists, but unlike Prabhupada, C&T try to use science, rather than metaphysical argument, to counter science. Again like some Creationist writers, they provide a recital of credentials or reputations, where these are thought to strengthen their case.[12] Supporters of the concept of Tertiary Man are represented as 'leading paleoanthropologists', 'famous scientists', and 'experts' all.

So *FA*'s target is not science *per se*, but the claimed deficiencies of orthodox evolutionary palaeoanthropology. The technique of argument is to comb the early literature in great—indeed impressive—detail and show how the old evidence was built up but, C&T claim, never really refuted. Also, by reconsidering the early documents it is claimed that a more secure factual base is obtained than can be provided by recent investigations; for in many cases it is not possible to replicate the original observations. (For example, the original specimens of *Homo erectus* found near Peking were lost during the War; and original archaeological sites have sometimes been badly damaged by years of rather careless investigations.) Throughout, C&T argue that 'facts' are what matter, not theory.

Yet while C&T state on several occasions that they want to rely on facts, and while they use the scientific literature to mount their case, in fact it is Darwinian evolution that they have in their sights. They claim, for example, that if evidence for very early humans in Europe be admitted, then the African origin of mankind would be in jeopardy. Then 'not only the idea of an African origin but the whole concept of an evolutionary origin of the human species becomes untenable. And if scientists are forced to give up an evolutionary explanation of human origins, what does that say about the whole theory of evolution?' (181). Without the 'African origins hypothesis...where would paleoanthropology be?' (ibid.).

The Historians' Role

If scientists have lost sight of the idea of 'Tertiary Man', perhaps historians bear some responsibility. Certainly some pre-*FA* histories of palaeoanthropology, such as Peter Bowler's,[13] say little about the kind of evidence adduced by C&T, and the same may be said of some texts published since 1993, such as Ian Tattersall's recent book.[14] So perhaps the rejection of Tertiary *Homo sapiens,* like other scientific determinations, is a social construction in which historians of science have participated.[15] C&T claim that there has been a 'knowledge filtration operating within the scientific community', in which historians have presumably played their part. The authors go so far as to claim that the 'filtration' amounts to 'unconscious fraud' (525).[16] Hence the word 'forbidden' in *FA*, and the promise to reveal the 'hidden history of the human race' in the book's subtitle. Hence, too, the packaging (literary technology!) of each book, announcing: 'Major scientific cover-up exposed!'.

However, regardless of what scientists think about the idea of very ancient humans, co-existing alongside what we regard as non-human evolutionary ancestors, one cannot say that all historians have ignored the issues raised by C&T. Grayson and Spencer, for example, consider the early ideas about Tertiary humans in some detail.[17] Even so, it must be acknowledged that *FA* brings to attention many interesting issues that have not received much consideration from historians; and the authors' detailed examination of the early literature is certainly stimulating and raises questions of considerable interest, both historically and from the perspective of practitioners of SSK. Indeed, they appear to have gone into some historical matters more deeply than any other writers of whom we have knowledge. So let us sample some of the book's specific arguments.

Eoliths

In the late nineteenth and early twentieth century, given the idea of the 'antiquity of man', evolutionary theory (which was applied to artefacts just as much as to fossil remains), and the existence of Neolithic and Palaeolithic implements, it seemed likely that there should be even more primitive stone tools, made by presumed precursors of the stone tools that were unquestionably man-made. In England, two amateur archaeologists,

Benjamin Harrison and James Reid Moir, were the most ardent seekers of what were called 'eoliths' or 'dawn stone tools'. On the Continent, workers such as l'Abbé Bourgeois,[18] de Mortillet and Rutot thought that they could find similar evidence. It is, of course, easy to pick up pieces of flint in southern Britain or northern France that look as if they *might* be primitive artefacts. And the search for such objects became a popular pastime among amateur archaeologists.[19] Harrison found what appeared to him to be very primitive implements in the *upper* gravels of the Kent Plateau, which were regarded by geologists as Pliocene.[20] He succeeded in convincing the distinguished, but aging, geologist, Sir Joseph Prestwich, that the eoliths were indeed artefactual, and the specimens were described at the Anthropological Institute and claimed to be Pliocene in a much discussed paper.[21]

C&T describe the whole story in considerable detail, without omission of the fact that there was plenty of opposition to the eolith hypothesis, as well as enthusiastic acceptance. They refer to a paper by Alfred Barnes of 1939,[22] which has for long been thought to provide criteria for distinguishing between natural and artefactual stones. However, they successfully 'problematize' Barnes' criteria, referring to recent dissenting voices such as that of L.W. Patterson.[23] In fact, as late as 1968, authorities such as R.G. West were willing to accept the possibility of the artefactual status of eoliths.[24] But in the post-war period, belief in eoliths seems gradually to have faded away,[25] and today students in archaeology are regularly taught how to make flint implements themselves, observe the characteristic features of such artefacts, and hence distinguish between the artefactual and the natural. Of course, in such a learning process they are being 'acculturated': they are learning how to work within or according to a paradigm. But if C&T want to show that this teaching is in error, it is incumbent on them to do so by their own practical methods or theoretical arguments, not by combing the literature and finding some items therein that speak with dissenting voices.[26] Indeed, the extent to which C&T can find such voices in the literature evidences the fact that there is not some conspiratorial 'cover-up' in palaeoanthropology. In any case, the controversial eoliths investigated by Harrison, Moir and others were *not* associated with organic human or hominid remains. So whatever one may make of the 'tools', they tell us rather little about the actual course of human evolution. (But then C&T don't hold with human evolution.)

The Reck Skeleton

Perhaps the most significant withdrawal from the idea of very ancient humans occurred in the 1930s when Louis Leakey, influenced by arguments of Imperial College geologist Percy Boswell, retracted his support for the idea of a high antiquity for the so-called Reck skeleton (now called OH1),[27] discovered at Olduvai Gorge by Hans Reck in 1913. The skeleton concerned was undoubtedly modern in appearance, but it was found in a stratum near the surface as a result of erosion. Was it where it was because it was perhaps over a million years old? Or was it in a relatively modern grave?

C&T explain the issues clearly, and point out that Reck himself was on the look-out for evidences of burial, but he and others could not find such evidence. Louis Leakey, studying the skull back in Berlin, was initially sceptical as to the skeleton's age; but when he visited the site himself, in company with Reck and others, be became convinced of the specimen's great antiquity. The issue was of considerable theoretical importance, for if the Reck skeleton was of great age—perhaps comparable to that of Java man and Peking Man—then it would seem unlikely that these Asian remains were ancestral to modern humans; for modern humans were apparently already in existence at about the same time as the Javanese and Chinese remains of 'primitive' appearance. The question attracted a deal of discussion in the 1930s.

Some of the sedimentary matrix associated with the Reck skeleton, including red pebbles, was examined by Boswell and J.D. Solomon in London. It appeared that materials from beds *above* that suggested intrusive burial. As a result, both Reck and Leakey eventually changed their minds about the age of the skeleton.[28] Yet, as C&T argue, it is somewhat odd that Reck and Leaky had not noticed the red pebbles when they earlier examined the skeleton in Germany.

A post-war radiocarbon dating of some fragments of the Reck skeleton was undertaken by Reiner Protsch in 1974, yielding a date of 16,920 years,[29] and this is usually regarded as the correct date for the specimen. However, C&T problematize the data and the date, pointing out that there are discrepancies between accounts of the matter given by Protsch himself and Leakey's biographer Sonia Cole. It was, they claim, uncertain whether the surviving fragments of the main skeleton that were analyzed were indeed part of the original specimen;[30] and they had been soaked in an organic preservative, contamination which could have influenced the dating result so as to make it appear younger than it actually was. However, C&T omit to mention Protsch's statement that the skeleton was found in a 'contracted position', suggesting burial; and the preservative lacquer 'was easily flaked off and removed'.

C&T also consider Leakey's possible motives in changing his views about the dating of the Reck specimen. They point out that at the time of Leakey's change of heart there was a conference at Cambridge to review his work at Kanam: 'It is quite possible Leakey thought it best to withdraw his reputation from somebody else's controversial fossil and thus pave the way for acceptance of his own better-dated finds at Kanam and Kanjera. After all, some of the most vocal opponents of Reck's skeleton, such as Boswell,...would be sitting on the committee that would review the Kanam jaw and Kanjera skulls (641). This is surely SSKish, with appeal to devotees of interest theory.

C&T also have a plausible reason to offer as to why opinion about the Reck specimen swung round. By the time Protsch's date was published, it was theoretically unacceptable to have a modern-looking human of the same age as Java and Peking Man. The 16,000 year date, however, fitted conveniently into the crystallizing post-war paradigm. So the case was conveniently closed by Protsch. However, to re-open it properly, C&T would have to do further dating work on fragments of the surviving skull. The historical research can do no more than problematize the consensus.

Java Man and Peking Man: Homo erectus

C&T also show that there has been a deal of confusion in the development of ideas about *Homo erectus*, ever since the work of Dubois in the 1890s. So far as Dubois' original work was concerned, the skullcap and femur he discovered were found some distance apart, and one could not be sure that they belonged to the same organism. From C&T's perspective (silently in accord with Vedic theory), the human-like femur and the skullcap of more primitive appearance suggest that humans and an ape-like creature may have co-existed.

Von Koenigswald went to Java in the 1930s, and found further remains. But the investigations were carried out in a thoroughly messy fashion. Some remains were found on the surface and their geological horizon could only be gauged in uncertain fashion by their adhering sediment. Native collectors were employed and paid per fragment of bone. Not surprisingly, with this inducement many fragments became yet more fragmented. C&T point out that apparently authoritative tabulations of the locations and ages of various Javan *erectus* specimens, such as that of Wilfred Le Gros Clerk,[31] rest on numerous doubtful assumptions.

In China, there was confusion too. Quite apart from the tragic loss during World War II of the original 'Sinanthropus' specimens, C&T draw attention to interesting discrepancies between the descriptions of the site at Choukoutien as made by the first discoverers and subsequent investigators. Notably, the initial investigations by Davidson Black (1927) made no mention of the evidence for the use of fire in the Choukoutien cave. But, by 1931, Teilhard de Chardin and Breuil both thought that they could see evidence for fire. And stone implements were also reported from the site, whereas previously they had not been noticed (or reported in print). Why might this be so? C&T hypothesize that in the 1920s fire and tools were regarded as indicators of modern humans or Neanderthals. So observers might have remained silent about indications of fire and tools associated with the small-brained Peking man. So far as C&T are concerned, they seem to be sympathetic to the idea expressed by Breuil, in 1932,[32] that humans contemporary with *erectus* might have been responsible for the fire and the tools with which Sinanthropus was stratigraphically associated.

The Laetoli Footprints

While mainstream palaeoanthropology—in the form of the supporters of Timothy White and Donald Johanson—has given much attention to the specimen known as 'Lucy' (*Australopithecus afarensis*) and has associated the creature (and others from the so-called 'First Family') with the famous footprints of seeming humans discovered in a layer of volcanic ash at Laetoli in Tanzania, this association is strained in that White had to reconstruct the missing portions of Lucy's feet with the help of analogies from more complete specimens of *Homo habilis*.[33] Yet the surviving pieces of Lucy's skeleton suggest at least a partial arboreal habit, and by no means a fully upright gait. C&T's solution to the problem is to suggest that Lucy was indeed a non-human creature; but one that co-existed with true upright-walking humans who made the Laetoli footprints—3.7 million years ago. Again the evidence is construed in the silent light of Vedic metaphysics.

Fraud?

There is a thread running through *FA* that the history of palaeoanthropology is one that has sometimes been associated with fraud, as well as confusion or controversy. Hardly surprisingly, the authors place considerable emphasis on the notorious Piltdown affair. While this has long since been shrugged off by palaeoanthropologists and is now chiefly of historical concern, it is interesting that C&T give the story a curious twist, seeking to use the case to argue once again for their hypothesis of very ancient humans. The skull, C&T suggest, may in fact have been found *in situ*, being a genuine fossil of a Pleistocene human of modern form.[34]

Referring again to the red pebbles associated with the Reck skeleton which were noticed by Boswell in his examination, C&T suggest that they may have been deliberately sent to Boswell from Germany by Professor T. Molleson, who had already expressed the belief that the skeleton was a

Masai burial. This might explain why Reck and Leakey had previously overlooked the red material. We are not in a position to form a judgment as to the occurrence of fraud in this case, but it must be conceded as a possibility.

As another example, C&T refer (558-61) to an article by two Chinese scientists, Wu Rukang and Lin Shenglong, published in *Scientific American* in 1983,[35] which suggests that there was a definite increase in cranial capacity of the hominids living in the Choukoutien cave during the period of its occupation. But, C&T argue, after checking back to the original data of Weidenreich, that due allowance was not made for the ages of the specimens; some specimens that might have been considered (and which conflicted with the claimed trend) were omitted from the cranial volume data, seemingly deliberately; and one specimen that might seem to have an anomalously large cranial capacity got 'lost' in an averaging process.[36]

We concede that the science of palaeoanthropology does not have an unblemished record, though that does not mean that it is riddled with fraud. It is true, however, that it is a branch of science that has been characterized by passionate controversies, and it has been peculiarly prone to the effect of the theory-ladenness of observations. It is also a subject where large stories ('scenarios') have to be told on the basis of often limited or slender evidence.

More Curious Claims

In their 750 pages of main text, C&T take readers through great tracts of the history of palaeoanthropology, and it is not possible to summarize or analyze all the material here. It should be remarked, however, that *FA* is not content to argue for the existence of intelligent beings (humans) in the Old World in the early Pleistocene, or even as far back as the Eocene. Evidence is adduced for the existence of implements and traces of fire in Argentinean deposits dated between three and five million years. (C&T bring forward evidence for a much greater time-span for the human occupation of the Americas than is generally admitted.)[37] Seeming metallic tubes are reported from the Cretaceous in France. A find of a gold chain is claimed from the Coal Measures of Illinois; hieroglyphic markings on coal in Ohio; a shoe print from Cambrian shale in Utah; and what appear to be grooved spherical artefacts from the Precambrian of South Africa. The authors do not rest much of their case on this strange anecdotal evidence, but its presence in the book should be noted. (Some of it is referred to in Cremo's radio interview with Laura Lee.)

On the other side of the coin, so to speak, C&T take seriously reports of yeti, 'bigfoot' and extant Neanderthals, as described by Myra Shackley.[38]

And we were not a little startled to read of a report in *Nature* in 1908 from a Mr K. Stolyhwo of the Cracow Academy of Sciences of a Neanderthal skull found as part of a skeleton in a tomb, along with a suit of armour and iron spearheads![39] In our view, giving credence to such claims weakens rather than strengthens the interesting historical arguments mounted by C&T. But it has been their decision to bring such claims to the attention of their readers.

The Argument Assessed

Though rather a lot is foreshadowed in the early parts of *FA*, few SSK ideas are actually explored or applied to C&T's historical reconstructions, and their theoretical analysis of 'the real processes of science' consists for the most part of examples of the theory-ladenness of observation. Certainly, the effects of theory-generated expectation upon interpretation can be shown leading the way in the history of palaeoanthropology and through to the present day. Perhaps because the origins in question are our own, it is quite easy to show that palaeoanthropology has always been prone to 'story-telling' and to subjective bias.[40] Expectation led on all sides in the debates about human origins. It fueled the efforts of amateurs in the field like Harrison and Reid Moir, as well as professionals like Sir Arthur Keith who wanted to put England on the palaeoanthropological map ahead of France (in terms of tool finds) and Asia (in terms of the 'distinction' of witnessing 'the earliest stages of man's evolution'—Reid Moir's words).[41] Indeed, according to Moir, palaeoanthropologists in the early years of this century regarded the existence of man in the Pliocene as almost a necessity. This same expectation clearly influenced Elliot Smith's confident 'reading' of simian features in the anatomically modern Piltdown skull, as well as Louis Leakey's indefatigable search for evidence of human antiquity early enough to exclude *Australopithecus* from any direct ancestral relationship.

But C&T's critique is also theory-driven. And although they claim that facts are all-important, quite often they have recourse to 'why not?' rhetoric. Because the geological record has so many gaps, they argue: 'Can we say with complete certainty that humans of the modern type did not exist in distant bygone ages?...Out of 6 million years, only 100,000 may be represented by surviving strata. In the unrecorded 5.9 million years there is time for even advanced civilizations to have come and gone leaving hardly a trace' (12-13). If there is no conclusive evidence to the contrary, why can't all sorts of strange and remarkable things have happened?

Of course, if eoliths are unconvincing, then some of the earliest tool finds in East Africa which had no clear human association may also be. 'Expert' opinion that these African objects *are* 'tools' is something agreed to consensually by palaeoanthropologists, and we concur that, under a different paradigm, it is possible that eoliths too *might* have been consensually awarded artefactual status.[42] But suppose they were. Would that mean that the creatures that might have made the eoliths were human? Today's palaeoanthropology is happy with the idea of creatures such as *Homo erectus*, *Homo habilis* and even the Australopithecines, as tool makers. Louis Leakey used to suppose that evidence of intentional tool making to 'a set and regular pattern' was a mark of humanity.[43] But not everyone holds such a view today. Some authors hold that 'anatomically modern humans' may have first appeared about 100,000 years ago, or maybe earlier, but the quite sudden advance in lithic technology and associated behavioural changes that seem to have occurred about 40,000 years ago accompanied the invention of grammatical speech; and that *this* event signalled the true origin of humans.[44]

Finally, C&T's dichotomous strategy of argument—or one-sided presentation of the evidence—needs comment. As said, their intention is to document evidence for the existence of modern humans many millions of years ago, and also for the present-day existence of ape-like and other hominid species. But evidence favourable to the *evolutionary* anatomical origin of humans is omitted. It includes, for example: several 'borderline' fossils occurring around 150,000 BP in southern and eastern Africa, and—some would claim—also in China at comparable times, until by 100,000 BP all the physical features characteristic of modern humans—'shorter, rounder skull, smaller brow ridge and face, prominent chin, and lighter skeletal build'—consistently appear together.[45] C&T also gloss over the sudden and geographically widespread appearance of completely modern humans around 40,000 BP, coinciding approximately with the last appearances of Neanderthals in the fossil record. On the whole, the work of recent palaeoanthropologists—in terms of both recent finds and reviews of current theory—is omitted from the discussion.

FA's presentation means that the reader is not offered a thorough review of the relevant evidence, despite the book's great length. It is as if the Aristotelian and Galilean cosmologies were presented, and evidence provided that not all Galileo's arguments were sound. *Ergo,* we should consider reviving Aristotelian theory. Needless to say, for us, it is modern astronomy that has to be put beside Aristotelian physics and cosmology. C&T should do the same with palaeoanthropology. To attempt to resurrect an old paradigm the way they do, by analyzing outmoded scientific texts, is one of the recognized marks of 'pseudo-science'.[46]

In any case, even if C&T's historical arguments are granted, while this would considerably alter current thinking about the time span of human history, it would not invalidate orthodox evolutionary *theory.*[47] The present co-existence of humans and chimpanzees is obviously no argument against evolutionary theory, and punctuated equilibrium theory is quite comfortable with protracted periods of co-existence of ancestral and descendent, or closely related, species. So there would be nothing contrary to evolutionary theory if (say) Neanderthals and humans were co-existing today.

So has *FA* made any contribution at all to the literature on palaeoanthropology? Our answer is a guarded 'yes', for two reasons. First, while the authors go in for overkill in terms of swamping the reader with detail—a strategy which may persuade readers who lack access to the relevant sources and no special expertise in the palaeoanthropology, and are therefore likely to assume that such a thorough exploration of the historical terrain must signify accuracy and equity[48]—much of the historical material they resurrect has not been scrutinized in such detail before. Second, while the form of their arguments is sometimes as much rhetorical as logical, C&T do raise a central problematic regarding the lack of certainty in scientific 'truth' claims, not so much because such claims are necessarily influenced or driven by vested interests but because the status of all knowledge is inherently a matter of degrees of probability and emerges as the result of social negotiation in concert with observation and inference. As Curtis has recently suggested in this journal,[49] when all that we can aspire to is a reduction in the level of doubt, recognition that this is indeed the case is a healthy state of affairs, and those scientists who insist that evolution is a fact might be better advised to recast this as 'highly probable theory'. This more modest claim might go some way to disarming today's increasingly vociferous opposition. However, if C&T do succeed in showing that in palaeoanthropology, as in other sciences, knowledge is in part socially constructed, *FA*, their own massive intervention in palaeoanthropological debates, via its historiography, can equally be seen as a social act, with its own deep and obscure(d) metaphysical roots.

In conclusion, we would say that palaeoanthropology is a branch of science that is most suitable for scrutiny by *aficionados* of SSK. But such scrutiny requires insights into the SSK literature, not a set of metaphysical assumptions that are totally out of line with those of modern science.

Notes

1. Ronald L. Numbers, *The Creationists: The Evolution of Scientific Creationism* (New York: Alfred A. Knopf, 1992).

2. While we acknowledge the useful role such perusal of the literature

can play both in theoretical evaluation and in historical reconstruction, it is noteworthy that neither 'orthodox' Creationism nor C&T's critique of evolutionary theory is based on archaeological research carried out by the authors themselves. Both are parasitic on the literature of science.

3. According to a Gallup poll taken in 1991, 47% of Americans, including one quarter of college graduates, believe that humans were specially created sometime within the last 10,000 years, a 3% increase since the last poll a decade earlier. In Canada, the figure is 53%. See Numbers, op. cit. note 1, 349.

4. Incidentally, it was Sankara, the first Vedantin, who supposedly planted the mystical stones that the 'thugs' tried to collect in *Indiana Jones and the Temple of Doom.*

5. See T.D. Singh, 'Vedantic Views on Evolution', in Singh and Ravi Gomatam (eds), *Synthesis of Science and Religion: Critical Essays and Dialogues* (San Francisco, CA & Bombay: The Bhaktivedanta Institute, 1987), 91.

6. We are indebted to Dr Max Harcourt for the information contained in this paragraph and his illuminating insights into the history of 'Vedic Science'.

7. Much energy has been spent distributing copies of *FA*, accompanied by glowing testimonials, and a leaflet accompanying a recent distribution of the *Newsletter* of the British Society for the History of Science contained an extract from a letter by one of us (Oldroyd) which, taken out of context, might suggest support for the views expressed in *FA*. The testimonials include a (non-committal) comment from that *aficionado* of the social studies of science, David Bloor, that the book could provide interesting material for students in the History and Philososphy of Science or Sociology of Science. This suggestion was followed by one of us (Wodak), who made a detailed study of *FA* in an honours dissertation. We have noticed, incidentally, that C&T are active in the popular media. For example, we have seen a published transcript of a radio interview between Cremo and American radio journalist Laura Lee, which discussed the arguments of *FA*, and the idea of a scientific 'cover-up': see L. Lee & M. Cremo, 'Forbidden Archaeology', *Nexus: New Times Magazine* (June-July 1995), 11-16. (We are indebted to Mark MacKenzie for this reference.) It is interesting that this 'New Age' magazine has of late published several papers representing the views of the American 'far right'. It would appear that one factor that the 'New Agers' and the 'loony right' hold in common is a suspicion of supposed conspiracies—whether perpetrated by palaeoanthropologists suppressing evidence or FBI agents blowing up buildings in Oklahoma! It is a strange mix, itself worth sociological investigation.

8. A.C. Bhaktivedanta Swami Prabhupada, *Life Comes from Life: Morning Walks with His Divine Grace* (Los Angeles, CA, London, Paris, Bombay & Sydney: The Bhaktivedanta Book Trust, 1984).

9. Other writings published by the Institute, especially their pamphlets such as *Origins* and *Atma*, reveal that the argument from design plays an important part in the Hare Krishna/Vedic critique of evolution: see Drutakarma dasa, Bhutatma dasa and Sadaputa dasa, *Origins: Higher Dimensions of Science* (Los Angeles, CA: The Bhaktivedanta Book Trust, 1984), which gives a synopsis of some of the main arguments of *FA*. Instead of Paley's clock, the analogue suggested in *Atma* (Sydney: Bhaktivedanta Books, 1984), 11, is that of the urban freeway. Just as traffic laws do not evolve but are 'established and enforced by the will and intelligence of the government administrators', so the laws of nature 'depend on a universal will and intelligence'. Likewise, even if we cannot see the traffic controller, 'we can logically conclude by observing the highly-ordered systems of nature that a supreme intelligence is directing them'. One might suggest that, given the traffic congestion and mayhem of many freeways, this analogy is by no means convincing. (But Paley held that even broken watches evinced design.)

A related institution in the US is the Maharishi International University (Fairfield, Iowa), which publishes a journal, *Modern Science and Vedic Science*, promulgating the ideas of the University's founder, Maharishi Mahesh Yogi. The journal contains chiefly articles on psychological matters, but includes such curious items as putative solutions to Fermat's Last Theorem. There is great interest in the 'unified field theory' of physics. The mind, when in a suitable state of quiet awareness, can supposedly apprehend the 'unified field'. Some exceedingly bizarre notions are expounded, for example: '[T]he Veda unfolds through its own commentary on itself, through the sequential unfoldment, in different sized packets of knowledge, of its own knowledge of itself. All the knowledge of the Veda is contained in the first syllable "Ak" of the *Rig Veda*, and each subsequent expression of knowledge elaborates the meaning inherent in the packet of knowledge through an expanded commentary. The phonology of that syllable, as analysed by Maharishi, expresses the self-interacting dynamics of consciousness knowing itself': see Kenneth Chandler, 'Modern Science and Vedic Science: An Introduction', *Modern Science and Vedic Science,* Vol. 1 (1987), 5-26, at 10. Given such strange ideas, we cannot see any prospect of a unification of the ideas of modern science and Vedic science, no matter what social benefits might (very likely) flow from a quieter, meditative existence for all of us. (*Modern Science and Vedic Science* is appropriately located in the 200s section [Dewey System] of the University of New South Wales Library.)

10. As well as Singh & Gomatam (eds), op. cit. note 5, see also *Mechanistic and Nonmechanistic Science: an Investigation into the Nature of Consciousness and Form, Vedic Cosmography and Astronomy, Alien Identities: Ancient Insights into modern UFO phenomena*, all authored by the prolific Richard Thompson, PhD (in Mathematics), Cornell University.

11. Like Creation Scientists, Bhaktivedantists are fond of credentials.

12. Perhaps we should be gratified that one of us has been cited in the BSHS advertisement!

13. Peter J. Bowler, *Theories of Human Evolution: A Century of Debate 1844-1955* (Baltimore, MD: The Johns Hopkins University Press, 1986; Oxford: Basil Blackwell, 1987).

14. Ian Tattersall, *The Fossil Trail: How We Know what We Think We Know about Human Evolution* (New York & Oxford: Oxford University Press, 1995).

15. There is here a residual positivist assumption on the part of C&T that because the outcome of scientific investigation is social in character it is thereby likely to be in error. Such a view would not be accepted by SSK practitioners. C&T refer to the existence of a kind of 'conspiracy' in palaeoanthropology. But presumably they would see no such conspiracy if the palaeoanthropological consensus were one that endorsed Tertiary—or even Precambrian!—humans. We suspect, therefore, that C&T would not support the strong programme's 'symmetry thesis'.

16. It is a nice question whether fraud can be unconscious.

17. Donald K. Grayson. *The Establishment of Human Antiquity* (New York: Academic Press, 1983); Frank Spencer's book on the Piltdown forgery has as its opening chapter: 'The Search for Tertiary man: A Prologue to Piltdown', which discusses the topic in some detail: see Frank Spencer, *Piltdown: A Scientific Forgery* (London, Oxford & New York: Natural History Museum Publications and Oxford University Press, 1990). Other texts which mention the debates include Glyn E. Daniel, *A Hundred Years of Archaeology* (London: Duckworth, 1950) and John Reader, *Missing Links* (Harmondsworth, Middx: Penguin, 1988). Too recent to benefit C&T is A. Bowdoin van Riper, *Men Among the Mammoths: Victorian Science and the Discovery of Human Prehistory* (Chicago, IL & London: The University of Chicago Press, 1993).

18. Bourgeois actually initiated the notion of eoliths.

19. This can be seen from the pages of the early volumes of the *Proceedings of the Prehistoric Society of East Anglia* (before it mutated in the 1930s into the more professional *Proceedings of the Prehistoric Society*). In the earlier period, it seemed natural to assume that there were human progenitors to the makers of the quite sophisticated Palaeolithic implements found in northwest Europe, and the idea of Palaeolithic man having migrated there form some less exalted spot seemed less attractive. After all, northwest Europe did seem to be the centre of the intellectual universe in the late nineteenth century, where everything should have begun!

20. In a system of river terraces the upper levels are older than the lower ones, contrary to the usual geological 'principle of superposition'.

21. Joseph Prestwich, 'On the Primitive Characters of the Flint Implements of the Chalk Plateau of Kent, with Reference to the Question of their Glacial or Pre-Glacial Age. With Notes by Messrs. B. Harrison and De Barri Crawshay' *Journal of the Royal Anthropological Institute of Great Britain and Ireland*, Vol. 21 (1892), 246-62.

22. Alfred S. Barnes, 'The Differences between Natural and Human Flaking on Prehistoric Flint Implements', *American Anthropologist*, Vol. 41 (1939), 99-112.

23. L.W. Patterson, 'Criteria for Determining the Attributes of Man-made Lithics', *Journal of Field Archaeology*, Vol. 10 (1983), 297-307.

24. R.G. West, *Pleistocene Geology and Biology With Special Reference to the British Isles* (London: Longmans, 1968), 343-45.

25. The fading acceptance of eoliths is to be seen, for example, in: Alfred L. Kroeber, *Anthropology: Race, Language, Culture, Psychology, Prehistory* (New York: Harcourt, Brace and Company, 2nd edn, 1948), 630-31.

26. To be sure, theoretical arguments are deployed by C&T against Barnes. But for scientific effectiveness, these need to be published in appropriate scientific journals.

27. Today, only the skull, which is modern in form, is preserved. The skull was separated from the rest of the skeleton before shipment to Germany.

28. L.S.B. Leakey, H. Reck, P.G.H. Boswell, A.T. Hopwood and J.D. Solomon, 'The Oldoway Human Skeleton', *Nature*, Vol. 131 (18 March 1933), 397-98.

29. Reiner Protsch, 'The Age and Stratigraphic Position of Olduvai Hominid I', *Journal of Human Evolution*, Vol. 3 (1974), 379-85.

30. But Protsch stated that they were labelled as such; and certain unspecified 'microanalytical tests' were run to show that the fragments belonged to the same skeleton as the skull.

31. Wilfred E. Le Gros Clerk, *The Fossil Evidence for Human Evolution* (Chicago, IL & London: The University of Chicago Press, 3rd edn, 1978), 94.

32. Henri Breuil, 'Le Feu et l'Industrie de Pierre et d'Os dans le Gisement du Sinanthropus' *L'Anthropologie*, Vol. 42 (1932), 1-17.

33. T.D. White & G. Suwa, 'Hominid Footprints at Laetoli: Facts and Interpretations', *American Journal of Physical Anthropology*, Vol. 72 (1987), 485-514.

34. However, this hypothesis is at odds with radiocarbon and AMS dates for Piltdown cranial remains, which come out at less than one thousand years: see H. de Vries and K.P. Oakley, 'Radiocarbon Dating of the Piltdown Skull and Jaw', *Nature,* Vol. 184 (25 July 1959), 224-25; J.C. Vogel and H.T. Waterbolk, 'Groningen Radiocarbon Dates V (Piltdown Series)', *Radiocarbon,* Vol. 6 (1964), 368; F. Spencer and C.B. Stringer in R.E.M.

Hedges et al, 'Radiocarbon Dates from the Oxford AMS System: *Archaeometry* Date list 9', *Archaeometry,* Vol. 31 (1989), 207-34, at 210.

35. Wu Rukang & Lin Shenglong, 'Peking Man', *Scientific American,* Vol. 248 (June 1983), 78-86.

36. It should be pointed out, however, that C&T give incorrect page numbers for the paper of Wu and Lin; and we have not, therefore, been able to verify all the claims that they make against the Chinese article.

37. They also refer to various cases where scientists have suffered career problems as a result of holding unorthodox views about the extent of American antiquity.

38. M. Shackley, *Wildmen: Yeti, Sasquatch and the Neanderthal Enigma* (London: Thames & Hudson, 1983).

39. 'Notes', *Nature,* Vol. 77 (23 April 1908), 587.

40. The mythic elements of evolutionary story-telling in this period have been well described by Misa Landau, *Narratives of Human Evolution* (New Haven, CT: Yale University Press, 1991)

41. James Reid Moir, *Pre-Palaeolithic Man* (Ipswich: W.E. Harrison, n.d.), 61.

42. Cf. a recent remark by Iain Davidson: 'We see much of the patterning of early and late stone industries as deriving from the ingenuity of archaeologists in seeing pattern' ('The Archaeology of Language Origins', *Antiquity,* Vol. 65 [1991], 39-48, at 41).

43. L.S.B. Leakey. *The Progress and Evolution of Man in Africa* (London, New York & Toronto: Oxford University Press), 37-38.

44. J.D. Clark, 'African and Asian Perspectives on the Origins of Modern Humans', in M.J. Aitkin, C.B. Stringer and P.A. Mellars (eds) *The Origins of Modern Humans and the Impact of Chronometric Dating* (Princeton, NJ: Princeton University Press, 1993), 148-78.

45. Christopher Stringer and Clive Gamble, *In Search of the Neanderthals* (London: Thames & Hudson, 1993). 129.

46. See Daisie Radner and Michael Radner, *Science and Unreason* (Belmont, CA: Wadsworth Publishing House, 1982), 27-52. However, we prefer Laudan's distinction between 'good science' and 'bad science', rather than Popper's 'science' and 'pseudo-science'. And we certainly acknowledge that C&T's historical work has yielded matter of considerable interest.

47. We refer here to the interesting historical evidence adduced concerning such matters as eoliths, Java Man, and so on, not such matters as the outlandish notions of gold chains in coal.

48. The same strategy can be found in some CS publications, such as Harold S. Slusher and T.P. Gamwell, *Age of the Earth*, Technical Monograph No. 7 (El Cajon, CA: Institute for Creation Science Research, 1978). The pages of this monograph fairly burst with complicated mathe-

matical formulae, raising questions as to the particular readership being targeted. Formulae like these can only be understood by the mathematically educated. What function do they serve in the strategy of CS apologetics? Are they designed to dazzle believer and non-believer alike with the scientific proficiency of the authors (whose academic credentials are of course cited on the front cover), or are they a means for CS's self-justification as a credible 'scientific' research institute? (Perhaps such publications aim to serve both functions.) The same question can be asked of *FA*. What sort of reader can be expected to part with A$60 to purchase the book, and to plough through hundreds of pages of detailed palaeoanthropological discussion? The converted presumably need no such persuasion, while the unconverted who have no prior palaeoanthropological interests or expertise are unlikely to persevere. Perhaps this publication, too, is primarily designed to make credible BI's status as a 'scientific' research centre.

49. Ron Curtis, 'Narrative Form and Normative Force: Baconian Story Telling in Popular Science', *Social Studies of Science*, Vol. 24 (1994), 419-67.

Related Correspondence

Dr. David R. Oldroyd and Ms. Jo Wodak have asked that I not publish any of their letters to me.

2.2.3.1. Letter to Dr. David R. Oldroyd, June 25, 1995.

I see in the *History of Science Society Newsletter* that you are organizing the annual conference of the Australian Association for the History, Philosophy, and Social Studies of Science in Sydney, July 8-11. I am not able to attend this year, but I would like to receive a conference program and be placed on the mailing list for next year's announcements.

I hope Jo Wodak's thesis is coming along (and would like to see that when it's finished), and I am of course waiting to see what you two will have to say about *Forbidden Archeology* in *Social Studies of Science*. The latest review is by W. Stoczkowski in *L'Homme* (Jan.-Mar. 1995). Tim Murray, head of the archeology department at Latrobe University down there, tells me his review in *British Journal for the History of Science* is coming out soon.

Tim also has my paper "Puranic Time and the Archeological Record," presented at the World Archeological Congress-3 in New Delhi last December, which he says looks like it will fit in the conference proceedings volume on time concepts, to be published by Routledge.

I also recently presented a paper on the impact of *Forbidden Archeology* at the Sixth Annual Interdisciplinary Conference on Science and Culture, Institute for Liberal Studies, Kentucky State University. Of course, I sometimes step out of mainstream channels. I've accepted an invitation to give a paper on *Forbidden Archeology* at the Ancient Astronaut Society world conference in Bern, Switzerland, this August.

Other than that, I'm busy with another book. I hope everything is going well with your work in your new field of interest, the history of paleoanthropology.

2.2.3.2. Letter to Jo Wodak and David Oldroyd, May 30, 1997.

The past year has been quite busy for me, and it is only now that I have had time to sit down and write to you about your review article on *Forbidden Archeology*. The book has received full reviews in *British Journal for the History of Science, L'Homme, L'Anthropologie, Geoarchaeology, American Journal of Physical Anthropology*, and other publications, but I regard your review article as the most significant scholarly response.

I am particularly grateful for your recognition that *Forbidden Archeology* does make a contribution to the literature on paleoanthropology (for its detailed examination of the historical material and for its raising "a central problematic regarding the lack of certainty in scientific 'truth' claims'").

I shall now offer a few comments on some of your statements. You wrote: "In their Preface, C&T acknowledge the Vedic inspiration of their research, but what this amounts to is not explained. They claim that such explanation is unnecessary, for their critique of current human evolutionary theory is so well supported by an 'abundance of facts' as not to need theoretical/philosophical justification. However, the authors promise that a forthcoming second volume will show how Vedic theory agrees with 'the facts' more completely than either 'Darwinian evolutionists' or 'Biblical creationists' (240-41). This, of course, is not much help to readers of *FA*, given that its conclusions are really only persuasive if the underlying Vedic assumptions are accepted."

When one encounters a massive critical review of the literature on human evolution that strongly suggests current ideas about human origins and antiquity are incorrect, one will naturally wonder about underlying motives. Such wonder is not a logical necessity, but it is a fact of human nature. Therefore, in my Introduction to *Forbidden Archeology* I wanted to clearly but briefly indicate the motive for the review.

Furthermore, logically speaking, one can offer a negative critique of current evolutionary accounts of human origins and antiquity without offering any alternative explanation. It is quite possible to say that the present explanation is contradicted by numerous facts and we need a new one. It may be desirable or customary to offer an alternative explanation when criticizing and rejecting an existing explanation, but it is not necessary. *Forbidden Archeology* does not offer an alternative explanation, but appreciating the customary desire for an alternative explanation, I indicated that one was indeed forthcoming.

I disagree that *Forbidden Archeology*'s conclusions are persuasive only if the underlying Vedic assumptions are accepted. The book's conclusion is that current ideas of human origins and antiquity are wrong. There is no logical necessity to accept any Vedic assumptions in order to be persuaded of this. And practically speaking, I have found that Christian and Islamic creationists, who do not accept the Vedic assumptions that inspired *Forbidden Archeology*, find the book persuasive. The same can be said of reviewers from the New Age and alternative science communities.

That said, I do not think it is wrong for reviewers, such as yourselves, to explore the underlying Vedic assumptions as deeply as you might desire. But there was no necessity for the authors of *Forbidden Archeology* to accompany their critique of modern concepts of human evolution with an alternative explanation based on those assumptions. They could have done so. But I will share with you my reasons for their not having done so. The first step, I thought, was to show, as convincingly as possible, that an alternative explanation, Vedic or not, is necessary. Furthermore, to offer a persuasive Vedic alternative explanation of human origins will itself take a rather large book, and it must incorporate many categories of evidence in addition to archeological evidence. So rather than offer a summary of that alternative explanation that would not in itself be persuasive, I thought it better simply to signal that such an alternative explanation was coming and let readers encounter it in its full expression.

In any case, I do find it gratifying to be pressed by you and others for the Vedic alternative explanation.

Some of the details of your explanation of *Forbidden Archeology*'s Vedic background are not entirely correct. (1) The Bhaktivedanta Institute is only one part of the Hare Krishna movement—the science studies branch. (2) The Bhaktivedanta Institute is at present a rather loose confederation of individuals and organizations. At the time you wrote the article, one of the BI centers, under the direction of Dr. Richard Thompson, was in San Diego. Richard Thompson has since moved his center to a Hare Krishna community in Florida. The main centers of Bhaktivedanta Institute activity are located in Oakland, California, and Bombay, India, and are under the direction of Dr. T. D. Singh. The Bombay center has recently started a graduate program in consciousness studies in association with an Indian university. (3) Sankara represents a nontheistic element in Indian philosophy, and is not normally considered a proponent of *bhakti* philosophy. (4) According to Puranic cosmology, the duration of the present creation system is 311 trillion years (the lifetime of the demigod Brahma in solar years) instead of 311 billion years. We are about 150 trillion years into the life of the current Brahma. During this creation system, universes are created during each day of Brahma, and are withdrawn during each night of Brahma. The day of Brahma lasts 4.3 billion years, and the night lasts an equal period of time. We are currently about 2 billion years into the current day of Brahma. Accordingly, the age of this universe, which is one of many millions of universes currently existing, is about 2 billion years. We would thus expect the fossil record to extend 2 billion years, which, interestingly enough, it apparently does, according to modern science.

You wrote: "although C&T certainly discuss social interactions between paleoanthropologists, they make little explicit effort to relate their work to the technical SSK literature." True. It is not that I am unaware of the technical SSK literature. It certainly informed the writing of *Forbidden Archeology*. But I noticed during the eight years I was working on *Forbidden Archeology* that the technical SSK literature reflected, even over that period of time, some fairly substantial, if arcane, changes of concerns and methodologies in the SSK community. Such change is ongoing, and any attempt to tie the argument of *Forbidden Archeology* to current (i.e. early 1990s) technical SSK literature would quickly have been dated and would also have restricted even the academic audience for the book.

Concerning the eolith issue, you wrote: "Today students in archaeology are regularly taught how to make flint implements themselves, observe the characteristic features of such artefacts, and hence distinguish between the artefactual and the natural. Of course, in such a learning process they are being 'acculturated': they are learning how to work within or according to a paradigm. But if C&T want to show that this teaching is in error, it is incumbent on them to do so by their own practical methods or theoretical arguments, not by combing the literature and finding some items therein that speak with dissenting voices. Indeed, the extent to which C&T can find such voices in the literature evidences the fact that there is not some conspiratorial 'cover-up' in palaeoanthropology. In any case, the controversial eoliths investigated by Harrison, Moir and others were *not* associated with organic human or hominid remains. So whatever one may make of the 'tools', they tell us rather little about the actual course of human evolution. (But then C&T don't hold with human evolution.)"

Several comments are in order: (1) As documented in *Forbidden Archeology*, the original discoverers and supporters of eoliths also experimented with making stone implements and used the results to arrive at conclusions about the artifactual nature of chipped stones. (2) I do not feel obligated to conduct my own experiments with making stone tools. *Forbidden Archeology* is a metaepistemological critique of archeology and paleoanthropology. It is founded on the idea that these disciplines exist mainly as corpus of texts, and that a critical study of texts and how they are manufactured, read, and archived can yield important insights into truth claims in these disciplines over time. (3) Your reference to "conspiratorial coverup" is misleading. First of all, it contradicts your own earlier correct statement that *Forbidden Archeology* posits a social process of "knowledge filtration." Second, in the Introduction to the book and several other places my coauthor and I point out that this process of coverup or knowledge filtration generally does not involve groups of scientists diabolically plotting to keep certain facts hidden. It is a social process that involves directing scientific attention to or from certain reports over time. (4) In some cases eoliths were associated with organic remains, as at the Foxhall site, which yielded a controversial human jaw, as well as signs of fire, in addition to eoliths. (5) The presence of crude stone tools in Europe, at the times indicated by the eolith discoveries of Harrison and Moir (up to 5 million years ago), would say *something* about the course of human evolution. According to current ideas, the earliest stone tools date back about 2 million years.

You wrote: "But evidence favourable to the *evolutionary* anatomical origin of humans is omitted.... *FA*'s presentation means that the reader is not offered a thorough review of the relevant evidence, despite the book's great length." The literature on the evolutionary anatomical origin of humans is so voluminous and accessible, to specialists and general readers alike, that we did not feel compelled to report it in detail.

You went on to say: "It is as if the Aristotelian and Galilean cosmologies were presented, and evidence provided that not all Galileo's arguments were sound. *Ergo,* we should consider reviving Aristotelian theory. Needless to say, for us, it is modern astronomy that has to be put beside Aristotelian physics and cosmology. C&T should do the same with palaeoanthropology. To attempt to resurrect an old paradigm the way they do, by analyzing outmoded scientific texts, is one of the recognized marks of 'pseudo-science'."

You neglect to mention that *Forbidden Archeology* did in fact extensively analyze the latest available reporting in paleoanthropology, especially in the latter half of the book. The book purported to be a historical survey of evidence for human antiquity, so of course it analyzes texts from different periods. Your references to "an old paradigm" and "outmoded" texts seem to display a kind of age bias. Old is bad. New is good. Old is wrong, New is right. Also "outmoded" means no longer fashionable, and fashion, although it certainly does have a role to play in scientific inquiry, may not be the best guarantee of epistemic superiority. Any kind of science, pseudo or otherwise, is based on a corpus of texts, some of them new, some old. Even orthodox paleoanthropology is based on a corpus of texts, some old and some new. But this corpus of texts is drawn from a larger body of texts. And there is some utility in from time to time going over the corpus of current texts, in the context of the complete archive, and seeing if everything is in order. To characterize this as pseudoscience is inaccurate. In general, accusations of pseudoscience strike me as a kind of pseudoscience, an attempt to monopolize discourse.

In a footnote you yourselves say, "We prefer Laudan's distinction between 'good science' and 'bad science', rather than Popper's 'science' and 'pseudo-science'. And we certainly acknowledge that C&T's historical work has yielded matter of considerable interest."

You wrote: " In any case, even if C&T's historical arguments are granted, while this would considerably alter current thinking about the time span of human history, it would not invalidate orthodox evolutionary *theory*."

We said the same thing in the book, but noted that the time scale would have to be altered so drastically that many might be hesitant to readily accept the new version of the theory. In practice, the evidence would therefore cause severe problems for evolutionary theory.

William B. Howells, one of the principal architects of modern human evolutionary theory, wrote to me on August 10, 1993, after receiving a review copy of *Forbidden Archeology*: " Most of us, mistakenly or not, see human evolution as a succession of branchings from earlier to more advanced forms of primate, with man emerging rather late, having feet developed from those of apes, and a brain enlarged from the same kind. Teeth in particular strongly indicate close relations with Miocene hominoids. This makes a clear phylogenetic pattern. To have modern human beings, with their expression of this pattern, appearing a great deal earlier, in fact at a time when even simple primates did not exist as possible ancestors, would be devastating not only to the accepted pattern. It would be devastating to the whole theory of evolution, which has been pretty robust up to now....The suggested hypothesis would demand a kind of process which could not possibly be accommodated by evolutionary theory as we know it."

Finally, you wrote: "In conclusion, we would say that palaeoanthropology is a branch of science that is most suitable for scrutiny by *aficionados* of SSK. But such scrutiny requires insights into the SSK literature, not a set of metaphysical assumptions that are totally out of line with those of modern science."

One of the things insight into the SSK literature prepares one to do is question the metaphysical assumptions of modern science. Perhaps some new ones would serve us better.

Thank you again for having the patience to go through *Forbidden Archeology* and give it such a detailed, fair, and courteous review.

Finally, how may I obtain a copy of Jo Wodak's honours dissertation on the book?

2.3. Forewords by Scholars

For the sake of completeness, I am including in this section the Forewords to Forbidden Archeology *and* Hidden History of the Human Race.

2.3.1. Dr. Pierce J. Flynn (1993) Foreword to *Forbidden Archeology*, by Michael A. Cremo and Richard L. Thompson.

Dr. Pierce J. Flynn is a sociologist at the California State University at San Marcos. He is the author of The Ethnomethodological Movement: A Semiotic Interpretation *(Mouton de Gruyter, 1991).*

I perceive in *Forbidden Archeology* an important work of thoroughgoing scholarship and intellectual adventure. *Forbidden Archeology* ascends and descends into the realms of the human construction of scientific "fact" and theory: postmodern territories that historians, philosophers, and sociologists of scientific knowledge are investigating with increasing frequency.

Recent studies of the emergence of Western scientific knowledge accentuate that "credible" knowledge is situated at an intersection between physical locales and social distinctions. Historical, sociological, and ethnomethodological studies of science by scholars such as Harry Collins, Michael Mulkay, Steven Shapin, Thomas Kuhn, Harold Garfinkel, Michael Lynch, Steve Woolgar, Andrew Pickering, Bruno Latour, Karin Knorr-Cetina, Donna Haraway, Allucquere Stone, and Malcolm Ashmore all point to the observation that scientific disciplines, be they paleoanthropology or astronomy, "manufacture knowledge" through locally constructed representational systems and practical devices for making their discovered phenomenon visible, accountable, and consensual to a larger disciplinary body of tradition. As Michael Lynch reminds us, "scientists construct and use instruments, modify specimen materials, write articles, make pictures and build organizations."

With exacting research into the history of anthropological discovery, Cremo and Thompson zoom in on the epistemological crisis of the human fossil record, the process of disciplinary suppression, and the situated scientific handling of "anomalous evidence" to build persuasive theory and local institutions of knowledge and power.

In Cremo and Thompson's words, archeological and paleoanthropological "'facts' turn out to be networks of arguments and observational claims" that assemble a discipline's "truth" regardless, at times, of whether there is any agreed upon connection to the physical evidence or to the actual work done at the physical site of discovery. This perspective, albeit radical, accords with what I see as the best of the new work being done in studies of scientific knowledge.

Forbidden Archeology does not conceal its own positioning on a relativist spectrum of knowledge production. The authors admit to their own sense of place in a knowledge universe with contours derived from personal experience with Vedic philosophy, religious perception, and Indian cosmology. Their intriguing discourse on the "Evidence for Advanced Culture in Distant Ages" is light-years from "normal" Western science, and yet provokes a cohesion of probative thought.

In my view, it is just this openness of subjective positioning that makes *Forbidden Archeology* an original and important contribution to postmodern scholarly studies now being done in sociology, anthropology, archeology, and the history of science and ideas. The authors' unique perspective provides postmodern scholars with an invaluable parallax view of historical scientific praxis, debate, and development.

2.3.2. Dr. Phillip E. Johnson (1994) Foreword to *The Hidden History of the Human Race* by Michael A. Cremo and Richard L. Thompson.

Dr. Phillip E. Johnson is a professor of law at the University of California at Berkeley. He is the author of Darwin on Trial.

Human prehistory is not something about which anyone ought to be dogmatic. A few years ago, the "Mitochondrial Eve" hypothesis was being presented to the public virtually as fact; now it is under a cloud. Only a few days before I wrote these words, newspaper stories reported the redating of a skull fragment in Java attributed to *Homo erectus.* Now said to be 1.8 million years old, the fossil seemingly places this claimed ancestral species in Asia long before it was supposed to have migrated from Africa.

Evidence of this kind could receive wide publicity because, although it disappoints the expectations of some paleoanthropologists, it excites others and does not threaten the coherence of the accepted picture of human evolution in any fundamental way. But what if an apparently modern human fossil were found in sediments dated two million years old? Would the astonishing finding receive credence? Possibly there would be irresistible pressure to recalculate the date, to reattribute the fossil to some prehuman species, to question the competence of the discoverer, and eventually to forget the whole thing.

According to Michael Cremo and Richard Thompson, something of that sort has happened before, and happened often. This is because of a dual standard that is applied to evaluate evidence. Evidence of early humans or their tools is readily accepted if it fits with the orthodox model of human evolution. Evidence that is just as reliable, but which does not fit the model, is ignored or even suppressed. It fairly quickly drops from the literature, and within a few generations is almost as invisible as if it had never been. As a result, it is virtually impossible for rival understandings of early human history to gain credence. The evidence that would have supported them is no longer available to be considered.

In their lengthy work titled *Forbidden Archeology*, Cremo and Thompson provided a stunning description of some of the evidence that was once known to science, but which has disappeared from view due to the "knowledge filter" that protects the ruling paradigm. The detective work required to unearth this evidence was impressive, and the authors reported what they had found and how they had found it in such careful detail, and with such thorough analysis, that they deserved to be taken seriously. Unfortunately, relatively few professional scientists are willing to consider evidence that upsets prevailing views and comes from a source out of the academic mainstream. This present work presents a summary of the larger work for the ordinary reader, and I hope it will attract the attention of fair-minded professionals, who may then be motivated to study the much more detailed presentation of the same evidence in the original volume.

The authors frankly acknowledge their motivation to support the idea, rooted in the Vedic literature of India, that the human race is of great antiquity. I do not share their religion or their motivation, but I also do not think that there is anything disreputable about a religious outlook which is candidly disclosed. Scientists, like other human beings, all have motives and biases that may cloud their judgment, and the dogmatic materialism that controls the minds of many mainstream scientists is far more likely to do damage to the truth because it is not acknowldged as a bias. In the end the important thing is not why the investigators were motivated to look for a certain kind of evidence, but whether they found something worth reporting, and worth serious consideration by the scientific community.

As far as I am able to judge, Cremo and Thompson have reported evidence that is very much worth that kind of serious consideration. I am not writing this Foreword to endorse their findings, but to encourage serious students of the subject to give them fair hearing. This is a very interesting book, which makes for exciting reading. I would like very much to see how well the evidence it reports stands up to fair-minded scrutiny from the best-informed readers, who may be glad to have the chance to examine evidence that was not included in the textbooks and review articles they were given in their college and graduate school classes.

3

Additional Reviews, Notices, and Related Correspondence

3.1. Proevolution Interest Group Publications

Forbidden Archeology *attracted reviews from several publications that identify themselves as skeptical, rationalist, and so forth. These are code words for a rather strict and cranky materialism, which came into vogue in the latter nineteenth century, matured in early decades of the twentieth, and now seems rather outdated to almost everyone except its current practitioners, who still consider it the height of intellectual fashion. The academic journals included in Chapter Two at least make a pretense of scientific objectivity. The journals included in this section are openly hostile to creationism and paranormal phenomena, subjecting claims favorable to either to enthusiastic ritual debunking. Their writers' style generally incorporates heavy doses of haughty sarcasm and ridicule. Works expressing unwanted conclusions are often ostentatiously tagged with the label pseudoscience, as if this itself were a sufficient response to any argument or evidence. Jonathan Marks's review of* Forbidden Archeology, *published in* American Journal of Physical Anthropology, *is similar in style and tone to the reviews found in this section, showing that there is some overlap between the partisan interest group publications and the more objective scientific and academic journals.*

3.1.1. Wade Tarzia (1994) "*Forbidden Archaeology*: Antievolutionism Outside the Christian Arena." *Creation/Evolution*, 14(1): 13-25. Reprinted by permission of the American Humanist Association.

Wade Tarzia has a doctorate in comparative literature and folklore. He works as a writer and editor at the School of Engineering of the University of Connecticut at Storrs. Creation/Evolution *is a publication of a small and highly partisan anticreationist group ambitiously titled the* National Center for Science Education.

Forbidden Archeology, a new Bhaktivedanta Institute book, argues that anatomically modern humans have existed for hundreds of millions of years, disproving the theory of human evolution; it makes no specific claims about other kinds of biotic evolution. The book also claims that archaeologists have become a "knowledge filter" (p. xxv ff.) since the 19th century, laboring under a predisposition to ignore evidence for anatomically modern humans having existed for millions of years. Sometimes the book develops a dishonesty theory—evidence is said to be "carefully edited" (p. 150) by scientists so that younger investigators do not see evidence that invalidates the theory of human evolution.

The authors have worked hard in collecting and quoting an enormous amount of material, most of it from the 19th- and early 20th-century, certainly interesting for its historical perspective. Their evidence is as diverse as it is detailed, including, for example, "eoliths" (crudely broken stones some have considered early tools), "wildmen" (Bigfoot, etc.), and even a fossilized shoe sole from the Triassic period.

Despite all this hard work, I think the book falls short of a scientific work primarily (but not entirely) because (1) its arguments abandon the testing of simpler hypotheses before the more complex and sensationalistic ones, and (2) the use of so many outdated sources is inadequate for a book that seeks to overturn the well-established paradigm of human evolution—scholars must not work in isolation, especially today, when multidisciplinary approaches are needed to remain on the cutting edge of knowledge. However, for researchers studying the growth, folklore, and rhetoric of pseudoscience, the book is useful as "field" data.

I confine my review to some basic categories of flawed scientific argumentation. I show a couple of examples in each category but by no means have exhausted the pool. Throughout the book, examples of "loose" science appear. I hesitate in judging the book to be utterly worthless from a scientific standpoint—as I said, various specialists need to compare notes on the book—but if good ideas exist in *Forbidden Archeology*, they are hidden under a mass of undisciplined details, lack of critical contextual information, leaps of logic, and special pleading. The authors would have done better to devote years of research to a smaller list of topics to allow themselves space to consider and test all of the implications of their hypotheses.

Forbidden Archeology is so expansive that it forms good ground on which to explicate the style of pseudoscientific writings, especially on the topic of archaeology. It is an exhaustive attack on the idea that humans have evolved. It is also a well-written example of pseudoscience—it *looks* like the real thing, a phenomenon discussed in Williams (1991:15)—and a quick review of the book is not possible. Serious treatment of new ideas, however much on the fringe they may be, is an appropriate venture in science. "The idea is not to attempt to settle such ideas definitely, but rather to illustrate the process of reasoned disputation, to show how scientists approach a problem that does not lend itself to crisp experimentation, or is unorthodox in its interdisciplinary nature, or otherwise evokes strong emotions" (Sagan 1979:82).

Mass of Details

The mass of details with attached analyses would require book-length responses from specialized reviewers to confirm or critique. This style is a common diversionary tactic in pseudoscience. Since the authors have not aired their arguments previously through professional journals, as many scholars do before writing such a huge synthesis of material, the task of validation becomes a career itself. Such a style burdens an analysis with long leaps between broad assumptions (i.e., scientific cover-up) to the detailed evidence (i.e., minutiae of strata and dating from obscure sites)—all on the same page.

In the process of amassing details, the book seems to go to great length on minutiae, while more important data are passed over. Example: in a discussion of a purportedly incised bone (p. 38-40), discussion of the nature of the cuts and the context of the bone in the site are given short shrift while the discussion focuses on the fauna appearing in the stratum of the find. Evidence from an electron microscope study is not yet forthcoming; additionally, a reference central to the issue is a personal communication, and other evidence in the form of drawings or photographs is lacking. We are diverted from the primary issue of *whether it is an artifact at all.* A discerning reader simply needs more than this to credit unusual claims for controversial artifacts.

Overuse of Old Sources

Quotations from the 19th-/early 20th-century are copious— comprising, I would guess, at least 25 percent of the book. A few examples: (1) a 1935 work of Weidenreich is cited as opposition to a 1985 work of Binford and Ho (p. 553); was there no current reference to refute Binford and Ho, and if not, what does this mean? (2) a question is raised about the geological time-scale, and the latest reference on the matter cited is a lecture given by Spieker in 1956 (p. 16); surely, additional and more recent work is available on a topic as important as this; (3) the authors cite a 1910 work of Osborn that mentions archaeological work done in 1863 and 1867, which seems like desperately searching for supportive evidence in old reports; (4) experts are cited—from ca. 1870—on the subject of shark teeth to suggest that these Pliocene fossils were drilled by Pliocene humans (p. 49-51); this case is conspicuous in its avoidance of modern sources on shark biology and paleontology, sources that might better elucidate the work of tooth decay, parasites, and fossilization at work on shark teeth.

I do not indict the sincerity and ground-breaking work of 19th century scholars. However, because knowledge seems to accumulate and research techniques seem to improve, assuming a blanket equivalency of research level between 19th and late 20th century science is just going too far.

Rusting Occam's Razor

A major flaw of *Forbidden Archeology* is its quick leaping toward sensational hypotheses (see in general Williams 1991:11-27). Sensational ideas are not intrinsically bad—plate tectonics was pretty astonishing at one point (Williams 1991:132), but also true. However, the cautious investigator hopes that less sensational, or simpler, hypotheses are first proposed and well tested before more complex or less likely explanations are considered.

This jumping over possible explanations is what Dincauze (1984:294) calls avoidance of alternatives in archaeological argumentation. Dincauze fairly draws her cases from an array of archaeologists, some professional, others on the fringe. Her cases are drawn from the controversial claims for pre-Clovis (pre-12,000 BP) Paleoindian occupation in the Americas, but her ideas perfectly suit this current review. Dincauze writes,

> Critical tests must be applied to each and every claim for great antiquity so long as there remains no supporting context of ancient finds in which the claims can be readily accepted....We have at hand an unprecedented number of powerful analytic techniques. Because of the expanded base of theory, data, and method, we should be able to define related series of contrastive hypotheses around any question. Given multiple hypotheses, we can proceed to exclude or disprove all of but a few of them, leaving those that are not contradicted.

Consider the famous fossilized foot prints at Laetoli, Tanzania, dated to about 3.6 million years BP. Most scientists agree they were made by hominids. Since the footprints are surprisingly familiar, Cremo and Thompson feel they are direct evidence for 3.6-million-year-old *modern* humans (p. 742). Yet, one can more easily see the footprints in light of a main point of evolutionary theory—if parts of an organism are well-adapted to certain uses, selection pressure for change would be absent. Thus human feet may be relatively well-adapted to walking and need not have changed rapidly over a few million years. So far this seems to be the simplest explanation. *Forbidden Archeology* has not offered an alternative that falsifies this concept nor proposed a better one.

Reference to reports of living ape-people (or "wildmen") caps my list of giant leaps. *Forbidden Archeology* uses this section to suggest the simultaneous existence of ancestral hominids with modern-type humans (cf. 622) (which would supposedly disprove the notion of human evolution, ignoring the possibility of shared common ancestry). The authors seem very credulous of reports of wild-folk sightings. Here the easiest explanation, in the absence of a caged abominable snowperson, is that Yeti, Sasquatch, etc. are manifestations of folklore about anthropomorphic creatures, which is spread worldwide and goes back quite far; the human-eating monsters Grendel and his mother in the 1,000+ year-old epic Beowulf are examples (see Donaldson 1967). In fact, some of the reports cited in *Forbidden Archeology* remind me of Beowulf when the theme of the report is an attack of an ape-man (examples on pp. 610, 611, 614, 618). The nature of some reports reminds me of contemporary legends in which the actual witness of a strange event is removed from the informant by space and time; one informant said, "Many years ago in India, my late wife's mother told me how her mother had actually seen what might have been one of these creatures at Mussorie, in the Himalayan foothills" (p. 607).

Discussing wildmen existing in folklore, the authors cite a reference that says, in part, that wolves appear in folktales because they are real; so if wildmen did not show up in folktales, then their reality could be doubted (p. 617). Well—dragons, giants, and vampires show up in folklore; are we to believe they are real? But chipmunks seldom appear in folktales, so perhaps they are mythical? Asking simple questions such as these help us make a "reality check" on arguments. As a folklorist, I need to see the folklore hypothesis first discussed and soundly falsified before I consider that Yeti is real. And as a person interested in science, I also need to see a sound ecological defense of their lifestyle; as William says, "[T]here is a worldwide belief in human-like monsters, often lurking in the unknown woods....we've got them everywhere we want them—but conveniently they don't take up much space and eat very little" (Williams 1991:17).

Missed Evidence

While presenting a voluminous amount of detail, sometimes *Forbidden Archeology* overlooks important points. For example, the book discusses the Timlin site in New York, where researchers reported finds of tools dated to 70,000 BP (p. 354). Yet *Forbidden Archeology* does not mention the responses to these claims by several professionals which cast the nature of these finds in doubt (Cole and Godfrey 1977; Cole, Funk, Godfrey, and Starna 1978; Funk 1977; Starna 1977; a reply to the criticisms is in Raemsch 1978). I found it interesting that a student created similar "eoliths" by rattling the same source material in a garbage can (Funk 1977: 543); this simple experiment has much to say about eoliths!

The authors have also missed Dincauze's (1984) work which analyzes the flaws in theorizing about bones and artifacts from alleged early man sites. The flaws in logic, artifactual context, and hypothesis testing (or lack of it) that she discusses are perfectly applicable to arguments on eoliths and alleged incised bones, and her discussions include some of the very sites referred to in *Forbidden Archeology* and the problems associated with them.

In addition, the book appears to miss the point that conclusions drawn from the paleoarchaeological record rely heavily on the context of evidence found from a variety of sites. When an artifact or fossil has a good context, it has been found among other evidence of cultural activity and has been dated by more than one method. The artifact might be found in concentrations of other artifacts at a butcher site comprising the bones of an animal. Such a context supports a claim that simple tools, comprising rather crudely chipped cores and flakes, were indeed tools. Similarly, the dating of the remains should rely not only on a chemical method but also on other contexts, such as datable fossil remains of other life (Dincauze [1984:301-305] discusses these issues; see Mania and Vlcek [1981:134] for an example in use: testing amino acid racemization, geological strata, and faunal analysis).

Problems of missing context plague eolith arguments. Thus, the authors state that crude eoliths are not accepted as tools whereas allegedly similar-looking artifacts (such as Oldowan and Acheulian industries) discovered by professional archaeologists are accepted as artifacts (p. xxvii). But many Acheulian artifacts and quite a few Oldowan artifacts are quite distinctively styled—impossible to confuse with randomly-broken eoliths.

Furthermore, Cremo and Thompson think that Oldowan tools cannot be accepted as tools because they were not found near hominid fossils (p. 154). This chain of logic continues: if one rejects eoliths as tools, then one must also reject Oldowan tools, which dismisses most of the tools from East Africa and Zhoukoutien in China (p. 188); or—take your choice!—in the absence of hominid remains, Acheulian artifacts could be attributed to *Homo sapiens* (p. 410). In some cases the authors may be correct—some of the early finds at Olduvai that have no supporting context may indeed be shaky evidence. Beyond this, however, *Forbidden Archeology* builds a shaky correspondence between the alleged evidence of eoliths and the accepted hominid and tool finds. First of all, archaeologists do not fail to question their data, a fact that *Forbidden Archeology* conveniently fails to mention at strategic points. A cursory library search introduced me to Walker (1981:198-201), who notes that the dating, surface-find context, and sample sizes of hominid finds present currently unsolved problems (although, on the other hand, Walker emphasizes that surface finds, under certain defined conditions of context, can offer reasonable evidence [p.200]. On the same stroll I found Rightmire (1984:298) observing that *Homo erectus* probably made the early Acheulian tools, but the association of the tools with these hominids is not clear in the southern African sites.

However, these cases do not make sites with better contexts disappear. Rightmire (1984:298,300) mentions sites at which fossil hominids and tools are found in more solid contexts. Mania and Vlcek (1981:133-151) also report a hominid site with associated hominid fossils, faunal remains, and tools. The Koobi Fora (Kenya) site is undoubtedly a butcher site replete with concentrations of stone tools; the only creature that could have made tools in that region is an early hominid species (Leakey and Lewin 1978:12). (See Isaac [1984:7-10] and Jones, et al. [1992] for further evidence.) And most would disagree with the authors about Zhoukoutien; tool-using *Homo-erectus* is most likely represented at that site near Beijing (Harrold 1990:6).

Archaeologists would love to find an Australopithecine or *Homo habilis* who choked to death on a classifiable bone on an extinct animal, an Oldowan utensil in hand, covered over by a layer of hardened, datable volcanic ash preserving the footprints of disappointed family and friends leaving the body. This hasn't happened. Yet, finds of tools in context with butcher sites or living sites, with hominid remains existing in the general region (near tools, in a few cases), are too strong to disavow in the absence of any other fossil of an intelligent creature that could produce tools and living floors. This evidence cannot be compared with eolithic evidence found out of context in the 19th century.

Acceptance of Poor Evidence

Any supporting evidence seems acceptable to the authors. I wish the Triassic "shoe sole" (p. 807) (also cited by many "scientific" creations) were held to some standard of documentation, with its blurry photograph and no sign of the stitching, etc., proving it to be a shoe fossil. The authors criticize the quality of Java Man and Zhoukoutien Cave finds even though the techniques and documentation of these finds cannot be compared to the unconvincing claims of other reports, such as those stemming from sworn testimonials, as can be found on page 376 ff., and they ignore the well-documented on-going discoveries at the Beijing site. Similarly, when the book documents a claim for a modern-type human skeleton (reported in an 1862 geology journal) in a coal deposit 90 feet deep, we learn the authors wrote the Geological Survey to date the coal to about 286 million years (p. 454). But we are not treated to contextual discussion of the bones—how they were found, who found them, what was the site like, and how these allegedly old bones came out of the earth with only a loose black coating that was easily scraped away to reveal nice white bone, etc. The impression left is that, if a tabloid reported Jimmy Hoffa's corpse was found in Triassic deposits, then the authors would no doubt perform rigorous research to date those deposits and then include the data in their next book.

The best example of reliance on poor evidence is an attempt to make negative evidence into support. The introduction to the wildman chapter tries to use lack of evidence for wildmen to support the existence of them. The argument begins by questioning how—for example—we can really trust that Johanson's Ethiopian hominid finds were discovered as reported in the literature; also, how do we know that those same fossils are actually in the museum now? (p. 592). This line of argument leads into the plea that, if (for example) scientists believe Johanson's words, his reports, and assurances that the actual fossils are in the museum, then scientists ought to believe in reports of ape-people, since these scientific data are no more trustworthy than reports of ape-people. Said simply: "If you trust evidence from professionals, which we believe to be doubtful, then please trust our doubtful evidence."

Too often, accepted evidence (and mainstream theory) is called into question by claiming that scientists are dishonest. The idea is a venerable two-edged weapon, because if you accept this view of science and of dishonorable or clumsy scholars, then how can this book be trusted, either? If the evidence of Johanson's (or others') excavations can be so easily lost, switched, lied about, then how much more could the 19th century evidence be warped, the evidence on which this book relies so heavily? And how can we trust the authors, who attempt to use this evidently untrustworthy science-stuff?

Faulty View of the Scientific Process

One of the most striking themes of *Forbidden Archeology* is the notion that scientists are slaves to tradition, which slows down or stops the adoption of new ideas. Yet, scientists have often turned over paradigms in the face of a social tradition or peer pressure that penalized them for it.

Galileo pushed his "wild" view of a heliocentric solar system until threatened by state-officiated torture. Modern cosmology is another example, a branch of knowledge under such motion revision that I suspect astronomers are giants among coffee drinkers. Similarly, paleoarchaeology is revised often in the face of new evidence (see Tuttle 1988 for a feel for the controversy). The "knowledge filter" would have to be impossibly acrobatic to span all this change.

Forbidden Archeology says that 19th-century scientists are to be trusted, however: they were open-minded about the nature of the artifacts they found in early strata, while today's scientists automatically explain away such finds (p. 90). (The authors don't notice the hard-hitting critics of eolith claims in the 19th and early 20th century [cf. Warren 1905])! The authors feel that the discovery of Java Man (one of the earliest pieces of direct evidence for human evolution) was a turning point that made scientists so narrow-minded. After the Java find, scientists became predisposed to the theory of evolution. I am not sure how this process works. If scientists ignore truth to be predisposed to tradition, then this paradigm would have favored the idea of the extreme age of modern human types because it is more easily worked into Biblical tradition than is evolution. (Perhaps this is why the Cardiff Giant hoax [see Feder 1990] worked so well on the public—they were predisposed to believe in a fossil "giant" because they were imbued in a Biblical tradition of antediluvian giants.) How could Java man change such a tradition by itself unless scientists eventually become disposed to consider new evidence? Dubois would have been given cement overshoes, otherwise! Scientists were indeed open-minded—eventually the theory of evolution was adopted despite all the penalties of challenging an entrenched social tradition of Biblical history.

A more specific complaint centers on the exploitation of uncertainty in science. Some people may perceive (perhaps envy) that scientists feel confident delivering "truth"—what else, from the people who enable moon landings and Tylenol? Of course, abundant mysteries exist to continually remind scientists of their limitations. However, an anti-science approach tries to turn this natural uncertainty into proof that mainstream science cannot be expected to get it right. For instance, *Forbidden Archeology* opens its case in the Introduction by citing that anthropologist Russell Tuttle saw a mystery in the fact that australopithecines existed around the same time as the human-like footprints at Laetoli. The citation ends there, and we don't know exactly what the mystery is that Tuttle sees (is it a mystery about two distinct species of hominid living simultaneously or how the curved big-toe of an ape-like creature could have left a modern-seeming footprint?). Let us be happy that Tuttle was mystified—this is proof that the curious and honest scientist in him is alive and kicking; but the authors have made a mystery in science into a crack in scientific process. Mysteries are everywhere, and when they disappear, so does science, because science is only a method for understand mysteries as reliably as possible.

It's Antievolution, but Is It Creationism?

I think so, as my title suggests. The authors state that they are followers of Vedic philosophy and aim to explain the history of the human race according to information preserved in Vedic texts and religion. They inform the reader that their religious affiliation should not matter if their ideas are solid (p. xxxvi), and I agree. Any person's work should be regarded on its merits. Religion and other cultural beliefs can bias an outlook, however, as the authors themselves would agree.

With this in mind, we can fairly ask if the authors are trying to force data into a mold shaped by Vedic religion. In his forthcoming review in *Geoarchaeology,* Feder mentions that the authors admit their religious affiliations but do not state their theoretical outlook. He writes, "Like fundamentalist Christian creationists, they avoid talking about the religious content of their perspective, so we can only guess at it." Feder tells us of the concept of the Vedic world cycle (manvantara) of 300,000,000 years in which the world with its humans is created and then destroyed in cycles. I think this concept is in keeping with *Forbidden Archeology*'s thesis of modern-type humans existing throughout Earth history. Feder says, "We all know what happens when we mix a literal interpretation of the Judeo-Christian myth with human paleontology; we get scientific creationism. It seems that we now know what happens when we mix a literal interpretation of the Hindu myth of creation with human paleontology; we get the anti-evolutionary Krishna creationism of *Forbidden Archeology*, where human beings do not evolve and where the fossil evidence for anatomically modern humans dates as far back as the beginning of the current manvantara." Actually, they push their artifactual evidence back 600,000,000-plus years to the Precambrian, where allegedly a grooved metallic sphere was found in South Africa and a metal vase in Massachusetts (p. 815).

I add that the Sanskrit epic, *Ramayana,* includes intelligent monkeys and bears who side with the Vedic gods against the demons. Has this narrative motif predisposed the authors to believe in modern-type humans living alongside intelligent animals (i.e., other hominids)? I can raise the hypothesis but know of no method for supporting it beyond the purely circumstantial evidence of the authors' stated religious affiliations, their broad theory of human existence alongside other hominids, and their belief in living apemen.

Conclusion

The authors posit a vast "knowledge filter" and often indict the honesty and biases of scientists. A fairer judgment is that scientists are human and have human potentials for failings; in my mind, this means that knowledge is accumulated at a slower rate than in a perfect world, but accumulate it does. At the most cynical point, I could posit that untruthful biases are uncovered because scientists eventually criticize loose thinking if only to further their careers. At their best, scientists—indeed, all scholars and artists—love truth and are driven to know how the world is made. Multiply these drives by the number of scholars living, and it all adds up to a normally self-corrective tradition (cf. Sagan 1979:82) that Cremo and Thompson reject with little basis.

Scientists have developed a rhetoric to report and, perhaps, to think about their studies as objectively as possible; however, this rhetoric can be used to further personal agendas even when the science is solid (see for example Halloran 1984:79)—the human and scientist are inseparable. But instead of using *Forbidden Archeology*, with its poorly supported claims, people interested in the problems associated with scientific reporting would do well to begin with professional work on the subject (for example, Coletta 1992; Fahnestock 1986; Gross 1990; Halloran 1984; Prelli 1989; Weimer 1977). Discussions of the history and nature of pseudoscience are available in Cole (1980), Feder (1990), Harrold and Eve (1987), and Williams (1991). Many of their characterizations will be recognized in *Forbidden Archeology*.

To close this discussion, I suggest that Cremo and Thompson have succumbed to a logical fallacy that can plague both professional and amateur or marginal archaeologists. Dincauze (1984:292) writes about the trap of possibilist arguments.

> The possibilist fallacy "consists in an attempt to demonstrate that a factual statement is true or false by establishing the possibility of its truth or falsity" (Fischer 1970:53)....The danger comes when possibilities are confused with demonstration, when "it could be" is followed by an unearned "therefore, it is." One cannot falsify possibilities, and most skeptics wisely eschew the effort. From the skeptics' refusal to face evidence....The only appropriate engaged response to a possibilist argument is a request for evidence, rather than assertion.

Dincauze reminds us that investigation must begin with possibilist ideals, with the following caution: "Possibilist arguments are only the first step toward knowledge; they indicate a problem domain where the method of multiple hypotheses might be applied." (1984:310). Possibilist ideals inherent in part of the scientific approach are, perhaps, one reason why some people seem to be excited about *Forbidden Archeology*.

The publisher included a notice of "advanced praise" along with the review copy. Some selections: Dr. Virginia Steen-McIntyre, a geologist, writes, "What an eye-opener! I didn't realize how many sites and how much data are out there that don't fit modern concepts of human evolution...[publisher's ellipsis] I predict the book will become an underground classic." *Fortean Times* said, "Cremo and Thompson have launched a startling attack on our whole picture of human origins and the way we have arrived at that picture: not only is the evidence impugned, but also the scientific method of handling it." Dr. Mikael Rothstein of the *Politiken Newspaper,* Denmark, remarks, "*Hidden History* (sic) is a detective novel as much as a scholarly *tour de force.* But the murderer is not the butler. Neither is the victim a rich old man with many heirs. The victim is Man himself, and the role of the assassin is played by numerous scientists." On the other hand, Richard Leakey replied to their request for a book blurb: "Your book is pure humbug and does not deserve to be taken seriously by anyone but a fool." In parentheses the publisher adds: "Representative of the scientific establishment's viewpoint."

This book, like other creationist texts that use similar techniques, is most useful as ethnographic data in studies of comparative religion, cult movements, popular movements, anti-science, fantastic archaeology, rhetoric, folklore—the book can be studied in any of these fields. With its emphasis on "secrets" and "hidden history" and "cover-up," the book participates in the popular genre of the conspiracy, akin to popular beliefs about the Kennedy assassination and crashed alien spaceships kept in guarded Air Force hangars. Sometimes the motifs of these modern legends are mixed with traditional motifs, as in the example of UFOs combined with traditional Irish fairy lore (Smith 1980:402), and a "scientific" explanation of why mermaids do not appear in Lake Michigan (Degh and Vazsonyi 1976:109, 112-113). These instances mark the relatively recent transition from agrarian to technological society, showing a need to react against mainstream science—or at least to dilute it—by adopting, re-inventing, or continuing traditional beliefs in the supernatural. The need for people to fantasize about such things is genuine; the behavior forms an aspect of Western industrialized culture (perhaps an aspect diagnostic of our particular pressure) well worth interdisciplinary study.

This folklore connection is suggested in the book's constant looking-backward toward a "golden age" of open-minded scholars, which reminds me of the function of myth, in which the past is formed in a mythological story tradition to legitimize the present. I am also reminded of the romance genre of literature: "Romance is the mythos of literature concerned primarily with an idealized world in which subtlety and complexity of characterization are not much favored and narrative interest tends to center on a search for some kind of golden age" (after Lee 1972:227). Much of *Forbidden Archeology* does read like a romance.

In any event, I have no evidence that people were or were not much more open-minded or golden a hundred years ago; but in the present I see *Forbidden Archeology* fantasizing about a past open-mindedness to legitimize a vast restructuring of our present understanding—without good evidence.

References

Cole, John R. 1980. Cult Archaeology and Unscientific Method and Theory. *Advances in Archaeological Method and Theory* 3:1-23.

______, and Laurie R. Godfrey. 1977. On "Some Paleolithic Tools from Northeast North America." *Current Anthropology* 18(3):541-543.

______, Robert E. Funk, Laurie R. Godfrey, and William Starna. 1978. On Criticisms of "Some Paleolithic Tools from Northeast North America": Rejoinder. *Current Anthropology* 19(3):665-669.

Coletta, W. John. 1992. The Ideologically Biases Use of Language in Scientific and Technical Writing. *Technical Communication Quarterly* 1(1):59-70, Winter.

Day, Michael H. 1986. *Guide to Fossil Man*. 4th edition. University of Chicago Press.

Degh, Linda, and Andrew Vazsonyi. 1976. Legend and Belief. *Folklore Genres.* Dan Ben-Amos, ed. Austin: University of Texas Press, pp. 93-123.

Dincauze, Dena F.. 1984. An Archaeo-Logical Evaluation of the Case for Pre-Clovis Occupations. *Advances in World Archaeology* 3:275-323. F. Wendorf and A.E. Close, eds. London: Academic Press.

Donaldson, E. Talbot, trans. 1967. Beowulf. *The Norton Anthology of English Literature,* Vol. 1. 3rd ed. New York: W.W. Norton & Co.

Fahnestock, Jeanne. 1986. Accommodating Science: The Rhetorical Life of Scientific Facts. *Written Communication* 3:275-296.

Feder, Kenneth L. 1990. *Frauds, Myths, and Mysteries: Science and Pseudoscience in Archaeology.* Mountainview, CA: Mayfield.

______.1994. Forthcoming Book Review for *Geoarchaeology.*

Fischer, D.H. 1970. *Historians' Fallacies*. NY: Harper and Row.

Funk, Robert. 1977. On "Some Paleolithic Tools from Northeast North America." *Current Anthropology* 18(3):543-545.

Gross, Alan G. 1990. *The Rhetoric of Science*. Cambridge: Harvard University Press.

Halloran, S. Michael. 1984. The Birth of Molecular Biology: An Essay in the Rhetorical Criticism of Scientific Discourse. *Rhetoric Review* 3(1):70-83.

Harrold, Francis B. 1990. Past Imperfect: Scientific Creationism and Prehistoric Archaeology. *Creation/Evolution* 27: 1-11.

Harrold, F.B., and R.A. Eve, eds. 1987. *Cult Archaeology and Creationism: Understanding Pseudo-scientific Beliefs about the Past.* Iowa City: University of Iowa Press.

Isaac, Glynn. 1984. The Archaeology of Human Origins: Studies in the Lower Pleistocene in East Africa, 1971-1981. *Advances in World Archaeology* 3:1-87. F. Wendorf and A.E. Close, eds. London: Academic Press.

Jones, Steve, R. Martin, and David Pilbeam, eds. 1992. *The Cambridge Encyclopedia of Human Evolution.* Cambridge: Cambridge University Press.

Leakey, Richard E., and Roger Lewin. 1978. *People of the Lake: Mankind and Its Beginnings.* New York: Doubleday.

Lee, Alvin A. 1972. *The Guest-Hall of Eden.* New Haven: Yale University Press.

Mania, Dietrich, and Emmanuel Vlcek. 1981. *Homo erectus* in Middle Europe: The Discovery from Bilzingsleben. *Homo erectus: Papers in Honor of Davidson Black.* B. Sigmon and J. Cybulski, eds. Toronto: University of Toronto Press, pp.133-152.

Prelli, L.J. 1989. *A Rhetoric of Science: Inventing Scientific Discourse.* Columbia: University of South Carolina Press.

Raemsch, Bruce. 1978. On Criticisms of "Some Paleolithic Tools from Northeast North America." *Current Anthropology* 19(1): 157-160.

Rightmire, G.P. 1981. *Homo erectus* at Olduvai Gorge, Tanzania. *Homo erectus: Papers in Honor of Davidson Black.* B. Sigmon and J. Cybulski, eds. Toronto: University of Toronto Press, pp. 189-192.

Sagan, Carl. 1979. *Broca's Brain: Reflections on the Romance of Science.* New York: Random House.

Smith, Linda-May. 1980. Aspects of Contemporary Ulster Fairy Tradition. *Folklore Studies in the Twentieth Century. Proceedings of the Centenary Conference of the Folklore Society.* Venetia J. Newall, ed. Rowman and Littlefield: D.S. Brewer, pp. 398-404.

Starna, William A. 1977. On "Some Paleolithic Tools from Northeast North America." *Current Anthropology* 18(3):545-546.

Tuttle, R.H. 1988. What's New in African Paleoanthropology? *Annual Review of Anthropology* 17:391-426.

Walker, Alan. 1981. The Koobi Fora Hominids and Their Bearing on the Genus *Homo.* *Homo erectus: Papers in honor of Davidson Black.* B. Sigmon and J. Cybulski, eds. Toronto: University of Toronto Press, pp. 193-216.

Warren, S.H. 1905. On the Origin of "Eolithic" Flints by Natural Causes, Especially the Foundering of Drifts. *Journal of the Royal Anthropological Institute* 35:337-364.

Weimar, Walter B. 1977. Science as a Rhetorical Transaction: Toward a Nonjustificational Conception of Rhetoric. *Philosophy and Rhetoric* 10:1-19.

Williams, Stephen. 1991 *Fantastic Archaeology: The Wild Side of North American Prehistory*. Philadelphia: University of Pennsylvania Press.

Related Correspondence

3.1.1.1. Letter to Dr. Wade Tarzia, August 22, 1996.

I have only this year seen your 1994 review of my book *Forbidden Archeology* in *Creation/Evolution* (14,1). A librarian friend of mine came across it in a search, and sent me a copy. I read it with some dismay; but considering the nature of the publication in which it appeared I was not surprised.

Creation/Evolution, as I am sure you will admit, is a journal with a pronounced ideological slant, its principal purpose being to discredit questioning of Darwinism, especially from the perspective of any of the world's great wisdom traditions. A more honest name for the journal would be *Anti-Creation/Pro-Evolution*. As you know, *Creation/Evolution* is the house organ of the National Center for Science Education. The name of this organization is itself misleading, giving the impression that it is connected with the government or that it is a branch of some large and truly national educational or scientific organization, with offices in every state. It is not. Instead, the NCSE is a small but vocal, and quite partisan, proevolution group, run out of a little office in Berkeley.

Taking all this into account, the rhetorical approach you have taken in your review is understandable but nevertheless regrettable. It is a textbook example of fundamentalist Darwinism. A fundamentalist Darwinian is someone committed to Darwinism more on ideological rather than scientific grounds. A fundamentalist Darwinian instinctively sees any scientific work that does not come to a Darwinian conclusion as "pseudoscience."

But even a fundamentalist Darwinian may sometimes hedge his absolutist accusations. Concerning *Forbidden Archeology*, you admit that you "hesitate in judging the book to be utterly worthless from a scientific standpoint." You say that "it *looks* like the real thing," and you find it "certainly interesting for its historical perspective." You acknowledge that "the mass of details with attached analysis would require book-length responses from specialized reviewers to confirm or critique" and that "a quick review of the book is not possible." Yet you go on to quickly review the book, and dismiss it as pseudoscience.

Blanket accusations of pseudoscience are themselves a hallmark of pseudoscience. Thus, your work, although it *looks* like a scientific review really is not. It would have been better if you had written your article without resorting to the overworked pseudoscience ploy and had gotten it published in a real academic journal, such as *American Anthropologist*, instead of an interest-group organ such as *Creation/Evolution.*

Forbidden Archeology is not pseudoscience. It is a controversial scientific work in support of a conclusion with which many scientists concerned with human origins, admittedly, disagree. One very good sign that *Forbidden Archeology* is, despite its controversial nature, a work of science is that it has been the subject of extensive scientific discourse within the scientific audiences to which it was directed. If *Forbidden Archeology* had drawn only the attention of interest-group publications such as *Creation/Evolution, Skeptic,* and other organs of fundamentalist Darwinism, your characterization of the book as being "not scientific" might seem more plausible.

Nonscientific, or pseudoscientific, works are seldom given serious reviews in academic journals. But *Forbidden Archeology* has drawn full reviews from *American Journal of Physical Anthropology, Geoarchaeology, L'Homme, L'Anthropologie, British Journal for the History of Science, Social Studies of Science,* and *Ecology, Ethology, and Evolution*, and notices in *Journal of Field Archeology, Antiquity,* and other publications. And while none of the reviewers have renounced the standard Darwinian views of human evolution, or endorsed *Forbidden Archeology* (quite the opposite), most have treated the book as one of genuine scholarship, even if inspired by Vedic writings and representing a knowledge tradition distinct from that of modern science.

For example, archeologist Tim Murray, in his review of *Forbidden Archeology* in *British Journal for the History of Science* (28: 377-379), says: "I have no doubt that there will be some who will read this book and profit from it. Certainly, it provides the historian of archaeology with a useful compendium of case studies in the history and sociology of scientific knowledge, which can be used to foster debate within archaeology about how to describe the epistemology of one's discipline."

Murray even goes so far as to admit that the Vedic outlook that inspired *Forbidden Archeology* might have some place in future developments in the discipline of archeology. He notes, "The 'dominant paradigm' has changed and is changing, and practitioners openly debate issues which go right to the conceptual core of the discipline. Whether the Vedas have any role to play in this is up to the individual scientists concerned." What I find encouraging about a statement like this is the recognition that traditional knowledge systems may have some role to play in reformulations of archeological paradigms.

This is not to suggest that Tim Murray endorses the book, or sympathizes with its purpose. He most definitely does not, and I am sure there is much in his review with which you would wholeheartedly agree (and with which I would wholeheartedly disagree). But the point is this: Tim Murray, a real representative of the discipline of archeology, writing in a real academic journal, has treated *Forbidden Archeology* in a more intellectually respectable fashion than you have in the house organ of the partisan NCSE. Tim Murray may be a committed Darwinist, but I do not regard him as a fundamentalist Darwinist. You could learn from Murray's review how to strongly disagree with a controversial scientific work without resorting to stale, ritual accusations of "pseudoscience."

Jo Wodak and David Oldroyd, in their review of *Forbidden Archeology* in *Social Studies of Science* (26: 192-213), ask (p. 207), "So has *Forbidden Archeology* made any contribution at all to the literature on palaeoanthropology?" They respond, "Our answer is a guarded 'yes,' for two reasons. First, while the authors go in for overkill in terms of swamping the reader with detail...much of the historical material they resurrect has not been scrutinized in such detail before. Second, while the form of their arguments is sometimes as much rhetorical as logical, Cremo and Thompson do raise a central problematic regarding the lack of certainty in scientific 'truth' claims." Wodak and Oldroyd go on to suggest that "those scientists who insist that evolution is a fact might be better advised to recast this as a 'highly probable theory.'"

In other words, these scholars take a stand against the kind of fundamentalist Darwinism represented by the NSCE, *Creation/Evolution*, and your review of *Forbidden Archeology*. Again, this is not to suggest that Wodak and Oldroyd endorse *Forbidden Archeology*, or that they are sympathetic to its purpose. But they do spend twenty pages seriously discussing the book in an academic journal (in a review article rather than the standard short book review), and they do acknowledge the book does make a scholarly contribution to the literature on paleoanthropology. This should demonstrate something to you. For your own growth as a scholar and human being, you might do some soul-searching about the nature of your commitment to Darwinism and your tactics in defending it. It would be helpful if you could elevate yourself above fundamentalist Darwinism and its creaky devices, such as characterizing all opposition as pseudoscience. As you yourself said (p. 14): "Serious treatment of new ideas, however much on the fringe they might be, is an appropriate venture in science."

In general, your whole approach is flawed, but let's get down to some specifics. In the very first paragraph of your review, you say, "Sometimes the book develops a dishonesty theory—evidence is said to be 'carefully edited' (p. 150) by scientists so that younger investigators do not see evidence that invalidates the theory of human evolution."

This is a dishonest caricature of what is actually said on page 150 of *Forbidden Archeology*. After documenting some cases of archeological discoveries no longer found in current texts, my coauthor and I said: "These discoveries are not well known, having been forgotten by science over the course of many decades or in many cases eliminated by a biased process of knowledge filtration. The result is that modern students of paleoanthropology are not in possession of the complete range of scientific evidence concerning human origins and antiquity. Rather most people, including professional scientists, are exposed to only a carefully edited selection of evidence supporting the currently accepted theory."

In the same chapter (p. 85), the "editing" process of knowledge filtration is described: "The hard facts of these discoveries, though disputed, were never conclusively invalidated. Instead, reports of these ancient stone implements were, as time passed, simply put aside and forgotten as different scenarios of human evolution came into vogue." In other words, many of the participants in the editing process may have been conventionally honest persons, who simply channeled their research efforts in certain directions rather than others, because of their theoretical commitments. The result, not necessarily deliberate on their part, was that certain categories of evidence were dropped from a discipline's field of active discourse.

You say (p. 13) that *Forbidden Archeology* "abandons the testing of simpler hypotheses before the more complex and sensationalistic ones." The example you give of this is FA's treatment of the Laetoli footprints. Here you are clearly wrong.

The Laetoli prints are anatomically modern in all details. My coauthor and I, *after considering other alternatives*, therefore suggested they could be taken, in the context of other evidence documented in *Forbidden Archeology*, as consistent with the presence of anatomically modern humans existing 3.6 million years ago in Africa.

You maintain it is easy to understand, from evolutionary theory, how an early hominid could have developed modern feet that were carried unchanged to the present. And then you say, "So far this seems to be the simplest explanation. *Forbidden Archeology* has not offered an alternative that falsifies this concept nor proposed a better one." There are several things wrong with your statements.

First, you indicate we did not even consider the "simpler" possibility you mentioned—that the footprints were made by a primitive hominid with advanced feet. That is not true. Either you were very careless in your reading of our work, or you deliberately misrepresented it to your readers. On page *xxiii* of the Introduction to *Forbidden Archeology*, I wrote: "In 1979, researchers at the Laetoli, Tanzania, site in East Africa discovered footprints in volcanic ash deposits over 3.6 million years old. Mary Leakey and others said the prints were indistinguishable from those of modern human beings. *To these scientists, this meant only that the human ancestors of 3.6 million years ago had remarkably modern feet.*" The same observation about a primitive hominid with modern feet is made on page 742, in the detailed discussion of the Laetoli case. It is obvious from both statements that these humanlike feet would be carried forward unchanged in the evolutionary process from the presumed Laetoli hominid to the modern human type. That possibility was considered and rejected by us, for very good reasons described in the book (and summarized below).

Concerning the idea that a primitive hominid with modern human feet could have made the Laetoli prints, you also say, "*Forbidden Archeology* has not offered an alternative that falsifies this concept." This statement shows a lack of understanding of the concept of falsification and remarkable inattention to what *Forbidden Archeology* says.

Concerning the whole issue of falsification, the concept was introduced by the German philosopher of science Karl Popper. He held that for a scientific theory to be truly useful, it should be falsifiable. By this he meant a scientific theory should be stated in such a way that *evidence* could conceivably contradict it. Interestingly enough, Popper wrote in his autobiography: "Darwinism is not a testable scientific theory but a metaphysical research programme."

In any case, it is not an alternative theory, or concept, that falsifies a certain theory but contrary *evidence.* Keep in mind that the theory in question must be based on factual evidence. While it is certainly *possible* that the Laetoli footprints were made by a primitive hominid with advanced feet, there is no direct physical evidence to support this theory. According to R. H. Tuttle (*Phil. Trans. Royal. Soc., B, 292*: 89-94), the foot bones of the hominids known to exist at that time do not fit the Laetoli prints.

T. White and G. Suwa (*Am. J. Phys. Anthrop.* 72: 485-514) made a highly speculative reconstruction of a *possible* hominid foot, using rescaled foot bones from three individuals from two different genera (*Homo habilis* and *Australopithecus afarensis*), and maintained that such a foot could have made the prints. But as of now, there has not been discovered an intact foot of any single known primitive hominid species that fits the print. So your whole line of reasoning is based on the *possibility* that such evidence might exist and be discovered.

I draw your attention to a statement you made in your review regarding "possibilist arguments." You cite Dena F. Dincauze as saying, "The danger comes when possibilities are confused with demonstration, when 'it could be' is followed by an unearned 'therefore it is.' One cannot falsify possibilities....The only appropriate engaged response to a possibilist argument is a request for evidence."

This certainly applies to the possibilist argument you give concerning the Laetoli footprints. You suggest that because *it could be possible*, according to evolutionary theory, that a primitive hominid could have had advanced feet, *therefore it is true* the advanced footprints found at Laetoli belonged to such a primitive hominid—thereby excluding other explanations. But as of now, there is not a primitive hominid foot that matches those prints.

After giving their speculative reconstruction of such a foot, White and Suwa were reduced to predicting that "the discovery of a complete foot skeleton at Hadar or Laetoli will conform in its basic proportions to the reconstruction," but this prediction remains to be fulfilled. Therefore, your conclusion that at Laetoli there must have been a primitive hominid with anatomically modern feet is *unearned* by the evidence. Hence, it is improper for you, by your own standards, to ask for falsification of your possibilist argument. The correct response for me, in this case, is to request *you* to give good fossil evidence for your proposal that there existed at Laetoli 3.6 million years ago a primitive hominid with feet that could have made the prints found there.

So your proposal is really not quite so simple. Actually, the simplest proposal is to match the Laetoli prints to whatever complete foot fossils from individual hominids now exist. When this is done, it is found that only a hominid of the anatomically modern human type possesses a foot that matches the Laetoli prints. This is not unreasonable. Your proposal is actually more "complex and sensationalistic," because it relies on foot bones that have yet to be discovered. I will add that in *Forbidden Archeology*, we document fossil evidence for anatomically modern humans existing at the time of the anatomically modern Laetoli prints (see FA Chapter 6).

Some of your complaints about *Forbidden Archeology* are so superficial that they hardly require a response, but just for the sake of thoroughness, I will answer. You complain that FA is too full of details; then in your discussion of a particular case, you ask for—guess what?—more details! And in your description of this case you make an inaccurate statement. You say (p. 14), "In a discussion of a purportedly incised bone (p. 38-40 [your citation]), discussion of the nature of the cuts and the context of the bone in the site are given short shrift while the discussion focuses on the fauna appearing in the stratum of the find." Actually, there is in the cited example *no discussion at all* of the fauna appearing in the stratum of the find. Such inaccuracy and misrepresentation are sadly typical of your review.

You complain (p. 15) that "quotations from 19th-/early 20th-century are copious—comprising, I would guess, at least 25 percent of the book." But what do you expect from a work that was conceived as a critical historical survey of evidence for human antiquity, from the time of Darwin to the present?

You complain (p. 15) that "a 1935 work of Weidenreich is cited as opposition to a 1985 work of Binford and Ho." The topic under discussion was the nature of markings on the original Beijing man fossils, discovered in the cave at Zhoukoudian in the 1930s. Binford and Ho, writing in 1985, 50 years afterward, suggested that animals brought the bones into the cave. Weidenreich thought the bones were brought into the caves by Beijing man, himself the killer and eater of his own kin. The most important evidence in this regard was the markings on the bones themselves. The bones were lost to science during the Second World War, although some photographs and casts remained. In this instance, it is entirely appropriate to contrast the views of one of the original discoverers of the fossils, who had a chance to study the actual bones closely and carefully over the course of several years, with views of later workers who did not have this opportunity.

Your pathological distrust of reports from earlier generations of scientists seems to be quite selective. You apparently distrust such reports if they go against your fundamentalist Darwinist beliefs. But what about Darwin himself? He published *Origin of Species* in 1859. It thus predates most of the older material cited in *Forbidden Archeology*. And in making his arguments, Darwin relied on material from the early nineteenth century and even the eighteenth century. Are you prepared to reject all of Darwin's arguments, on the grounds that he made use of old reports, older than those cited in *Forbidden Archeology*? Furthermore, today's standard archeology texts are full of reports from earlier generations of scientists, often praised for the accuracy of their observations.

Actually, according to your proposal ("knowledge seems to accumulate and research techniques seem to improve"), even today's reports are not to be trusted, because in the future even more knowledge will accumulate and research techniques will improve even more. That would mean that scientists operating in the year 2096 will automatically be dismissing reports from 1996, the ones you deem so entirely credible. Talk about folklore! The myth of eternal progress in science is certainly worth investigation by folklorists. According to you, the best way to arrive at the truth would be to project yourself into the most distant future, because knowledge is always accumulating and research techniques are always improving. Practical reality is somewhat more complicated than your naive folklore belief in the eternal cumulative progress of science. Observational techniques developed centuries ago persist, retaining their utility, and new research methods often introduce new sources of error.

In suggesting that new reports may be better than old ones, you are again falling into the possibilist fallacy that Dincauze warns against. Citing specific cases from *Forbidden Archeology*, you claim it is *possible* that reports from earlier scientists are not as trustworthy as those from today's scientists. While that may indeed be true, it is up to you to provide concrete evidence for that in these specific cases. You don't do it. For example, you suggest that modern shark biologists *might possibly* contradict the opinions of earlier experts about humanlike marks found on Pliocene shark teeth. This might be so. But I have found no such evidence of it. Have you? Give me some citations.

Finally, my coauthor and I are not unaware that modern researchers might have different, although not necessarily better, opinions than earlier ones. In a book that is primarily concerned with the geological antiquity of human remains and artifacts, the age of the strata in which the objects were found becomes one of our principal concerns. In this regard, we noted (*FA*, p. 14): "In some cases, the geological periods assigned to certain strata in the nineteenth century have been revised by modern geologists....In general, whenever strata in a given locality have been identified, we have tried to look up the periods assigned to them in current geological literature. We have then given dates to these strata on the basis of the modern period assignments." If you go through the book, you will see that we have followed this practice very systematically.

And the same is true in other areas of concern. We acknowledge the possibility that there might be variation in research and reporting over time, but when we make statements about such variation we give supporting evidence. You do not. Falling into the possibilist fallacy, you simply say it is possible the later reporting is not only different but better, and conclude the older reporting is untrustworthy, without giving any real evidence why this is so. To make matters worse, you do not even cite any later contrary reports. You simply invite your readers to believe that such reports exist, that they are different, and must be better. That is asking quite a bit.

You are highly critical of the concept of living ape-men, suggesting that the simplest explanation is that "Yeti, Sasquatch, etc., are manifestations of folklore." You ignore the fact that such evidence emanates not only from folk reports. John R. Napier, one of England's most famous and influential physical anthropologists, wrote an entire book about this topic (*Bigfoot: The Yeti and Sasquatch in Myth and Reality*, 1973), in which, after considering all available physical and observational evidence, he stated: "One is forced to conclude that a man-like life-form of gigantic proportions is living at the present time in the wild areas of the north-western United States and British Columbia." A similar conclusion was reached by physical anthropologist Grover S. Krantz of the University of Washington. *Forbidden Archeology*'s chapter on this topic contains many more reports by scientists who have studied the physical evidence for Bigfoot and reached conclusions different from yours. You may happen to disagree with them, but I put more confidence in their reporting and judgment than I do in yours. I note that you did not give any hint of their reports or conclusions in your review.

You do, however, manage to twist some of the statements my coauthor and I made about Bigfoot in the book. For example, you give the false impression that we subscribe to the view that because wildmen appear in folklore, therefore they must exist. What we actually say on the page you cite is the following. In the 1950s, Y. I. Merezhinski, a senior lecturer on anthropology at Kiev University, reported seeing a live wildman (called the Almas) at a distance of a few yards in the mountains of Azerbaijan. The report is found in anthropologist Myra Shackley's book *Wildmen: Yeti, Sasquatch, and the Neanderthal Enigma* (1983, p. 110). My coauthor and I then stated, "It is reports like this that tend to dispel the charge that the Almas is a creature that exists only in folklore. *And as far as folklore is concerned, accounts of the Almas and other wildmen are not necessarily a sign that the Almas is imaginary*." We then cited Dmitri Bayanov of the Darwin Museum of Moscow, who pointed out that real animals like wolves and bears also appear in folklore. Bayanov then offered his own opinion that if wildmen did not appear in folklore, this might cause us to doubt their existence. That is, however, his own opinion. The only thing that my coauthor and I said is that accounts of the Almas and other wildmen in folklore are *not necessarily a sign that the Almas is imaginary*. I am sure you would agree with that.

In common with some other reviewers, you seem highly upset that some cases of extremely anomalous evidence, not as well documented as other cases in *Forbidden Archeology*, were included in the book. You cite as an example an account of a Triassic shoe print. You complain (p. 18), "*Any* supporting evidence seems acceptable to the authors."

You neglect to tell your readers that this case and others like it were segregated in an appendix to the book, with the following introduction (*FA*, p. 795): "The reports of this extraordinary evidence emanate, with some exceptions, from nonscientific sources. And often the artifacts themselves, not having been preserved in standard natural history museums, are impossible to locate. We ourselves are not sure how much importance should be given to this highly anomalous evidence. But we include it for the sake of completeness and to encourage further study."

Furthermore, you apparently fail to grasp the nature of the general argument presented in *Forbidden Archeology*. The book is not simply a catalog of anomalous evidence, each piece of which is meant to be taken as sufficient reason to reject the current evolutionary account of human origins. The book is instead an epistemological critique of the scientific treatment of evidence related to human origins and antiquity. It is the entire pattern of selective reporting of anomalous evidence that is significant.

If the earth holds, as *Forbidden Archeology* proposes, some evidence for extreme human antiquity, one would expect that this evidence would be found over the years, and that it would be reported by different categories of researchers. Given the tendency of theoretical preconceptions to influence the treatment of evidence, a concept accepted by almost all historians, philosophers and sociologists of science, one might predict that the quality of the reporting would vary, as follows:

(1) Scientifically trained observers in the past, whose theoretical preconceptions did not prevent them from seriously considering such evidence, would find it and report it in some detail. In *Forbidden Archeology* we show that many nineteenth-century scientists published detailed reports documenting evidence of an ape-man (or even human) presence as far back as the Miocene and Eocene (i.e. in the time period ranging from 5 to 55 million years ago). Most of these scientists were evolutionists, but until the discovery of the Java ape-man in the 1890s (in a geological context less than a million years old), there was not a very firm time line for the human evolutionary process. So before the 1890s, these scientists were prepared to encounter evidence for primitive tool-using hominids in the Miocene and earlier. It is not, as you suggest, that we consider these early researchers to have been more enlightened and honest than later ones. They were simply operating according to different theoretical frameworks, which were also evolutionary.

(2) Scientists today, with theoretical preconceptions that prevent them from seriously considering such evidence, would find it and perhaps report it without considering its radical implications for the current paradigm. The Laetoli footprints provide an example of this. Scientists discovered footprints consistent with the presence of anatomically modern humans living 3.6 million years ago but, because of theoretical preconceptions, could not even entertain this possibility.

(3) In rare instances, "maverick" scientists today might find anomalous evidence and report it in detail, giving attention to how it contradicts the dominant paradigm. This reporting, one might predict, would not attain wide circulation and the reporting scientists might be subjected to inhibiting social pressures. The case of Virginia Steen-McIntyre, who reported anomalously old dates for the Hueyatlaco site in Mexico and suffered professionally because of it, provides an example (FA, pp. 354-366).

(4) Nonprofessionals might also encounter such evidence, and give reports, published in nonscientific literature, that are somewhat incom-

plete. We might also expect these reports to include the most extreme anomalies, ones that have always been beyond the limits of scientific acceptability, even in the past.

Reports from all four categories may be found in *Forbidden Archeology*, as part of a systematic epistemological critique of the handling of anomalous evidence in archeology and anthropology. Reports of extreme anomalies, found in nonscientific publications, have, in this context, some value in confirming the hypothesis that there is in fact some physical evidence for extreme human antiquity. In any case, by taking reports of this category out of their clearly stated context, you deliberately misled your readers about the quality of evidence reported in *Forbidden Archeology*.

Of course, reports of the most extremely anomalous evidence sometimes do come from scientific literature. As you point out, we included in *Forbidden Archeology* (p. 454) a report from an 1862 edition of a geology journal about a human skeleton found in coal 90 feet below the surface in Macoupin County, Illinois. The report, as we admitted, was all too brief, and lacked much crucial information. But even today it is common for scientific journals to include brief notes of new discoveries and events.

In any case, the task we set before us in writing *Forbidden Archeology* was to survey all categories of evidence for extreme human antiquity. Generally speaking, none of this evidence can be found in current literature, hardly even a hint of its existence. To account for this, we proposed a process of knowledge filtration. Given that the filtration process has worked so effectively, we felt it useful to give readers as complete a picture as possible of the evidence for extreme human antiquity. We especially wanted to counter the impression that evidence for extreme human antiquity, if reported at all, was not reported in scientific literature. This meant including not only the most exhaustively documented cases from the earlier scientific literature, but also cases from this scientific literature that are less well-documented, for the sake of completeness and as a guide to future research efforts.

In introducing the report of the Macoupin County skeleton, we noted: "One is tempted not to mention such finds as these because they seem unbelievable. But the result of such a policy would be that we discuss evidence only for things we already believe. And unless our current beliefs represent reality in total, this would not be a wise thing to do."

As brief as the report of the Macoupin County skeleton was, the details of the skeleton's position (90 feet below the ground, beneath a two-foot layer of slate rock) seem to rule out simple burial. Your response to this report was to write: "The impression is left that, if a tabloid reported Jimmy Hoffa's corpse was found in Triassic deposits, then the authors would no doubt perform rigorous research to date those deposits and then include the data in their next book." No, they would not. But this author will include your ridiculous statement as data of another kind in his next book.

Here is another completely false statement from your review: "The authors don't notice the hard-hitting critics of eolith claims in the 19th and early 20th century [cf. Warren 1905]! This leads me to believe that you did not read the chapter on the very crude stone implements called eoliths or were deliberately misrepresenting the book to your readers. The eolith chapter contains many lengthy discussions of the hardest critics of the eoliths, including Warren. References to Warren's negative views on eoliths may be found on pages 129-132, 134, 138-139, 142, 144, 154, 163, 166 and 174 of *Forbidden Archeology*. The book also contains a whole section on the views of H. Breuil and A.S. Barnes (*FA*, pp. 151-175), two of the most prominent critics of eoliths. And this does not exhaust the references to eolith critics.

Your smug observation that a student once created "eoliths" by rattling stones around in a garbage can is uninformed. Such superficial arguments were considered and dealt with by advocates of eoliths long ago. There is ample discussion of this issue in *Forbidden Archeology*, as in pages 175-178. Opponents claimed that chipped stones resembling eoliths (crude stone tools) were to be found in cement mills and other places where stones were thrown against each other in random fashion. But supporters of eoliths pointed that the resemblance was superficial. This view is supported by findings of modern researchers, such as Patterson, who say it is generally possible to distinguish random chipping on stone from deliberate human action (*FA*, p. 176).

Here is another absolutely incorrect statement from your review (p. 17): "The authors state that crude eoliths are not accepted as tools whereas allegedly similar looking artifacts (such as Oldowan and Acheulian industries) discovered by professional archeologists are accepted as artifacts (p. *xxvii*). But many Acheulian artifacts and quite a few Oldowan artifacts are quite distinctively styled—impossible to confuse with randomly broken eoliths."

First of all, the Acheulian industry, as noted in *Forbidden Archeology* (pp. 189, 354, 410), is an advanced stone tool industry associated with *Homo erectus*. It is characterized by forms easily recognizable as tools, whereas eoliths may appear to an untrained observer as simply broken pieces of stone. Nowhere in *Forbidden Archeology* do we say that eoliths are similar to Acheulian tools. Certainly, this is not stated on the page of the Introduction that you cited. What is stated there is this: "Significantly, early stone tools from Africa, such as those from the lower levels of Olduvai Gorge, appear identical to the rejected European eoliths." The reference is to the Oldowan industry, which is in fact similar to the European eoliths, as documented by illustrations and comparative study of reports (*FA*, pp. 190-193). For example, Mary Leakey (*FA*, p. 191) described among the Oldowan tools "various fragments of no particular form but generally angular, which bear a minimum of flaking and some evidence of utilisation."

Acheulian industries do occur in the *upper* levels of Olduvai Gorge. Similarly, some of the discoverers of European eoliths also found more advanced tools of the Acheulian type. They called them paleoliths, and they recognized them as belonging to a much later period than the eoliths. Therefore we said on page 190 of *Forbidden Archeology*, "The Acheulian type of Olduvai appears to correspond with the Paleolithic implements described by Harrison and Prestwich, while the Oldowan type, especially its unifacially flaked specimens, appears to roughly correspond with the flint implements described as eoliths." Nowhere do we say that eoliths correspond to Acheulian implements.

You claim: "The authors have also missed Dincauze's (1984) work which analyzes the flaws in theorizing about bones and artifacts from alleged early man sites." Dincauze is not the only author to prescribe stringent criteria for site evaluation. James B. Griffin made similar proposals, about the need for identifiable geological context, multiple dating methods, presence of animal fossils, human skeletal remains, and so on. As we noted in our discussion of Griffin in *Forbidden Archeology* (pp. 350-351): "The problem here is that practically none of the locations where major palaeoanthropological discoveries have been made would qualify as genuine sites....For example, most of the African discoveries of *Australopithecus, Homo habilis,* and *Homo erectus* have occurred not in 'clearly identifiable' geological contexts, but on the surface or in cave deposits, which are notoriously difficult to interpret geologically. Most of the Java *Homo erectus* finds also occurred on the surface, in poorly specified locations. At none of the places of these discoveries can one find the combination of factors Griffin deemed necessary for a proper site." So Dincauze and Griffin can make their requirements for site evaluation as stringent as they like. The problem comes in the application of such standards, and there does appear to be a double standard.

What, for example, do you make of the discoveries of stone tools in the Pliocene formations of Monte Hermoso in Argentina, reported by the respected paleontologist F. Ameghino (*FA*, pp. 292-294)? There the artifacts were found along with hearths, charcoal, and split animal bones—exactly the type of evidence you say is required. A human bone was also found at the same site, which has been dated, by more than one method, as Pliocene—about 3-4 million years old.

Another false statement in your review (p. 20): "The authors feel that the discovery of Java Man...was a turning point that made scientists so narrow-minded. After the Java find, scientists became predisposed to the theory of evolution." Throughout *Forbidden Archeology* we make a somewhat different point. Here is just one example, from the Introduction (*FA*, p. *xxxii*): "Historically the Java man discovery marks a turning point. Until then, there was no clear picture of human evolution to be upheld and defended. Therefore, a good number of scientists, *most of them evolutionists,* were actively considering a body of evidence indicating that anatomically modern humans existed in the Pliocene and earlier. With the discovery of Java man, now classified as *Homo erectus*, the long-awaited missing link turned up in the Middle Pleistocene. As the Java man find won acceptance among evolutionists, the body of evidence for a human presence in more ancient times gradually slid into disrepute. This evidence was not conclusively invalidated. Instead, at a certain point scientists stopped talking and writing about it."

So our point is quite different from what you state in your review. *Before and after* the Java man find, most scientists were disposed to evolutionary ideas. Furthermore, it is not that before Java man scientists were necessarily more open-minded in any absolute moral sense. They were just operating under theoretical concepts that did not place such firm boundaries on human antiquity.

Stripped of its fundamentalist Darwinian rhetoric, about 90 percent of your "review" is simply a collection of misstatements, relying on quotes taken out of context, questionable assumptions, and logical fallacies. In this letter, I have pointed out a good many, but by no means all of them.

You did, however, manage to get some things right.

You wrote that *Forbidden Archeology* should have mentioned the reports critical of the geological context of Raemsch's stone tool discoveries at the Timlin, New York, site. You are absolutely correct about this, and I thank you for pointing it out.

You wrote: "The Sanskrit epic, *Ramayana,* includes intelligent monkeys...who side with the Vedic gods against the demons. Has this narrative motif predisposed the authors to believe in modern-type humans living alongside intelligent animals (i.e., other hominids)?" The answer is yes. And there is quite a bit of physical evidence to support this idea, as amply documented in *Forbidden Archeology*.

Now let me ask you the following. Folklore contains accounts of dividing, shape-changing demons and gods. Has that disposed you to accept the primary myth of Darwinism, that of the little cell that divides and changes shape, resulting in many new forms? Keep in mind that the central "fact" of Darwinism—one species transforming into another—has never been directly witnessed, but is merely inferred from different categories of indirect evidence, subject to multiple interpretation and, according to Popper, is not really testable as a falsifiable scientific theory.

You wrote: "With its emphasis on 'secrets' and 'hidden history' and 'cover-up,' the book participates in the popular genre of the conspiracy, akin to popular beliefs about the Kennedy assassination and crashed alien spaceships kept in guarded Air Force hangars." You are partially right about that. In radio and television interviews, and other communications with the general public, I find considerable interest in this kind of thing, and there is some degree of truth to it. But what we are clearly talking about in *Forbidden Archeology* is a "knowledge filter," which has its academic equivalents in the postmodernist critiques of science. As I said in my Introduction *(FA,* p. *xxvi*), "When we speak of suppression of evidence, we are not referring to scientific conspirators carrying out a satanic plot to deceive the public. Instead, we are talking about an ongoing social process of knowledge filtration that appears quite innocuous but has a substantial cumulative effect. Certain categories of evidence simply disappear from view, in our opinion unjustifiably."

Of course, there are examples of deliberate fraud. The Piltdown case is the best example. The forgery of fossils, including accompanying faunal remains, was so exacting that scientists were fooled for forty years. Almost everyone who has investigated the fraud has concluded the perpetrator had to be a professional scientist. Concerning the motive for this fraud, J. S. Weiner, one of the scientists who exposed it, suggested (*FA*, p. 523): "There could have been a mad desire to assist the doctrine of human evolution by furnishing the requisite 'missing link.'" The most recent suspect is Dr. Hinton, then head of the zoology department at the British Museum. So it appears that evolutionists themselves have admitted that at least one of their number was so bold as to manufacture fake evidence in support of Darwinism.

You wrote: "If the evidence of Johanson's (or others') excavations can be so easily lost, switched, lied about, then how much more could the 19th century evidence be warped, the evidence on which this book relies so heavily?" In other words, both sets of reports, if subjected to intense negative scrutiny, can be distrusted. I have no objection to that line of reasoning. What I object to is the application of a double standard, whereby one set of reports is spared from intense negative scrutiny. If both sets are held to the same standard, and both fail, that is fine.

You wrote: "This book...is most useful as ethnographic data in studies of comparative religion, cult movements, popular movements, anti-science, fantastic archaeology, rhetoric—the book can be studied in any of these fields." Actually, I would be happy to cooperate with you in introducing *Forbidden Archeology* into those areas of scholarship. Even if you were personally representing the book to students and scholars in those areas, and doing your best to characterize it as you have in your pseudoreview, I am confident that at least some of those actually reading the book would begin to seriously question what they have been taught about human origins and antiquity. Anyway, let's do it. If you provide me with a list of book review editors for appropriate journals, as well as a list of possible course instructors, I will be happy to send them review copies of *Forbidden Archeology*, along with a cover letter from you, plus a brief note from me (the contents of which can be reviewed by you).

3.1.2. Colin Groves (1994) "Creationism: The Hindu View. A Review of *Forbidden Archeology*." *The Skeptic* (Australia), vol. 14, no. 3, pp. 43-45.

Dr. Colin Groves is a paleoanthropologist, and Reader in Biological Anthropology at the Australian National University. The Skeptic *is a publication of the Australian Skeptics. I found his surveys of paleoanthropological literature valuable. For example, he offered opinions of the australopithecines that challenge some dominant views. I therefore cited his work in* Forbidden Archeology. *His review manifests many of the features typical of writing in this kind of publication, starting with a large dose of sarcasm and ridicule.*

When a big square package, weighing over 3.5kg, arrived in my pigeonhole, a number of thoughts flitted across my mind. Which student hates me enough to send me a letter bomb? Will the postman sue me because of his hernia? After the package, when unwrapped, proved to contain a 914-page book, I felt like the Prince Regent on being presented by Edward Gibbon with a copy of his "Decline and Fall of the Roman Empire": "Another great damn thick square book! Always scribble, scribble, scribble, eh, Mr. Gibbon?" And then that final, heart rending, cry, "Why me?"

There is a letter from the senior author, Michael Cremo, accompanying the book. "Because your work, or that of your colleagues, is discussed in my new book *Forbidden Archeology*, I am sending you an advance copy." Can this be conspiracy theory as applied to archaeology by someone who feels that The Truth has been suppressed by The Establishment? It can. The letterhead is "Bhaktivedanta Institute, San Diego." Can this be a representative of that other fundamentalism, the Hindu variety? It can.

Remind ourselves what fundamentalist Hindus believe. Like fundamentalist Christians and Jews, they dismiss evolution. Unlike the latter, who believe the world has existed only six to ten thousand years, fundamentalist Hindus believe it has been going for billions and billions of years—far more than geology allows, in fact. And human beings, and indeed all living creatures, have been here all along. But in the event, it is going to make little difference; an apologia will consist of a recital of long-forgotten (long-suppressed, in their view) "evidence" of humans coeval with trilobites and dinosaurs, and arguments that supposed ape/human intermediates really aren't that at all.

But this time we get nearly a thousand pages! Gish, Bowden and Lubenow, the Christian creationists, can't raise even half of this between them. The difference is that Cremo and Thompson have read much, much more of the original literature than the other creationists, and their survey is correspondingly more complete. Yet I can't really say that their understanding is much greater, for all that; their tone of argument is as perverse, they are just as biased.

The fossil and archaeological evidence for human and cultural evolution is not all of consistently high quality. In the nineteenth century, human remains and artefacts were usually found by accident and by amateurs; they would be dug up, removed from context, and presented with a flourish to the nearest "expert". Controlled excavation was not a widely practised art; photography of a find in situ was an unusual occurrence. The finds' stratigraphy was often vague in the extreme; those re-examining their significance in later times had to rely on the fading memories of untrained workmen who had been enlisted by the finder.

This state of affairs improved as archaeology and palaeontology developed, and contextual information came to be recognised as crucial. Today, accidental discoveries are rarities; usually specimens turn up because someone has an idea where to look, given the prevailing geology and landscape, and an excavation is mounted with all kinds of specialists—geomorphologists, geochemists, taphonomists, above all photographers—riding along to ensure that everything about the site and its contents is recorded.

Cremo and Thompson seem not to understand this; they seem to want to accord equal value to all finds. One of many, many "out-of-context" human fossils which they discuss is the Foxhall jaw, a specimen of modern *Homo sapiens* discovered in 1855 and commonly ascribed at the time to the Late Pliocene, when (as we now believe) the human lineage was represented by just a bunch of near-apes called the australopithecines. The jaw was found by workmen, one of whom sold it to Dr. Collyer, a passing American physician, for the price of a glass of beer, and Collyer showed it to the luminaries of the day —Owen, Prestwich, Huxley, Busk— who expressed a variety of opinions, that it could or could not have come from the site and level claimed for it, and so that it could or could not be an example of "Pliocene Man". The jaw not long afterwards disappeared.

The authors quote the palaeoanthropologists Boule and Vallois in 1947: "It requires a total lack of critical sense to pay any heed to such a piece of evidence as this", and I can only agree; but, oddly, Cremo and Thompson disagree. Their opinion has nothing to do with the obvious fact that the whole case for the specimen's Pliocene origin was based on hearsay and supposition, and because the fossil has since disappeared, but because the stratigraphic provenances of other, nowadays widely accepted, fossils—"Java Man" and the Heidelberg jaw—were likewise based on flimsy evidence, and the original "Peking Man" fossils have likewise disappeared!

One has only to turn to their accounts of these fossils, and to read between the lines, to see why these other fossils are today taken seriously whereas Foxhall is not: other "Java Man" and Heidelberg-like fossils are known, whose stratigraphy has been exhaustively studied; excellent photographs, radiographs and casts survive of the lost "Peking Man" fossils, and others exactly like them have turned up since. But the same sort of non-evidence (Galley Hill, Clichy, Castenedolo, Calaveras, all *Homo sapiens* fossils briefly famous in their day because their finders thought they were Miocene, Pliocene or whatever) is taken seriously by the authors, who then completely miss the point when they imply, or claim boldly, that the evidence for the australopithecines, habilines and so on is also somehow flimsy.

There is an Appendix on the dating of fossils, mainly radiocarbon; Potassium-Argon dating is given the hatchet job in the main text (section 11.6.5). Devastating "exposure" of the alleged deficiencies of radiometric dating is obligatory in all creationist texts on fossils, and this one is no different. There they all are: the 160 million to 2.96 billion year dates for Hawaiian lava flows known to be less than 200 years old; the supposed "cover-up" of discrepant dates; the arguments over the correct date of the KBS Tuff at Koobi Fora, whether it was laid down 2.6, 2.4 or 1.88 million years ago. It is as if Cremo and Thompson think that an invention, as soon as it is made, either works or it doesn't; of course, the understanding of new methodologies—potassium-argon dating like any other—improves as its practitioners make mistakes (and, alas, are often embarrassed enough about their mistakes to keep quiet about them) and learn from them.

Potassium-argon dating and its now more generally used successor, the Argon/Argon method, are by now rather well understood. It is understood, for example, that mineral erupted from a volcano will release its store of radiogenic argon, resetting the "clock", only if it reaches a high enough temperature, and that the lava from deep-sea eruptions is chilled and does not usually reach this temperature; so that if you measure argon in an undersea lava flow (say, for the sake of argument, in Hawaii) you will be measuring what has been stored up over millions and millions of years, not just what has accumulated since the eruption.

It is understood, too, that tuffs are volcanic products brought down by water and deposited alongside other, much older sediments; so that if you simply pick up some grains from a tuff (say, for the sake of argument, at

Koobi Fora) you are very likely to get some very ancient ones along with your recent volcanic ejecta, and unless you clean the sample very carefully you will get anomalously high readings because of this mixture. This all seems very obvious nowadays, but the earlier practitioners of the method had to learn it the hard way. And in the main it is not suppressed: their errors are in the literature for all to see, and for creationists to point out with a delighted "see, it doesn't work!".

Now, palaeoanthropology is a specialty of mine, but archaeology is not, so I showed the book to a couple of colleagues whose specialty it is. Dr. Andrée Rosenfeld was not highly delighted, but offered some comments on the book's long, long, discussion of Eoliths. These are (no, were) supposed stone tools from extremely ancient deposits, believed in by many archaeologists in earlier generations but now universally discounted.

"The problem," Andrée explained, "lies in their selective emphasis and choice of language; have they not heard of semiotics? For example, on p 106 they quote an early objector to eoliths, Worthington Smith in 1892, and totally misunderstand its significance; eoliths can be extracted from any gravel from any period, whether with or without other artefacts, and with any range of patina—eoliths in fact only occur, as far as I am aware, in gravel or similar deposits." That is to say, in any deposit with lots of small stones in it, you are going to find some stones that by chance resemble crude artefacts! "They have not examined eoliths, but present a value laden discussion of the literature. The question is not 'could such fractures arise from hominid action' but could such fractures (or other marks) arise naturally—and if so, they cannot be taken as evidence for hominid presence."

Eoliths are not commonly featured in creationist texts—after all, here are Hindu not Judaeo-Christian creationists—but there are other bits and pieces in the book which I have met with before. On p. 811 we have the famous "Meister print", a supposedly shoe-like print, associated with trilobite fossils, in Cambrian deposits in Utah. The junior author, Thompson, examined the print in 1984 and (p. 812) saw "no obvious reason why it could not be accepted as genuine" despite the careful arguments to the contrary by a geologist, Stokes, quoted in two previous paragraphs.

Where I had met the Meister print before was in the first edition of a (Christian) creationist pamphlet, *Bone of Contention* by Sylvia Baker, MSc, and where I failed to meet it again was in the second edition of said pamphlet; presumably Ms. Baker learned of Stokes's analysis and quietly dropped it.

Another bit and piece and which I have met with before is a "carved shell from the Red Crag, England (Late Pliocene)", a period long before art was supposed to have existed, of course. This is a shell with what looks like two little round eyes, a simple triangular nose and a slit of a mouth carved into it; it resembles a Halloween pumpkin. Where I had met this one before was in an issue of *Creation Ex Nihilo* some four or five years ago, and I must say that when I saw it there I laughed out loud. Here it is again, just as chuckleworthy, on pp. 71-72. See above, under Eoliths.

Andrée Rosenfeld again: "What is curious is that an essentially religious organisation feels the need to justify themselves by recourse to science—but their discourse is scientistic, not scientific." In this, they are no different from any other creationists. Try to think ourselves into the mindset of a religious fundamentalist: "I believe in my sacred texts. I am aware that science does not support their veracity. My belief is not wrong—that is axiomatic—therefore science must be. I must look into this science business, to find out where it went wrong."

The fundamentalist convinces him/her/itself as supposed holes in the scientific fabric turn up, and wow! this can be used to convince others too! It's a kind of top-down learning experience; what is missing is what students get as they learn their science bottom-up: context. That, really, is why it is so difficult to actually open a dialogue with the creationist: why it is that scientists debating with creationists are effective mainly when they are pointing out their opponents' ignorance, stupidity or outright lies. Their opponent—let alone the audience—simply has no conception of context.

A book like this, simply because it is superficially scholarly and not outright trash like all the Christian creationist works I have read, might indeed make a useful deconstructionist exercise for an archaeology or palaeoanthropology class. So it's not without value. You could do worse, too, than place it in front of a Gishite with the admonition "Look here: these guys show that human physical and cultural evolution doesn't work. Therefore it follows that the Hindu scriptures are true, doesn't it?".

Related Correspondence

3.1.2.1. Letter to Dr. Colin Groves, June 1, 1997.

What I most missed in your "review" of *Forbidden Archeology* in *The Skeptic* (which only recently came to my attention) was your opinion on how your own work was cited in the book. That was, after all, the reason I sent a copy to you.

In your review you noted, somewhat sarcastically, that you expected to find in *Forbidden Archeology* "arguments that supposed ape/human intermediates really aren't that at all." The book does contain such arguments. In fact, some of them are based on statements from your book *A Theory of Human and Primate Evolution* (1989).

In many scientific and popular accounts, *Australopithecus afarensis* (represented most famously by "Lucy") is depicted as an intermediate between early apes and modern humans. The fossil evidence for this supposed intermediate was gathered mostly at the Hadar site in Ethiopia. You, along with others, have questioned the integrity of the fossil evidence for *Australopithecus afarensis,* suggesting that this supposed intermediate species, as defined by its discoverers, was mistakenly manufactured from the bones of several species.

Forbidden Archeology (p. 741), after reviewing the case for *Australopithecus afarensis*, states:

> This brief review does not, however, exhaust the various opinions about the phylogenetic status of *A. afarensis.* "For Ferguson (1983, 1984) the Hadar sample contains three different taxa: *Sivapithecus* sp., *Australopithecus africanus,* and *Homo antiquus* (new species)," noted Groves (1989, p. 262). Groves himself (1989, p. 263), in his comprehensive taxonomic survey of the hominids, said: "Certainly the post-cranial data are absolutely clear, and split the Hadar sample into two divisions." Groves (1989, p. 263) classified one Hadar group as early *Homo* and the other as an unnamed new hominid genus. Under the species designation *Australopithecus afarensis*, he kept only the Laetoli jaws. So Groves, like Ferguson, found three species instead of one in the *A. afarensis* fossils of Johanson and White.

This serious uncertainty about attributions of early hominid fossils tends to lessen our confidence that in particular cases (of *Australopithecus this* or *Homo that*) we are dealing with real species, let alone intermediate species, ancestral to modern humans.

Orthodox evolutionists nevertheless consider the genus *Australopithecus* to be a clear evolutionary intermediate between the early apes and modern humans. They generally say *Australopithecus* walked exclusively upright on two legs in modern human fashion, thus distinguishing it from the early apes. But your own work seems to suggest otherwise.

Forbidden Archeology (p. 727) states: "An author of a recent survey (Groves 1989) takes the side of Stern, Susman, Tardieu, Oxnard, and others who have argued for a substantial component of arboreality in *Australopithecus afarensis* and the australopithecines generally. Groves (1989, p. 310) said that in the australopithecines 'bipedal locomotion was only part of a pattern which also incorporated sophisticated climbing ability.'"

Forbidden Archeology (pp. 721-722) also states: "Groves (1989, p. 307), after reviewing studies by Oxnard and others, agreed that 'the locomotor system of *Australopithecus africanus* was unique—not simply an intermediate stage between us and apes.' He found the same to be true of other species of *Australopithecus.* This fact, along with other aspects of the hominid fossil record, caused him to suggest that 'non-Darwinian' principles were required to explain an evolutionary progression from *Australopithecus* to modern human beings (Groves 1989, p. 316)."

So it would seem that your work also presents "arguments that supposed ape/human intermediates really aren't that at all." I anticipate that you will say that *Australopithecus* really is an intermediate, although not in a way that standard Darwinian evolutionary principles can explain. That means, however, we are in substantial agreement on one point: standard Darwinian principles do not account very clearly for an evolutionary link between *Australopithecus* and modern humans. Of course, you still favor some kind of evolutionary link, and in your 1989 book posited some unspecified non-Darwinian principles to account for it. I favor some kind of nonevolutionary relationship between the two, and in a forthcoming book will offer some Vedic principles to account for it.

You said in your review, "The fossil and archaeological evidence for human and cultural evolution is not all of consistently high quality." This is quite true, and the same point is consistently made throughout *Forbidden Archeology*. Your statement that "Cremo and Thompson seem not to understand this; they seem to want to accord equal value to all finds" is therefore grossly incorrect. In fact, given the many explicit statements about variation in the quality of evidence found throughout *Forbidden Archeology*, I would have to say your statement is not only incorrect but deliberately misleading. Deliberate misrepresentation, sadly, is a common feature of writing such as yours and publications of the kind in which your review appeared.

In the first chapter of *Forbidden Archeology*, there is an entire section explaining the book's epistemological approach, which is based principally on evaluation of the quality of reports. In that section may be found the following statement: "A collection of reports dealing with certain discoveries can be evaluated on the basis of the thoroughness of the reported investigation and the logic and consistency of the arguments presented. One can consider whether or not various skeptical counterarguments to a given theory have been raised and answered. Since reported observations must always be taken on faith in some respect, one can also inquire into the qualifications of the observers. We propose that if two sets of reports appear to be equally reliable on the basis of these criteria, then they should be treated equally. Both sets might be accepted, both might be rejected, or both might be regarded as having an uncertain status."

The problem is not that the quality of evidence and reporting varies. That is granted. The problem is that application of standards for evaluating the quality of evidence and reports tends to vary unevenly. *Forbidden Archeology* argues not for equal acceptance of all evidence but for equal application of standards for evaluation of evidence. There is a big difference, but it appears to have escaped your notice.

Furthermore, it is not true, as you suggest, that all archeological and paleoanthropological evidence reported before recent times was uniformly poor. After all, modern textbooks are full of early discoveries that happen to be consistent with modern evolutionary accounts of human origins. As early as the nineteenth century, geologists, paleontologists, anthropologists, and archeologists had developed rather exacting techniques for determining the relative ages of geological formations. Generally speaking, these relative age determinations are still accepted, as can be seen from any comprehensive study of geology or paleontology.

In some of these early cases, objects were, as desired by you, photographed in situ. For example, *Forbidden Archeology* contains a detailed description of some artifacts found in a Pliocene formation at Miramar in Argentina. These were left in place by the original excavator to be examined and photographed (*F. A.*, pp. 329-324).

Furthermore, not all evidence for human origins found in current textbooks, including recently discovered evidence, meets the high standards that you set. None of the Java man discoveries, ranging from the original ones made by Dubois in the 1890s to those of the late twentieth century, were made in controlled excavations, photographed in situ, etc. The great majority were surface finds, made by farmers and paid collectors, who later brought them or sent them to scientists. This occurred not only in the nineteenth century, but as recently as the 1970s. Also, many of the important discoveries of australopithecines, beginning with the very first one (Dart's "Taung baby"), were not made in controlled excavations.

You say that today "an excavation is mounted with all kinds of specialists—geomorphologists, geochemists, taphonomists, above all photographers—riding along to ensure that everything about the site and its contents is recorded," implying that under such conditions no evidence grossly contradicting orthodox views of human origins ever turns up. But *Forbidden Archeology* documents, among others, the case of Virginia Steen-McIntyre, a tephrachronologist (an expert in dating volcanic ash), who in the 1970s arrived at dates of about 300,000 years for an archeological site at Hueyatlaco, Mexico—far too old for humans to be in North America, or anywhere else in the world.

Regarding the Foxhall jaw, I stand by my judgment that it is worthy of serious consideration as evidence for a human presence in the Pliocene. Whatever else might be said about the fossil, testimony indicates it was taken from the coprolite beds at the 16-foot level in the Red Crag formation at Foxhall and the physical condition of the bone (heavily impregnated with iron) supported this judgment. The Red Crag formation has been extensively studied and is well dated. The fossil was examined by leading authorities of the day, who did, as you say, have different opinions about it. Although the fossil is now missing, a good contemporary drawing of it exists. From the drawing and the testimony of the experts who examined the actual specimen, it is apparent that it is anatomically modern.

That the fossil was originally found by a workman should not automatically disqualify it from consideration. Many fossils accepted by orthodox evolutionists were also found by workmen, such as the Heidelberg jaw and almost all of the *Homo erectus* specimens from Java (not just the original Java man). If the provenance of the Foxhall jaw is suspect, then the provenance of these well-known finds still present in current textbooks is also suspect.

The brings us back to the working method employed in *Forbidden Archeology*. If two sets of evidence have a similar key attribute, they must be treated similarly—on the basis of this attribute both may be accepted, both may be rejected, or both may be considered doubtful. If we propose that the provenance of any fossil not found by professional scientists in a controlled excavation is suspect, then we must apply that judgment equally. If you want to say that the provenance of the Foxhall jaw is based on hearsay and supposition, that judgment applies equally to the Heidelberg jaw and many other fossils currently accepted by orthodox evolutionists. You, however, insist upon applying a double standard.

That the original fossil is now missing should not completely disqualify it either. The original Beijing man (*Homo erectus*) fossils are also missing. You point out that other fossils and signs of Beijing man's presence were later found at the Zhoukoudian site. And other fossils of *Homo erectus* fossils have been found elsewhere in the world. This does, as you say, increase our confidence in the original finds. But the same thing is true of the Foxhall jaw. Several decades after the discovery of the jaw, stone tools, debris from stone tool manufacturing, and signs of fire were discovered in the 16-foot level at Foxhall, thus confirming a human presence. And *Forbidden Archeology* documents many other well-dated discoveries of fossils and artifacts indicative of a human presence in the same Pliocene period.

Taken in this overall context, the Foxhall jaw is worthy of consideration as evidence for extreme human antiquity. However, I have no objection to anyone rejecting it, as long as the standards for such rejection are applied equally to evidence now considered acceptable by orthodox evolutionists.

Regarding your rejection of *Forbidden Archeology*'s suggestion that the evidence for the australopithecines and habilenes is flimsy, you yourself, as shown above, have contributed to the suggestion of flimsiness. For example, you yourself have expressed the opinion that Don Johanson's version of *Australopithecus afarensis* was mistakenly put together from fossils of three separate hominid species. Furthermore, physical anthropologists such as T. J. Robinson and Bernard Wood say the same thing about the traditional picture of *Homo habilis*. It is not a real species; it was mistakenly put together from bones of two or more species (*F. A.*, p. 710). So it appears the evidence for the habilenes is also somewhat flimsy.

You say, "Today accidental discoveries are rarities; usually specimens turn up because someone has an idea where to look, given the prevailing geology and landscape." A true skeptic should be able to quickly discern the problem with this procedure. Researchers are looking only where they expect to find certain kinds of fossils, given not only "the prevailing geology and landscape" but also the prevailing ideas about human evolution. And thus they find and report only what they expect to find. Even so, a surprising amount of anomalous evidence has turned up. One very good example is the anatomically modern human footprints, in volcanic ash 3.7 million years old, that turned up at the Laetoli site in Tanzania in 1979.

Concerning radiometric dating, you admit all the problems we described in *Forbidden Archeology*, so there is not too much to add. Apparently, you think all the mistakes and deficiencies occurred in the past, and today everything is just fine. But the history of these methods does not inspire such general confidence.

Your colleague Dr. Andrée Rosenfeld makes some very naive statements about eoliths, probably gathered not from a study of eoliths or the original literature on eoliths, but from received traditional knowledge in his discipline.

Rosenfeld, with you in apparent agreement, claims the authors of *Forbidden Archeology* totally misunderstood a statement made by Worthington Smith in 1902 (*F. A.*, p. 106). On the contrary, Rosenfeld and you have misunderstood the statement, probably because neither of you carefully read section 3.2.5 of *Forbidden Archeology*, which establishes the context for the statement. For someone who talks so much about context, you pay remarkably little attention to it.

The topic of the statement in question is the relative antiquity of eoliths (very crude stone tools) and more advanced implements found in the same deposits. Sir John Prestwich, a famous English geologist, and Benjamin Harrison, a local collector, had maintained that the crude eoliths were older than the more advanced stone tools found along with them in certain deposits.

Addressing this issue, Worthington Smith said, "I don't attach much importance myself to the dubious and disputed forms [the eoliths], because such forms occur with genuine implements in all paleolithic gravels. The very rudest forms can never mean anything, unless such forms are exclusive, and pertain only to certain deposits." Rosenfeld wants to interpret this statement to mean (in your words) "in any deposit with lots of small stones in it, you are going to find some stones that by chance resemble crude artefacts."

That is not at all what Worthington Smith was saying in this case, although he might have said it elsewhere. The context for this particular statement is as follows. Prestwich and Harrison found two kinds of tools in very old, high-level river gravels—some of advanced type and some of a crude type. In river valleys, the gravels found at terraces high on the valley slopes are older than the gravels on the lower slopes, in some cases considerably older. This is because the gravels at the high levels were deposited when the river first started cutting its valley, whereas the gravels at the lower levels of the valley were deposited more recently, as the river continued cutting the valley deeper and deeper. What is one to make of tools found in the high-level gravels? Prestwich and Harrison considered that some very sophisticated tools (neoliths and paleoliths), showing little sign of weathering and wear, might have been dropped in the high-level gravels during the past few thousand years, making them far younger than the gravels in which they were found. Found in the same gravels were cruder tools (eoliths) with signs of considerable weathering and wear, suggesting they were far older than the more advanced tools, perhaps as old as the early Pleistocene or the Pliocene. These eoliths, according to Prestwich and Harrison, would have been made by primitive humans or human ancestors, whereas the neoliths would have been made more recently by anatomically modern humans with a fairly high level of culture.

But Prestwich himself posed this question about the eoliths: "Could these implements, like the neolithic implements which occur on the same ground, have been dropped on the surface where they are now found, at some later date?" In other words, the tools might all have been of the same fairly recent age.

It is against this background that Worthington Smith's statement must be understood. What he is saying is this: "Even if we grant that the much disputed eoliths are genuine crude tools, that does not establish that they are older than the more advanced paleoliths that accompany them in the high level gravels. In all stone tool industries, from gravels of whatever age, we find that more sophisticated tools are accompanied by less sophisticated ones because people make different kinds of tools for different purposes. So if you want to make a case for the greater antiquity of the eoliths, it would be best if you could find particular gravel deposits of great age where they occur by themselves (i.e., "exclusively"), with no other more advanced tools."

If you go carefully through the discussion of eoliths in *Forbidden Archeology*, you will find there are cases that satisfy Worthington Smith's requirement. And you will also find arguments by Prestwich showing that his requirement does not apply in the case of the high level specimens. As mentioned above, Prestwich found he could make a distinction between the advanced tools and the eoliths, found in the same deposits, in terms of the latter's signs of greater wear and weathering, indicating greater antiquity.

The suggestion offered by you and your colleague that eoliths were in all cases just stones that looked like tools displays a truly remarkable degree of ignorance about the issue. You smugly proclaim that "in any deposit with lots of small stones in it, you are going to find some stones that by chance resemble crude artefacts!"—as if this rather obvious objection occurred neither to the original discoverers of eoliths nor to the authors of *Forbidden Archeology*.

Over the years, from the nineteenth century to the present, a great deal of careful research has been carried out on how to distinguish natural breakage from human action on stone.

For example, *Forbidden Archeology* (p. 97) says: "A modern expert in lithic technology, Leland W. Patterson, also believes it is possible to distinguish even very crude intentional work from natural action. Considering 'a typical example of a flake that has damage to its edge as a result of natural causes in a seasonally active stream bed,' Patterson (*J. Field Arch. 10*: 303) stated: 'Fractures occur randomly in a bifacial manner. The facets are short, uneven, and steeply transverse across the flake edge. It would be difficult to visualize how random applications of force could create uniform, unidirectional retouch along a significant length of a flake edge. Fortuitous, unifacial damage to an edge generally as no uniform pattern of retouch.' Unifacial tools, those with regular chipping confined to one side of a surface, formed a large part of the Eolithic assemblages gathered by Harrison and others."

As an advocate of "bottom up" research, it would be better for you to study either the eoliths themselves or the original reporting about them before deciding the entire issue on a "top down" basis, by blindly accepting tattered old disciplinary proverbs, reported to you by your colleague.

Speaking of tattered old disciplinary proverbs, which signify a stubborn unwillingness to consider the real technicalities of the eolith question, here is another one from Rosenfeld, who probably heard it from his professors, who probably heard it from theirs: "Eoliths in fact only occur, as far as I am aware, in gravel or similar deposits."

Rosenfeld is suggesting that eoliths made of flint, for example, occur only in deposits of flint gravel, from which one picks out a few flint specimens that appear to be tools. That eoliths occurred in gravel deposits may be true in some cases, but not all. About one set of eoliths, Geologists J. Hamal-Nandrin and Charles Fraipont reported: "At Thorington Hall, the rarity of stones and their dispersal does not permit us to suppose that the flints have been naturally retouched by impact or pressure. One can observe that in the level where the flints are found one does not find any worn and fractured flints other than the ones appearing to be the result of intentional work" (*F. A.*, p. 134). They also say, "At Thorington Hall the detritus bed lacks many rocks. It contains coprolites, phosphate nodules, and only some small flint pebbles." Thus this does not appear to be a big flint gravel deposit; and the only fractured flints are the ones identified by contemporary experts (not the original discoverer) as the result of intentional work.

I could give other examples. At Foxhall, the worked flints are found in layers of black sediment and are quite rare (*F. A.*, p. 142-43). There are horizontally stratified clean sand deposits above and below these layers. J. M Coles, a modern English archeologist, argued that the relative rarity of flints in the black layers, as well as the fact that it is hard to account for their presence by natural processes, indicates that they arrived in these layers by human agencies.

When one objectively studies the eolith question one quickly learns that the simplistic slogans one has heard for ages , and accepted as gospel ("eoliths occur only in gravel deposits"), do not always apply. In any case, even if eoliths did occur only in gravel deposits, it would still be possible to make reasoned judgments as to whether they were of human manufacture or the result of natural breakage.

About the authors of *Forbidden Archeology*, Rosenfeld asserts: "They have not examined eoliths, but present a value laden discussion of the literature."

First of all, the authors have examined eoliths. For example, my coauthor and I visited Dr. Ruth D. Simpson, an expert in lithic technology, at the San Bernardino County Museum of Natural History, and there examined eolithlike implements from the Calico, California, site. With Simpson and her coworkers we carefully looked at specimens of crude work on stone, and compared the actual specimens with illustrations and photos of eoliths from European eolith sites. We also discussed with them some of the technical literature on distinguishing natural breakage from intentional human work on stone. Also, during the time I was writing *Forbidden Archeology*, I spent many hours studying natural breakage of stone on the beaches and in the canyon stream beds of San Diego County. I conducted my own stone flaking experiments and also looked at stone tools in local museum collections. My coauthor visited Herbert L. Minshall, a local collector and expert on crude stone tools, and examined his collection of eolithlike implements.

Concerning the value-laden nature of our discussion of the literature, all scientific writing is laden with value, for if nothing else scientific writing is, or should be, concerned with the value of true knowledge and honest reporting. I take it that Rosenfeld meant to say that our discussion of the literature was one-sided. But we presented in detail the views of those who were opposed to accepting eoliths as actual human artifacts. Indeed, after a thorough discussion of the pros and cons of the eolith question, *Forbidden Archeology* (p. 188) says: "Of course, even after having heard all of the arguments for eoliths being of human manufacture, arguments which will prove convincing to many, some might still legitimately maintain a degree of doubt. Could such a person, it might be asked, be forgiven for not accepting the eoliths? The answer to that question is a qualified yes. The qualification is that one should then reject other stone tool industries of a similar nature. This would mean the rejection of large amounts of currently accepted lithic evidence, including the Oldowan industries of East Africa and the crude stone tool industry of Zhoukoudian in China."

Overall, I would have to say that *Forbidden Archeology*'s discussion of the eolith controversy was fairer and more objective than your representation of that discussion to your readers. Anyone who actually takes the trouble to read the book will immediately see that the caricature you have presented in your review is so inaccurate as to suggest deliberate misrepresentation.

Rosenfeld also said: "The question is not 'could such fractures arise from hominid action' but could such fractures (or other marks) arise naturally—and if so, they cannot be taken as evidence for hominid presence." I believe the question is, "Did Rosenfeld read the eolith chapter in *Forbidden Archeology* or not?" I can only conclude that he did not, as there is abundant discussion of how to distinguish natural fracturing of stone from intentional human or hominid action. Or perhaps he did read the chapter, in which case the matter of deliberate misrepresentation again arises.

In any case, the answer that *Forbidden Archeology* gives to Rosenfeld's question, with extensive references to the technical literature on the subject, is that natural action cannot explain the fracturing on eoliths. For example, I've already provided in this letter Patterson's criteria for ruling out natural fracturing. Many of the eolith meet these criteria, and would thus appear to be intentionally manufactured.

Considering the Meister footprint, the geologist W. L. Stokes offered the following objections to its authenticity as a genuine human print: (1) It was just a single print instead of a set of prints in sequence, and he knew of no instance in which a scientific journal had accepted a solitary print as genuine; (2) there was no sign of squeezing aside of the matrix; (3) spalling, a natural fracturing of rock, offered a good explanation.

Forbidden Archeology (p. 812) answered these objections as follows: (1) we provided an example from *Scientific American* of a single footprint being accepted as genuine; (2) we noted that shoeprints of the kind found by Meister do not always leave obvious signs of squeezing aside of the matrix and suggested that it would be advisable to microscopically examine the grain structure of the shale for signs of displacement by pressure from above; (3) we noted that spalling is a weathering phenomenon that normally occurs on the exposed surface of rocks, whereas the Meister print was from the freshly split interior of a block of shale.

Forbidden Archeology's final statement on the Meister print (pp. 812-813) was not one of uncritical acceptance: "The Meister print as evidence for a human presence in the distant past is ambiguous. Some scientists have dismissed the print after only cursory examination. Others have rejected it sight unseen, simply because its Cambrian age puts it outside the realm of what might be expected according to evolutionary theory. We suggest, however, that the resources of empirical investigation have not yet been exhausted and that the Meister print is worthy of further research."

Since you are concerned that variation of the quality of reports be recognized, it would have been appropriate for you to note in your review that the Meister footprint appeared, with similar extremely anomalous cases, in an appendix to the main text of *Forbidden Archeology*, with this introduction: "The reports of this anomalous evidence emanate, with some exceptions, from nonscientific sources....We ourselves are not sure how much importance should be given to this highly anomalous evidence. But we include it for the sake of completeness and to encourage further study."

Regarding the "carved shell from the Red Crag, England (Late Pliocene)," you give no reasons for rejecting it other than that an article about it once appeared in a Christian creationist journal and that it personally struck you as amusing. *Forbidden Archeology* (pp. 71-72) says about the shell: "In a report to the British Association for the Advancement of Science in 1881, H. Stopes, F.G.S. (Fellow of the Geological Society), described a shell, the surface of which bore a carving of a crude but unmistakably human face. The carved shell was found in the stratified deposits of the Red Crag." The Red Crag is at least 2.0-2.5 million years old. Later reporting in *The Geological Magazine* gave other reasons for the carved shell's authenticity.

Regarding your references to Hindu beliefs about the age of the earth, yes, "it has been going on for billions and billions of years." But when this is properly understood, it is not necessarily contradicted by modern geology. Hindu cosmology speaks of cycles of creation and destruction. The basic unit of these cycles, which are of vast duration, is the day of Brahma. During the day of Brahma, which lasts about 4.3 billion years, the earth is manifest. During the night of Brahma, which also lasts about 4.3 billion years, the earth is unmanifest. According to the Hindu cosmological calendar, we are about 2 billion years into the present day of Brahma. That would be the age of the current earth. Because life has, according to Vedic accounts, been here since the beginning of the current manifestation of earth, we should therefore expect the fossil record to extend back about 2 billion years. And according to modern paleontology it apparently does go back about 2-3 billion years, with the age of the earth itself being about 4 billion years. In short, there is a rough equivalence between Hindu cosmology and modern geology regarding the age of the earth and the extent of the fossil record.

Yes, *Forbidden Archeology* is an exercise in apologetics, with the Vedic texts of India the object of the apologia. The Vedic texts do contain accounts of human origins and antiquity that differ from those of orthodox science. It is not so much that believers need physical evidence from science to uphold their faith in what the texts say. Believers, some of them scientists and scholars, instead resent the constant stream of statements from sources within the scientific world that the accounts of the ancient texts are absolutely incorrect because all the available evidence, it is said, shows that Darwinian evolution is absolutely true. Those scientists and scholars who accept the Vedic texts feel obligated to make a response. The nature of the response is dictated by the nature of the audience, which consists of scientists who denigrate creation accounts and persons they have influenced, both inside and outside the scientific community. The intended audience places great faith in the accumulation of physical evidence and evaluating it in ways that are called scientific. This dictates, therefore, the nature of the reply.

In any case, Rosenfeld's description of the mentality of a fundamentalist certainly applies to fundamentalist evolutionists like yourself. "I believe in evolution. I am aware that authors identified with certain religious traditions write books that discredit evolution using evidence from scientific sources. My belief in evolution is not wrong—that is axiomatic—therefore the scientific evidence reported by these authors, or their use of it, must be wrong." As you say, you refuse to discuss ideas and evidence, but confine yourself to pointing out your "opponents' ignorance, stupidity or outright lies." Yes, only you are intelligent. All your opponents are stupid, ignorant liars. Talk about "value-laden discussion"!

I might add, the term scientism is often applied to writings, such as yours, which reflect a doctrinaire ideological commitment to anticreationism and materialism in the name of science. There is no necessary link between science and materialism. Actually, the dominance within science of the kind of fundamentalist Darwinism and materialistic scientism represented by your review, and publications such as *The Skeptic*, is a fairly recent, and hopefully short-lived, aberration.

3.1.3. Gordon Stein (1994) Review of *Forbidden Archeology: The Hidden History of the Human Race. The American Rationalist,* vol. 39, no. 1, May/June, p. 110. Reprinted by permission of the Rationalist Association.

This massive (914 page) book has one singe purpose: to convince you that the entire chronology of the human race on earth, as pictured by archeology and anthropology, is incorrect. Humans have been here far longer than the 500,000 or so years that science currently alleges. The proof, these authors argue, has long been available from many scientific studies, especially those done in the last century. However, traditional scientists have suppressed and denied the results of those studies. It is an improbable scenario, but let's look at the evidence offered.

The first 400 pages of this book is all about paleoliths (stone tools used by early humans). It is more than you would ever want to know about paleoliths, and is (I thought) rather boring reading. Next comes 200 pages about bones that might be those of humans. This I found moderately interesting. Next comes a section of about 40 pages on apemen (bigfoot, sasquatch, the almas), of considerable interest to me. There follows 170 pages on *Australopithecus* and Leakey's findings, which are also interesting, I thought. Finally, there is a section on radio-carbon dating and on anomalous objects found in archeological digs, both of which I thought highly interesting. If only the book's balance had been different! A lengthy set of appendices containing bibliography, illustration sources, and tables summarizing everything follows.

In short, this book is overkill on a grand scale. Certainly the findings of any science are tentative. When evidence builds to a sufficient amount, any view that contradicts the established view *should* get recognition. To postulate a vast conspiracy does not seem a likely explanation to me. We must remember that [the sponsors of this book and indeed the author's motivation] is to show that the Hindu view of the age of man, as given in the Indian Vedas, is the correct one. In other words, the authors have a *religious* motivation behind their writing and research. That, in and of itself, should merely raise our suspicions, but we should allow the *evidence* they present to decide the truth or falsity of their case. Here, I think they raise certain evidence that should make us cautious to accept the conclusions of archeology, but not enough to refute those conclusions. Something to keep in mind would be that this book is the mirror image of the creationist literature, namely it is a supposed examination (really a critique) of the literature of science with a view to showing that the chronology of science is wrong. In the case of the creationists, they seek to show that man has only been around for a short time (less than 10,000 years). In the case of the present book, they seek to show that man has been around for a longer time (millions of years). In each case, there is the same motivation, namely the confirmation of a religious position.

Related Correspondence

3.1.3.1. Letter to Gordon Stein, June 1, 1997.

Please forgive this much delayed response to your review of my book *Forbidden Archeology* in *The American Rationalist.*

I am grateful for your proposal that *Forbidden Archeology* raises "certain evidence that should make us cautious to accept the conclusions of archeology." At this stage of the game, that is progress. I am also grateful for your recognition that although the book is inspired by religious commitment, it should be the evidence that decides the case.

In general, I found your review fair, with the exception of your statement that the book posits a massive conspiracy to suppress evidence. That is not exactly what the book says. *Forbidden Archeology* instead posits a social process of knowledge filtration, whereby certain categories of evidence are set aside, ignored, and eventually forgotten.

3.1.4. Bradley T. Lepper (1996) "Hidden History, Hidden Agenda. A Review of *Hidden History of the Human Race.*" *Skeptic*, vol. 4, no. 1, pp. 98-100. Reprinted by permission of the Skeptics Society.

Dr. Bradley T. Lepper is Curator of Archaeology at the Ohio Historical Society, an occasional visiting Assistant Professor in the Department of Sociology and Anthropology at Denison University in Granville, Ohio, and editor of the journal Current Research in the Pleistocene.

The Hidden History of the Human Race, by Michael A. Cremo and Richard L. Thompson, is an ideologically motivated assault on the conventional view of human evolution and prehistory. The authors claim "various humanlike and apelike beings have coexisted for long periods of time" (hundreds of millions of years, in fact) and that scientists have "systematically suppressed" the evidence for this incredible notion (p. xvii, 133).

The Hidden History of the Human Race is an abridged edition of *Forbidden Archeology*, published by the Bhaktivedanta Institute in San Diego, and dedicated to "His Divine Grace A. C. Bhaktivedanta Swami Prabhupada," the implications of which will be apparent below. In the preface to the abridgment Michael Cremo states the rationale for this leaner version: it's "shorter, more readable, and more affordable." In other words, they hope to reach a wider audience with their message that human evolution didn't happen the way the textbooks claim, and that generations of archaeologists and paleoanthropologists have conspired to conceal the truth from the public.

The original book has been reviewed in various places (Feder, 1994; Marks, 1994; Tarzia, 1994) and, as the substance of the work has not changed, the interested reader might want to consult these other reviews for different, if concordant, perspectives. It is worthwhile to consider the new abridgment because it is likely to be more widely read than its rather ponderous predecessor (in fact, it can be found in many mainstream bookstore chains, including Barnes and Noble).

The Hidden History of the Human Race is a frustrating book. The motivation of the authors, "members of the Bhaktivedanta Institute, a branch of the International Society for Krishna Consciousness" (p. xix), is to find support in the data of paleoanthropology and archaeology for the Vedic scriptures of India. Their methods are borrowed from fundamentalist Christian creationists (whom they assiduously avoid citing). They catalog odd "facts" which appear to conflict with the modern scientific understanding of human evolution and they take statements from the work of conventional scholars and cite them out of context to support some bizarre assertion which the original author would almost certainly not have advocated. Cremo and Thompson regard their collection of dubious facts as "anomalies" that the current paradigm of paleoanthropology cannot explain. Sadly, they offer no alternative paradigm which might accommodate both the existing data and the so-called anomalies they present; although they do indicate that a second volume is planned which will relate their "extensive research results" to their "Vedic source material" (p. xix). Kuhn noted that "To reject one paradigm without simultaneously substituting another is to reject science itself" (1970, p. 79); and that is precisely what Cremo and Thompson do. They claim that "mechanistic science" is a "militant ideology, skillfully promoted by the combined effort of scientists, educators, and wealthy industrialists, with a view towards establishing worldwide intellectual dominance" (p. 196).

The work is frustrating because it mixes together a genuine contribution to our understanding of the history of archaeology and paleoanthropology with a bewildering mass of absurd claims and an audaciously distorted review of the current state of paleoanthropology.

Cremo and Thompson are quite right about the extreme conservatism of many archaeologists and physical anthropologists. While an undergraduate at a prominent southwestern university, I participated in classroom discussions about the claims for a very early occupation at the Timlin site (in New York) which had just been announced. The professor surprised me when she stated flatly that, if the dates were correct, then it was "obviously not a site." This dismissal of the possibility of such an ancient site, without an examination of the data or even a careful reading of the published claim, is dogmatism of the sort rightfully decried by Cremo and Thompson. George Carter, the late Thomas Lee, and Virginia Steen-McIntyre are among those whose claims for very early humans in America have been met with unfortunate ad hominem attacks by some conservative archaeologists; but, regardless of how shamefully these scholars were treated, the fact remains that their claims have not been supported by sufficiently compelling evidence. Cremo and Thompson are wrong, however, when they condemn scientists for demanding "higher levels of proof for anomalous finds than for evidence that fits within the established ideas about human evolution" (p. 49). It is axiomatic that extraordinary claims demand extraordinary evidence.

Cremo and Thompson have little understanding of history and almost no understanding of the disciplines of paleoanthropology and archaeology. In the Introduction, Thompson is identified as a generic "scientist" and "a mathematician," while Cremo is "a writer and editor for books and magazines published by the Bhaktivedanta Book Trust" (p. xix). Their naive approach to history is revealed in their discussion of the alleged discovery of broken columns, "coins, handles of hammers, and other tools" quarried from limestone in France between 1786 and 1788 (p. 104). In order to establish the credibility of this report they note that it was published in the *American Journal of Science* in 1820. They attempt to support their charge that modern scientists are dogmatic by observing that "today, however, it is unlikely such a report would be found in the pages of a scientific journal" (p. 104). The *American Journal of Science* in the 1820s published many reports that would not be found in modern science journals. Mermaids (Shillaber 1823), sea serpents (*American Journal of Science and Arts*, 1826), and the efficacy of divining rods for locating water (Emerson, 1821) were topics of interest to scientists of that era. That such material was presented in a 19th century journal with "Science" in the title is no measure of its reliability or its relevance to modern science; likewise, that modern marine biologists no longer consider mermaids a worthy subject for research is no measure of their dogmatism. Cremo and Thompson might disagree, however, for they devote an entire chapter to reports of "living ape-men" such as Bigfoot, which, even if true, contribute nothing to their thesis that anatomically modern humans lived in geologically ancient times. Chimpanzees are "ape-men" of a sort, sharing 99% of our genetic makeup, and their coexistence with *Homo sapiens sapiens* does no violence to evolutionary theory.

Cremo and Thompson's ignorance of the basic data of archaeology is exemplified by their reference to the Venus of Willendorf as a work of "Neolithic" rather than Paleolithic art (p. 84) and their mistaken identification of a nondescript stone blade from Sandia Cave as a "Folsom point" (p. 93). Folsom points are highly specialized and distinctive artifacts and, although the excavators of Sandia Cave did recover several from that site, a Folsom point is not what is depicted in the photograph reproduced by Cremo and Thompson (p. 93). Moreover, although they have plumbed the depths of 19th-century literature in search of crumbs of data that support their rather vague notions about the extreme antiquity of *Homo sapiens,* they are not abreast of the latest developments in the field of archaeology. They refer to claims of great antiquity for artifacts from the Calico, Pedra Furada, Sandia Cave, Sheguiandah, and Timlin sites, but are apparently unaware of recent (and some not so recent) work concerning these sites which substantially refutes (or calls into serious question) the claims of the original investigators (e.g., Cole and Godfrey, 1977; Cole et al., 1978; Funk, 1977; Haynes and Agogino, 1986; Julig et al., 1990; Kirkland, 1977; Meltzer et al., 1994; Preston, 1995; Schnurrenberger and Bryan, 1985; Starna, 1977; Taylor, 1994).

This is a book designed to titillate, not elucidate. The authors discuss a weathered rock more than 200 million years old which they identify as a fossilized partial shoe sole (p. 115-116). They allude to "microphoto magnifications" of the fossilized stitches which allegedly show "the minutest detail of thread twist and warp" (p. 116), but do not present these magnified images. Instead, they reproduce a somewhat blurred photograph of the weathered outlines which do not, at least to this reviewer, resemble any portion of a shoe sole.

Cremo and Thompson discuss the three to four million year old fossilized footprints discovered at Laetoli, and note that scholars have observed "close similarities with the anatomy of the feet of modern humans" (p. 262). Cremo and Thompson conclude that these footprints actually are the tracks of anatomically modern humans, but they offer no explanation for why these individuals were not wearing the shoes which supposedly had been invented more than 296 million years earlier.

Cremo and Thompson are selectively credulous to an astonishing degree. They accept without question the testimony of 19th-century gold-miners and quarrymen, but treat with extreme skepticism (or outright derision) the observations of 20th-century archaeologists. That Von Koenigswald purchased *Pithecanthropus* fossils from native Javanese causes Cremo and Thompson "uneasiness" (p. 164); but they blithely accept Taylor's purchase of the "Foxhall Jaw" from "a workman who wanted a glass of beer" (p. 133) without similar unease. The authors are critical of archaeologists for rejecting the very early radiometric dates for technologically recent stone artifacts at Hueyatlaco, Mexico (pp. 91-93), but they are as quick to reject radiometric dates which do not agree with their preconceived interpretations (pp. 125, 139-140).

Cremo and Thompson's claim that anatomically modern *Homo sapiens sapiens* have been around for hundreds of millions of years is an outrageous notion. Accepting that there is a place in science for seemingly outrageous hypotheses (cf. Davis, 1926) there is no justification for the sort of sloppy rehashing of canards, hoaxes, red herrings, half-truths and fantasies Cremo and Thompson offer in the service of a religious ideology. Readers who are interested in a more credible presentation of the overwhelming evidence for human evolution should consult Ian Tattersall's wonderful recent book *The Fossil Trail: how we know what we think we know about human evolution*.

References

American Journal of Science, and Arts, 1826. "Sea Serpent." *American Journal of Science, and Arts*, 11:196

Cole, J. R., R. E. Funk, L. R. Godfrey, and W. Starna. 1978. "On Criticisms of 'Some Paleolithic Tools from Northeast north America': rejoinder." *Current Anthropology,* 193:665-669

Cole, J. R. and L. R. Godfrey. 1977. "On Some Paleolithic Tools from Northeast North America." *Current Anthropology*, 18(3):541-543.

Davis, W. M., 1926. "The Value of Outrageous Geological Hypotheses." *Science,* 63:463-468.

Emerson, R. 1821. "On the Divining Rod, With Reference to the Use Made of it in Exploring for Springs of Water." Oct. 23, 1820. *American Journal of Science and Arts,* 3:102-104.

Feder, K. L. 1994. "Review of *Forbidden Archaeology: The Hidden History of the Human Race." Geoarchaeology*, 9(4):337-230.

Funk, R. E. 1977. "On Some Paleolithic Tools from Northeast North America." *Current Anthropology,* 18(3):543-544.

Haynes, C. V., Jr. and G. A. Agogino. 1986. "Geochronology of Sandia Cave." *Smithsonian Contributions to Anthropology, No. 32.*

Julig, P. J., W. C. Mahaney, and P. L. Storck. 1991. "Preliminary Geoarchaeological Studies of the Sheguiandah Site, Manitoulin Island, Canada." *Current Research in the Pleistocene, 8*:110-114.

Kirkland, J. 1977. "On Some Paleolithic Tools From Northeast North America." *Current Anthropology, 18(3)*:544-545.

Kuhn, T. S. 1970. *The Structure of Scientific Revolutions*. 2nd edition. International Encyclopedia of Unified Science, Vol. 2, No. 2. University of Chicago Press.

Marks, J. 1994. "Review of *Forbidden Archaeology: The Hidden History of the Human Race.*" *American Journal of Physical Anthropology, 93(1)*:140-141.

Meltzer, D. J., J. M. Adovasio, and T. D. Dillehay. 1994. "On a Pleistocene Human Occupation at Pedra Furada, Brazil." *Antiquity, 68(261)*:695-714.

Preston, D. 1994. "The Mystery of Sandia Cave." *New Yorker*, 12 June 1994, pp.66-83

Schnurrenberger, D. and A. L. Bryan. 1984. "A Contribution to the Study of the Naturefact/Artifact Controversy." In *Stone Tool Analysis*, M. G. Plew, J. C. Woods, and M. G. Pavesic, (eds.) pp.133-159. Albuquerque: University of New Mexico Press.

Shillaber, J. 1823. "Mermdid." (sic) *American Journal of Science and Arts, 6*:195-197

Starna, W. A. 1977. "On Some Paleolithic Tools from Northeast North America." *Current Anthropology, 18(3)*:545.

Tarzia, W. 1994. "Forbidden Archaeology: Antievolutionism Outside the Christian Arena." *Creation/Evolution, 14(1)*:13-25.

Taylor, R. E. 1994. "Archaeometry at the Calico Site." *The Review of Archaeology, 15(2)*:1-8.

Related Correspondence

3.1.4.1. Letter to Dr. Bradley T. Lepper, June 1, 1997.

I've only recently seen your review of my book *The Hidden History of the Human Race*. I thank you for acknowledging that the book "makes a genuine contribution to our understanding of the history of archaeology and paleoanthropology." I also thank you for your statement that the authors "are quite right about the conservatism of many archeologists and physical anthropologists" and for your admission that archeologists sometimes dismiss ancient dates for sites "without an examination of the date or even a careful reading of the published claim."

Furthermore, I am grateful to you for listing the articles that are not mentioned in *Hidden History*'s discussion of the Timlin, New York, site. It was a mistake not to include them. They do raise important questions about the artifacts recovered from the site and about the geological interpretation of the age of the site.

Nevertheless, in one of the papers cited by you, Bryan and Schnurrenberger (p. 149) concluded that at least five of the Timlin artifacts were genuine. From new studies of the geology of the site (p. 147), they concluded it was more recent than the original discoverers (Raemsch and Vernon) claimed. Raemsch and Vernon thought the Timlin artifacts were found in glacial till deposits, laid down by glaciers 60 or 70 thousand years ago. According to Bryan and Schnurrenberger, the glacial deposits had been reworked by a stream in early postglacial times. They thought the Timlin artifacts dated to this period. But it seems to me that if the tools were found in reworked glacial deposits, they could have come from those glacial deposits. This is a possibility that must at least be considered.

I now want to offer some comments on the parts of your review that do not accurately reflect the content and purpose of *Hidden History*.

The methodology employed in *Hidden History* was not borrowed from fundamentalist Christian creationists. As acknowledged in the Introduction to *Forbidden Archeology*, the major methodological influences on the authors, particularly this author, were recent work in the history, philosophy and sociology of science, as well as the Sanskrit historical literature of India, as interpreted by His Divine Grace A. C. Bhaktivedanta Swami Prabhupada, whose translations and commentaries have drawn favorable reviews from Sanskrit scholars and Indologists worldwide.

In my study of Christian creationist books, I found some of higher quality than others. Not many of them, however, were directly related to human origins and antiquity, and those that were did not treat these topics in sufficient depth. They were not, therefore, extensively cited in *Hidden History*.

Furthermore, I sensed that most Christian creationist work was directed more toward other Christian creationists than to mainstream scientists and scholars. *Hidden History*'s parent book, *Forbidden Archeology*, was consciously aimed at mainstream scholars. It was intended to open a genuine scholarly dialogue or debate, and this effort has been successful. The large number of serious reviews the book has received in academic journals is a sign of that. To my knowledge, such journals do not normally give this kind of attention to other kinds of antievolutionary literature.

As might be expected, some mainstream scholars succumbed to the temptation to force this book into familiar categories and responded accordingly; but many resisted that temptation, at least partially, and looked at the book somewhat objectively.

The distinction between *Hidden History*'s parent book (*Forbidden Archeology*) and ordinary creationist literature has been noted by many scientific reviewers, including Kenneth Feder, whom you cited in your article.

Feder wrote in *Geoarchaeology* (9:338): "While decidedly antievolutionary in perspective, this work is not the ordinary variety of antievolutionism in form, content, or style. In distinction to the usual brand of such writing, the authors use original sources and the book is well written. Further, the overall tone of the work is far superior to that exhibited in ordinary creationist literature."

Hidden History is not simply a "catalog" of "odd 'facts' which appear to conflict with the modern scientific understanding of human evolution." It develops a refined epistemological argument, which you were unable or unwilling to follow. It appears that you, like others, instinctively forced *Hidden History* into a familiar category (poorly contrived creationist tracts). And then you automatically repeated the customary set of accusations—"catalog of odd facts," "quoting out of context," etc.

But these conventional maledictions do not apply to *Hidden History*, which presents a thorough and systematic survey and critique of evidence relevant to human origins and antiquity. The facts in the book are deployed within the framework of a well-articulated analysis of the quality of archeological and paleoanthropological reporting.

About *Hidden History*'s parent book, *Journal of Field Archeology* (*21*: 112) said: "This volume combines a vast amount of both accepted and controversial evidence from the archeological record with sociological, philosophical, and historical critiques of the scientific method to challenge existing views and expose the suppression of information concerning history and human origins."

And archeologist Tim Murray wrote in *British Journal for the History of Science* (28: 379) that the book "provides the historian of archeology with a useful compendium of case studies in the history and sociology of scientific knowledge, which can be used to foster debate within archaeology about how to describe the epistemology of one's discipline."

This is not to suggest that the writers of these statements endorsed the book's conclusions; they did not. But it is apparent that they thought the book was something more than a collection of poorly documented odd facts and quotations taken out of context. They could recognize something of the book's epistemological framework and intellectual integrity, whereas you could not.

Hidden History does not quote out of context. *Hidden History* examines particular cases in considerable detail, with long quotations from original sources. The reasons for the detailed treatment are outlined in *Hidden History*'s parent book (*Forbidden Archeology*, p. 35): "It would of course be possible to more briefly summarize and paraphrase reports such as these. There are two reasons for not doing so. The first is that paleoanthropological evidence mainly exists in the form of reports...and we shall therefore take the trouble to include many selections from such reports, the exact wording of which reveals much....A second consideration is that the particular reports...are extremely difficult to obtain....A final consideration is that proponents of evolutionary theory often accuse authors who arrive at nonevolutionary conclusions of 'quoting out of context.' It therefore becomes necessary to quote at length, in order to supply the necessary context."

Admittedly, some of the context may have been lost in abridging the 900-page *Forbidden Archeology* to the 300-page *Hidden History*. But the preface I wrote to *Hidden History* explicitly refers readers desiring more complete context to the unabridged version of the book.

Regarding quoting an author in support of a conclusion the author himself would not have advocated, there is nothing wrong with that if the quotation is accurate and the meaning of the quotation is taken as intended by the author. For example, Richard Leakey reported that the ER 1481 femur, found isolated from other bones, was anatomically modern and about 1.8 million years old (*Hidden History*, p. 253). It is not wrong for me to cite this information in support of the idea that the femur could have come from an anatomically modern human living in Africa 1.8 million years ago, even though Richard Leakey would probably not entertain this idea himself.

Speaking of taking quotes out of context, you yourself are not sinless. You lifted the quote about mechanistic science being a militant ideology from its context, which deals with the activities of the Rockefeller Foundation in China in the first decades of the twentieth century, and presented it as the authors' general indictment of today's science.

What *Hidden History* (pp. 195-196) actually says is that the Rockefeller Foundation, the board of which included educators like Charles W. Eliot (formerly president of Harvard University), scientists like Dr. Simon Flexner, and industrialists like John D. Rockefeller, wanted to open an independent secular university in China, for the purpose of introducing Western science. This was opposed by both the Chinese government, which wanted control over it, and Christian missionaries in China, worried about an influence detrimental to their own educational activities. To get around this opposition, Eliot suggested opening a hospital and medical school. According to a Foundation official, Eliot thought "there was no better subject than medicine to introduce to China the inductive method of reasoning that lies at the basis of all modern science." He apparently felt that China was being held back by its attachment to traditional Buddhist and Taoist ways of knowledge. The ploy was successful. The Foundation accepted Eliot's idea, as did the Chinese government and the Christian missionary establishment. It was in this context that *Hidden History* (p. 196) said: "Here mechanistic science shows itself a quiet but nevertheless militant ideology, skillfully promoted by the combined effort of scientists, educators, and wealthy industrialists, with a view toward establishing worldwide intellectual dominance." This particular statement is thus tied to a very specific event in history, and to a very specific group of educators, scientists, and industrialists. To make a case that the statement applies in a more general way to today's entire scientific enterprise might be worth attempting, but that would take a book in itself.

In short, you took out of context a limited statement that was reasonable in terms of its supporting evidence and deliberately gave your readers the misimpression that *Hidden History* was making an unsupported wild generalization of the kind your readers are properly conditioned to reject.

You also took out of its clearly stated context the report of evidence for extreme human antiquity discovered in France, published in *American Journal of Science and Arts*. This case was included in a chapter containing *Hidden History*'s most extreme anomalies. The chapter introduction (p. 103) clearly stated: "The reports of this extraordinary evidence emanate, with some exceptions, from nonscientific sources....We ourselves are not sure how much importance should be given to this highly anomalous evidence. But we include it for the sake of completeness and to encourage further study." This statement of context is so clear that your omission of it from your disparaging discussion can only be characterized as deliberate and, hence, intellectually dishonest.

The following paragraph from your review illustrates just how far you are willing to descend into the realm of silliness and triviality: "Cremo and Thompson discuss the three to four million year old fossilized footprints discovered at Laetoli, and note that scholars have observed 'close similarities with the anatomy of the feet of modern humans' (p. 262). Cremo and Thompson conclude that these footprints actually are the tracks of anatomically modern humans, but they offer no explanation for why these individuals were not wearing the shoes which supposedly had been invented more than 296 million years earlier." *Hidden History* proposes that scientists who hold firmly to orthodox views on human origins often employ ridicule as their weapon of choice when confronted with challenging evidence. I see you are no exception to this rule. The real question is this: how do you explain the occurrence of anatomically modern footprints in rock 4 million years old? The foot bones of the australopithecines, supposedly the only hominids then in existence, could not, according to physical anthropologists such as Russell Tuttle, have made those prints.

Regarding the report of the Nevada shoe print, the stimulus for your attempt to ridicule your way out of considering the obvious implications of the Laetoli footprints, it was included in *Hidden History*'s chapter on extreme anomalies, with a very clear statement of its context. And you insisted on taking it out of this context. It was duly acknowledged that reports such as this, from nonscientific publications, leave much to be desired but were included in the book for the sake of completeness and to encourage further study. The photograph you complained about is of value in that it to some degree confirms the existence of the object in question. The report also offers opportunities for pursuing further investigation of this object. It might be possible, for example, to track down the object itself or to find the original microphotographs, said to have been taken by an employee of the Rockefeller Institute in New York.

In addition to blatantly taking the reports of the Nevada shoe print and the above mentioned discoveries in France out of their clearly stated context and improperly suggesting that the "naive" authors were accepting of them to a degree that they were not, you were also unable or unwilling to see how this particular category of reports fits into the overall epistemological argument presented in the book. For your benefit, I shall put it as simply and briefly as I can.

Assuming that there is evidence for extreme human antiquity in the earth's strata, we can make the following predictions. Some of the evidence will be close to conventionally accepted limits for human antiquity and some will far exceed these limits. Some of the evidence will be found by scientists, who will react to it according to their theoretical preconceptions and report it according to their professional standards. Some of the evidence will be found by nonscientists with few theoretical preconceptions and reported in nonscientific literature. It is likely that the evidence that most radically departs from conventionally accepted limits will be reported by nonscientists in nonscientific literature. In terms of this approach, evidence of the kind reported in the chapter on extreme anomalies does have some value in confirming the hypothesis that evidence for extreme human antiquity does exist and has been reported by various categories of researchers, ranging from professional scientists whose findings are published in academic journals to nonscientists whose findings are reported in newspapers and magazines.

You said that just because reports of unusual phenomena were published in a 19th-century journal that happened to have the word "science" in its title is no measure of the reports' "reliability or relevance to modern science." Neither is this, in itself, any measure of their unreliability or irrelevance to modern science.

You have misunderstood and taken out of context Thomas Kuhn's statement that "to reject one paradigm without simultaneously substituting another is to reject science itself." Kuhn did not intend this to mean that any individual who introduces evidence contradicting a reigning paradigm must himself immediately introduce a new paradigm.

If you carefully study Kuhn's entire description of scientific revolutions, you will find the following development. In the beginning of a science there is no reigning paradigm. Individual scientists gather evidence from nature and use it to build competing paradigms. Eventually, one of these may triumph, and at this point a mature science, a research community with a common program, develops. This research community is guided by a single dominant paradigm, through which it structures its research goals and methods. The paradigm does not resolve all questions and problems. If it did, there would be no need for further research. Instead it gives a systematic approach for solving various puzzles suggested by the paradigm. The solving of these puzzles constitutes the activity of normal science. In the course of normal science, it may happen that anomalies begin to accumulate. Some of these may be set aside for future research. Some may be dismissed as irrelevant. But if a sufficient number of anomalies accumulates, anomalies which resist solution by the paradigm or incorporation into it, a crisis develops. As the crisis intensifies, scientists begin to offer and promote new paradigms capable of accommodating the anomalies. If one of these paradigms attracts the attention of a sufficient number of members of the research community, a scientific revolution takes place. The research community learns to see things in a different way. It develops a new set of methods and concerns.

Kuhn points out that unless there is a recognizable crisis, provoked by an accumulation of crucial anomalies, there will be no movement to a new paradigm. The first step toward movement to a new paradigm is thus recognition of anomalies, of counterinstances to the current paradigm.

In the 1970 edition of *The Structure of Scientific Revolutions* (pp. 93-94), Kuhn compares scientific revolutions to political revolutions: "Initially it is crisis alone that attenuates the role of political institutions as we have already seen it attenuate the role of paradigms. In increasing numbers individuals become increasingly estranged from political life and behave more and more eccentrically within it. Then, as the crisis deepens, many of these individuals commit themselves to some concrete proposal for the reconstruction of society in a new institutional framework. . . .The remainder of this essay aims to demonstrate that the historical study of paradigm change reveals very similar characteristics in the evolution of the sciences."

The purpose of *Forbidden Archeology* is to confront the community of human evolution researchers with the massive number of unassimilated crucial anomalies in their field, and thus provoke a sense of crisis in at least some small section of this community. This effort has been to some degree successful, but the sense of crisis must be intensified. Only when the sense of crisis becomes intense will researchers give serious consideration to adopting a new paradigm. Kuhn noted (p. 76) that "retooling is an extravagance to be reserved for the occasion that demands it."

In any case, I can assure you that I will be offering a new paradigm in a forthcoming book, as promised in *Forbidden Archeology*. In my opinion, the occasion demands it.

I suppose we shall have to disagree on whether or not the claims of Carter, Lee, and Steen-McIntyre are supported by sufficiently compelling evidence. But I think your admission that they have been shamefully treated and subjected to unfortunate ad hominem attacks should cause researchers interested in North American archeology to take a second look at their findings, or, perhaps, a first look, and make up their own minds.

I also have to disagree with your insistence that evidence that goes against current ideas of human origins must be subjected to a much higher standard of proof than evidence supporting current ideas. I agree with George Carter, who said in his book *Earlier Than You Think* (1980, p. 318): "When a new idea is advanced, it necessarily challenges the previous idea.... The new idea is then attacked, and support of it is required to be of a high order of certainty. The greater the departure from the previous idea, the greater the degree of certainty required, so it is said. I have never been able to accept this. It assumes that the old order was established on high orders of proof, and on examination this is seldom found to be true."

I also agree with Alfred Russell Wallace, cofounder with Darwin of the theory of evolution by natural selection, who said (*Nineteenth Century,* vol. 22, p. 679) that "the proper way to treat evidence as to man's antiquity is to place it on record, and admit it provisionally wherever it would be held adequate in the case of other animals, not, as is too often now the case, to ignore it as unworthy of acceptance or subject its discoverers to indiscriminate accusations of being impostors or the victims of impostors."

You are correct that, even if true, the chapter on living ape-men does not directly contribute to the book's thesis that anatomically modern humans existed in the very distant past. But if true, the chapter would support the book's general picture of the coexistence of humans and more apelike hominids from the distant past until the present. While evidence of the coexistence of anatomically modern humans with more apelike hominids today does not do any violence to evolutionary theory, their coexistence in the distant past would do some violence to it. And the evidence documented in *Hidden History* suggests they did coexist in the distant past.

You say that *Hidden History* offers "a mistaken identification" of a stone tool from Sandia Cave as a Folsom point and cite this as an example of the authors' "ignorance of the basic data of archaeology." The identification is, however, not that of the authors, but of the archeologists who took the photograph of the stone tool and published the photograph along with their identification of the stone tool as a Folsom point in an archeological publication of the Smithsonian Institution, duly cited in the permission credits on the copyright page of *Hidden History*. The tool is shown cemented in the cave breccia. The references you gave in your review (Haynes and Agogino, Preston) also refer to the tools found cemented in the cave breccia as Folsom tools.

I find it somewhat unusual that you faulted *Hidden History*, a book published in 1994 (as an abridgment of a work published in 1993), for not citing reports published in 1994 and 1995. I suppose the author of any book-length work on any scientific subject could also be accused of not being totally up to date on the latest work in his or her field. But that is the nature of the book writing process. You write a book, it goes to press, and comes out a year later. So automatically the book is going to be a year or two behind the times as soon as it becomes available for sale. In any case, I have already admitted that I should have been aware of the 1977 and 1978 reports on the Timlin site.

Concerning the other reports you have cited (and I do thank you for calling them to my attention), I have the following comments.

Haynes and Agogino think the implements found at Sandia Cave are no more than 14,000 years old. But their report on the Sandia Cave discoveries does not rule out the possibility that the human artifacts found there are perhaps as much as 300,000 years old.

At Sandia Cave two kind of implements were found—Folsom implements and Sandia implements.

The sequence of layers at Sandia Cave were as follow, from top to bottom. First came a layer of recent dirt and debris (Unit J). Under this was a layer of dripstone (calcium carbonate). This layer of dripstone (Unit I) yielded carbon 14 ages of 19,100 and 24,600 years. Haynes and Agogino found these ages hard to accept and proposed they must be wrong. They proposed that the dripstone had been contaminated with old carbon, and that this had caused the tests to yield ages that were falsely old. But even Haynes and Agogino were mystified by this. They acknowledged that contamination of samples is usually with younger carbon instead of older carbon (p. 26). Furthermore, over 80 percent of the carbon in the samples would have had to have been introduced by contaminants. They admitted that they could see no visible sign of any such contamination, and were reduced to speculating that the old carbon must have come from old carbon dioxide in the air trapped in the cave (p. 27).

Haynes and Agogino (p. 7) said the ages of 19,100 and 24,600 "cannot be correct because of the archaeology and the dating of more reliable materials in the underlying units." In referring to the archaeology, they mean that tools identified as Folsom tools were found under the dripstone. Folsom tools are generally considered to be 10 or 11 thousand years old. Therefore, they reasoned, the carbon 14 dates of the overlying dripstone must be wrong. This is exactly the kind of reasoning that *Forbidden Archeology* criticizes. As for younger carbon dates on rock and bone below the dripstone being more reliable, this is hard to believe. For one thing, Haynes and Agogino (p. 26) admitted that none of their carbon 14 dates were reliable because they had been obtained by methods now considered obsolete. Furthermore, if they could attribute the ages of 19,100 and 24,600 years for the dripstone to contamination by old carbon (even though no contaminants were visible), they should also be able to attribute the younger ages of other samples to contamination by recent carbon.

Below the dripstone, which could be 24,000 years old, was a deposit of cave breccia (Unit H). A bone from this deposit gave a uranium series age of 73,000 years (p. 7). Haynes and Agogino (p. 7) said that "the U-series date cannot be supported archaeologically." As mentioned above, Folsom implements, generally thought to be 10 or 11 thousand years old, were found cemented in the cave breccia of Unit H. Therefore the U-series date must be wrong. One could also propose that the generally accepted upper age limit for Folsom implements is wrong, and that the Folsom implements from Unit H of Sandia Cave are about 70,000 years old.

Haynes and Agogino cited carbon 14 dates of 9,100 years for carbonate rock from the Unit H breccia and 12,830 years for bone fragments. But they have admitted that these dates are unreliable. In other words, the dates could be correct, or they could be falsely old or young. As we have seen, Haynes and Agogino have felt free to adjust the carbon 14 and uranium series dates to fit their conviction that the Folsom implements found in Unit H could not be more than 10 or 11 thousand years old. But there is another way to adjust things. We can accept the carbon 14 date of 24,000 years for the dripstone of Unit I and the uranium series age of 73,000 years for the bone found in the cemented breccia of Unit H. This would mean the Folsom implements of Unit H would be between 24,000 and 73,000 years old.

Below Unit H is found another layer of dripstone called Unit G. This dripstone yielded a carbon 14 age of 30,000 or more years. In other words, the Unit G dripstone could be of any age more than 30,000 years. For example, Unit G could be 100,000 years old. Haynes and Agogino concluded this date must be wrong. But it fits into the sequence that we have established. The dripstone of Unit I yielded a radiocarbon date of 24,000 years, bone from the Unit H breccia yielded a uranium series age of 70,000 years, and the radiocarbon date of Unit G could be in excess of that.

Below the dripstone of Unit G was another layer of cemented breccia called Unit F. Organic carbon from a cave wall gave a carbon 14 date of 12,000 years, but this could have resulted from contamination by younger carbon.

Then comes a gypsum crust (Unit E), followed by another layer of dripstone, Unit D. This lower dripstone gave a carbon 14 age of 32,000 years, but Haynes and Agogino thought it was not correct. Wanting it to be younger, they proposed it had been contaminated with older carbon. The same dripstone yielded a uranium series age of 226,000 years. According to Preston (p. 74), another uranium series test gave an age of 300,000 years.

Haynes and Agogino dismissed these ages and revised the radiocarbon date downward to 27,000 years. I would propose dismissing the radiocarbon age and keeping the uranium series dates, which fit nicely into the sequence of radiocarbon and uranium series dates that I have established.

Beneath the dripstone of Unit D is a deposit of yellow ocher. When burned, yellow ocher yields a reddish substance used as a cosmetic. Haynes and Agogino suggested the ocher deposits had been mined by early Indians.

Beneath the yellow ocher, the original discovers found a loose deposit of rock and dirt (Unit X), containing in some places stone tools of a type different from the Folsom implements found in the upper breccias. They called these stone tools Sandia implements. According to Haynes and Agogino, rodents carried all of these tools down from the upper levels of the cave deposits (Units F and H), starting about 14,000 years ago. To support this hypothesis, they pointed to the presence of rodent bones yielding carbon 14 dates ranging from 8,000 to 14,000 years. If the positions in which they were found in Unit X were their original positions, then the tools would be at least 300,000 years old (if we accept the uranium series date for the overlying Unit D dripstone).

There are several points to consider here. First, Haynes and Agogino (p. vii) admitted that their carbon 14 dates were unreliable: "There is no foolproof method of positively isolating indigenous bone carbon from contaminant carbon in leached bone." So the rodent bones could have been much older than their maximum carbon 14 dates of 14,000 years, and the rodents could thus have moved the tools to their positions in Unit X much earlier than 14,000 years ago.

The effective range of the carbon 14 dating method used by Haynes and Agogino was about 40,000 years. This means that if the rodent bones were 300,000 years old, then even a small amount of contamination would have caused them to yield a carbon 14 date of 40,000 years. More extensive contamination would have brought the ages down even further.

Furthermore, there are some difficulties with the idea that rodents moved the Sandia tools from higher levels in the cave down to positions in and below the ocher deposits at any point in time, whether 14,000 or 300,000 years ago. Only Sandia implements are found in Unit X. None are found in Units F and H, which according to Haynes and Agogino were the most likely source of the Sandia implements. Haynes and Agogino (p. 28) proposed that the Sandia implements were deposited in Unit F between 14,000 and 11,000 years ago, before Unit F was consolidated into a hard cave breccia between 11,000 and 9,000 years ago. The Folsom implements would have been deposited in Units F and H during this latter period of consolidation.

But Haynes and Agogino (p. 28) noted a problem: "If Sandia occupation was before Folsom occupation, it is surprising that no diagnostic Sandia artifacts were found in the [Unit F] breccias." Preston (p. 75) asked Haynes this question: "How was it possible that *all* the Sandia points—nineteen of them—were somehow carried by rodents to the bottom layer only?" Rodent tunnels are found in many of the layers, not just the bottom layer. Haynes replied, "Don't think we didn't ask ourselves that same question. It's very, very strange."

Haynes and Agogino also considered the possibility that the Sandia implements were younger than the Folsom implements found in the cave breccias of Units F and H. This means they would have to be from the loose cave debris (Unit J) that started accumulating on the cave floor after 9,000 years ago. But Haynes and Agogino (p. 28) note that this "appears unlikely because no Sandia artifacts are reported from the upper loose debris (Unit J)."

A more reasonable possibility would be that the Sandia artifacts below the yellow ocher are in their original positions and that they were not moved by rodents from the upper levels of the cave. Just because there is evidence that rodents made tunnels into the levels containing the implements does not rule out the possibility that the implements were in their original positions. That the Sandia implements were found only in and beneath the ocher argues for them being in their original positions. In that case, the implements would be over 300,000 years old, about the same age as the crude stone tools from the Calico, California, site and the advanced stone tools from the Hueyatlaco, Mexico, site.

Here is another reason for suspecting that the Sandia implements were in their original positions, arriving their as a result of human action rather than transport by rodents. The first Sandia implement was found on the level of the cave floor, alongside a hearth made from four stream-rounded cobble stones, charcoal, and a jaw of a large mammal (Haynes and Agogino, figure 6). Haynes and Agogino (p. 28) noted: "Apparently all witnesses considered the point, four rounded cobbles, and a bovid mandible to be in situ and associated with the hearth."

According to Preston, some archaeologists have suggested that the Sandia cave discoveries were all fraudulent. Haynes and others disagreed. Of course, if the discoveries were fraudulent, that would be significant. We would have another case, in addition to Piltdown, in which professional scientists manufactured evidence in support of their own theories.

To me, however, the most likely interpretation of the Sandia evidence is that you have Folsom implements in Units F and H that could be anywhere from 24,000 to 73,000 years old and Sandia implements in Unit X that are at least 300,000 years old.

Taylor's report on the Calico site is a review of published literature and does not give any new evidence. Taylor (p. 7) admitted that the age of the artifact-bearing sediments at Calico is "in excess of 100,000 years and perhaps as much as 200,000 years old." He doubted, however, that the objects found in these sediments are the result of human work. In this regard, Taylor cited a report by Payen, which analyzed the Calico artifacts in terms of the Barnes platform angle method. According to Barnes, at least 75 percent of the platform angles should be acute (less than 90 degrees) for the object to be of human manufacture. Payen found that the Calico implements did not satisfy this requirement. But Taylor neglected to mention a later report by Leland W. Patterson, an expert in lithic technology, and his coworkers that appeared in *Journal of Field Archeology* (vol. 14, pp. 91-106). Patterson and his coauthors (p. 97) reported, "Acute platform angles were found on 94.3% of the Calico flakes with intact platforms...The average platform angle of the Calico flakes was 78.7%....This is consistent with the usual products of intentional flaking." A more detailed discussion can be found in *Forbidden Archeology* (pp. 166-175). Patterson disputed Payen's conclusions and methodology.

The paper by Meltzer, Adovasio, and Dillehay about the Pedra Furada site in Brazil raises questions about the evidence for a human occupation there at 30,000 years. The original discoverers reported hearths with charcoal at levels of this age, along with stone tools. Meltzer, Adovasio, and Dillehay expressed some doubts about the charcoal (was it from wildfires rather than hearths?) and the stone tools (were they really of human manufacture?). The original discoverers have, of course, already given attention to such doubts. Meltzer, Adovasio, and Dillehay (pp. 695-696) themselves acknowledged that they "are not experts on the data and evidence recovered from Pedra Furada" and that they did not "expect our opinions will be shared by our colleagues (even those who viewed the site with us)." I would not therefore characterize their report as a refutation of the claims of the original discoverers.

Regarding the report of Julig, Mahaney, and Storck on the Sheguiandah, it is, as the title indicates, a very brief preliminary study. The original investigators, T. E. Lee and J. T. Sanford, found stone tools in deposits they characterized as glacial till. This would give them considerable antiquity. Other investigators have challenged the identification of the Sheguiandah deposits as till. But the preliminary studies of Julig, Mahaney, and Storck (p. 111) revealed that "the so-called 'till' deposits are clearly non-sorted and may be till or colluvium." They noted that "of the 27 samples analyzed from the so-called 'till' deposits and underlying sediments, 20 exhibit curves which are characteristic of nonsorted sediment such as till or colluvium." Furthermore, Julig, Mahaney, and Storck stated that "crescentic gouges...which are widely considered to be the effect of transport by continental ice, were...observed on grains of several samples." So your suggestion that their report clearly contradicts the earlier work of Lee and Sanford is mistaken. In fact, the report tends to confirm the judgments of Lee and Sanford.

Returning to general methodology, you accuse the authors of being "selectively credulous to an astonishing degree." You find it objectionable that we "accept without question the testimony of 19th-century goldminers and quarrymen, but treat with extreme skepticism (or outright derision) the observations of 20th-century archaeologists." Of course, we did not accept without question the testimony of anyone. But I can understand how you could have gotten an impression of selective credulity. Missing from *Hidden History*'s discussion of epistemological principles is the following paragraph from *Forbidden Archeology* (p. 25), which may help you understand something about the methods we employed in evaluating reports:

"In discussing anomalous and accepted reports...we have tended to stress the merits of the anomalous reports, and we have tended to point out the deficiencies of the accepted reports. It could be argued that this indicates bias on our part. Actually, however, our objective is to show the qualitative equivalence of the two bodies of material by demonstrating that there are good reasons to accept much of the rejected material, and also good reasons to reject much of the accepted material. It should also be pointed out that we have not suppressed evidence indicating the weaknesses of anomalous findings. In fact, we extensively discuss reports that are highly critical of these findings, and give our readers the opportunity to form their own opinions."

In a final flourish of rancor, you hurl at *Hidden History* a veritable barrage of curses, practically exhausting the fundamentalist Darwinian's stock of clichÈs, calling the book a "sloppy rehashing of canards, hoaxes, red herrings, half truths and fantasies." But others have passed a different final judgment. In their review article about the unabridged version of *Hidden History*, historian of science David Oldroyd and his graduate student Jo Wodak wrote in *Social Studies of Science* (*26*: 107): "So has *Forbidden Archeology* made any contribution at all to the literature on palaeoanthropology? Our answer is a guarded 'yes', for two reasons. First, while the authors go in for overkill in terms of swamping the reader with detail...much of the historical material they resurrect has not been scrutinized in such detail before. Second,...Cremo and Thompson do raise a central problematic regarding the lack of certainty in scientific 'truth' claims."

Earlier in the same review article (p. 196) they noted: "It must be acknowledged that *Forbidden Archeology* brings to attention many interesting issues that have not received much consideration from historians;

and the authors' detailed examination of the early literature is certainly stimulating and raises questions of considerable interest, both historically and from the prospective of practitioners of sociology of scientific knowledge."

Regarding Ian Tattersall's book, the following quote from Wodak and Oldroyd is perhaps relevant: "If scientists have lost sight of the idea of 'Tertiary Man', perhaps historians bear some responsibility. Certainly some pre-*FA* histories of palaeoanthropology, such as Peter Bowler's, say little about the kind of evidence adduced by C&T, and the same may be said of some texts published since 1993, such as Ian Tattersall's recent book. So perhaps the rejection of Tertiary *Homo sapiens*, like other scientific determinations, is a social construction in which historians of science have participated."

Anyone who is really interested in learning the complete story of how we know what we think we know about human origins and antiquity cannot afford to ignore *Hidden History* or, better yet, *Forbidden Archeology.*

In the end, I cannot judge you too harshly. After all, like many of your generation, you probably grew up believing in Darwinism and were conditioned to regard such belief as one of the characteristics of scientific and intellectual respectability. You were also conditioned to regard opposition to Darwinism as a symptom of religious intolerance or irrational credulity, deserving of righteous contempt. Given all that, your fundamentalist reaction to *Hidden History* is understandable. Even so, I detect in your review some signs that you may someday rise to the platform of virtuous scientific impartiality to which you now pretend.

3.2. Alternative Science Publications (General)

3.2.1. John Davidson (1994) "Fascination Over Fossil Finds." *International Journal of Alternative and Complementary Medicine* August, p. 28. Reprinted by permission of *International Journal of Alternative and Complementary Medicine.*

When a scientific theory gathers the status of a dogma, the possibilities of new research being conducted in that area and the room for new theories on the matter become severely restricted.

Those who try to break through such barriers run the risk of castigation and prejudice. They find few champions in academia. The going is all uphill. Mentally they are pushing against the habits and collective unconscious of many powerful minds and help even of the simplest kind from the 'establishment' is rarely forthcoming.

Their task is not an easy one and many do not have the character, the time, the funds or the other necessary resources to do justice to their thesis. It then becomes easy for others to criticise their work, dismissing it from the viewpoint of 'established opinion' as the work of a misguided enthusiast without giving the real consideration it deserves.

Michael Cremo and Richard Thompson are therefore to be congratulated on spending eight years producing the only definitive, precise, exhaustive and complete record of practically all the fossil finds of man, regardless of whether they fit the established scientific theories or not. To say that research is painstaking is a wild understatement. No other book of this magnitude and calibre exists. It should be compulsory reading for every first year biology, archaeology and anthropology student—and many others too!

The 914 excellently produced pages of *Forbidden Archeology* take us through so many anomalies of fossil man—anomalies only according to modern theories—that unless every single one of these finds is incorrectly dated, documented and observed, man's present scientific theories of his own origins must now be radically re-assessed. If only one human fossil or artifact of the 50 or so meticulously documented and discussed from the Miocene or early Pliocene is correctly dated, then everything concerning the theories of human origins must return to the melting pot. And the evidence is that a large proportion of them are entirely credible.

Why then have they not been previously considered? Because the roller-coaster of habituated mind patterns and dogma has simply brushed them aside, as do creationists who—faced with all the evidence of ancient times—still insist that the world was created in 4004 BC, according to a preconceived opinion. The psychological processes are in both instances the same.

We are treated to Pliocene bones, including a skull from middle Pliocene strata near Castenedolo in Italy, maybe five million years old. Bones found in Carboniferous coal in Pennsylvania, at least 286 million years old and capped by two feet of slate rock, 90 feet below the surface. Footprints of human-like, bipedal creatures who lived in Carboniferous Kentucky and Pennsylvania and Missouri, too. Flint tools from the Miocene, 10 to 12 million years old, found in Burma and the same from even older Late Oligocene sands in Belgium. Hundreds of metallic spheres with three parallel grooves running around their equator, found in recent decades by South African miners in Precambrian mineral deposits 2.8 billion years old. And a great deal more.

The book is both entertaining and scholarly—a rare combination. It rolls along presenting its information in a logical and coherent fashion, making honest comment and assessment as it goes. There is nothing long-winded about it—only thoroughness. Data is not pressed into the service of any particular doctrine but presented and left to tell its own story. Words like 'possible' and 'not sure' are used quite commonly, a practice that demonstrates an intellectual honesty and integrity that would with profit benefit many proponents of the more conventional points of view.

Cremo and Thompson also describe the processes by which data gets suppressed consciously and unconsciously and discuss all the evidence upon which modern theories are founded.

Forbidden Archeology deserves to provoke discussion and controversy. It should not be swept aside or ignored. If the general scientific community once again puts their heads in the sand until the furor passes by, they will be guilty of negligence in their duty to the world at large as self-professed seekers of the truth of things.

3.2.2. William Corliss (1993) Notice of *Forbidden Archeology. Science Frontiers Book Supplement*, no. 89, September-October, p. 1. Reprinted by permission of William Corliss.

William Corliss has edited compilations of anomalous evidence in many fields of science. The following notice of Forbidden Archeology *is from a newsletter in which Corliss draws the attention of his readers to books he considers of special interest.*

Forbidden Archeology has so much to offer anomalists that it is difficult to know where to start. One's first impression is that of a massive volume bearing a high price tag. Believe me, *Forbidden Archeology* is a great bargain, not only on a cents-per-page basis but in its systematic collection of data challenging the currently accepted and passionately defended scenario of human evolution.

The first half of the book assembles "anomalous" evidence, by which the authors mean "rejected and suppressed" observations. Here are fat chapters on incised bones, eoliths, crude tools, and skeletal remains—all properly documented and detailed, but directly contradicting the textbooks and museum exhibits. The book's second half deals with "accepted" evidence, such as Java Man and the recent African and Asian skeletal finds. While anthropologists have not suppressed these data, they often interpret them to fit reigning paradigms. Cremo and Thompson provide alternate interpretations of Lucy, *Homo habilis*, and the famous skull ER 1470. There is even a 35-page chapter entitled "Living Ape-Men?" The salient theme of this huge book is that human culture is much, much older than claimed.

3.2.3. Steve Moore (1993) Review of *Forbidden Archeology. Fortean Times,* vol. 59, no. 72, p. 59. Reprinted by permission of Mike Dash, publisher of *Fortean Times.*

Charles Fort (1874-1932) was an American skeptic who spent most of his adult life compiling collections of facts contradicting accepted scientific ideas. The Fortean Society was created to carry on his work, and Fortean organizations are still active today.

Once upon a time there was a new idea called 'evolution' and all the brave scientists and academics set off to find the evidence with which they could trace the origins of the human race. It was the 19th century and the evidence soon mounted up, got reported and interpreted; some of it was explained and some of it explained away. These days, of course, we have a nice cozy picture of the way it all happened and if we can just find a few more pieces of the jigsaw, then we can all live happily ever after...

It isn't often that one can say that a book is truly imbued with the Fortean spirit, but *Forbidden Archeology* undoubtedly is, in spite of the fact that Fort's name is not mentioned once. Cremo and Thompson have launched a startling attack on our whole picture of human origins and the way that we've arrived at that picture: not only is the evidence impugned, but also the scientific method of handling it.

The book falls into two main sections, the first dealing with anomalous evidence (finds which seem to contradict the dominant view of human evolution), the second with defects in the accepted evidence on which that view is based. But just as important as the evidence itself are the preconceptions of the interpreters, and C & T manage to demonstrate how, as the prevailing dominant built up, those preconceptions became more important than the quality of the evidence, thus consigning the anomalous to oblivion.

C & T have taken on an enormous task. Much of the anomalous evidence dates from the 19th century and exists now only as reports; the extensive quotations from those reports accounts in large part for the vast size of the book. Beyond that, the controversy over each piece of evidence is fully covered, and we're allowed to see the process of exclusion at work, with its accusations of misinterpretation, its hints of cheating and its simple "I'm a more famous scholar than you..." And yet, of course, many of the adherents of the anomalous were among the most famous scholars of their day, and such technical matters as the stratification of deposits were as well understood then as now. Indeed, more than a century later, there still remain serious questions about our modern dating methods, as is pointed out in a lengthy appendix. There are occasions where the book strays into the more familiar (and possibly sensational) Fortean territory by dealing with the notion that the yeti / sasquatch / wild men reports suggest co-existence of species rather than linear evolution, and with more outrageous anomalies (though handled with suitable reserve) such as apparent shoeprints in ancient strata and puzzling artefacts found in solid coal. For the most part, however, it deals scrupulously with matters of palaeoanthropological record. It might be thought that 900 pages of detailed material on anomalously dated flint tools, broken bones and fragmentary skeletons might not make the most absorbing reading, but by providing the human controversies as well, the authors tell a gripping story. Thus, with a clear conscience, one is able to say "I couldn't put the book down", rather than simply "I couldn't pick it up"!

C&T avowedly believe that modern man has been upon this earth longer than currently believed and that our picture of human origins is seriously flawed. Be that as it may, they appear to have given a commendably even-handed treatment of the material and have certainly made a compelling case for its re-examination, both accepted and anomalous. Their own ideas on human origins are deferred to a promised second volume. If you have any interest in fossil humanity or, perhaps more importantly, the way scientists put together our picture of the universe, then you ought to read this vast and surprisingly cheap book...

3.3. Alternative Science Publications (Archeology)

3.3.1. Jean Hunt (1993) "Antiquity of Modern Humans: Re-Evaluation." *Louisiana Mounds Society Newsletter*, no. 64, Nov. 1, pp. 2-4. Reprinted by permission of Jean Hunt.

Jean Hunt, now deceased, was the founder of the Louisiana Mounds Society and editor of its journal.

A teeth-rattling new hypothesis has been proposed in *Forbidden Archeology* by Michael A. Cremo (writer) and Richard L. Thompson (researcher, Ph.D., mathematics, Cornell), a formidable (914 pages) book published in 1993 by Govardhan Hill Inc. and available from David Childress, Adventures Unlimited, Box 74, Kempton, IL 60946, $40 plus $4 s&h.

As soon as I read the introduction written by Michael, I knew that they had stumbled upon the same information I had, and at about the same time (1981 for me, 1984 for them), about the censorship and suppression practiced during the past hundred years by the academic establishment regarding human history. I predicted the discovery of multiple previous sophisticated civilizations, currently unsuspected, and my prediction has been supported by the increasing number of anomalous bits of evidence that kept cropping up, like the Ica Stones, which establishment figures, like Glyn Daniel and Stephen Williams, kept having to stomp on to get them back into the obscurity to which the establishment had consigned them years ago.

By its "stomping," the establishment had worked itself over the years into a tar pit with no exit, and the increased bellowing and lunging had only succeeded in working them deeper into the mire. At this point, in 1993, its credibility even with the least educated and informed elements of the general public is almost non-existent.

Here, in *Forbidden Archeology*, a very low-key and kind statement identified the phenomenon as it first began to appear: "When we speak of suppression of evidence, we are not referring to scientific conspirators carrying out a satanic plot to deceive the public. Instead, we are talking about an ongoing social process of knowledge filtration that appears quite innocuous but has a substantial cumulative effect. Certain categories of evidence simply disappear from view, in our opinion unjustifiably" (p. xxvi).

Later, the phenomenon is given a less favorable odor with the account of what happened when:

> ...Thomas E. Lee of the National Museum of Canada found [in the 1950s] advanced stone tools in glacial deposits at [a Canadian site]. Geologist John Sanford of Wayne State University argued that the oldest...tools were at least 65,000 years old and might be as much as 125,000 years old. For those adhering to standard views on North American prehistory, such ages were unacceptable. Thomas E. Lee complained: 'The sites discoverer [Lee] was hounded from his Civil Service position into prolonged unemployment; publication outlets were cut off; the evidence was misrepresented by several prominent authors...; for refusing to fire the discoverer, the Director of the National Museum, who had proposed having a monograph on the site published, was himself fired and driven into exile; official positions of prestige and power were exercised in an effort to gain control over just six...specimens that had not gone under cover; and the site has been turned into a tourist resort...[The site] would have forced embarrassing admissions that the Brahmins did not know everything. It would have forced the rewriting of almost every book in the business. It had to be killed. It was killed." (pp. xxix-xxx).

I knew all that already, having read about countless other similar cases in my own research. Most books written in the last 50 years are so flawed by their foundations of layer upon layer on honest error and dishonest self-interested censorship, that most simply perpetuate an accepted mythology; the exceptions are those dealing with input from other disciplines, such as the radiocarbon testing techniques from physics and DNA testing from medicine. It was necessary, for Cremo and Thompson, as it had been for me, to go back to older archeology and "antiquarian" books and papers presented before 1900 to get to the date before it had been twisted and corrupted by the academic interest groups.

This is what Cremo and Thompson did, and they have collected a mass of data about the earliest evidences of modern humans, the discussions and reasoning that took place in regard to them, and the forks in the road not taken which would have resulted in accurate assessment of the data and a true picture of human history. They begin their presentation with Darwin and Neanderthals, and end with a 13-page table (pp. 815-828) listing the evidence, item by item, which actually carries modern human existence back millions of years.

Here is an admirable statement of their working method:

> In discussing the anomalous and accepted reports...we have tended to stress the merits of the anomalous reports, and we have tended to point out the deficiencies of the accepted reports. It could be argued that this indicates bias on our part. Actually, however, our objective is to show the qualitative equivalence of the two bodies of material by demonstrating that there are good reasons to accept much of the rejected material, and also good reasons to reject much of the accepted material. It should also be pointed out that we have not suppressed evidence indicating weaknesses in the anomalous findings. In fact, we extensively discuss reports that are highly critical of these findings, and give our readers the opportunity to form their own opinions (p. 25).

The difficulties faced by most readers, casual or professional, are described in this paragraph:

> The case of Wright and Sollas shows how researchers who share a certain bias (in this case a prejudice against evidence for Tertiary humans) cooperate by citing a poorly constructed "definitive debunking report" (in this case by Breuil) as absolute truth in the pages of authoritative books and articles in scientific journals. It is a very effective propaganda technique. After all, how many people

> will bother to dig up Breuil's original article, in French, and, applying critical intelligence, see for themselves if what he had to say really made sense? (p.164).

Occam's Razor would have us accept that hypothesis which covers the most evidence with the least use of assumption. The Cremo-Thompson hypothesis covers *all* the evidence, even the "Fortean" anomalies which have previously embarrassed the establishment so: beginning on page 795, they discuss the coins, hammer handles and other fragments of tools buried deep within quarries; a nail in Devonian sandstone from North Britain; a gold thread in carboniferous stone, England; a metallic vase from precambrian rock at Dorchester, Massachusetts; a tertiary chalk ball from Laon, France; a coinlike object from the middle pleistocene in a well boring in Illinois; a petrified triassic shoe sole from Nevada; a cambrian shoe print in shale from Utah; a precambrian grooved metallic sphere from South Africa; and others.

To those examples must be added the Ica Stone art: Cremo and Thompson plan to write a second, companion volume, and I'm sure they'll want to include the inscribed pictures showing people attacking dinosaurs, people using telescopes and magnifying glasses, people operating and performing Cesarean sections and heart transplants. Since Ica Stones have been found in South American tombs sealed more than a thousand years ago and undisturbed since, no charge of "modern hoax" or "fraud" can be made. If someone wants to call them "antique frauds," then who were the people who knew about dinosaurs and sophisticated surgery a thousand years ago? (The photograph below, copyright 1993 Bill Cote, was provided by him for this use.)

The Cremo-Thompson hypothesis even offers a "home" for the most outstanding anomaly discovered in recent times: the face and pyramids of Mars. Mars was last habitable (if we may assume that this information is correct and not censored) 500,000 years ago; it is therefore quite within the scope of this hypothesis that people either went from Earth to Mars and built these structures—or that they came here and became our ancestors.

The exact places of discovery of the Ica Stone art objects have not been revealed by the principal researcher involved with them. This makes it very difficult to determine their age, even if the objects are genuine (and there is some dispute about this).

Related Correspondence

3.3.1.1. Letter from Jean Hunt, August 3, 1993.

David Childress has told me that you have written a book, *Forbidden Archeology*, which I should read and review for our *Louisiana Mounds Society Newsletter*. David said that his *Adventures Unlimited* will be selling the book at a price of $40, although it was published by Govardhan Hill Publications. I would, of course, give price and ordering information along with the review.

If you will send me a review copy, I will be glad to do so. Since you probably don't know us from Adam's housecat, I'm enclosing a copy of our current *LMSN* so you can judge for yourself what our interests are.

3.3.1.2. Letter to Jean Hunt, September 5, 1993.

I am very pleased to send you a review copy of *Forbidden Archeology*, and I hope you and your members will enjoy it—forbidding though it is in size and price.

You say that we "probably don't know you from Adam's housecat." Well, not quite. I had run across your name, and I was intending to get in touch with you. But, as fate would have it, you got to us first. I think we were meant to link up.

The research and writing of *Forbidden Archeology* took about 8 years altogether. And now that it is in print it has been very exciting to see it find its audience, which is not easy to reach through the normal bookstore channels. It really seems to be a matter of networking with all kinds of organizations such as yours and with people like Dave Childress and Adventures Unlimited. If you have in mind any other people or groups who might like to receive review copies, please let me know.

P. S. May I ask you to clear up for me the gender-ambiguity of your name?

3.3.1.3. Letter from Jean Hunt, September 18, 1993.

I am now a little less than half-way through your book, and I haven't found anything yet with which I disagree; your and Richard's approach to the material and use of it is exactly what I hoped it would be after my first, superficial overview. I am so happy now to know that there are people in the world who are actively campaigning for the same thing I am—truth in scholarship!

I am enclosing a blind copy of a letter to George Carter, which will explain the prediction I made over the phone—that he would not be receptive to your hypothesis—which seemed to disappoint you. It may be that his response to your book will be different...[*respecting the wishes of Jean Hunt, I've deleted some sentences about a dispute with George Carter*] I forgot to give you Dr. Cabrera's address in my last letter [*address deleted*].

You have my permission to quote anything I write you or anything in the *Louisiana Mounds Society Newsletter* (the Carter correspondence is the only exception). If you look in the "Invitation to Join" paragraph on the first page of the *LMSN,* you'll find blanket permission to quote given there, provided that the author and publication are properly credited.

I'm looking forward to reading the rest of your book this weekend. My suspense is already killing me: am I permitted to know your hypothesis about the origin of humans at this point? You are aware, no doubt, that the people who accept your viewpoint will find it very difficult, not only to continue their positive attitudes toward either Darwinism or creationism, but also toward any of the religions based on the Bible? Since I dismissed the currently-known (to me) religions long ago, that's no problem for me, but I can understand that lots of people would not agree.

3.3.1.4. Letter from Jean Hunt, September 19, 1993.

I finished reading your book last night; I felt punchy and cross-eyed. If I felt that way after just a marathon of reading it, then how did you and your co-workers feel about writing it, proofing it and all the rest of the dog-work that goes into getting such a book ready for publication? It is an admirable work, and you are all to be congratulated.

What you have done is to prove the case by giving "chapter and verse"; the rest of us—those who have tried to use the scientific method with existing archeological reports—*knew* what had been happening, but none of us had proved it on the massive scale that you have. I tried, in my small way, but I was not nearly as successful or comprehensive.

We are still a small and outnumbered group, and I have no doubt that it will be years before our positions will become generally accepted; but I have no doubt that that will happen in time. The growth of my own LMS shows the potential: from 6 members in 1986 to 400 in 1993, and all recruited by articles, book reviews and word-of-mouth.

I asked a friend, Ingo Swann, to check into the possibilities of getting a review published in *Fate* magazine, the only commercial publication that is likely to print it. *Fate* has a readership of about 50,000, and, when Ingo reviewed my book, the sales that came from it were unbelievable! I suggested that he contact you direct, if the editor approves, and that way the money can go directly to you, rather than being diverted to a "middle man." Since David Childress told me about it, I feel honor bound to list his organization in my review.

Ingo objected at first, because he said he wasn't expert enough in the field; so I asked him why he reviewed mine. He said he did it because he knew nobody else would! So then he realized what I meant and said he would try.

I have not revised my review as much as I'm going to. I changed the "600 million years" to "several million years" for political reasons: I don't want to take any chance on "scaring" people away in a preliminary review. Let them see your evidence first. I have added two quotes, which are highly identifiable because I've offset them into smaller type, along with one long quote I had already chosen.

Please let me know if you have any objection as soon as possible, because I plan to start sending out advance copies singly to certain people in private correspondence. Colin Renfrew is one of those to whom I plan to mail a copy. I have been urging him to attempt a "clean up" for quite some time now; it's a wonder that he continues to read and respond to my letters. If he decides to do so, then we will have a major breakthrough—but don't hold your breath!

The *LMSN* is going to be a treasure trove of "anomalies" for you. Your item about the dinosaur-and-human footprints in Russia (p. 458) can now be combined with the same thing in Texas. The Volume II you have contains an index by subject in the back. We had several articles on the dinosaur-and-human footprints in Texas, now enthusiastically publicized by the creationists.

The names of the people who provide information and follow up on the items are given there, and I have their mailing addresses and phone numbers.

I hope that, by the time you receive this, you will have received Bill Cote's videos on Mars and the Ica Stones. His group will be seeking funding for the making of the second documentary (after the one on the Sphinx) which will be about the Ica Stones and Marcahuasi. It is fortunate that your book and his television plans have coincided in time.

3.3.1.5. Letter to Jean Hunt, September 27, 1993.

This letter was in response to a telephone conversation with Jean Hunt, during which she suggested I send a copy of Forbidden Archeology *to her friend Bill Cote, a New York television producer in exchange for copies of his videos. Cote was later responsible for the NBC television special* Mysterious Origins of Man, *which featured material from* Forbidden Archeology *and interviews with its authors. I also sent a review copy of* Forbidden Archeology *to George F. Carter, author of* Earlier Than You Think, *a controversial work about evidence for a human presence in North America predating that allowed by conventional archeologists.*

I very much appreciated being able to speak to you on the phone recently. Thank you for the introduction to Bill Cote. I am sending him a copy of *Forbidden Archeology*.

If you have an address for George Carter, I would like to send him a copy as well. Probably he won't like everything about the book, but since I did talk about him and his discoveries he might like to see it.

Please find enclosed a check for my membership in the Louisiana Mounds Society.

3.3.1.6. Letter to Dr. Barry and Reneé Fell from Jean Hunt, October 1, 1993.

In a phone conversation, Jean suggested that I personally deliver a review copy of Forbidden Archeology *to Dr. Barry Fell, the Harvard biologist who devoted the latter part of his life to studying inscriptions and other evidence suggesting that Old World peoples had visited the Americas before the voyages of Columbus and Leif Erikson. Jean wrote to Fell to prepare the way for a meeting. Both Fell and I were living in San Diego, California at the time.*

You'll be delighted to know that I wrote George Carter yesterday, apologizing to him for losing my temper with him a year ago—and telling him what he did to make me lose it—and trying to get back into a friendly relationship. When there is a need to "circle the wagons," like right now, it's best to bury old conflicts.

Part of what I wrote him—and John Dan's death—sparked some thinking about my own wishes and role in life, and I got up early this morning to write a rather whimsical piece on the subject. I don't know if it's appropriate for publication in the next ESOP, but you're welcome to use it if you think so. The more I re-read it, the better I like it; I think I'll use it myself in the next LMSN. I think I'll also send a copy to Jim Guthrie and Jim Whittall.

I spoke to Michael Cremo about his book, *Forbidden Archeology*, and he agreed to deliver to you a complimentary copy—you're one of his heroes, too. I know that you and Reneé will be delighted to have a 900+ page book to occupy your "leisure" time!

3.3.1.7. Letter to Jean Hunt, October 9, 1993.

Thanks for everything. I don't know if I shall get to visit Barry Fell before I go to India. I have been having problems with visas, and there is a whole stack of other things that I absolutely have to get done before I go. If I get through them all, there is some chance I can still deliver the book, but if not I'll have to try when I get back.

3.3.1.8. Letter from Jean Hunt, October 11, 1993.

I know you're away, so this letter will just sit there and "age" until you get back; there's nothing in it that will spoil on the shelf!

Ingo Swann has agreed to review your *Forbidden Archeology* for *Fate* magazine and its 300,000 readers. Nobody knows in which month it will be printed, but Ingo will send me an advance copy and I'll send it on to you.

I'm enclosing a letter and photograph from Evan Hansen, together with a copy of my letter responding to him. I sent Evan an advance copy of my review of your book, and this is in reply to that. Evan lives in Utah, knows the countryside well, and has a deep and abiding desire to "discover" something important. He is not well educated and does not do much "homework." He has made claims in the past that proved out to be fantasy rather than reality; he is not a "liar," just imaginative. Once in a while he comes up with a real gem, though, and it is a mistake to write him off entirely. I examine his claims and sometimes check them out with people more expert than I am in whatever the subject is, and sometimes publish his articles; but mostly they don't check out.

As it happens, your book is probably going to get more good attention than expected: a major confrontation has developed centering on Barry Fell. A book, *Ancient American Inscriptions*, was published recently which turned out to be a malicious, dishonest "hatchet job" on Barry and some of his closest associates. I decided to call for rebuttal information, to be printed in *LMSN,* starting with this issue (#64); the first "errors of fact" are about copyright violation and plagiarism, and the plagiarism is documented by Barry. I think our membership will increase greatly, because Barry has ordered (and paid for) 600 extras copies to be inserted in his book, *ESOP* [*Epigraphic Society Occasional Papers*], which will be mailed out in early November. All this developed over a two-week period, and I had no idea of the "happy" outcome for us when I started my project. Of course, I may also get sued, but I doubt it; the material I'm printing is truthful, and that's an absolute defense against a successful libel suit.

I'm looking forward to your return.

3.3.2. Bill Welch (1995) Review of *Forbidden Archeology*. *Stonewatch*, vol. 13, no. 2, Winter, p. 10. Reprinted by permission of David Barron, editor.

This is one of the most remarkable books to appear in many years, and is recommended to anyone with the fortitude to read 800 pages. Even merely sampling the material would be rewarding. Although repetitious, it is so for an obvious reason. Equally obvious is the scholarly research and impeccable documentation of the authors' surprising thesis. The secondary thesis will be familiar to most readers of *Stonewatch*, since we have dealt with the same problem regularly since the emergence of evidence for cultural diffusion.

Cremo and Thompson began their research for religious reasons, but that is insignificant compared to the prodigious task they took on and the conclusions to which their work inescapably leads. They have pointed out fully documented scientific evidence for the astonishingly early existence of "anatomically modern" humans. In many cases, the evidence was examined and accepted by commissions or committees of respected members of the professional scientific community. Most was produced by respected professionals, and withstood the critical challenges of peers.

The conclusions are astonishing because of the currently prevailing view most of us have been taught, that humans evolved as recently as 100,000 years ago. Yet there is that persuasive and once-accepted evidence that our species [have] been around many times longer. Unmistakable artifacts have been found imbedded in stone and in coal. Manufactured tools have been found in geologically dated strata of great age.

Not surprising is the clear case made for the systematic dismissal of the evidence, and omission of it from professional discussion and texts, an all-too-familiar practice of contemporary professional scientific communities in other disciplines also of interest to Gungywamp Society members. It is clearly shown that some of the technical tests used to establish younger ages for remains are seriously flawed and unreliable. Stratigraphic dating is cited as more reliable, properly so, although not for absolute dating in more than very approximate terms.

One little recognized fact of science called to our attention by this book is that most scientific information is inferential, not directly observed. Many professionals cite scientific theories and opinions as fact, creating a widespread public confidence in their pronouncements. This is unfortunate, often obscuring vital information and misleading the general public.

The authors demonstrate repeatedly that two different sets of criteria have been applied in accepting or rejecting evidence by the contemporary professional scientific community. The preferred evidence has been accepted on very liberal criteria, while contrary evidence has been rejected on far more critical bases, which in many cases can be easily challenged on technical grounds. Specific cases show the "establishment" has gone to extreme lengths to suppress evidence which disputes the prevailing views. Overall, the views of these authors and the evidence they have compiled warrant serious consideration.

Related Correspondence

3.3.2.1. Letter to David P. Barron, August 20, 1994.

David Barron is president of The Gungywamp Society and editor of Stonewatch, *the Society's journal.*

Having recently read over a copy of your newsletter, I am sending you a review copy of my book *Forbidden Archeology*. Although it is not specifically about your focused area of interest, I think it does address some larger concerns about the treatment of archeological evidence by the establishment scholars. If you do decide to review the book, please send a copy of the review. The book is generating quite a bit of controversy, as you might imagine.

3.3.2.2. Letter from David P. Barron, September 1, 1994.

I am writing to confirm the safe arrival of your book *Forbidden Archeology* and to thank you for your generous kindness. Having published a bit myself, I know how expensive it is and how selective one must be in offering copies for review.

In view of the sheer size and scope of your fine publication I strongly doubt if I can get it read and properly reviewed for our Fall issue of *Stonewatch,* but review it I will. I think it is too important an addition to the literature to pass off onto one of our other three reviewers and I will undertake the project myself. The Winter issue comes out in December, 1994 and I anticipate that timing for the review.

My late friend Jean Hunt, with whom I had many fine debates and disputes, raved about your book and I had intended to secure a copy in the future, just to see what she was so enthusiastic about. You have solved that problem neatly and I will try to do my most positive best to present it to our 650+ readers.

Should you have a small illustration cut suitable for accompanying the review, I would be glad to insert it, otherwise I could reduce the front cover on my Canon. The choice is yours, OK?

Barron did pass the book to one of his other reviewers.

3.3.3. Colonel W. R. Anderson (1993) Notice of *Forbidden Archeology*. *Vikingship: Bulletin of the Leif Ericson Society*, vol. 29, no. 4, p. 6. Reprinted by permission of *Vikingship*.

The Leif Ericson Society is concerned with evidence for a Norse presence in North America before Columbus.

This book doesn't have a word about Vikings, but I found it extremely gratifying. Primarily, it explains and denounces the vicious efforts of the Academic Establishment to protect its outmoded status quo. It vindicates two good friends: George Carter of Texas A&M, and the late Tom Lee, professor of Laval University and longtime editor of the *Anthropological Journal of Canada*, both of whom have made important contributions to these pages. It exposes the outrageous methods of Phuddy Duddies to protect their seats on their fat er, cushions, and their intense efforts to bury—literally and figuratively—evidence of their outdated rote. I found hilarious the tale about a student of Tom Lee, who wrote a book with Tom's findings in Sheguiandah, Ontario, indicating that the Phuddies were out-of-date by about 100,000 years. The offending chapter was deleted en masse—*but they forgot to exclude the extensive bibliography.*

It is over 900 pages of fascinating, the depressing, tales of academic misdeed.

3.4. Reviews from Religious Perspectives and Publications

3.4.1. Peter Line (1995) Review of *The Hidden History of the Human Race*. *Creation Research Society Quarterly*, June, p. 46. Reprinted by permission of the Creation Research Society, P.O. Box 8263, St. Joseph, MO 65408-8263.

The Creation Research Society Quarterly *is the main journal of young-earth Christian creationists.*

This book is must reading for anyone interested in human origins. I was surprised to come across the book in a major U.S. book chain store, given that its contents strongly contradict the standard view of human origins. Although the authors do not explicitly state how they believe the human race came into being, they make it clear that their theoretical outlook is derived from the Vedic literature in India, which supports the idea that the human race is of great antiquity.

Phillip E. Johnson writes a favorable foreword to the book, which is a condensed version of a much larger unabridged volume, but makes it plain that he does not share the authors' religion or motivation, as neither do I. The authors present evidence for the existing of man in the Pliocene, Miocene, and even earlier periods, if one assumes that the evolutionary time scale and dating methods are valid. A lot of the evidence is based on stone tools, but there is also anomalous human skeletal remains mentioned as well.

The authors have made accessible to others valuable information that is not readily available due to the existence in the scientific community of "a knowledge filter that screens out unwelcome evidence." (p. xvii). Particularly informative is the authors' discussion involving "morphological dating," where the morphology of a fossil often determines the date assigned to it. Also informative are the examples showing the ease at which contradictory evidence is dismissed or discredited by members of the scientific establishment, whereas evidence in support of the orthodox evolutionary view is accepted all too readily.

This book presents ample evidence, most of which is open to dispute no more than "orthodox" evidence for evolution is, that humans have always existed alongside their supposedly ape-men ancestors. I found the most controversial part of the book the chapter on living ape-men, which incorporates many "wildmen" stories, some of the stories being too far-fetched.

As a recent earth creationist, I would not accept the evolutionary time scale that the authors appear to accept. However, the authors have shown that even if you accept the evolutionary view of a vast age for the earth, the theory of human evolution cannot be supported.

3.4.2. Dick Sleeper (1995) Notice of *The Hidden History of the Human Race*. Dick Sleeper Distribution, Inc. Catalog, Winter/Spring, p.1 Reprinted by permission of Dick Sleeper.

Dick Sleeper is a wholesaler of books to Christian bookstores.

While not written from a Christian perspective, this book will provide comfort to those creationists who have been frustrated by the prejudices of the scientific community. The authors clearly demonstrate that massive evidence attacking the widely held views of evolutionists has been suppressed, ignored, or forgotten. Deploying a great number of convincing facts, deeply illuminated with critical analysis, the authors challenge us to rethink our understanding of human origins, identity and destiny.

"A stunning description of some of the evidence that was once known to science, but which has disappeared from view due to the 'knowledge filter' that protects the ruling paradigm."

Philip E. Johnson, Author of *Darwin on Trial*

3.4.3. Salim-ur-rahman (1994). "Spanner in the works." *The Friday Times*, April 21, 1994. Reprinted by permission of Salim-ur-rahman.

The Friday Times *is an English-language newspaper in Pakistan. I was grateful for this review from an Islamic author.*

How old is man, factually speaking. Not very, if we subscribe to current scientific theories about his origin. At present it is believed that anatomically modern humans evolved fairly recently, about 100,000 years ago in Africa and perhaps elsewhere.

From a scientific point of view the scenario is straightforward and evolutionary correct. At first there were ape-like creatures. Then, about a million years ago, up the evolutionary ladder, came *Homo erectus*. He was succeeded by *Homo habilis* or *Homo Sapiens.* As simple as that. A cut and dried case. The discovery of Java man, now classified as *Homo erectus*, served as a clincher. The stratum in which the skull of the Java man was found was considered to be 800,000 years old, that is, belonging to the Middle Pleistocene. According to the scientific circles, the main argument was as good as settled and only a few loose ends remained to be tied up.

Unfortunately, as Cremo and Thompson's book amply and ably demonstrates, nothing should ever be taken for granted. It is unscientific to be cocksure. The element of *hubris* which has crept into scientific thinking is as inconvenient as it is dangerous.

But back to the book in question. The authors claim, and go to great lengths to show, that the story of man's evolution or of his presence on earth is too involved and the evidence too controversial. Nothing can be settled to a nicety. Their book makes it plain that suppression of enigmatic data is not only common but also has a long history. If the facts do not agree with a favoured theory, then the scientists and researchers simply discard such facts, even an imposing array of them. In other words, the motto among research circles is a pragmatic one: save the theory you believe in at all costs and to hell with facts which don't fit in.

For instance, after Darwin introduced his theory, numerous scientists discovered incised and broken bones and shells suggesting that tool-using humans or human precursors existed in the Pliocene (2.5 million years ago), the Miocene (5-25 million years ago) and even earlier. To accept this evidence would have left the neat evolutionary theory in shambles. Therefore all such data was either suppressed or ignored or, worse still, ridiculed.

Forbidden Archeology is more than a well-documented catalogue of unusual facts. It is also a sociological, philosophical and historical critique of the scientific method, as applied to the question of human origins and antiquity. From the evidence gathered by them the authors conclude that the now dominant assumptions about human origins are in need of drastic revision.

They have identified two main bodies of evidence. The first is a body of controversial evidence (A) which shows the existence of anatomically modern humans in the extremely distant past. The second is a body of evidence (B) which can be interpreted as supporting the currently dominant views that anatomically modern humans evolved fairly recently, about 100,000 years ago.

They also identify standards employed in the evaluation of palaeo anthropological evidence. After detailed study they found that if these standards are applied equally to A and B, then we must either accept both A and B or reject both. They have also recounted in detail how the controversial evidence has been systematically suppressed, ignored or forgotten, even though it is qualitatively (and quantitatively) equivalent to evidence favouring currently acceptable views on human origins.

There is a very strong reason why so much uncertainty prevails in this particular field and why it would continue to plague research work even in future. All attempts to show the evolution of species (the human species in particular) must rely on the interpretation of fossils and other remains found in the earth's strata.

I had the impression that the geological record was, barring a few exceptions here and there, generally intact. After going through this book I stand disabused of my belief. The drastic incompleteness of the fossil record remains a critical factor in palaeontology. Geological record is exceedingly incomplete. An observation in the USA showed that in a particular instance, out of the 6 million years record, as many, as 5.9 million years of strata were missing. It is assumed that about 4.1 million fossillisable marine species have existed since the Cambrian period some 600 million years ago. Yet only 93,000 fossil species have been catalogued. Conservatively, the ratio is one out of every 44 species. But there is reason to believe that the ratio is one out of every 100. When we turn from marine organisms to the totality of living organisms the situation only gets worse. It is estimated that 982 million species have existed during the earth's history, compared with the 130,000 known fossil species. This is only .013 of one per cent.

Forbidden Archeology is a serious and thought-provoking book, reminding us that the history of the human race may be far more older than we are led to imagine, that anatomically modern man may have co-existed with more ape-like creatures millions of years ago. Perhaps these ape-like creatures still exist. There is a whole chapter about them. Also one about strange artefacts which are occasionally found in strata unimaginably ancient.

In a way all this ties up with the remarks attributed to the Holy Prophet (PBUH) Hadhrat Ali and Imam Jafar Sadiq in which they said that the Adam we are descended from was preceded by numerous Adams and their progeny.

3.4.4. Mikael Rothstein (1994) "*Forbidden Archeology:* Religious Researchers Shake the Theory of Evolution." *Politiken,* January 1, Section 3, page 1. Reprinted by permission of *Politiken* and Mikael Rothstein.

Mikael Rothstein is assistant professor of religious history at the University of Copenhagen, Denmark. This article appeared in the science section of Politiken, *Denmark's largest newspaper.*

Human beings like us have lived on the Earth for millions of years. In order to substantiate the theory of evolution scientists who study the origin of man have deliberately suppressed or even destroyed evidence, claim two researchers in a remarkable book.

When Charles Darwin published his pioneering work *The Origin of Species* in 1859, he suspected his theory of evolution would cause a stir. According to his theory of evolution, man was reduced to a creature on par with all other living beings.

The theory of evolution also disputed the religious thesis of the simultaneous creation of the species, appearing at the time of Darwin as they had done since the time of creation.

The theory of evolution challenged the unique qualities of man and the divine nature of the creation; Darwin himself became a victim of the vicious teasing and humiliation of his times. He was pictured as a monkey in the newspapers and his relation to worms and maggots was made clear because he said the species evolved from each.

However, the theory of evolution was not that easy to subdue, and with persistent support the new view survived and became within the next decades a real alternative to the theory of creation of the church.

Later on, especially when paleontology and paleoanthropology (the study of the evolution of man) set sail, great parts of theology yielded and formulated an adjusted creation theory. Now it was said that the species probably had evolved somehow or other, but God was behind.

The Roles Are Changed

So now today it is 1994, and the scientific worldview has cemented its influence in all ways. While previously science had to justify itself through religion, religion today has to prove that its dogmas and conceptions are scientifically valid, if it wants to be taken seriously. The roles have been changed. Therefore, the provocations of today come from religion.

By the way, regarding the discussion of the origin of man, there is a new book, *Forbidden Archeology*, which was published last year, which is exciting reading material. The book, which takes up more than 900 compact pages, is a *tour de force* through innumerable archeological and paleoanthropological facts. The authors aim to prove that it is not possible to maintain the theory of evolution when the facts are examined in their totality.

The two authors, Richard Thompson and Michael Cremo, have spent around eight years researching the book before sitting down at the typewriter. The result is in principle just as provoking as *The Origin of Species*.

Hard to Turn Down

The conclusion is hard as stone: the authors claim that the established group of scientists, who deal with the descent of man, has, in order to make the theory of evolution fit, consciously suppressed or even destroyed evidence.

Now, dishonesty in that field is not an unknown phenomena. In the dawn of archeology and paleoanthropology, there were quite a few examples of fraud. On the other hand, it is not unusual that those who are up against the dominating views see conspiracies everywhere.

Objectively speaking, however, it is difficult to turn down great parts of the evidence that the authors present: Why has a long list of problematic findings not been treated scientifically? Why are there purges in the scientific collections? Why do they ignore imbalances between chronological and geological assumptions?

Thompson and Cremo take the reader through several hundred years of research, mention hundreds of cases, end up in all continents, and on the basis of an impressive file, unearth documents which were forgotten, but which conclusively affect the present position of science.

Among other things, they claim that men like us have been on the Earth for millions of years—generally man (*Homo sapiens*) is considered to be around 200,000 years old.

Forbidden Finds

The proof for the significantly older age should be—among other things—the so-called out-of-place-artifacts, that is, objects made by men, which appear at places where they should not—e.g. a shoeheel from the Triassic (i.e. 200 million years old), a nicely-made gold thread in sediments more than 320 million years old, a metallic jar more than 600 million years old in rocks from the Precambrian.

Most solid is, however, the analysis of the fossil finds and the archeological procured objects, which are analyzed by the same method that paleoanthropology and archeology normally employ.

What is new, Thompson and Cremo say, is not the method, but the material that is under study. If the established group of scientists delved into the suppressed material with an open mind, they would, it is claimed, reach the same results.

Since the arguments rest on a long list of examples, the authors expect that the serious critic will systematically refute them all; because that is what Thompson and Cremo have done, to systematically question the time-honored conceptions bit by bit.

Read Critically

On the one hand, the authors have written an interesting history of science. On the other hand, a genuine thriller.

Part of the story is, however, that both the authors belong to the Bhaktivedanta Institute in San Diego, which is the academic center for ISKCON (The International Society for Krishna Consciousness), a part of the Vaisnava religion from India (a branch of Hinduism).

The authors—who are trained not only as scientists and mathematicians, but as monks—thus also have a missionary and theological project. Their otherwise thorough academic argumentation can thus find support in the Vaisnava mythology, which actually describes the history of man and the geological development of the Earth in a way that is compatible with the conclusions Thompson and Cremo provide.

One therefore has to read critically, but that one has to do always, anyway.

A number of reviewers abroad have praised the book to the skies. Others have condemned it as nonsense. It seems, however, difficult to refute the concrete documentation as wrong quotations, manipulation with facts, etc. More difficult is it—for me, also—to accept that the evolutionistic faith of our childhood is not so safe and sure anyway, even though it can be proved.

Some, however, delight in the provocation. Also, in Denmark there are religious scientists, who argue creationistic views (Christian thought). More people would have a hard time to change horses midstream and ride along the creationistic path.

By reading Thompson and Cremo's book one can thus get a glimpse of the feeling the people of the church experienced when Darwin's theory was presented. The discussion is not only about facts. It is about our self-perception as human beings and about our ideas of the world we live in.

Related Correspondence

3.4.4.1. Letter to Richard Thompson and Michael Cremo from Mikael Rothstein, November 11, 1993.

The other book referred to in this letter is Richard Thompson's Alien Identities, *which examines the modern UFO phenomenon from the standpoint of the ancient Sanskrit writings of India.*

You were so kind as to send me your most recent books for review some months ago. I have read both, although your vast account on the *Forbidden Archeology* is somewhat difficult to read from cover to cover! I find both books most interesting, well written and—each in its own way—important. A review article on the *Forbidden Archeology*-book will appear in Denmark's largest and most important newspaper (*Politiken*) within a month or so. It has been accepted for the weekly "science-page".... My review (among other things) states (translated from the Danish): "The book is a detective novel as much as a scholarly *tour de force.* But the murderer is not the butler. Neither is the victim a rich old man with many heirs. The victim is Man himself, and the role of the assassin is played by numerous scientists. The book takes the case to court, and asks the reader to judge for himself. What can we do? How do we avoid questioning otherwise infallible truths after reading this minute and well-documented account? It takes more courage to recognize one's [own] misjudgment than to challenge other's. The burden lies with the reader. I personally must admit that the argumentation provoked me thoroughly. I certainly don't feel comfortable. After all, I always thought that Man was about 100,000 years old, but now sincere people, with good intellectual and academic arguments, tell me that I am wrong. The book presents many disturbing facts, but, of course, these facts only become really meaningful if you are willing to accept that current scientific notions about Man basically are wrong. I shall look forward to the reply from more traditional scientists in this field...."

This kind of religious argumentation, I believe, reveals important element[s] in the relation between religion and secular society in general, and in this case Vaisnava-bhakti and the surrounding intellectual milieu in particular. As a historian of religions this obviously interests me.

3.4.4.2. Letter to Mikael Rothstein, December 30, 1993.

I am very pleased to learn that you have written a review of my book *Forbidden Archeology*, coauthored with Richard Thompson.

When the review comes out in *Politiken,* please send us a copy for our files. I sent review copies of *Forbidden Archeology* to some of the academic journals in religious studies. Hopefully some reviews will be forthcoming. If you have any knowledge of a journal that might be likely to review the book, please let me know and I will send them a review copy, if I have not already done so. I think you are correct that our book "reveals an important element in the relation between religion and secular society in general, and, in this case, Vaisnava-bhakti and the surrounding intellectual milieu in particular." Anything that you can do to bring this point to the attention of scholars in your field would be quite valuable, in my humble opinion. If you know of any upcoming conferences in history of religion or related disciplines that might have panels or presentations on this topic, please let me know.

3.4.4.3. Letter from Mikael Rothstein, February 2, 1994.

I hereby have the pleasure to forward you my brief article on your work *Forbidden Archeology*. As it may appear to you—although the article is in the strange Danish language—it has been published in Denmark's largest, and most influential, newspaper, *Politiken.* The editors did not want a genuine review after all, but rather an article which set your claims in perspective. The text refers your points through examples and compares your message to that of the evolutionists at Darwin's time in order to demonstrate how the positions have changed: Today the creationists deliver the provoking news. Previously this was the function of the evolutionists. The article acknowledges your solid argumentation which is often more than hard to refute, but I do not present any judgment as to whether you are right in your conclusions (but then, I *am* a real *karmi*!). However, I find the book amazing in many ways and hopefully I have made my modest contribution to get it sold.

ISKCON in Denmark has responded to the article with appreciation!

For the record: The article is placed in the specific science-section of the paper, and it is entitled (as you may understand after all) *Forbidden Archeology*. The subtitle reads: "Religious scientists provoke the theory of evolution". "Religious" because I mention your affiliation with ISKCON and state that you have a religious interest in your otherwise scholarly enterprise.

3.4.4.4. Letter to Mikael Rothstein, April 24, 1994.

Thank you for sending the copy of your *Politiken* article about *Forbidden Archeology*. One of the ISKCON members from Copenhagen recently wrote to me that a Danish publisher wants to do an edition of the book and that Danish national television is interested in interviewing me and Richard Thompson. So your article had quite an impact.

I do not, however, read Danish. Even my fairly good reading knowledge of German and a passing acquaintance with Icelandic have not allowed me to completely penetrate the article, although I can pick out some words and sentences.

In an earlier letter, you included a very quotable passage from the review you said would be published in *Politiken*. You wrote that *Forbidden Archeology* was "a detective novel as well as a scholarly tour de force." You then developed the idea of the book as a murder mystery, with man the victim and many scientists the murderers. Am I correct that these passages do not appear in the article that was finally printed?

3.4.4.5. Letter from Mikael Rothstein, May 16, 1994.

I am glad that my article (*article* rather than review) on your book has had an impact. I have also had a number of reactions to the article myself. A week ago, while I was in Paris, France, I also met people who had read it.

I don't blame you for being unable to penetrate the strange language of Danish. Just wait until you hear how it is spoken! However, your impression is correct. The sentence you mention did not appear. What I sent you was a comment to the article as I presented it to the paper. However, because of the editors' desire for unforgettable pictures, I had to cut the article with some 40%. As a matter of fact the editors suggested a revised text which I accepted. A sentence you may be interested in (for PR purposes?), which is in the text, is the following (last paragraph, line 1-4 under the heading "Las kritisk" i.e. "read critically"): *On the one hand, the authors have written an interesting history of science. On the other hand a genuine thriller.*

Towards the end of the text I point to the fact that you (the authors) are associated with the Bhaktivedanta Institute/ISKCON, which, of course, means that you have scholarly as well as religious motivations behind your work. The first third of the article uses the old theological reaction against Darwin's theory to show how you are suggesting just as provocative an idea today, only the positions have changed completely. The last sentence of the article reads: "The discussion is not only about facts. It is about our self-perception as human beings and about our concepts/notions/ideas of the world we live in."

3.4.5. Richard Heinberg (1995) "The Lost History of Humankind." *The Quest*, Winter, pp 24-31. Reprinted by permission of *The Quest*.

The Quest *is a publication of one of the major Theosophical organizations. The passage below is from pages 28-29 of "The Lost History of Humankind," which deals with alternative views of human civilization.*

Perhaps the most shocking unorthodox new book having to do with the human past is Michael Cremo and Richard Thompson's *The Hidden History of the Human Race* (Govardhan Hill, 1994), a condensation of their daunting 952-page *Forbidden Archeology* (1993). In both books, the authors collect the evidence that mainstream archaeologists have rejected—bones of anatomically modern humans in geological formations tens or even hundreds of millions of years old; artifacts recovered from mines and coal beds; signs of human presence in the Americas up to 750,000 years ago. They also reevaluate the accepted evidence of the human evolutionary past—the bones of *Australopithecus, Homo erectus,* and Neanderthal, and show convincingly that this evidence has passed through a "knowledge filter" whose purpose is to perpetuate a reigning paradigm. Whatever evidence fits the paradigm, no matter how flimsy, is accepted; whatever doesn't, no matter how solid and unequivocal, is suppressed.

Along the way, Cremo and Thompson compare all of the Australopithecine/*Homo erectus* data with modern reports of living apemen (the Yeti of the Himalayas, the Sasquatch of the Pacific Northwest, and the Yeren of southern China). Perhaps, they suggest, the apemen who lived a couple of million years ago were not our ancestors; they were merely other primate species who co-existed with *Homo sapiens* then, just as the Yeti and Sasquatch do to this day. The authors do far more than push the temporal borders of civilization back a few thousands years; they question the basic premises on which we have based all our ideas about the prehistoric human past. They don't offer an alternative theory; they merely show that the one that is dominant today is based on an extreme form of intellectual tunnel vision.

3.4.6. Ina Belderis (1995) "Will the Real Human Ancestor Please Stand Up!" *Sunrise*, April/May, pp. 111-118. Reprinted by permission of the Theosophical University Press.

Sunrise *is a publication of the Theosophical Society. Ina Belderis is the librarian for the Theosophical Society Library in Pasadena, California.*

Most of us have seen the classic illustration of human origins: a pictorial sequence of apelike beings progressing from left to right, with each succeeding "apeman" showing more and more human traits, and finally ending up with an anatomically modern human being. This is the scientific picture of human evolution put forward for many decades. Does the image still stand: Paleoanthropological finds of the last 25 years seem to challenge it thoroughly, but its most serious challenge may very well come from anthropology's own historical records: when these are critically analyzed, they reveal a widespread evolutionary prejudice.

The most extensive analysis of this kind is *The Hidden History of the Human Race* by Michael A. Cremo and Richard L. Thompson. The authors pose two hypotheses based on the Vedic scriptures: the human race is much older than now generally accepted, and various human-like and apelike beings coexisted for long periods in the past. They claim that anomalous finds that do not fit the accepted theory of human evolution have been reported for years, especially in the late 19th century and early 20th. Yet these have been ignored and suppressed by scientists to such an extent that most are now virtually unknown. According to the authors, scientific preconceptions have caused such anomalous finds to be categorically rejected and severely criticized.

The book starts by explaining several key limitations of paleoanthropological research. First, discoveries are fairly rare and have often been made under questionable circumstances. As soon as something is found, dug up, and taken elsewhere, essential elements—such as its exact position in the strata—are destroyed, and afterwards one is dependent on the testimony of the discoverers. Many of their statements depend on various observations and conclusions about geological layers and disturbances within them. The testimony of one person may be different from that of another. Then there is the temptation to cheat, which can be systematic (as in the Piltdown case), or less premeditated (as in reports omitting research material which does not quite fit the desired conclusions). Also, modern chemical and radiometric dating is not without its limitations. Contamination may influence the result, or preliminary calculated dates are sometimes rejected or accepted on the basis of arguments that are not always clearly stated or published. Since paleoanthropological reports tend to provide incomplete information about "complex, unresolvable issues," the authors decided to compare the *quality* of different reports. The first part of the book discussed numerous reports of so-called anomalous finds of artifacts and human skeletal remains. The second part describes reports of finds that have been accepted by scientists to support the prevailing ideas on human evolution.

It should be noted that in Darwin's time there was no established theory of human descent—no sequence of apelike beings and dates—because there had been no discoveries of hominid fossil remains except for two Neanderthal skulls and some finds of modern morphology. It was not until the early 1890s when Eugene Dubois discovered "Java man" that a sequence with dates was theorized. These protohuman fossils later became known as *Homo erectus,* and since they were found in Middle Pleistocene deposits, they were said to be 800,000 years old. This discovery functioned as a benchmark, according to Cremo and Thompson: "Henceforth, scientists would not expect to find fossils or artifacts of anatomically modern humans in deposits of equal or greater age. If they did, they (or someone wiser) concluded that this was impossible and found some way to discredit the find as a mistake, an illusion, or a hoax" (p. 7).

Thus the *anomaly* was born...as nothing was anomalous before the established theory came into being. A number of anomalies had been discovered in the 19th century by reputable scientists, who found skeletal remains of anatomically modern humans in fairly old geological layers (Pliocene and Miocene). In addition they found numerous stone tools and bones that showed signs of the activities of human beings. Since the human evolution theory took shape, these finds have been ignored and rejected. "Knowledge filtration" has prevented scientists from (re)examining these reports. Also scientists have more or less stopped looking for artifacts and remains in older layers that lie outside the possible scope of the theory. When anomalies turn up, they are judged by very strict standards, while finds that *do* fit the theory are judged by very lenient ones. Some of the stricter standards have been described by anthropologist James B. Griffin: a proper site must have a clearly identifiable geological context (no possibility of intrusion); it must be studied by several expert geologists (and there must be substantial agreement among them); there must be a range of tool forms, well-preserved animal remains, pollen studies, macrobotanical materials, human skeletal remains, dating by radiocarbon and other methods. The authors of *Hidden History* point out that

> By this standard, practically none of the locations where major paleoanthropological discoveries have been made would qualify as genuine sites...most of the African discoveries of *Australopithecus, Homo habilis*, and *Homo erectus* have occurred not in clearly identifiable geological contexts, but on the surface or in cave deposits....Most of the Java *Homo erectus* finds also occurred on the surface, in poorly specified locations. —p. 89

The book goes on to discuss six kinds of anomalous finds: broken and incised bones, eoliths (chipped flints), crude paleoliths, advanced paleoliths and neoliths, extreme anomalies, and human skeletal remains. After the discovery of Java man and Peking man, scientists believed that the transition to toolmaking humans had taken place in the Early to Middle Pleistocene, so they no longer searched for Pliocene or older tools, or even investigated alleged finds. Cremo and Thompson discuss over forty anomalous cases of incised bones, as well as different eoliths, paleoliths and neoliths from various parts of the world, and present the arguments of the so-called debunkers. The authors point out a pattern of standard approach toward controversial evidence. "One mentions an exceptional discovery, one states that it was disputed for some time, and then one cites an authority...who supposedly settled the matter, once and for all. But when one takes the time to dig up the report that...supposedly delivered the coup de gr,ce, it often fails to make a convincing case" (p. 81).

The last category of anomalous finds are human skeletal remains. These finds may vary from bone fragments to partial or entire skeletons. They have been found in several places in the Americas and in Europe. Among these are also extreme anomalies, for instance a skeleton found on a coal bed capped with two feet of slate rock, or skeletons found in various Oligocene, Eocene, or early Miocene layers. More frequent finds involve skeletons and fragments from late Miocene (10 to 5 million years old), Pliocene (5 to 2 million years old), and the Pleistocene era (from 10,000 to 2 million years old).

Those who challenge the finding of human remains in very old geological strata often claim that it involves a recent burial in old layers, especially if anatomically modern humans are found. For this to be the case, the layers above the remains would have to be disturbed, yet scientists have discovered many human remains in very old strata where the layers above the remains were undisturbed. Another argument used by opponents is that the remains ended up where they were found because of mudslides out of more recent layers. Mudslides, however, are also traceable in the geological stratigraphy, and they have proven not to be a factor in those cases. Moreover, skeletal remains often take on the color of the ground in which they were housed for long geological periods, which is another argument against intrusion from more recent, differently colored soils.

The second part of *Hidden History* focuses on those finds of human skeletal remains that have been accepted by science as contributing to the human evolutionary theory. The first major discovery was made in Java by Dubois, who found a molar and skull cap in 1891, and a fossilized human femur in 1892. He believed that these belonged together and were the remains of an extinct giant chimpanzee. Only after corresponding with Ernst Haeckel, who postulated the existence of a missing link (*Pithecanthropus*), did Dubois consider his find a specimen of this ape-man. After initial opposition, *Pithecanthropus* took firm root in the minds of scientists as an early ancestor of man. In the 1930s, G. H. R. von Koenigswald traveled to Java to continue the search for *Pithecanthropus*, and hired scores of Javanese workers. Local villagers, promised money in exchange for discoveries, were paid by the piece, so that bone pieces were broken up to earn more money. With this approach the strict criterion of exact site description became impossible, yet *Pithecanthropus* is now known as *Homo erectus* and still considered an accepted link in our ancestral history.

As anthropologists found more hominid remains, they developed a procedure of morphological dating which Cremo and Thompson consider very questionable. For example, when two hominid specimens of different morphology are found in the same stratum and in the presence of similar fauna, they must both be placed within the same geological period. This period, however, may span many hundreds of thousands of years. In some cases, when different paleomagnetic, chemical, and radiometric methods give a wide spread of conflicting dates within this period,

> scientists will decide, solely on the basis of their commitment to evolution, that the morphologically more apelike specimen should be moved to the early part of its possible date range, in order to remove it from the part of its possible date range that overlaps that of the morphologically more humanlike specimen. As part of the same procedure the more humanlike specimen can be moved to the later, or more recent, part of its own possible date range. Thus

> the two specimens are temporally separated....It would look bad to have two forms, one generally considered ancestral to the other, existing contemporaneously.... With this maneuver completed, the two fossil hominids, now set apart from each other temporally, are then cited in textbooks as evidence of an evolutionary progression. —pp. 204-5

Currently, Africa appears to be the arena where scientists have their arguments and disagreements on the theory of human evolution. The line of descent promoted till not long ago was roughly *Ramapithecus* (fossil ape), *Australopithecus, Homo habilis, Homo erectus*, and finally ending up with *Homo sapiens*. In this sequence *Australopithecus* and *Homo habilis* were given rather humanlike characteristics, such as an apelike head or face on a modern-looking human body. Paleoanthropological discoveries of the last 25 years have done much to undermine this fairly simple picture. Besides these, various other kinds of research on fossil hominids have disturbed the picture even more.

Homo sapiens is believed to have first appeared about 100,000 years ago. *Homo erectus* supposedly goes back to about 1 million years, while *Australopithecus* is several million years old. Recent finds, however, date these hominids back beyond their allotted time periods and sometimes make them contemporaneous, which of course destroys the notion that they are each other's ancestors. The fossil remains of *Homo habilis* are so heterogeneous that some scientists wonder if perhaps some should really be ascribed to *Australopithecus*, and others to *Homo erectus*. Because of dimorphism (the male being about twice as big as the female), in some of these early species, it is also possible that remains were ascribed to different types, while in reality they represented different genders within one species.

Donald Johanson, discoverer of Lucy, an *Australopithecus* dated at 3.5 million years old, continues to "maintain that *Homo* came directly from *Australopithecus afarensis*" (p. 265). "Louis Leakey held that *Australopithecus* was an early and very apelike offshoot from the mainline of human evolution. Later, his son Richard Leakey took much the same stance" (p. 257). Presently at least four different types of *Australopithecus* are known: *A. afarensis, A. africanus, A. robustus*, and *A. boisei*, of which the last two represent more robust types with bigger jaws. Scientists do not agree, however, about their lineage. Some think one descended from the other, while others believe that the more robust types were a sideline that specialized. Then the so-called Black Skull was found by Alan Walker at Lake Turkana in 1985. This skull had larger teeth, a bigger jaw, and a sagittal crest, resembling *A. boisei*, but it turned out to be 2.5 million years old, older than that oldest robust *Australopithecus.* This made the theory of specialization in the robust types very doubtful.

Considering the results of their investigations on anomalous finds, accepted discoveries, and what has lately come to light in Africa, Cremo and Thompson draw the following conclusions.

> (1) There is a significant amount of evidence from Africa suggesting that beings resembling anatomically modern humans were present in the Early Pleistocene and Pliocene. (2) The conventional image of *Australopithecus* as a very humanlike terrestrial biped appears to be false. (3) The status of *Australopithecus* and *Homo erectus* as human ancestors is questionable. (4) The status of *Homo habilis* as a distinct species is questionable. (5) Even confining ourselves to conventionally accepted evidence, the multiplicity of proposed evolutionary linkages among the hominids in Africa presents a very confusing picture. —pp. 265-6

When they combine these observations with what they discovered by studying reports of anomalous finds, they conclude that the evidence as a whole (bones and stones) "is most consistent with the view that anatomically modern humans have coexisted with other primates for tens of millions of years" (p. 266).

This conclusion is in line with what H. P. Blavatsky maintained over a hundred years ago—in theosophical writings which also referred to the Vedic scriptures. In the viewpoint of theosophy, human beings did not descend from ape-ancestors: humanity forms the main stock from which all beings are derived. Monkeys came into being after unions between early mindless humans and primitive mammals tens of millions of years ago. At that time matter was more plastic than it is now, and the barriers between species were not as pronounced. Later this act was repeated by degenerate (but no longer mindless) beings of the human stock and the descendants of the earlier hybrids (monkeys). The result was a variety of semi-human beings with more or less apelike traits. The ancient scriptures describe these beings as "apes" who resembled humans much more than our present anthropoid apes do. They also tell us that humans eventually waged war on these semi-humans and exterminated most of them, letting only the most beastlike ones live. Our present anthropoid apes are allegedly the descendants of these beastlike hybrids.

An evolutionary descent of this nature would explain why the physical makeup of man is decidedly primitive, while those of the animals, including mammals and anthropoid apes, is increasingly specialized. G. de Purucker points out a few of these primitive traits in his *Man in Evolution* (1977), mentioning the human skull, nasal bones, facial features, skeleton, muscles, tongue, vermiform appendix, great arteries, premaxilla, and foot, as examples of primitive mammalian simplicity in human beings. The ape's foot compared to the human foot is a clear instance of specialization in the ape—it has developed into a hand. Scientific research concerning the dental development and the position of the larynx in humans and apes has brought to light a marked difference. This research has been applied to fossil hominids and is described by Richard Leakey in his *Origins Reconsidered* (1992). It turns out that *Australopithecus, Homo habilis*, and early *Homo erectus* have traits that are more apelike, while the patterns in later *Homo erectus*, *Neanderthals* and *Homo sapiens* are human.

If the hypotheses brought forward in *The Hidden History of the Human Race* are correct, that humans are far older than commonly believed, and that humanlike and apelike beings coexisted for long periods of time, then who were *Australopithecus, Homo habilis*, and *Homo erectus*? Some scientists have already acknowledged that we really do not know where *Homo sapiens* came from. Could all these early hominids be the mixed forms between humans and apelike beings as described in theosophical literature? Perhaps the search for the first apeman who stood up and behaved like a human is irrelevant. Could it be that man is his own ancestor?

Related Correspondence

3.4.6.1. Letter to Ina Belderis, June 13, 1995.

Thank you for your thoughtful review of my book *The Hidden History of the Human Race*, which appeared in the April/May issue of *Sunrise: Theosophic Perspectives*. Although I am striving for detachment in the face of dualities such as praise and blame, I must say that it is helpful to sometimes get some praise to be detached from! Thus far there has been no shortage of negative reviews from the scientific establishment (mostly for the unabridged version). The book has, at least, not been ignored.

If you do not mind telling me, how did you encounter *The Hidden History of the Human Race*? I always find it interesting to see by what pathways the book is circulating.

3.4.6.2. Letter from Ina Belderis, June 24, 1995.

Thank you for your kind letter. I enjoyed reading *Hidden History of the Human Race* in the abridged version, and I also have *Forbidden Archeology*, which until now I've only scanned. I intend to read it this summer. I decided to make a review of the shorter book, because I thought the 900 page book may be too intimidating for many people. On the other hand, I did not want them to miss out on the information.

You asked me how I encountered *Hidden History*. One of my colleagues here at the Theosophical Society saw an ad for the book in an Atrium Distributors catalog and pointed it out to me as something we might want to consider for our library. I am one of the librarians here at the Theosophical Library Center, and also the bookbuyer for both the library and Theosophical University Press. When the book came in, it intrigued me—I have always been interested in paleoanthropology and related subjects—so I usurped it right away to read it. That experience led to the review. All in all, it resulted in my brushing up on the subject. A lot of new material has come out in the field and obviously paleoanthropologists have increasing problems fitting everything into their theories. It was a real treat to me to read about these old forgotten reports about anomalies and their comparison with reports about accepted finds and see that everything is not as clear cut as scientists sometimes would like to believe.

I heard through the head of our society that you are working on another book. Will this be a continuation of the things you dealt with in *Hidden History* or something entirely different? I am very curious about it, and I'm looking forward to its publication.

3.4.6.3. Letter to Ina Belderis, July 16, 1995.

Than you for telling how you acquired *Forbidden Archeology* and *The Hidden History of the Human Race*. Yes, I am working on another book, *Human Devolution*, and in it I will give an alternative to current scientific ideas of human origins. I start with a chapter summarizing the physical evidence for extreme human antiquity catalogued in *Forbidden Archeology*. I will point out that this evidence suggests a definite need for an alternative explanation of human origins, as any evolutionary hypothesis appears to be ruled out. In the remainder of the book I will show that just as evidence of a physical nature has been suppressed (stones and bones) much evidence of a psychical nature has also been suppressed. This suppressed evidence shows that humans have a spiritual component that needs to be addressed in any account of human origins. All of this, along with cross cultural accounts of human origins, suggests that we have devolved from a spiritual position off of this planet (rather than having evolved from apes on this planet). I hope to have the writing finished by the end of the year, and hopefully the book could be in print by the end of 1996. How did you start on your spiritual path? I was always attracted to writing and spiritual matters as I was growing up around the world, and joined the Krishna consciousness society in my early twenties in 1973, staying with it till now.

The writing of my new book extended beyond the estimate I gave in this letter.

3.4.6.4. Letter from Ina Belderis, August 1, 1995.

Thank you for your letter. Your new book sounds very interesting. I also think that our spiritual component has to be addressed when we are talking about human origins, and that our spiritual origins are far more encompassing than what happens on this planet. The problem of paleoanthropology is that it is very difficult when you find stones and bones to know exactly what you are looking at. Premises and conclusions about what scientists think they see and saw through the history of that field reflect very strongly—more so than in any other scientific field perhaps—their own ideas and prejudices about man and life. I recently read a book called *The Neandertals* by Trinkaus and Shipman, which I enjoyed in particular because it gave a background history of every single scientist whose finds and theories they described. Perhaps you are familiar with the book? Knowing their background and their motivation or agenda was often a real eye-opener as to their conclusions. I am also looking forward to Richard Leakey's newest book, *The Sixth Extinction*, which is due in October. I like his writing style though I do not share all his conclusions.

You asked how I started my spiritual path. After visiting Pakistan in 1976 on a student's excursion for my geography degree, and seeing ruins of Buddhist monasteries in Taxila, I concluded that I really did not know very much about Buddhism, so I wanted to read something about it. I came across a book on Zen Buddhism and tried to read it. At that point in my life, however, that was too far out to interest me. On the bookshelf that harbored the book on Zen Buddhism I also found Christmas Humphreys' *Karma and Rebirth*. So I decided to read that. That did interest me a lot. Then I came across *Isis Unveiled* by Blavatsky, which I found really fascinating. It was like a recognition of certain truths that I had always harbored, but that I had never seen so clearly in writing before. After that

I read more books on theosophy and I contacted The Theosophical Society. This all happened in The Netherlands in Europe, and I joined the Society in 1980. I came to this country in 1987 on the invitation of the Head of the Society to work at our Headquarters here.

It is our Summer break from classes and public meetings now, and I am using the time to prepare a talk on the subject of human ancestors that I promised to give here in December. That will keep me busy for quite some time. What other work do you do besides writing your new book?

3.4.6.5. Letter to Ina Belderis, July 16, 1995.

I have also read *The Neandertals* by Trinkaus and Shipman. Like you, I found it fascinating because of the background they gave on the anthropologists and archeologists. Their admission that evidence is gathered and interpreted by human beings with motives and preconceptions was rather honest, in comparison with most other books of this kind.

Thank you for sharing with me how you came to your spiritual path. About two years ago, I visited the Taxila area of Pakistan. My purpose, however, was not archeological but paleobotanical. After landing in Islamabad and finding a hotel in Rawalpindi, I hired a car and driver and went to the salt mines at Khewra, in the Salt Range mountains. When Pakistan was still part of India, European and Indian scientists found evidence for advanced plants, including angiosperms, in the Salt Range formation, at Khewra and elsewhere. According to one group, the Salt Range formation is Cambrian and the plant remains were taken as evidence that the history of life on earth was different than most researchers thought. According to another group, the presence of the advanced plant remains was evidence that the Salt Range formation was not Cambrian but Eocene. The debate was never conclusively resolved, but modern geologists say the Salt Range formation is definitely Cambrian. And that to me suggests the plant remains are a genuine evolutionary anomaly. That is part of another book I am researching. I went to Khewra not to do any gathering of specimens but just to get a spiritual feel for the place—to walk up the rugged sunny gorges and see what vibrations from the geologists and botanists who worked there were still reverberating.

You ask what I do in addition to writing my new book. One thing is that I talk about the old one. I have just returned from Bern, Switzerland, where I gave a talk at the world conference of the Ancient Astronaut Society (Erich von Daniken's organization). My talk was titled "Forbidden Archeology—Evidence for Extreme Human Antiquity and the Ancient Astronaut Hypothesis." After the talk, I spent a few restful days at the

Hare Krishna temple in Zurich, then came home to Florida. As a practicing member of the Krishna consciousness movement, I devote a good part of each day to my spiritual discipline, which involves rising early, going to the temple, studying Sanskrit literature, chanting my mantra meditations, etc.

3.4.7. Anonymous (1995) "*Hidden History* Reveals Major Scientific Cover-Up." *Hare Krishna Report*, No. 5, March, p. 2. Reprinted by permission of ISKCON Communications Global.

The Hare Krishna Report *is a publication of ISKCON Communications, the public relations office of the International Society for Krishna Consciousness. It is circulated to religion reporters, professors of religious studies, and other select audiences.*

Over the past two centuries, researchers have discovered bones and artifacts showing that people like ourselves existed on earth millions of years ago. But the scientific establishment has suppressed, ignored, or forgotten these remarkable facts. Why? Because, argues a well-documented and controversial new book, *The Hidden History of the Human Race*, they contradict dominant views of human origins and antiquity.

Authors Michael A. Cremo and Dr. Richard L. Thompson, both members of the Bhaktivedanta Institute, the scientific branch of the Hare Krishna movement, claim that evolutionary prejudices, deeply held by powerful groups of scientists, have acted as what they call a "knowledge filter," leaving us with a radically incomplete set of facts for building our ideas about human origins.

The 300-page *Hidden History* (Govardhan Hill Press, 1994) is the product of eight years of research, and is a condensed version of the more formidable 900-page *Forbidden Archeology*, first published in 1993.

The authors call for a "change in today's rigid scientific mindset" and challenge us to rethink our understanding of not only human origins, but the accepted methods of science itself. Not surprisingly, response from the scientific community has been mixed.

Humbug

Responding on behalf of the scientific establishment, noted anthropologist Richard Leakey wrote, "Your book is pure humbug and does not deserve to be taken seriously by anyone but a fool."

Tour de Force

Others were more impressed. "The book is a detective novel as much as a scholarly tour de force," wrote Dr. Mikael Rothstein (*Politiken* newspaper, Denmark). "But the murderer is not the butler. Neither is the victim a rich old man with many heirs. The victim is Man himself, and the role of the assassin is played by numerous scientists..."

The Journal of Field Archeology commented that, "This volume combines a vast amount of both accepted and controversial evidence from the archaeological record with sociological, philosophical, and historical critiques of the scientific method to challenge existing views and expose the suppression of information concerning history and human origins."

David Heppell, of the Department of Natural History of the Royal Museum of Scotland, wrote, "A very comprehensive and scholarly compilation....Whether one accepts the evidence presented or not, it certainly looks as if there will no longer be any excuse for ignoring it."

3.4.8. Gopiparanadhana Dasa (1993) "Book Review: *Forbidden Archeology.*" *ISKCON World Review*, vol. 12, no. 2, August/September, p. 4. Reprinted by permission of *Hare Krishna Today.*

ISKCON World Review, *now renamed* Hare Krishna Today, *is a bimonthly newspaper of the International Society for Krishna Consciousness.*

What is it that makes us "modern"?

Our belief that we are better than our predecessors, that we are the culmination, at least so far, of the mysterious, historical process of progress.

Even more than all our curious artifacts, our adherence to this strange religion is the criterion for labeling us "modern man." The doctors of the faith, speaking from their cathedral seats in the institutes of science, have declared *our credo*: This ordered world came from chaos. It fell into place by natural evolution. By natural evolution the atoms became molecules. The molecules became proteins. The proteins became amoebas and apes. And the apes became man.

From time to time there will be dangerous skeptics, like the authors of *Forbidden Archeology*, daring to question the holy relics of bones and rocks which witness the accepted truth of natural selection. Such heresy may be short-sighted and dogmatic, in which case the faithful will easily defend themselves by belittling it with counterproof or just ignoring it. But the challenge becomes much more formidable when, as in this book by the unrepentant heathens Michael Cremo and Richard Thompson (Drutakarma and Sadaputa Prabhus), the expression of doubts is elevated to the strict standards of high science itself.

According to the official version of human evolution, about two million years ago certain apes turned into the first toolmakers who walked erect. A million years later, in the opinion of most experts, the descendants of this earliest kind of human first left their original home, Africa. Anatomically modern humans appeared only 100,000 years ago, and they entered the American continent, at the earliest, 30,000 years ago. To substantiate this doctrine, paleontologists present fossil evidence of the ape-men and early men which locates them when and where they were supposed to have lived.

It is the sacredness of this evidence that *Forbidden Archeology* takes over eight hundred pages to criticize. With remarkably focused persistence, this book pursues a single idea: The "anomalous" evidence of man's presence too early in the wrong places is too copious and relatively verifiable for a fair observer to simply ignore. This evidence has been kept stigmatized and forgotten by a biased and largely invisible effort to filter out data contradicting the official model of human evolution. Over and over again the discoverers of unwanted evidence have been discredited and their ideas conveniently buried in oblivion so that later generations will never find them.

When some of the obscured facts of archeology's history in this century and the last are unearthed from their archival graves, however, we find a lot of "anomalous" fossil evidence that was collected with just as much careful scientific rigor as the few fossils which supposedly prove the official account of human development. This unaccepted "anomalous" data includes bone and stone tools which, as primitive as they may be, show man in some form intruding on the scene a few million to fifty million years or more ahead of schedule. More sophisticated human artifacts and human skeletal remains have also been recovered from unsettlingly ancient fossil strata.

Archeology is never as exact a science as its proponents might hope it to become. Its data is too scarce and too easily misidentified. Taking advantage of this inherent uncertainty, the defenders of the official faith demand perfect verification of heretical theories and, of course, never get it. But the accepted version also can never be proven beyond a reasonable doubt. This is *Forbidden Archeology*'s main claim: The "anomalous" evidence is just as good, if not better, than the official evidence. If one of these two is unacceptable as insufficiently verified, then, to be fair, they both are.

The authors of this book lead us quite a distance through the dry bone fields of human archeology to show us how good the evidence of man's early presence is and how sparse and questionable the evidence of the

Darwinian version of man's origin is. Some of us will ask whether it is worth the trouble to join this painstaking safari, and wonder who else is going to bother. After all, the book is so long and technical that only experts may have much taste for it and the capacity to digest it; and then, being critical scholars, they will manage to find reasons to dismiss it. As your humble reviewer, I can only give my own opinion on this. I'm not a scientist or even a very clear thinker, yet I managed to read the whole book and found it interesting and convincing. The readers of this book are going to have to work hard, but they don't have to be qualified as specialists.

Messrs. Cremo and Thompson, and their assistant, Stephen Bernath, have done some valuable original research into the history of the science of human archeology. The facts and arguments they have assembled here are solid and thorough; whoever does read them through is going to have to take them seriously. And I think it will have been well worth the eight years' work it took to write *Forbidden Archeology* if anyone starts to see the modern myth of progress for the unrealistic superstition it really is.

3.5. New Age Publications and Publications Featuring Paranormal Phenomena, UFOs, and Conspiracy Theories

3.5.1. Douglas J. Kenyon (1996) "Ancient Mysteries: Exposing A Scientific Coverup." *Atlantis Rising,* No. 6, pp. 16-18, 46. Reprinted by permission of *Atlantis Rising.*

In 1966 respected archeologist Virginia Steen-McIntyre and her associates on a U.S. Geological Survey team working under a grant from the National Science Foundation were called upon to date a pair of remarkable archeological sites in Mexico. Sophisticated stone tools rivaling the best work of Cro-magnon man in Europe had been discovered at Hueyatlaco, while somewhat cruder implements had been turned up at nearby El Horno. The sites, it was conjectured, were very ancient, perhaps as old as 20,000 years, which, according to prevailing theories, would place them very close to the dawn of human habitation in the Americas.

Steen-McIntyre, knowing that if such antiquity could indeed be authenticated, her career would be made, set about an exhaustive series of tests. Using four different, but well-accepted, dating methods, including uranium series and fission track, she determined to get it right. Nevertheless, when the results came in, the original estimates proved to be way off. Way *under*—as it turned out. The actual age was conclusively demonstrated to be more like a quarter of a million years.

As we might expect, some controversy ensued.

Steen-McIntyre's date challenged not only accepted chronologies for human presence in the region, but contradicted established notions of how long modern humans could have been anywhere on Earth. Nevertheless, the massive reexamination of orthodox theory and the wholesale rewriting of textbooks which one might logically have expected did *not* ensue. What *did* follow was the public ridicule of Steen-McIntyre's work and the vilification of her character. She has not been able to find work in her field since.

More than a century earlier, following the discovery of gold in California's Table Mountain and the subsequent digging of thousands of feet of mining shafts, miners began to bring up hundreds of stone artifacts and even human fossils. Despite their origin in geological strata documented at 9 to 55 million years in age, California state geologist J.D. Whitney was able subsequently to authenticate many of the finds and to produce an extensive and authoritative report. The implications of Whitney's evidence have never been properly answered or explained by the establishment, yet the entire episode has been virtually ignored and references to it have vanished from the textbooks.

For decades miners in South Africa have been turning up—from strata nearly three billion years in age—hundreds of small metallic spheres with encircling parallel grooves. Thus far, the scientific community has failed to take note.

Among scores of such cases cited in the recently published *Forbidden Archeology* (and in the condensed version *The Hidden History of the Human Race*) it is clear that these three are by no means uncommon. Suggesting nothing less than a "massive coverup," co-authors Michael Cremo and Richard Thompson believe that when it comes to explaining the origins of the human race on earth, academic science has cooked the books.

While the public may believe that all the real evidence supports the mainstream theory of evolution—with its familiar timetable for human development (i.e. *Homo sapiens* of the modern type going back to only about 100,000 years)—Cremo and Thompson demonstrate that, to the contrary, a virtual mountain of evidence produced by reputable scientists applying standards just as exacting, if not more so, than the establishment has been not only ignored but, in many cases, actually suppressed. In every area of research, from paleontology to anthropology and archeology, that which is presented to the public as established and irrefutable fact is indeed nothing more, says Cremo, "than a consensus arrived at by powerful groups of people."

Is that consensus justified by the evidence? Cremo and Thompson say no.

Carefully citing all available documentation, the authors produce case after case of contradictory research conducted in the last two centuries. Included are detailed descriptions of the controversy and ultimate suppression following each discovery. Typical is the case of George Carter who claimed to have found, at an excavation in San Diego, hearths and crude stone tools at levels corresponding to the last interglacial period, some 80,000-90,000 years ago. Even though Carter's work was endorsed by some experts such as lithic scholar John Witthoft, the establishment scoffed. San Diego State University refused to even look at the evidence in its own back yard and Harvard University publicly defamed him in a course on "Fantastic Archeology."

What emerges is a picture of an arrogant and bigoted academic elite interested more in the preservation of its own prerogatives and authority than the truth.

Needless to say, the weighty (952 page) volume has caused more than a little stir. The establishment, as one might expect, is outraged, albeit having a difficult time ignoring the book. Anthropologist Richard Leakey wrote, "Your book is pure humbug and does not deserve to be taken seriously by anyone but a fool." Nevertheless, many prestigious scientific publications including *The American Journal of Physical Anthropology, Geoarchaeology*, and the *British Journal for the History of Science* have deigned to review the book, and while generally critical of its arguments, many conceded, though grudgingly, that *Forbidden Archeology* is well-written and well-researched. Some indeed recognize a significant challenge to the prevailing theories. As William Howells wrote in *Physical Anthropologist*, "To have modern human beings...appearing a great deal earlier, in fact at a time when even simple primates did not exist as possible ancestors, would be devastating not only to the accepted pattern, it would be devastating to the whole theory of evolution, which has been pretty robust up until now."

Yet despite its considerable challenge to the evolutionary edifice, *Forbidden Archeology* chooses not to align itself with the familiar creationist point of view nor to attempt an alternative theory of its own. The task of presenting his own complex theory—which seeks, he says, to avoid the "false choice" usually presented in the media between evolution and creationism—Cremo has reserved as the subject of a forthcoming book *Human Devolution*. On the question of human origins, he insists, "we really do have to go back to the drawing board."

As the author told *Atlantis Rising* recently, "*Forbidden Archeology* suggests the real need for an alternative explanation, a new synthesis. I'm going to get into that in detail. And it's going to have elements of the Darwinian idea, and elements of the ancient astronaut theory, and elements of the creationist nature, but it's going to be much more complex. I think we've become accustomed to overly simplistic pictures of human origins, whereas the reality is a little more complicated than any advocates of the current ideas are prepared to admit."

Both Cremo and Thompson are members of the Bhaktivedanta Institute—the Science Studies Branch of the International Society for Krishna Consciousness. Cremo and Thompson started their project with the goal of finding evidence to corroborate the ancient Sanskrit writings of India which relate episodes of human history going back millions of years.

"So we thought," says Cremo, "if there's any truth to those ancient writings, there should be some physical evidence to back it up but we really didn't find it in the current textbooks." They didn't stop there though. Over the next eight years Cremo and Thompson investigated the entire history of archeology and anthropology, delving into everything that has been discovered, not just what has been reported in text books. What they found was a revelation. "I thought there might be a few little things that have been swept under the rug," said Cremo, "but what I found was truly amazing. There's actually a massive amount of evidence that's been suppressed."

Cremo and Thompson determined to produce a book of irrefutable archeological facts. "The standard used," says Cremo, "(meant) the site had to be identifiable, there had to be good geological evidence on the age of the site and there had to be some reporting about it, in most cases in the scientific literature." The quality and quantity of the evidence—they hoped—would compel serious examination by professionals in the field, as well as by students, and the general public.

Few would deny that they have succeeded in spectacular fashion. Much in demand in alternative science circles, the authors have also found a sympathetic audience among the self-termed sociologists of scientific knowledge, who are very aware of the failure of modern scientific method to present a truly objective picture of reality. An upcoming NBC special, "The Mysterious Origins of Man", draws heavily upon Cremo and Thompson's suggestion that there is a "knowledge filter" among the scientific elite which has given us a picture of prehistory which is largely incorrect.

The problem, Cremo believes, is both misfeasance and malfeasance. "You can find many cases where it's just an automatic process. It's just human nature that a person will tend to reject things that don't fit in with his particular world view." He cites the example of a young paleontologist and expert on ancient whale bones at the Museum of Natural History in San Diego. When asked if he ever saw signs of human marks on any of the bones, the scientist remarked, "I tend to stay away from anything that has to do with humans because it's just too controversial." Cremo sees the response as an innocent one from someone interested in protecting his career. In other areas, though, he perceives something much more vicious, as in the case of Virginia Steen-McIntyre. "What she found was that she wasn't able to get her report published. She lost the teaching position at the university. She was labeled a publicity seeker and a maverick in her profession. And she really hasn't been able to work as a professional geologist since then."

In other examples, Cremo finds even broader signs of deliberate malfeasance. He mentions the activities of the Rockefeller Foundation, which funded Davidson Black's research at Zhoukoudian (in China). Correspondence between Black and his superiors with the Foundation shows that research and archeology was part of a far larger biological research project, (from the correspondence) "thus we may gain information about our behavior of the sort that can lead to wide and beneficial control." In other words, this research was being funded with the specific goal of control. "Control by whom?" Cremo wants to know.

The motive to manipulate is not so hard to understand. "There's a lot of social power connected with explaining who we are and what we are," he says. "Somebody once said 'knowledge is power.' You could also say power is knowledge. Some people have particular power and prestige that enables them to dictate the agenda of our society. I think it's not surprising that they are resistant to any change."

Cremo agrees that scientists today have become a virtual priest class, exercising many of the rights and prerogatives which their forebears in the industrial scientific revolution sought to wrest from an entrenched religious establishment. "They set the tone and the direction for our civilization on a worldwide basis," he says. "If you want to know something today you usually don't go to a priest or a spiritually inclined person, you go to one of these people because they've convinced us that our world is a very mechanistic place, and everything can be explained mechanically by the laws of physics and chemistry which are currently accepted by the establishment."

To Cremo it seems the scientists have usurped the keys of the kingdom, and then failed to live up to their promises. "In many ways the environmental crisis and the political crisis and the crisis in values is their doing. And I think many people are becoming aware that (the scientists) really haven't been able to deliver the kingdom to which they claimed to have the keys. I think many people are starting to see that the world view they are presenting just doesn't account for everything in human experience."

For Cremo we are all part of a cosmic hierarchy of beings, a view for which he finds corroboration in world mythologies. "If you look at all of those traditions, when they talk about origins they don't talk about it as something that just occurs on this planet. There are extraterrestrial contacts with gods, demigods, goddesses, angels." And he feels there may be parallels in the modern UFO phenomenon.

The failure of modern science to satisfactorily deal with UFOs, extrasensory perception or the paranormal provides one of the principal charges against it. "I would have to say that the evidence of such today is very strong," he argues. "It's very difficult to ignore. It's not something that you can just sweep away. If you were to just reject all the evidence for UFOs, abductions and other kinds of contacts coming from so many reputable sources, it seems we have to give up accepting any kind of human testimony whatsoever."

One area where orthodoxy has been frequently challenged is in the notion of sudden change brought about by enormous cataclysm, versus the gradualism usually conceived of by evolutionists. Even though it has become fashionable to talk of such events, they have been relegated to the very distant past, supposedly before the appearance of man. Yet some like Immanuel Velikovsky and others have argued that many such events have occurred in our past and induced a kind of planetary amnesia from which we still suffer today.

That such catastrophic episodes have occurred and that humanity has suffered from some great forgettings, Cremo agrees. "I think there is a kind of amnesia which when we encounter the actual records of catastrophes, it makes us think, oh well, this is just mythology. In other words, I think some knowledge of these catastrophes does survive in ancient writings and cultures and through oral traditions. But because of what you might call some social amnesia, as we encounter those things we are not able to accept them as truth. I also think there's a deliberate attempt on the part of those who are now in control of the world's intellectual life to make us disbelieve and forget the paranormal and related phenomena. I think there's a definite attempt to keep us in a state of forgetfulness about these things."

It's all a part of the politics of ideas. Says Cremo, "It's been a struggle that's been going on thousands and thousands of years and it's still going on."

3.5.2. Christina Zohs (1996) "An Interview with Michael A. Cremo." *Golden Thread*, vol. 1, no. 5, June, pp. 13-16. Reprinted by permission of *Golden Thread*.

G.T.: Michael, you and Richard Thompson have written a book called Forbidden Archeology. *I believe it required an enormous amount of research. Could you tell us a little about this book?*

Michael: *Forbidden Archeology* grew out of some discussions that Richard and I had around 1984, being both members of The Bhaktivedanta Institute—which is a branch of the International Society for Krishna Consciousness that studies the relationship between modern science and the worldview expressed in the Vedic literature. We also got our inspiration from the ancient Sanskrit writings of India, among which is one group of writings called the Puranas. The Puranas are histories that tell of human civilizations going back millions of years on this planet, which is quite different from what our modern scientific worldview is telling us. From the mainstream scientists of today we learn that human beings like ourselves have only been around on this planet for about 100,000 or 200,000 years at most, and that civilization has only been around for about 10,000 years.

So our question was "Is there any evidence to back up the accounts that are found in these ancient Sanskrit writings?" And when we looked in the current textbooks: we didn't find any such evidence. All you find in the current textbooks is evidence that goes along with the idea that humans like ourselves have come about fairly recently—having evolved from more ape-like creatures. So it looks very solid when you look at these textbooks: we came into being about 100,000 to 200,000 years ago and here is all the evidence, so it must be true! But we decided not to stop our research with just the information from the current textbooks.

What we did was we looked into the entire history of archeology and anthropology that had been uncovered over the past one hundred fifty years. That took about eight years of solid research, getting documents from all over the world in many different languages, getting many rare books and many rare out-of-print journals in archeology and anthropology. What we found was literally hundreds of examples of human skeleton remains and human artifacts of various kinds that place human beings like ourselves on this planet—going back millions and millions of years, which is just as the ancient Sanskrit writings of India say.

So what we did in *Forbidden Archeology* was gather all of this evidence and put it all together and we also wanted to document why this evidence *isn't* in the textbooks today—why it isn't displayed in the museums, why our children aren't taught about it in schools and why we don't see it, for example, on television specials. The reason the evidence is no longer current is that it is systematically eliminated by what we call a process of knowledge filtration.

Now the first thing that occurred to us was—maybe there is some very good reason why all of this very interesting and intriguing evidence that puts human beings back millions of years on this planet has been eliminated from the current scientific literature. On careful study we couldn't see any good reason why it would have been eliminated. We were able to document in many cases exactly why this evidence was rejected, and in many cases it was rejected by scientists simply because it went against the ideas that they favored. It went against the orthodox idea that began to crystallize in the latter part of the 19th century, which was we had evolved fairly recently from the more ape-like creatures.

G.T.: *Could you give us any examples concerning the suppressed evidence you discovered?*

Michael: There is one amazing example that I really like a lot because it shows two things that we tried to document in *Forbidden Archeology*.

In 1849 gold miners were coming to the gold mining regions of California in Tuolumne County. At first they were panning for gold in the streams but afterwards they started digging tunnels in the sides of mountains. Inside these tunnels, sometimes hundreds and over a thousand feet, they were finding human skeletons, spearpoints, stone mortars and other items at dozens of different locations....A geologist named J.D. Whitney, who was the state geologist of California then, became aware of these discoveries and collected many of the artifacts and wrote a massive book about them that was published by Harvard University's Peabody Museum of Natural History.

The amazing thing about these discoveries is that the rock in which they were found is anywhere from 10 million up to 55 million years old. We learned this by consulting with geological reports from modern scientific literature. That area is still being extensively studied, due to it still being economically viable for its gold, and the age of the rock formation from which these artifacts and human skeletal remains were taken is quite well known. So this is amazing to know—we *have* had a human presence, say, 55 million years ago in California.

There was a very powerful anthropologist, William Holmes, who worked at The Smithsonian Institution in Washington, D.C. He was in a more powerful position than Whitney in the scientific world and he was committed to the idea that human beings have come about fairly recently by a process of material evolution from more ape-like creatures. So what he said in the report he wrote was essentially this: "That if Dr. Whitney had understood the theory of human evolution as we understand it today then he would have hesitated to announce his conclusions despite the opposing array of testimony with which he uses to confirm it." In other words, if the facts don't go along with the theory, then the facts, even an opposing array of them, have to be thrown out and discredited—and essentially that is what happened. So, simply because these very well-documented facts did not agree with the emerging theory of human evolution, these very intriguing facts were put aside and that is why you don't see these objects on display in museums today with their proper dates attached to them.

Earlier this year, Richard Thompson and I were featured guests on an NBC television special called *The Mysterious Origins of Man,* narrated by Charlton Heston. The story that I just told you about these California gold mine discoveries was part of that show, and the producers tried to get permission from the University of California in Berkeley to film these artifacts that are still there, placed by J.D. Whitney back in the 19th century. But the director of the museum told the producers that they could not let them film the objects because they were understaffed and would have to pay overtime and they just couldn't do it. So the producers came back with an offer to pay any amount of money that would be required to have people out there to help do this, however they still refused us permission to film these artifacts.

So, this is how this whole process of knowledge filtration really works. It is not really a field for open-minded inquiry, because you can't get access to this evidence sometimes, it is kept out of sight and people just aren't allowed to see these things.

If you look at archeology and anthropology—it is kind of a big museum, and it's as if the public is only allowed into one room in that museum and they only see the evidence that agrees with the accepted theories. In that big museum there are many other rooms, locked rooms, and what we have tried to do in *Forbidden Archeology* is take the public on a tour of those locked rooms in the museum of human origins and history.

***G.T.:** Why do you feel it is important for us to know our beginnings?*

Michael: Well, ultimately what is involved is the whole direction of human civilization, because how we define ourselves in our origins to a large sense determines what kind of life we are going to lead, what kinds of institutions we are going to have and what kinds of goals we are going to have as a society. The evidence we have documented in these books puts a human presence so far back in time, literally hundreds of millions of years, that no evolutionary explanation really works. This means that you have to start looking for an alternative idea of human origins. We don't present that alternative idea in *Forbidden Archeology* or *The Hidden History of The Human Race* (which is an abridged version of *Forbidden Archeology*), but it is something I am going to present in a new book in progress, called *Human Devolution.*

The essential idea that I am going to be presenting in this new book is that if we want to understand where we came from, we can't look for some material explanation. In other words, we can't suppose on the basis of the actual evidence documented in *Forbidden Archeology*, which is that we evolved upward from more ape-like creatures on this planet simply by some physical process. It appears that if we want to really trace out our origins, we have to look to extraterrestrial dimensions and we have to take into account different non-material aspects of the human essence. So what really appears to be the case is that we devolve from some higher spiritual position and that is how we have come to this planet. I think that if you take one or the other of these views, you are going to get a whole different way of life, a different outlook—and a new set of goals will grow out of each outlook. So I think that is what really is at stake, and it is a conflict that has been going on for thousands of years.

Human Devolution will be presenting a spiritual alternative account of human origins—but for that to be taken seriously, first Richard and I wanted to establish that we really do need an alternative explanation, and we wanted to show that the current explanation, based on the physical evidence, simply does not work and make sense. It only makes sense if you throw out literally hundreds of very well-documented cases of scientific evidence that show that human beings like ourselves have been on this planet for hundreds of millions of years—that puts us so far back in time that no materialistic evolutionary explanation can account for our presence, so we have to look to other principles and alternative accounts.

G.T.: *When will your next book* 'Human Devolution' *be out?*

Michael: Well, I don't want anyone to hold their breath or put out an order for it now. Just as putting together *Forbidden Archeology* took eight years, putting together this next book is also going to take some time. There is another whole area of evidence to do with the human essence that has been suppressed by a process of knowledge filtration. This is evidence that tends to show that there is more to the human essence than biochemistry and matter. There is something else there—consciousness—which is just as real and just as fundamental as matter, and perhaps even more so. You find evidence for this in all kinds of paranormal research, out of body experiences, past life memories, subtle healing, extrasensory perception and things of this sort. All of this evidence tends to demonstrate that if we really want to understand what a human being is you have to take consciousness into account.

There is also a tremendous amount of evidence showing that we are not alone in the universe but are with other human life beings populating the universe—with us only being part of a whole cosmic hierarchy. This hierarchy is arranged in different levels and there may even be beings much more powerful than we are in this universe. It would appear that at the top of the hierarchy there is some kind of overall guiding intelligence—messengers who are attempting to transmit knowledge to us from the higher levels and attempting to improve our mental and spiritual state. There is a tremendous amount of evidence for this in the great traditions of the world—the ancient Hindus, Egyptians, Aztecs, Aborigines, Polynesians, African tribal people, and so on. They all had this idea of the cosmic hierarchy of beings and a spiritual dimension to the human essence. Much of this traditional wisdom is being supported by different researchers, and uncovering all that and putting it together in a very systematic way is going to take some time.

G.T.: *How else did we evolve from early man?*

Michael: Well, here is something to consider along those lines, which will be discussed in my next book. If we look at how Darwin attempted to establish his materialistic view of evolution, what he did was observe what people involved in breeding animals such as pigeons, dogs and horses can do. By certain breeding techniques you can direct the changes in these animals in a certain direction—by making them smaller or bigger, changing their colors or inducing changes by this process of breeding. Now there are limits to that, and in breeding you can only take an animal so far—you can breed dogs but they are still dogs, you can breed cats and get many different kinds of cats but they are still cats. Darwin's suggestion was, let's imagine that the process went on for millions of years and see what might happen—now I would take a similar approach.

There have been some very good studies done based on the idea that there is more than matter to life, that there are different subtle energies involved. There are cases where you have healers who by projection of subtle energies have been able to make physical changes in organisms by, for example, stopping cancerous growths. There have even been laboratory studies done where you would have people direct their attention to a petri dish full of cells—and they have apparently been able to induce changes in these cells. So in these cases of subtle healings there *does* appear to be some visible physical effects of mind over matter that we can observe today. So if ordinary humans among us have such powers—imagine beings higher up on the hierarchy who may be able to structure not only your different physical organisms and physical organs, but by their powers they might be able to even work *larger* changes in physical organisms by completely reconstructing them or even constructing them from their component elements using their mental powers. From the traditional wisdom sources of the world we see that this is often the case, and there does appear to be some evidence for such a thing to be possible—and the evidence comes from the whole area of subtle energies and different miraculous healings that researchers are now documenting.

So that's where we may have to look for an explanation of how human beings and, indeed, other kinds of creatures have originated. There may be some guided process of genetic engineering going on and I think it would involve many things. People have tended to want very simplistic explanations for human origins, whether they are talking about a very simple creation story or an equally simple Darwin story, but it may be much more complicated than that. There may be such things going on as some kind of guided genetic engineering by beings higher up in the cosmic hierarchy, there may be extra-terrestrial visitations going on and there may be processes of reproduction going on. So it may be a little more complicated than either our more simplistic religious or scientific explanations would have us believe.

G.T.: *So we've been tampered with or had some deliberate engineering?*

Michael: Actually from the ancient Sanskrit writings of India it would appear that there has been some deliberate engineering of the life forms by beings higher in the hierarchy. There have been many visitations by extraterrestrials, and there would be some overall guiding intelligence involved in the process that would even be something that resembles part of Darwinism. In other words, some kind of descent with modification, some kind of reproduction that goes on but it would not be accidental as in the Darwinian model. But there could be reproductive processes that these higher beings would engage in that would result in the production of new life forms. All these elements are part of a whole complex tapestry that is really needed to explain the human origin.

G.T.: *You have focused a lot on the ancient Sanskrit writings and teachings from India.*

Michael: Well, I think the information does not only come from India. For example, the ancient Greeks and Romans from which our own civilization springs have a similar idea that there was a multi-dimensional universe.

In other words, today we have a picture of the universe that largely says there are atoms in the void and that's it. But they had a hierarchical universe with material levels towards the bottom, more subtle realms in the middle and ultimately a spiritual realm at the top. Even the Greeks and Romans said that each level was inhabited by beings that were adapted to the life of each level. In other words, there were beings in the spiritual realms, there were beings in the ethereal realms called the demigods, and then there were humans and other living entities inhabiting the terrestrial realms. So besides the humans and demigods, there was a sort of creator god for the universe and then beyond that: spiritual beings. So if you look at all the great ancient wisdom traditions, you will see their universal origins, all with a similar pattern.

Now, why I focused on the ancient Sanskrit writings of India is because in those writings there is a vast literature that explains the system that you see reflected in cultures all over the world in a very systematic and complete way. The Sanskrit writings of India have also been preserved for thousands of years. While the Greek and Roman civilization broke down, the ancient civilization of India has persisted up to the present day. Today you don't see Egyptian priests and priestesses performing ancient rites, or Greek priests and priestesses—you don't find that. In India you can go to the temples and they are just as they were thousands of years ago, with the brahmans performing the ancient sacred rites and they *still* preserve their teachings, despite so many invasions. So it is not that this knowledge is something unique to India, but what I do find unique and inspiring about these writings is that they lay out the system in great detail and give much explanation. I just find that the whole cosmic hierarchical system is explained more thoroughly and completely in these ancient Sanskrit writings.

G.T.: *How old are the Sanskrit writings?*

Michael: Well, the writings themselves say that they are timeless, and that they have been passed down from higher dimensions since the very beginning of the universe—which would mean that they are billions and billions of years old. According to these ancient Sanskrit writings, there is a system of universal time—cycles of time—where in certain ages people have shorter memories, so they have to depend more on writing and different systems for preserving knowledge rather than the oral tradition. We are in such an age now in which these teachings—that have been passed down since time began—now exist in written form. And that process of writing teachings down in documents began about 5,000 years ago.

G.T.: *You have also written about the Great Pyramid and Sphinx. Do you have any ideas on how these were erected?*

Michael: I have always found it very interesting about one description saying that there were genies—spirit beings—that erected buildings for King Solomon. I have always found that intriguing and in the ancient Sanskrit writings of India you also find evidence that beings from higher levels have sometimes manifested monuments on the earth. If you go around the world into some of the ancient structures, sometimes you find these huge masses of architecture hard to explain. So I don't think that we can just look to human technology to understand and explain some of these monuments of past. It is not something that I have really studied extensively or looked into extensively.

G.T.: *What do you feel is mankind's destiny?*

Michael: In my book *Human Devolution*, I am making the point that we have not evolved upward by some material process on this planet from lower beings. If we accept that idea, then we would have to say that our destiny is just to invent more and more advanced kinds of technology and bring more and more of the material domain under our control for exploitation in various ways. That would take us beyond this planet to use our technologies to bring other planetary systems and resources and even other galaxies under our control. That is one idea, and I think that is the wrong idea and the wrong approach—we are never going to be satisfied if we take that path.

So in *Human Devolution* what I am putting forward is that we haven't evolved upward by some material process from lower beings, rather we have descended from some original spiritual position and now we find ourselves sort of encased by matter. What we really have to do is re-evolve—make a revolution—to go to our original spiritual position and that would be our destiny. To become free from the contact of matter and go back to our original spiritual position from which we've come—that is the message of the great wisdom traditions down throughout history that I think very significant. According to that view, the whole universe is designed to help us achieve that destiny; but if we reject that and accept another kind of destiny, then we have to live with the results of that. I think that many people are becoming dissatisfied with the results of what has happened to our planet and what has happened to the human spirit as we've pursued the destiny of trying to increase more and more our domination and exploitation of matter—rather than the destiny of cultivating the spirit.

***G.T.:** Michael, thank you very much.*

Michael: You're most welcome.

3.5.3. W. Ritchie Benedict (1995) Review of *Forbidden Archeology. The X Chronicles*, vol. 1, no. 5, November, p. 1. Reprinted by permission of W. Ritchie Benedict.

The X Chronicles *has ceased publication.*

The trained scientist regards archeology as a fascinating subject, but not so the average person who thinks it is as frozen in time as the bones and fossils it deals with. There doesn't appear to be much room for controversy about the accepted facts of human origin. Or is there? The authors have attempted to demonstrate in this exhaustively researched book (the bibliography alone run[s] to 30 pages) the notion that there are gaping holes in anthropological doctrines and Darwin's theory of evolution. As "Origins of Species" was published in 1859, it may seem strange his ideas have undergone few changes since the Victorian age. It is pointed out that the odder the discoveries that are made, the greater is the possibility they will be rejected by professional scientists. For example, in the past century, incised and broken bones have suggested a human presence in the Pliocene, Miocene and earlier periods. In 1872, shark teeth were recovered from a formation indicating an age of 2 to 2.5 million years. What was troubling was that each tooth had a hole bored in it, as is done by South Sea Islanders to make necklaces. Another incredible discovery was made in 1881. From the same formation and of the same age, a shell was found carved with a crude but undeniably human representation of a face on it.

To be fair, it is understandable why so many pieces of evidence have been rejected, as 19th-century scientists were often fooled by hoaxes, such as the famous Cardiff giant, an apparently petrified body of great antiquity.

An outstanding example of a peculiar artifact was found in Argentina in 1915 and featured a stone point in a Toxodon femur bone. A Toxodon was an extinct South American animal, resembling a furry, short-legged rhinoceros. This hunt must of happened some 2 to 3 million years ago.

The authors take a new look at the infamous Piltdown Man skull, now widely regarded as a manufactured fake. Charles Dawson was suspected of forgery—staining and filing the Piltdown jaw from a modern orangutan jaw. Cremo and Thompson feel Dawson may have been unjustly accused—it is possible the jaw was planted by one William Sollas, a professor of geology at Cambridge, and the real villain.

In Appendix Two, the authors tread on really dangerous ground, as far as concerns acceptance from science. They present a number of well-documented instances of an advanced culture (from earth or the stars?) in distant past ages. A coin-like object was retrieved from a well 144 feet deep in Illinois in 1881. Ten years later, in the same strata, a gold chain was found in carboniferous coal. While doing some research myself in the microfilm files of the University of Calgary, I came across mention of one of the peculiar items mentioned in this book. It was a metallic vase taken from Precambrian rock at Dorchester, Massachusetts, in June, 1852. A blast threw out an immense mass of rock, and two pieces of the vase were recovered from the debris. On the sides of the vase were six figures and/or a flower inlaid with pure silver. It is as much of a mystery today as it was 150 years ago, as well as being an affront to our thinking that something so advanced could exist 600 million (!) years ago.

I understand that a condensed popular version of the book is now available. This hardcover is designed for academic consumption rather than the general book-reading public, but it is to be hoped that both will open some closed minds about a topic we have all wondered about at one time or another. Who were these people who left advanced technology in coal seams? Or were they people at all? It takes some patience and some knowledge of science to comprehend all of the theories presented here, but it is well worth the time and effort. Amazing insights into the past.

3.5.4. Ingo Swann (1994) Review of *Forbidden Archeology. FATE Magazine*, January, p.106. This article appears courtesy of *FATE Magazine*. © January 1994, *FATE Magazine*, P. O. Box 64383, St. Paul, MN 55164-0303 U.S.A. Reprinted by permission.

Ingo Swann is a parapsychologist and author. Excerpts from my correspondence with him can be found in Chapter Four.

On April 26, 1993, *Time* published an article entitled "The Truth About Dinosaurs," saying "surprise, just about everything you believe is wrong." Conventional scientific theories about dinosaurs have long held that they were cold-blooded reptiles—lumbering, stupid beasts. Some earlier paleontologists object, saying certain evidence indicated otherwise, but were suppressed, debunked, or ignored.

Then came Robert T. Bakker's *The Dinosaur Heresies* (1986), reiterating old evidence combined with new technologies that showed many dinosaurs were mammalian, warm-blooded, agile, and surprisingly intelligent—as portrayed in *Jurassic Park*. The prevailing wisdom had been wrong all along.

Widely accepted evolutionary theories taught that, anatomically, humans evolved 100,000 years ago from earlier, ape-like primates. Not all scientists agreed, but again their objections were denied. Now this book comes and again, surprise.

Recently, near Glenrose, Texas, clearly preserved in deep-rock Cretaceous limestone were found:

1) a nearly complete Acrocanthosaurus dinosaur
2) 203 dinosaur footprints
3) 57 identifiable *Homo sapiens* footprints
4) a confirmed human finger in the same limestone
5) an iron hammer encased in nearby Cretaceous sandstone

The human footprints are our species, which has anatomically unique feet different from all other primate species. The geological Cretaceous Age occurred not 100,000 years ago, but 144 million years ago.

The Glenrose finds are not in the book, but hundreds of similar finds are—a manufactured vase in Precambrian rock (more than 600 million years old [m.y.o.]); a shoe print in Cambrian strata (550 m.y.o.); a *Homo sapiens* skeleton in carboniferous rock (320-260 m.y.o.); human footprints in Jurassic rock (150 m.y.o.); manufactured metal tubes in Cretaceous strata (65-144 m.y.o.); more human skeletons in Eocene (38-45 m.y.o.) and Oligocene (33-55 m.y.o.) strata, etc.

The authors have compiled all available evidence concerning human artifacts older than 100,000 years as presented and evaluated in earlier reports of paleontologists and archaeologists. Sociological details are discussed in these reports, along with the scientific debates they aroused, and the machinations utilized by mainstream scientists to exclude or debunk them. The evidence didn't fit in with the prevailing theories of how and when humans originated on earth.

Reasons for rejecting anomalous evidence are interesting. The process does not usually involve careful scrutiny of the evidence by the scientists who reject it. Most scientists prefer positive research goals, rather than scrutinizing unpopular claims. In the scientific community, the word goes out that certain findings are bogus, and most scientists avoid the rejected material. Then we are surprised when it becomes necessary to accept rejected evidence because of its sheer accumulation.

In the geological record near Glenrose Texas, and in Australia and Russia, human and dinosaur tracks have been preserved—together. For $13, the Creation Evidence Museum (PO Box 309, Glenrose, Texas 76043) will provide a 150-page booklet entitled *Dinosaurs: Scientific Evidence That Man and Dinosaur Walked Together*—while *Forbidden Archaeology* provides an understanding of the greater scientific and cultural issues involved.

3.5.5. George W. Earley (1994) Review of *Forbidden Archeology. The Gate*, July, p. 11. Reprinted by permission of George W. Earley and *The Gate*, P. O. Box 43516, Cleveland, OH 44143, U.S.A.

The Gate *describes itself as a "contrarian" publication.*

As is generally recognized, the Prehistory of Humankind is largely arrived at by consensus. After digging up old bones and miscellaneous ancestral rubble and cogitating over them at some length, the Scientific Establishment arrives at conclusions it can live with and which it then presents to the rest of us.

But there are always dissenters...mavericks who interpret the data differently, who do not accept the consensus and who present a contrarian view challenging conventional wisdom.

And sometimes the contrarians are right and the consensus changes: continental drift, for example.

Contrarian adherents now have a big (really big!) new book to marvel over and quote from: *Forbidden Archeology: The Hidden History of the Human Race* by Michael A. Cremo and Richard L. Thompson (Bhaktivedanta Institute; xxxviii + 914 pgs; $39.95).

As it is unlikely either your library or neighborhood book store will have a copy (tho you should try them), you may need to order directly from the publisher.

The authors believe that anatomically modern humans existed on Earth millions of years before they are currently believed to have done so. To 'document' their beliefs, they offer (mostly anecdotal) accounts of the findings of a wide range of what Ivan Sanderson likes to call OOPARTS, out-of-place artifacts such as 'modern' bones, implements, weapons, jewelry, etc. reportedly found while folks were excavating foundations, digging for coal or various metal ores. As the bulk of these discoveries date from the 18th and 19th centuries and were made accidentally, they weren't documented as the scientific community prefers, hence their refusal to accept them as valid.

Those familiar with William Corliss' *Source Book Project* will find such similar material here. It's fascinating, no argument about it, and hopefully will encourage both layfolk and professionals to take a hard look at the information the authors have turned up.

Unfortunately, in one instance with which I feel reasonably conversant, the authors have scanted their evidence. Their chapter 'Living Ape-Man?' recounts the sighting and photographing of an alleged Yeti in the Himalayas by a man named Wooldridge. Mr. Wooldridge wrote up his sighting for *Cryptozoology* [Vol. 5, 1986] the Journal of the International Society of Cryptozoology. Unfortunately, Cremo and Thompson ignore the commentary in Vol. 6 wherein considerable doubt is cast upon the photo...even Mr. Wooldridge has come to the conclusion that he neither saw nor photographed a real Yeti. Exclusion of this information makes one wonder what else might have been omitted from *Forbidden Archeology* to make its claims appear stronger. Such caveats aside, it's a fascinating book and should fuel hours of speculative thinking about our past.

3.5.6. Laura Lee (1995) "Notes from the New Edge: OOPS, Our Anomalies are Showing." *Common Ground*, vol. 10, no. 1., January/February. Reprinted by permission of *Common Ground* and Laura Lee.

Common Ground *is a Seattle, Washington, New Age newspaper. Laura Lee hosts a nationally syndicated radio program featuring alternative science. I've appeared as her guest several times.*

A stone mortar and pestle found in a California mine tunnel in 33 million year old rock.

Anatomically modern human bones and footprints, stone tools, cut wood, and carved stone; all found in geological strata dating back 55 million years.

What do they have in common? They are called "OOPARTS" by some (short for 'Out of Place Artifacts'), hoaxes by others, anomalies by all.

The surprise is that they are found in every old geological strata, where they're not supposed to be. After all, in the standard view of human evolution, intelligent humans are only a million years old; and the missing link between man and monkey is a 3 million year old fossil named Lucy.

But there were more surprises when I talked with researcher Michael Cremo, who, inspired by ancient Indian texts that talk of human civilization going back many millions of years, went back to the original field reports of such finds.

OOPARTS are often presented as being objects without documentation. But Cremo says many OOPARTS were as meticulously documented as the accepted artifacts.

And I'd always assumed anomalies were in short supply, the odd thing or two that didn't fit in. So I was surprised by their quantity. According to Cremo, in terms of artifacts and remains relating to human antiquity, the number of accepted pieces about equals the number considered anomalous. (Curious math, that. Like calling women who comprise 52% of the population a "minority.")

Yet, you won't find OOPARTS displayed in museums or mentioned in textbooks, because to study them would challenge the accepted scenario. So they are 'reburied' so as not to disturb consensual harmony.

Cremo says it isn't always so harmonious—that science projects this image of uniform consensus, but there is tremendous dissension behind the scenes. He points to the paleontologists who question the human-like attributes of Lucy—they say she was just a small ape and no missing link.

Science used to be more boisterous, before the turn of the century, before today's theories were 'written in stone.' Back then, OOPARTS were reported regularly in the major science journals, closely examined by leading authorities, and hotly debated.

Contrast that to today. "The pressure to conform to the standard view," says Cremo, "has resulted in a 'knowledge filter' that screens out anomalies. Today they are ignored and suppressed. . . .The logic goes: 'evidence that doesn't fit the accepted model must have been hoaxed or accidentally got into older strata,' and is therefore discounted. But that amounts to circular reasoning, not logic, not open and fair science."

Cremo's two books co-authored with Richard Thompson, *Forbidden Archeology* and *The Hidden History of the Human Race,* are currently creating a stir in archeological and anthropological circles.

Cremo and Thompson spent ten years examining hundreds of anomalous artifacts deserving of inclusion. What new picture then emerges from the fossil record?

Cremo's interpretation: "*Homo sapiens* and various apes co-existing on the planet millions of years ago, as they do today."

The ancient Indian texts from which Cremo gets his inspiration also portray human history as a roller coaster ride of alternating simple and advanced civilizations, offering a context for OOPARTS such as these.

Metallic tube found in France, in a 65 million year old chalk bed. Gold chain found in a 260 million year old chunk of coal.

The mere presence of anomalies is either disquieting or exciting, depending on your frame of reference, or your sense of adventure. OOPARTS are enigmas that tweak the comfort zone of conventional views. They serve as outermost 'signposts,' marking new territory for expanding paradigms, and pointing the way to the big picture.

3.5.7. Katharina Wilson (1996) Review of *Forbidden Archeology. The Observer*, vol. 7, no. 3 July-September, pp. 5-6. Reprinted by permission of *The Observer.*

The Observer *is a UFO publication from Oregon. Katharina Wilson is an abductee. She has written extensively about her alien abduction experiences in her book* Alien Jigsaw.

Forbidden Archeology is a new book that questions current beliefs about human evolution. Part I of *Forbidden Archeology* (which covers 458 pages) is based on what the authors call 'anomalous evidence' and "provides a well-documented compendium of reports absent from many current references and not otherwise easily obtainable." The authors discuss how scientific evidence has been "systematically suppressed, ignored, or forgotten...not through a conspiracy organized to deceive the public, but through an ongoing social process of knowledge filtration that appears quite innocuous but has a substantial cumulative effect."

Chapter One discusses information that has been overlooked, suppressed, or forgotten even though a lot of the evidence was discovered immediately after Darwin published *The Origin of the Species*. This chapter explains the basics about archeology, such as the geological timetable and the incompleteness of the fossil record.

The authors' thesis is based on the premise that anomalous finds should be studied and possibly accepted along with currently accepted 'evidence.' Perhaps as is the case with other types of controversial information, "One prominent feature in the treatment of anomalous evidence is what we could call the double standard...evidence agreeing with a prevailing theory tends to be treated very leniently....In contrast, evidence that goes against an accepted theory tends to be subjected to intense critical scrutiny, and it is expected to meet very high standards of proof."

There is a section in Chapter One titled "The Phenomenon of Suppression" which perfectly describes what experiencers and those involved with experiencer research are faced with. "...There are some observations that so violently contradict accepted theories that they are never accepted by any scientists. These tend to be reported by scientifically uneducated people in popular books, magazines, and newspapers."

Chapter Two covers detailed descriptions of reports involving intentionally cut and broken bones of animals. In other words, bones which have been altered by man. Some of this evidence points toward a theory that there was a human presence in the Americas far earlier than was originally believed (thought to be between 12 thousand and 25 thousand years ago).

However, many serious scientists of the nineteenth and early twentieth centuries reported that marks on bones as old as 25 *million* years old were indicative of human work. This chapter illustrates that when 'unbelievable' information arises and people are convinced that it cannot exist, the evidence pointing to such conclusions is overlooked or ignored by the scientific community.

Chapters Three, Four, and Five continue with extremely detailed studies of anomalous old stone tools and industries. Chapter Six closes out the authors first section of *Forbidden Archeology* with a discussion of anomalous human skeletal remains. In their conclusion of Part I, they write: "[the evidence] suggests the existence of anatomically modern humans as far back as the early Tertiary [the first period of the Cenozoic era; 65-37 million years ago]." A partial review of this anomalous evidence is listed at the end of this report.

Part II of *Forbidden Archeology* involves discussions of 'accepted evidence,' beginning with a review and discussions of "Java Man" and continuing with "The Piltdown case" and "Peking Man" (very interesting). A highly recommended read is Chapter Nine, "Peking Man and Other Finds in China".

The authors go into detail about how the Rockefeller Foundation funded many of the digs in Peking [Beijing]. If I may quote liberally from this chapter (page 534): "It thus becomes clear that at the same time the Rockefeller Foundation was channeling funds into human evolution research in China, it was in the process of developing an elaborate plan to fund biological research with a view to developing methods to effectively control human behavior. [Canadian physician Davidson Black] Black's research into Peking man must be seen within this context in order to be properly understood."

From pages 537 and 538, in reference to a new beginning of philanthropy, "All programs in various Rockefeller charities 'relating to the advance of human knowledge' were shifted to the Rockefeller Foundation, which was organized into five divisions...Each division was run by a highly competent academic and technical staff, who advised the trustees of the Foundation where to give their money...It was not to be five programs each represented by a division of the Foundation; it was to be essentially one program, directed to the general problem of human behavior, with the aim of control through understanding...the Foundation also saw itself engaged in a kind of thought control. Fosdick (1952, p. 143) said: "The possession of funds carries with it power to establish trends and styles of intellectual endeavor.'"

In a discussion about Beijing man, when actual physical evidence is not available for study, some reports are believed while others are dismissed: "[The authors] propose that reports about evidence conforming to the standard view of human evolution generally receive greater credibility than reports about nonconforming evidence. Thus deeply held beliefs, rather than purely objective standards, may become the determining factor in the acceptance and rejection of reports about controversial evidence."

Bigfoot?

Chapter Ten is titled "Living Ape Men?". This chapter reviews and discusses many descriptions of what we term in the Pacific Northwest as "Bigfoot." The term used by the authors most often is "Wildmen." This chapter is highly recommended. It increased my knowledge and awareness about the prevalence of reports concerning this type of creature or Being.

Indeed, after pages and pages of descriptions and discussions of evidence, the authors write, "Despite all the evidence we have presented, most recognized authorities in anthropology and zoology decline to discuss the existence of wildmen. If they mention wildmen at all, they rarely present the really strong evidence for their existence, focusing instead on the report least likely to challenge their disbelief.

Later in the book, during another discussion about skeletal remains discovered in Africa, the authors write [on page 649]: "...Most of the discoveries scientists have used to build up their picture of human evolution are similarly ambiguous, their significance obscured by professional rivalries and imperfect investigative methods."

A Sample of Anomalous Evidence

From [pages 795-814] Appendix 2, which is titled "Evidence for Advanced Culture In Distant Ages." A sample of anomalous evidence follows:

> "Raised letter-like shapes found inside a block of marble from a quarry near Philadelphia, PA. The block of marble came from a depth of 60-70 feet, which suggests the letters were made from intelligent humans from the distant past."
>
> "A metallic bell-shaped vessel that was blown out of pudding stone now called the Roxbury conglomerate, is over 600 million years old...by current standards, life was just beginning to form on this planet...but this vessel indicates the presence of artistic metal workers in North America over 600 million years before Leif Erikson."
>
> "A chalk ball was found...and based on its stratigraphic position, it can be assigned a date of 45-55 million years ago."

The appendix has a long list of other anomalous evidence, but since I've already over-quoted from this text, I will leave the remainder of the secrets to be discovered by the reader.

Related Correspondence

3.5.7.1. Letter to Katharina Wilson, undated

Thank you for sending me the copy of *The Observer* with the review of *Forbidden Archeology*. The review was unsigned, but I assume that you were the author. You are one of the few reviewers to pick up on the Rockefeller Foundation material that was in the book. For me, that was one of the most intriguing things that I turned up in my research. In any case, I was pleasantly surprised when members of the UFO community started reading and talking about *Forbidden Archeology*. Your review is another sign of that. May I ask how you learned about *Forbidden Archeology*? It is useful for me to know by what channels the book is circulating. In my next book, which I am calling *Human Devolution*, I am presenting an alternative view of human origins (*Forbidden Archeology* establishes the real need for an alternative). In this new book, I will be introducing some of the UFO/alien entity material, as this is an essential element in the whole picture.

3.5.7.2. Letter from Katharina Wilson, August 5, 1996.

It was so nice to receive your message today! I'm glad the newsletter was forwarded to you. Yes, I am the person who wrote the review, not an easy thing to do for someone who is not educated in the field of archeology. I purchased the book for my father (a professor of Geology and Geography) for his retirement, of all things. I thought it would be a 'round about way' to introduce him to some new 'theories.' Of course, I read the book before I gave it to him—so far, he has had positive things to say about it.

I heard about *Forbidden Archeology* on a television program I was watching this Spring (in Portland, Oregon). I can't remember what the name of the program was, but I tend to watch TLC and Discovery regularly. I also watch programs that carry information about the UFO phenomenon, so it may have been *Sightings* or possibly even *Paranormal Borderline.* After the show, I think the next day, I began calling bookstores. Powell's Books said they had sold out and had more on order, Barnes & Noble said they had sold out as well. Strangely, B&N told me the book didn't have an ISBN# and would have to be ordered from England and it would take many, many weeks to get it in. That sounded a little strange to me, so I talked with someone else in the bookstore, and they gave me the publisher's phone number. That's how I got the book.

I am the author of *The Alien Jigsaw* and a companion book *TAJ Researcher's Supplement.* The books are the publications of my journals, and the Introduction was written by Budd Hopkins. The book is available in major bookstores but I sell the *Supplement* via mail order. If you would be interested in the *Supplement* or some of my articles (for your research on your next book) I could send you some information that might be helpful. If you are interested please email me your mailing address. *Forbidden Archeology* sounds very interesting. I wish you all the best with this project. Please let me know when it's available.

I'm in the process of creating a web site and I will have information on how to order current and back issues of *The Observer* newsletter. I hope this helps to spread awareness about your book.

3.5.7.3. Letter to Katharina Wilson, August 24, 1996.

Thank you for your detailed response to my queries. At the moment, I am up in the Northwest, on a vacation/writing retreat, staying with some friends of mine near Bellingham, Washington. Looking out the window here I can see Mt. Baker rising over the hills. In a few days, I am leaving on a trip to Australia. Of course, I would be interested in hearing your father's reactions to *Forbidden Archeology*—if he wouldn't mind writing a few lines. It is gracious of you to offer to provide me some research materials. I would be happy to receive them.

3.5.7.4. Letter from Katharina Wilson, August 29, 1996.

I hope you are having a great vacation. Australia sounds wonderful!

I will send you copies of my books in case they will help you while researching UFO/abduction information for your new book. I think I'll include an article or two as well. (I hope this isn't too presumptuous of me.)

My father is on a month-long vacation, but when he returns, I'll ask him if he would write a few lines, as you requested.

3.5.7.5. Letter to Katharina Wilson, September 1996.

I spent most of my time in Australia on a working vacation—staying at a house owned by an English pop star on an island near the Great Barrier Reef. Encountered lots of exotic wild life right in the gardens of the home—giant fruit bats that explode out of the trees as you walk by in the early morning, wallabies (miniature kangaroos) that go crashing away through the underbrush, and those kookaburra birds with their wild, hooting calls, screeching parrots, etc.

I've been reading your book, and I think it is very interesting how we have been brought in touch with each other. I sense some kind of arrangement behind it.

I am a follower of the ancient Sanskrit teachings of India, and in those teachings I've learned that we are part of a cosmic hierarchy of beings. At the top of the hierarchy, there is a supreme conscious intelligent being, existing in a nonmaterial level of existence, along with other liberated beings. This material world is a place for beings who have departed that higher realm. Some are striving upward (*devas*) and others are striving downward (*asuras*). The entire universe is inhabited, and is divided into different levels, according to states of consciousness. The *devas* are good (godly) beings, and there are many levels of them, ranging from minor gods and goddesses to very powerful gods and goddesses, in charge of universal forces. The *asuras* are ungodly beings, and there are also many levels of them. Some of them are minor ghostly beings, and some of them are very powerful demonic beings. All of these beings are capable of interplanetary travel. Sometimes they travel by mystic power, sometimes by subtle spaceships, and sometimes by mechanical spaceships. The *asuras* in particular use the mechanical kinds, although they are also capable of the other kinds of travel. The *devas* do not visit this planet very much. Before the time of Buddha, about five thousand years ago, they would come a lot. But around the time of Buddha, people stopped performing the sacrificial rituals for the demigods. Even in Western culture, if we go back to the times of the Greeks, Romans, and Vikings, we can see that people were directly interacting with demigods. Of course, this was also true in India, Egypt, China, etc. Also, the universal time goes in cycles of four ages, the first one being very spiritual, and each of the others progressively more degraded. We are in the fourth age of a cycle. When you put it all together, you can predict that our contacts with other beings would most likely be unpleasant. In other words we would not be seeing very many contacts with demigods. According to the Sanskrit writings, there is located about 800 miles above the earth a zone inhabited by many kinds of asuric creatures. In the Sanskrit writings, there are many descriptions of them, and sometimes they abduct people. They also say that some of the effects are bad dreams, missing time, etc. My colleague Richard Thompson has compared the Vedic accounts with the modern UFO/abduction phenomena in his book *Alien Identities*. In any case, although there may be some unpleasant experiences as a result of contacts with asuric beings, we are confident that the whole system is under the control of the higher forces of pure goodness. In addition to the *devas* and *asuras*, there are liberated sages who travel through the universe, or take birth on different planets, to explain the whole system to people who really want to know. These sages are different than the demigods. The demigods, although basically good, are still infected by a desire to control forces in the material world. The liberated sages have no interest in the material world whatsoever, except to guide the beings there back to their original home.

What I have found especially interesting about your experiences is the strong paranormal element to them. Many in the UFO/abduction field seem to be very averse to that, and want to keep things on a strictly technical level, consistent with known physical laws. Your experiences seem to transcend that.

I also very much appreciated your teachings about not killing animals. As part of my spiritual practice, I have been a vegetarian for over twenty years.

Thanks again for sending your wonderful book and other very interesting materials. Of course, they will be helpful with my work, but I am mainly grateful for the glimpse they give into your personality.

3.5.8. Walter J. Langbein (1994) Notice of *Forbidden Archeology. PARA,* January. Reprinted by permission of Walter J. Langbein.

PARA *is an Austrian publication featuring paranormal phenomena. The following excerpt from the short notice is translated from the German.*

If we imagine the history of humanity as a giant museum, containing all knowledge on this topic, then we shall find that several of the rooms of this museum have been locked. Scientists have locked away the facts that contradict the generally accepted picture of history. Michael A. Cremo and Richard L. Thompson have, however, opened many of the locked doors and allowed laymen as well as scientists to see inside. Their massive work hit the USA like a bomb. Even scientists have been influenced, and rightly so. *Forbidden Archeology* compels the world of science to enter new territories and calls into question many revered theories about humanity and human history.

3.5.9. Robert Stanley (1994) Review of *Forbidden Archeology. UNICUS* vol. 3, no. 4, p. 38. Reprinted by permission of *UNICUS.*

Finally there is a reference book for serious students of ancient humanity that lists many of the politically and socially unacceptable archaeological discoveries of the past few hundred years. The text is scholarly and highly informative including clear line drawings of some of the featured artifacts.

The chapters dealing with wildmen, found nearly everywhere on earth, were based on credible witnesses and artifacts, analyzed under scientific methods. The evidence offered clearly shows that it is possible that man's missing link, a type of apeman humanoid, has lived near to but not directly with mankind for thousands if not millions of years.

This book is a fresh perspective on a very old subject. It carefully rethinks the known fossil records and the methods used for dating them. *The data is clear and powerful.* There are no final conclusions, regarding man's origins, drawn for the reader. You must make up your own mind after examining this massive manuscript of nearly 1,000 pages. *It is thoughtfully designed to give a much broader view of mankind's past.*

3.5.10. Jerry Watt (1996) "The Dawn of Humankind: A Review of *The Hidden History of the Human Race*." *Life In Action*, vol. 22, no. 5, September/October, pp. 169-171. Reprinted by permission of The Great School of Natural Science.

Life in Action *promotes the Philosophy of Individual Life as revealed through the teachings of The Great School of Masters.*

Mainstream paleoanthropologists will not like the *Hidden History of the Human Race* by Michael Cremo and Richard Thompson. Neither will many "new age" types be eager to embrace the latest speculative theory about human origins. While it is interesting reading, the *Hidden History of the Human Race* is more like an unauthorized biography of a science with more than a few skeletons in its closet (pun intended).

Those who are interested in the Philosophy of Individual Life will find *Hidden History* interesting for, perhaps, a reason not imagined by the authors. In the *Great Message,* John E. Richardson wrote:

> "The Great School of the Masters came into existence as a definite organic entity. The exact time of its birth as an organization is so remote that it is not within the range of historic certainty, and can therefore truthfully be said to be prehistoric. The remoteness of its antiquity, however, may be safely assumed from the suggestion that it is said to have a record(ed) history of more than 100,000 years."
>
> [John E. Richardson; *The Great Message: The Lineal Key of the Great School of the Masters*; The Great School of Natural Science (1950), p. 23]

How could there be an organization with a recorded history of 100,000 years? According to paleoanthropologists, just 10,000 years ago humankind existed as bands of hunter/gatherers. The bow had not yet been invented, let alone the written word. Cremo and Thompson make no claims about unknown ancient civilizations, but they cite one hundred and fifty or so pieces of evidence that suggest (and strongly so) that humankind is much more ancient than now thought. Perhaps millions of years older than now accepted. And, of course, this would allow for the existence of a 100,000 year old educational organization dedicated to the advancement of mankind.

According to Cremo and Thompson, the primary point they make is "...that there exists in the scientific community a knowledge filter that screens out unwelcome evidence." And this is not a radical view. It has often been noted that it takes a generation or so for new discoveries to be accepted into science. Often it isn't until the "old guard" passes away that new evidence is properly evaluated. The old guard busily tries to squeeze "square" evidence into "round" holes so they can keep alive the theories they've grown up with. As my instructor, Robert Johnson, recently wrote:

> ...many times in my scientific career I have seen this, with supposedly objective scientists twisting and/or ignoring anything not agreeing with their preconceived notions, or with a famous person's theory. Scientists are not as smart, wise, and objective as many non-scientists think them to be. Most are just "part-smart." [Robert W. Johnson, personal correspondence, June 23, 1996]

An example of the "knowledge filter" Cremo and Thompson write about is the so-called Java Man. Discovered in the early 1890's, this creature was surmised based on a cranium (skull cap) and femur (thigh bone) found about 45 feet apart. The cranium is clearly apelike while the femur is indistinguishable from modern humans. These bones were dated as 800,000 years old. For many reasons, the Java Man is now discredited. It was the product of paleoanthropologists attempting to create the "missing link." The apelike skull cap actually belonged to a large gibbon. But here's the interesting thing. We are left with a modern human thigh bone existing in Java about 400,000 years before humans are thought to have migrated to that area.

Scientists had no problem with a modern type human femur existing 400,000 years too soon in Java as long as it was part of their missing link. Today, however, it is not viewed as evidence for a greater antiquity of the human race. It doesn't fit the accepted theories. Quoting from adverse criticism in the front of their book (yes, they included adverse criticism as well as the favorable), quoting from this criticism: "Your book is pure humbug and does not deserve to be taken seriously by anyone but a fool." And that is from Richard Leakey.

Those who have [a]n interest in paleoanthropology will enjoy *Hidden History* the most. This is not a popularization. Some subjects are more than a little technical and, for example, deal with eoliths, paleoliths and neoliths—the first stone tools of mankind. Other sections deal with anomalous human skeletal remains, the Piltdown Man controversy and Beijing Man.

Perhaps the most controversial chapter of *Hidden History* is the one dealing with "Evidence for Advanced Culture In Distant Ages" (Chapter Six). Cremo and Thompson detail the unearthing of letters found *inside* a marble block found in Pennsylvania, a clay image found in Idaho, a fossilized shoe print from Nevada and a block wall discovered in an Oklahoma mine. They also discuss grooved spheres found in South Africa recovered from deposits said to be 2.8 billion years old. It is important to note that Cremo and Thompson make no special claims for the material in Chapter Six; rather, they include the information for completeness and to help promote further inquiry into these fascinating artifacts.

Hidden History, the abridged and more affordable version of *Forbidden Archeology*, makes a good case for the coexistence of anatomically modern humans and the very hominids we are supposed to have evolved from. Cremo and Thompson cite example after example of finds that never make their way into the mainstream literature because they do not support the prevailing theories. Even if just a handful withstand the rigors of scientific examination, they will have succeeded in showing that scientists are not the impartial investigators they present themselves as, and they will have shown that it is at least possible for the human race to be vastly older than now accepted.

And the Great School of the Masters may be a part of that greater antiquity of mankind.

3.6. Other Reviews and Notices

3.6.1. Anonymous (1994) Notice of *The Hidden History of the Human Race. Sci Tech Book News*, November.

A condensation of *Forbidden Archeology*, 1993, in which the authors, in keeping with their Vedic beliefs, argue that modern man's origins go back millions of years before the dates set by the establishment.

3.6.2. Diane C. Donovan (1993) Review of *Forbidden Archeology. Midwest Book Review*, September. Reprinted by permission of James A. Cox, publisher.

Michael Cremo and Richard Thompson's *Forbidden Archeology* is a weighty, eye-opening exposÈ of scientific cover-ups regarding evolution. Over a thousand pages document the real evidence about human origins, with a researcher and scientist joining forces to examine how inherent prejudice has affected the research establishing evolution. The authors gather a wealth of arguments and facts to help readers rethink human origins and history: chapters are surprisingly in-depth and detailed as they probe the key moments of archeological discovery and how these findings were regarded. Over eight years of research results in a controversial challenge to conventional thinking, making for an impressive, scholarly work.

3.6.3. Anonymous (1994) Notice of *The Hidden History of the Human Race. The Bookwatch,* September, p. 3. Reprinted by permission of James A. Cox, publisher.

Over the past two centuries scientists have found bones and artifacts showing that people like ourselves existed millions of years ago: facts suppressed by the scientific community. Evolutionary prejudices have hidden the truth and created a false story of prehistory, the authors maintain: chapters build an argument for an alternative viewpoint which rethinks human origins.

3.6.4. Lori Erbs (1993) Unpublished review of *Forbidden Archeology.* Reprinted by permission of Lori Erbs.

At the time she wrote this review, Lori Erbs was director of the United States Forestry Service Science Research Library in Juneau, Alaska.

Forbidden Archeology is a masterful examination of the vast body of research on human evolution, articulated with intrigue, precision, and sound logic. Authors Cremo and Thompson probe the "sociology" of scientific reasoning, contending that consistent standards have not been applied to the evaluation and acceptance of evolutionary evidence, also asserting that filtering effects have excluded findings that do not support prevailing theories of evolution. This comparative analysis of the voluminous expanse of research on evolution is conducted with objectivity and thoroughness. Of particular interest is documentation of suppressed and fraudulent investigations. Although unique and challenging views are presented, readers are allowed to form their conclusions on the basis of arguments given.

Written for the nonspecialist and specialist alike, *Forbidden Archeology* is bound to become a landmark in the literature on human evolution. Scrupulously researched and referenced, with a good index, its 814 pages of text are expertly crafted in a flowing style that invites readers onward in their exploration of "the hidden history of the human race." Recommended for academic and public libraries, as well as reference collections, for it contains an extensive bibliography and wealth of research on the subject and balances collections with an alternative view of human origins.

3.6.5. Joni Tevis (1995) "Krishna scientist debunks archaeology." ***Florida Flambeau*****, vol. 80, no. 122, Thursday, March 9, p. 3.**

The Florida Flambeau *is the student newspaper of Florida State University at Tallahassee, Florida.*

A process of "knowledge filtration" is causing the age of the human race to be underestimated by hundreds of millions of years, according to Michael Cremo, researcher for the Bhaktivedanta Institute in Gainesville.

"Imagine archaeology as a big museum. We can imagine ourselves going up the steps, hearing our footsteps echo down the hallways, seeing paintings of human evolution as we have always heard it. We find a room with fossils to support what we've heard. We would also find rooms that the public is not usually allowed to enter, rooms with artifacts that tell a different story. The effect is that the story we are accustomed to hearing isn't correct," Cremo said during a guest lecture in a Human Evolution class at Florida State University Wednesday.

Cremo, also known as Drutakarma Dasa, will also present his ideas on evolution tonight at 7 p.m. at the International Student's Center.

Cremo, along with coauthor Richard L. Thompson, dug through foreign journals and reports from the 19th and 20th centuries to find material for their book *Forbidden Archeology*. Along the way, Cremo claims to have found "literally hundreds and thousands of artifacts that provide contrary evidence to the accepted paradigm.

"We found a process of knowledge filtration, whereby evidence was set aside because it didn't fit with a theory accepted by the scientific community," he said. "The conflicting evidence was subjected to very intense negative scrutiny, and when holes were found in it, as holes can be poked in any piece of evidence, the conflicting evidence was put through the knowledge filter and rejected."

Cremo said his reason for beginning his controversial research stemmed from studies of ancient civilizations.

"My underlying motive was the inspiration of ancient Sanskrit writings of India and its accounts of ancient civilizations," Cremo said.

As evidence for his theory, Cremo cited the discoveries of human skeletons, mortars and pestles and obsidian spear points by California gold rush miners in a shaft dug into basalt-covered rock layers. The rock that the artifacts were found in dated from the Eocene, 38 million years ago.

Cremo said that definite proof showed that the artifacts could not have simply fallen into the mine shaft.

"Scientists stated that the artifacts came from stable layers of rock," he said. He also claimed that the artifacts are now at the University of California at Berkeley, but they are not on display for the public to view.

Cremo also cited the example of a stone tool found in Mexico. The site, dated by four different methods, is over 250,000 years old. This date contradicted the entire picture of New World archaeology, since the tool was assumed to have been made by fully modern man. Before the site was officially dated, estimates were put at 20,000 years. Virginia Steen-McIntyre, the geologist who dated the site, was labeled a "publicity-seeker" after her out-of-the-ordinary dating of the site.

Cremo displayed a poster detailing the differences between the "conventional scenario" and his "suppressed evidence." The poster showed an iron pot dated 300 million years old, far earlier than a stone tool 10 million years old. When asked about the apparent degression of human technology, Cremo said that the accepted concept of "linear time" should be replaced with "cyclical time."

"We have an unexamined preconception of linear time; it's culturally derived. Different levels of culture coexist even today, such as the MIT computer scientist and the Australian aboriginals," Cremo said.

When presented with the idea that the "knowledge filter" he decries actually screens material and provides dependable information, Cremo asked for a "degree of fairness" in dealing with facts contrary to the accepted hypothesis.

Cremo will give a lecture at 7 p.m. tonight at the International Student's Center, S. Wildwood Dr., followed by a free vegetarian feast.

3.6.6. Andrew M. Mayer (1995) Review of *The Hidden History of the Human Race. Small Press*, Winter. Reprinted by permission of *Small Press.*

Richard Thompson, a scientist and mathematician, has collaborated with Michael Cremo, a writer and editor for Bhaktivedanta Book Trust, and researcher Stephen Bernath to present an alternative version of the history of human evolution.

Many creditable scientists have little patience for such theories that do not credit Holocene and Pleistocene development of the appearance of modern man (Neanderthal, *Homo erectus, Australopithecus*). The authors take exception to these and other credential findings from eight eras.

Cremo and Thompson find difficulty in understanding the basis of evidence collected since the publication of Darwin's *Origin of the Species* and later *Descent of Man.* They feel the account of skeletal remains and fossilization fails to responsibly correspond to the genealogical chronology presented by scientists to justify evolutionary time. In effect, as practitioners of the Vedic literature of India, they have varied ideas of skeletal antiquity. They present enlarged versions of what they include as anomalous evidence and accepted evidence of man's evolution. They see deception of evidence from Piltdown man to present day.

Java Man, Beijing Man, 'Living Ape Men' and new evidence out of Africa (late 20th Century) are all pontificated upon in questionable fashion.

Anthropologist Richard Leakey states:

"Your book is pure humbug and does not deserve to be taken seriously by anyone but a fool. Sadly, there are some, but that's a part of selection and there is nothing that can be done."

William W. Howell, physical anthropologist, says:

"To have modern human beings...appearing a great deal earlier, in fact at a time when even simple primates did not exist as possible ancestors, would be devastating to the whole theory of evolution, which has been pretty robust up to now."

Yet Philip E. Johnson, law professor at the University of California, Berkeley and author of *Darwin on Trial*, concedes:

"In the end the important thing is not why the investigators were motivated to look for a certain kind of evidence, but whether they found something worth reporting, and worth serious consideration."

I would agree with Dr. Johnson that we deserve the right of current interpretation of data that long has been regarded as 'basic' for evolutionary science, in the event further evaluation justifies changes in the chronological approach to evolutionary theory and practice.

3.6.7. Ed Conrad (1993) "New book claims man existed on earth long before the apes." *Hazleton Standard-Speaker*, Wednesday, November 17, p. 13. Reprinted by permission of the *Hazleton Standard-Speaker* and Ed Conrad.

The Hazleton Standard-Speaker *is a daily newspaper published in Hazleton, Pennsylvania.*

This brand new book arrived in the mail for the *Standard-Speaker* to review and, since its title is *Forbidden Archeology: The Hidden History of the Human Race*, it came as no surprise that I wound up with the assignment.

After all, for more than a decade I've been insisting that I've discovered petrified human bones and soft organs between anthracite veins in Carboniferous strata dated at a minimum of 280 million years.

If I'm correct—and I assure you I am—then man *couldn't possibly* have evolved from the lowliest, earliest primates since they didn't pop up until multimillions of years later. And, of course, the bottom line is that Darwin and the evolutionists are dead wrong about man's animal ancestry.

Forbidden Archeology lends credence to my findings because its co-authors contend that they have come across evidence, during eight years of research, that the human fossil record has been suppressed by members of the evolutionary establishment to back up their theory of man's inhuman origin.

"Over the past two centuries, researchers have found bones and artifacts showing people like ourselves existed on earth millions of years ago," it states in the first paragraph of the cover overleaf. "But the scientific establishment has suppressed, ignored or forgotten these remarkable facts. Why? Because they contradict dominant views of human origins and antiquity."

Michael A. Cremo and Richard L. Thompson, the co-authors/researchers, insist that artifacts discovered in strata predating the earliest date given for the initial emergence of the apes substantiates that humans existed millions of years earlier.

But they charge that the scientific establishment has totally ignored a collection of remarkable findings.

They say evolutionary prejudices, deeply held by powerful groups of scientists, have been responsible for the coverup which they more politely call "a knowledge filter."

Cremo, a researcher specializing in the history of science, and Thompson, a scientist with published works in the field of mathematical biology, use an abundance of convincing facts.

In the most interesting final chapter, Evidence for Advanced Culture in Distant Ages, *Forbidden Archeology* reveals the following:

- A nail was discovered embedded in stone in a sandstone quarry in Scotland in strata dated between 360 and 408 million years.
- Gold thread was found inside stone in England in strata 320 to 360 million years old.
- A gold chain was discovered between bituminous coal veins near Morrisonville, Ill., in strata 260 to 320 million years old.
- A carved stone was found in a bituminous mine near Webster, Iowa, in strata 260 to 320 million years old.
- An iron cup was discovered inside a bituminous coal mine near Thomas, Okla., in strata 260 to 320 million years old.

Established science claims that the earliest primates date back 60-65 million years ago and that the earliest hominids—manlike creatures that walked upright—first emerged about 14 million years ago, if anywhere near that far back.

Forbidden Archeology already is receiving glowing reviews.

Dr. Siegfried Scherer, a biologist in Munich, Germany, observed, "If it stimulates professional re-investigation of reports not fitting the current paradigm on human evolution, *Forbidden Archeology* will have contributed to the advancement of knowledge of the history of mankind."

Diane C. Donovan of the Midwest Book Review calls *Forbidden Archeology* "a weighty, eye-opening exposé of scientific cover-ups regarding creation. Over a thousand pages document the real evidence about human origins, with a researcher and scientist joining forces to examine how inherent prejudice has affected the research..."

Meanwhile, I can only wonder why I'm not even mentioned in the book because I've certainly had enough publicity for Cremo and Thompson to at least have heard of me.

After all, lengthy articles about my discoveries have appeared in *The Congressional Record* on two different occasions; two multi-page features have been published in *The Spotlight,* the international populist newspaper; and numerous other stories have appeared in more than 500 newspapers from coast to coast.

Despite my specimens' rock-like appearance, two of the nation's most prestigious laboratories have examined the cell structures of one of them microscopically and have confirmed the presence of Haversian systems, the tell-tale sign of bone.

In addition, two of the world's foremost experts on human anatomy—the late Dr. Raymond Dart and the late Wilton M. Krogman, author of "*The Human Skeleton in Forensic Medicine*"—had physically examined several specimens and stated in writing that they dramatically resemble the contour of bone.

I'd like to add—for the record—that my discoveries of petrified human remains include, in addition to approximately three dozen bones, a gall bladder and a hemisphere of a human brain. I also have found an ax handle that has turned to a substance resembling coal; a dinosaur tooth and bones; and a portion of a giant prehistoric scorpion.

A number of the petrified bones, including a portion of a hominid mandible, are on permanent exhibit in the Greater Hazleton Historical Society Museum.

I am aware of Ed Conrad's discoveries, and my coauthor Richard Thompson visited him. Our standard for including discoveries in Forbidden Archeology *was that the original reports should include the locations where artifacts and bones were found, so that we could consult modern geological studies to check the age of the strata. Conrad did not reveal the exact locations and circumstances of his discoveries, and for this reason we decided not to include them. This does not imply that the discoveries are not genuine—only that the reporting did not conform to* Forbidden Archeology'*s standard. If the bones Conrad found are actually human (or even simply hominid), skeptics and sympathizers alike will naturally want to know how and where they were found. It is therefore important to show that the bones were found in strata of known great age and that there was little likelihood they could have been introduced into the deposit in more recent times. Conrad's discoveries are, however, worthy of attention and study.*

3.6.8. Philip Lipson (1995) "Michael Cremo and *Forbidden Archeology*." Unpublished review of *Forbidden Archeology*.

Philip Lipson is director of The As-You-Like-It Metaphysical Library in Seattle, Washington. I gave a lecture at the Library during a visit to Seattle. His article was submitted to the publication New Times *but was not printed.*

Who we are and where we come from as human beings has been the major question which everyone from ordinary people to philosophers, scientists, historians and archaeologists have been pondering over since the beginning of history, whenever that was.

And the key question of finding out "whenever that was" has been the major bone of contention between scientists, particularly archaeologists on the one hand and occultists, open-minded thinkers and simply the curious person who in his heart feels that the age of humankind is nearly infinite or at least extends back millions of years farther than most scientists have led us to believe.

The standard view of archaeologists regarding the origin of humankind puts the date of the origin as being not more than 100,000 years old. Yet much occult and ancient literature, particularly the Vedas and Puranas of Indian origin, as well as famed spiritual scientists such as Rudolf Steiner and Madame Blavatsky, have given much older dates for the origins of civilization going back millions of years.

Michael Cremo and Richard Thompson in their book *Forbidden Archeology* and *The Hidden History of the Human Race* (an abridged version of *Forbidden Archeology*) systematically question and in a sense debunk standard archeological consensus reality by citing case after case where artifacts and items, such as tools and skeletal remains, have been found in strata which is much older than when humankind as we know was supposed to have existed. This disinformation has apparently been going on for over 150 years, since around the time of Charles Darwin, father of evolutionary theory and developer of what is considered a kind of baseline for modern archeological thought.

Cremo and Thompson in the above-mentioned book expose some of the "myths" which at the present time are in the "everybody knows that" category and provide extremely well-documented evidence that question them.

One of these commonplace ideas is that "civilization began not more than 10,000 years ago in the Middle East" where there is actually documented evidence that "signs of civilization, such as gold chains, iron pots, and stone walls have been found in coal deposits hundreds of millions of years old."

Other ideas of which there is some evidence is that man and dinosaurs may have co-existed and that evidence for the existence of living ape manlike creatures such as Bigfoot have been seen by scientists and trained observers in all parts of the world.

Cremo points out that trying to get the truth about a scientific controversy from a scientist is like trying to get the truth about a political scandal from a politician.

A chart which accompanied the book presents a bold depiction of the difference between evidence the authors have found and the conventional scenario. The conventional viewpoint shows that between 10 and 100 million years ago, fairly advanced tools such as mortar and pestles have been found. Where convention says that the first birds and mammals appeared between 100 million and a billion years ago, in the same strata, humanoid footprints and gold chains have been discovered.

Cremo, who lectured at the As-You-Like-It Library on November 19th of this year, ascribes this lack of acceptance to what he calls "knowledge filtration." In other words, archaeologists agree that a certain explanation of humankind's origin makes sense to them. Therefore they force all evidence to fit this viewpoint by rejecting contradictory evidence and by making it difficult for opposing viewpoints, even by academically qualified archaeologists, to get published.

Cremo in his talk related an experience which he had while talking to a curator at the British Museum, who when asked what a certain file was about, was told that this is the file where we put reports we don't like to be sent out for "demolition jobs"—in other words sent to scientists who will make sure that these reports either get thoroughly debunked or else never see the light of day.

Cremo mentions cases of scientists who disagreed with the prevailing paradigm losing their academic positions and being made the victim of "whispering campaigns" meant to spread rumors that they were going "off the deep end."

Cremo also made it clear that archaeology wasn't the only "forbidden" subject, but that practically every academic subject has this aspect to it. A case in point is alternative medicine where literally thousands of treatments have been suppressed, often largely because they simply do not fit the prevailing medical model. Another example would be forbidden science, such as the lack of attention paid to the inventions of Nikola Tesla.

My speculation on part of the reason that this standard viewpoint is so steadfastly adhered to is simply that scientists feel it is necessary to maintain a coherent, progressive, linear universe which can be easily thought about and handled. Scientists would be loath to admit that their vast education is unable to explain everything. It may be a similar situation to why the government or scientists refuse to acknowledge the possibility of extraterrestrial life. It would just be too upsetting to their world view and the idea that the universe is going along business as usual.

Cremo posits the idea that archeology and history itself are actually cyclic, and that civilizations have arisen, flourished and died leaving behind their various artifacts, and that we may find evidence at various levels of development in the same areas, putting a lie to the myth of so-called progress.

Cremo's forthcoming book will be entitled *Human Devolution*, in which he speculates that we might have to go "off the planet" to get at the true story of humankind's origin. Much of thinking of mainstream scientists Cremo describes as "cognitive colonialism", and suggests that we expand our horizons and consider the universe in all its vastness and glory and infinite possibilities.

Richard Thompson is also the author of several remarkable and highly recommended books, *Alien Identities* and *Vedic Cosmography and Astronomy*. These books, as well as *Forbidden Archeology*, are available at bookstores and at the As-You-Like-It Library.

Michael Cremo will be appearing February 5th in the Rainier Room at The Seattle Center from 1 to 5 PM. Tickets are $10.00 in advance and $12.00 at the door. Call 206-726-3935 or contact Flashpoint Productions POB 295, Seattle, WA 98111 for more information.

4

Selected Correspondence

Forbidden Archeology *resulted in a correspondence of several hundred letters with persons ranging from professional scientists to ordinary readers. I have included here a representative sample of this correspondence, registering responses that range from enthusiastic endorsement to caustic rejection to polite noncommitment. I am hopeful that anthropologists and sociologists will find this material useful for studies in how controversial ideas circulate (and do not circulate) across different intellectual boundaries. Some will find profoundly unsettling the mix of professional scientists, journalists, talk show hosts, UFO enthusiasts, alien abductees, New Age practitioners, conspiracy theorists, fundamentalist Christians, parapsychologists, and collectors of archeological anomalies among my correspondents. I find it refreshing and a signal of the ongoing (and necessary) renegotiation of our conventional categories of authority and perception. I have identified all communications as letters, although a good many were email messages and faxes. Unless otherwise noted, all correspondence in this chapter is printed with the written permission of the authors. Unless otherwise noted, all letters are to or from Michael A. Cremo.*

4.1. Alan F. Alford

Alan F. Alford is an author in England with an interest in contacts between humans on earth with extraterrestrials.

4.1.1.Letter from Alan F. Alford, October 14, 1996 .

Gene Phillips [*president of the Ancient Astronaut Society*] gave me your number, as he thought you might be interested in joining forces with me to get some publicity in the UK.

You may recognize my name from *Ancient Skies* [*newsletter of the AAS*], as I'm a regular contributor, and I am also due to share a platform with you in Orlando in August 1997 [*at the AAS world conference*]. I have recently written and self-published a book entitled *Gods of the New Millennium,* which I am trying to promote in England. Like your own book, it is extremely controversial and has so far been given a wide berth by the national book reviewers...

However, I have recently been speaking to one of the UK's leading quality daily papers, who are interested in running a feature on the idea of a challenge to Darwinism as it applies to man. They are not prepared to stick their neck out on the back of one book/one author, but they are interested in running a 2-part feature comprising your conclusions and my conclusions. The two articles would complement each other well, as my book focuses on the most recent leap to *Homo sapiens* 200,000 years ago, whereas your own book goes back much further in time. How these two perspectives on our origins fit together is an intriguing subject that I would very much like to discuss with you.

Please let me know asap if you are interested in promoting your book in the UK, as outlined above.

I responded positively to Alford's invitation. As of the time of this writing, the articles have not been published.

4.2. Dr. Patricia A. Ariadne

Dr. Patricia A. Ariadne is a scholar of religion.

4.2.1. Letter from Dr. Patricia A. Ariadne, April 15, 1995.

I have just purchased your book, *Forbidden Archeology*, and find it fascinating. But I am especially interested in acquiring your second volume when it is written, as I am a scholar of religion. I am currently writing a book which speculates on the origins and processes of initiation (I am a Fourth Initiate in Eckankar so like both of you, I come to my writing with and from a spiritual foundation and perspective).

Please send me any journal writings or small publications that you might put together in the process of completing Vol. 2. Of course, I would be glad to pay for these. (And by all means, let me know when Vol. 2 is available).

4.3. Celine Armenta

Celine Armenta is the daughter of Juan Armenta Camacho, a Mexican archeologist who discovered signs of a human presence at Hueyatlaco, Mexico, in deposits over 200,000 years old.

4.3.1. Letter to Celine Armenta, January 3, 1995.

My good friend Virginia Steen-McIntyre has told me much about your late father's work *Vestigios de Labor Humana en Huesos de Animales Extintos de Valsequilla Puebla, Mexico.*

I would be most grateful if you could send me a copy of this work. In our book, *Forbidden Archeology*, Richard Thompson and I document many cases of good archeological work that has been suppressed or ignored by mainstream scientists. Most of this evidence tends to show a greater antiquity for the human race than the current paradigm of human origins allows. It seems to me that your father's work is in this category, and therefore I would be very much interested in seeing it and using his evidence in future editions of our book.

4.3.2. Letter from Celine Armenta, February 7, 1995.

I received your letter a month or so ago, but I did not reply earlier because I have been looking for the copy of my father's work you asked me for and unfortunately I have to tell you that there is not any available copy.

Some weeks ago I sent to Virginia Steen McIntyre my own copy (my very only one) and I still do not get any other one. The book was published by one Governor of Puebla state and I though[t] there were hundred of copies at some storing place, but nobody knows where they are and since the book was never sold at any book store I do not know how to get it.

Virginia sent me a copy of your book. It is amazing! I am reading it right now and of course I will keep trying to get a copy of my father's book to send it you.

I will keep in touch. Thank you for being interested in my dad's work.

4.3.3. Letter to Celine Armenta, April 11, 1995.

Thank you for searching for a copy of your father's book. I am sorry not to have my own copy, but fortunately I was able to make a copy of the book owned by Virginia Steen-McIntyre. I assume that you and Virginia have met recently there in Mexico. I hope you had a good time together. I have read through your father's book and find it very valuable. I hope to make use of the material in future writings.

Dr. Virginia Steen-McIntyre, who dated the Hueyatlaco site at over 200,000 years, has translated Juan Armenta Camacho's book from Spanish into English and is seeking to have it published.

4.4. John E. Barrett

John E. Barrett has a degree in archeology and classical studies from the University of Reading, England.

4.4.1. Letter from John E. Barrett, April 6, 1994.

He enclosed with his letter a clipping from The Sunday Express *of June 26, 1977. The short article by John Payne reported the discovery of three metallic disks, each about five feet in diameter, in an Australian coal mine.*

I am writing to congratulate you on an excellent introduction to some of the shortfalls in current palaeoanthropology. Your book is also an interesting contribution to the discipline of epistemology and the philosophy of science. I graduated from the University of Reading in 1980 with a degree in Archeology and Classical Studies. I enclose for your perusal a photocopy of a news clipping from one of Britain's biggest Sunday newspapers from 1977. I hope that you will find it of interest and of possible value for future editions of your book. I am not sure about the 'UFO' angle, but the objects themselves must be representative of some facet of history. Have you any information on them? I am always intrigued by the fact that one often hears of astounding discoveries or developments and then never hears anything else about them!

Congratulations on your excellent work, once again, and I hope sincerely that it runs into further editions. It certainly deserves to reach a wider audience.

4.4.2. Letter to John E. Barrett, April 25, 1994.

Thank you for your kind words about *Forbidden Archeology*. The article you sent us about the large metal discs found in coal in Australia was quite intriguing. I had not heard of this case previously. I think it is definitely worth investigation. It may indeed be useful for future editions of the book.

Richard Thompson and I are now absorbed in the research and writing of our next book. And, at this moment, we do not have a researcher to whom we could assign this investigation. Eventually, I myself might have time to work on it.

But I thought I might ask if you would be interested in following up on this very interesting account yourself. I can understand that you might be quite busy with other things, but if you do have the time and inclination I could offer some hints about how to proceed.

In any case, thank you again for sending the article, and if you come across anything else like that, please send it to me.

4.5. Dr. Henry H. Bauer

Dr. Henry H. Bauer is a chemist at the Virginia Polytechnical Institute and State University in Virginia. He is also editor of the Journal of Scientific Exploration *and a director of the Society for Scientific Exploration.*

4.5.1. Letter from Henry H. Bauer, September 28, 1994.

I would appreciate receiving a copy of *Forbidden Archeology* by Cremo & Thompson, which has been suggested for review by the *Journal of Scientific Exploration.*

The *Journal* (ISSN 0892-3310) has been published by the Society for Scientific Exploration since 1987 (2 issues per year until 1991, since then 4 issues per year). We review books on the nature of science and books that describe current scientific knowledge as well as works on unorthodox scientific claims. With respect to the latter, the *Journal* contains material that seeks to "advance the study of...any aspect of anomalous phenomena, including ...1)Phenomena outside the current paradigms of one or more of the sciences such as the physical, psychological, biological, or earth sciences. 2) Phenomena within scientific paradigms but at variance with current scientific knowledge. 3) The scientific methods used to study anomalous phenomena. 4) The...impact of anomalous phenomena on science and society....".

We strive for objectivity in reviewing: the Society's charter is to provide a forum for discussion, not to promote or deny the validity of specific claims. Since anomalous phenomena are characteristically subject to controversy, sometimes we publish more than one review of a given book, or carry short editorial annotations to the reviews to indicate other possible viewpoints. We are also sensitive to suggestions from publishers or authors that certain individuals might not be appropriate reviewers of certain books.

Copies of book reviews will be sent to you promptly after publication.

4.5.2. Letter to Dr. Henry H. Bauer, November 7, 1994.

My publishers recently passed on to me your request for a review copy of my book *Forbidden Archeology.* I of course recommended that they send you a copy. If you have not received it by now, please let me know, and I will do what I can to catalyze their reaction. Although you did not mention it in your letter, I think you are connected with the STS [*Science and Technology Studies*] program at VPI&SU, along with Steve Fuller, and others. I am a member of the HSS [*History of Science Society*], and attended the joint annual meeting of the HSS/PSA/4S [*History of Science Society/Philosophical Society of America/Society for Social Studies of Science*] this year in New Orleans. Although somewhat outside the academic mainstream (being a member of a small institute that is connected with a Hindu-oriented spiritual organization), I am nevertheless quite fascinated by the development of mainstream science studies in the overlapping areas of the history, philosophy, and sociology of science. I am interested in finding ways to interact more closely with members of that knowledge community. If there are any opportunities for this at the VPI&SU STS program (conferences, colloquia, etc.), please let me know. I would also like to receive membership information for the Society for Scientific Exploration.

As of May 1997, a review of Forbidden Archeology *had not appeared in* Journal of Scientific Exploration.

4.6. Dr. David Bloor

Dr. David Bloor, of the science studies unit in the sociology department of the University of Edinburgh in Scotland, is one of the leading figures in the sociology of scientific knowledge. He asked that I not publish his correspondence with me. I will, however, include some of my letters to him. I can only speculate as to Dr. Bloor's reasons for wanting his brief and rather noncommittal letter to me to remain unpublished. Sociologists of scientific knowledge appear to be a rather skittish lot. Perhaps he felt that to be included in any way in this work would be detrimental to his reputation as a conventional scholar. This same comment applies to the entry on Dr. Michael Mulkay.

4.6.1. Letter to Dr. David Bloor, February 7, 1993.

Since you are one of the core set of the sociology of scientific knowledge community, I thought I would send you an advance, prepress copy of my book *Forbidden Archeology: The Hidden History of the Human Race.* I am gathering some prepublication reviews and would welcome your comments. If you can think of anyone else who might like to receive an advance copy, please let me know.

4.6.2. Letter to Dr. David Bloor, May 10, 1993.

A few months ago I sent you an advance review copy of my book *Forbidden Archeology.* I can well imagine that your academic duties limit your reading time, but if you have had a chance to go through any of *Forbidden Archeology,* I would welcome your comments. I am not (necessarily) looking for endorsements of the book's central thesis, which is somewhat radical. But I would be interested in your impressions of the book's general level of scholarship and whether or not you think the book would be of interest to members of your discipline's research community.

In his response, Dr. Bloor indicated he could not comment on the book but did think it might be of interest to students in his discipline.

4.7. Dr. Geoff Brown

Dr. Geoff Brown is an organic chemist and director of undergraduate studies in Science and Society at the University of Greenwich, England.

4.7.1. Letter to Bhaktivedanta Books Limited from Dr. Geoff Brown, 1995 undated.

I greatly enjoyed reading what is surely one of the most comprehensive compilations ever published on the discoveries which might reasonably be taken to support the authors' claims.

4.8. Dr. Jean Burns

Dr. Jean Burns is a physicist with an interest in consciousness studies and scientific anomalies. She offered some useful comments on the manuscript of Forbidden Archeology.

4.8.1. Letter from Dr. Jean Burns, January 29, 1993.

I am delighted to hear that *Forbidden Archeology* is on its way to the printer. This is a careful piece of scholarship about a fascinating subject, and I am confident that it will become a classic, in print for many years.

I think the Introduction will be very helpful in giving an overview of the book.

It sounds to me like you have removed most of what I would call "polemics," by which I mean belaboring or accusatory comments which go beyond factual description. I think it is always appropriate to point out that an important line of thought has been overlooked and to describe particular circumstances of the way it happened.

I want to commend you again on the really excellent writing. It is always clear, even when discussing very technical material, and the descriptions of researchers of earlier years carry a sense of "aliveness" that holds the reader's interest.

I will keep in mind any possibilities for reviews. Since you are an excellent writer, you might want to propose an article to *Omni* magazine or *Fate* which would give an overview of the book for the layman (high-school level). I am more familiar with *Fate,* which I read regularly, and I am confident it would welcome an article. But *Omni* has a larger circulation. You probably could get book reviews from both magazines.

I am sure many people will be interested in this book. I recently mentioned it to a friend of mine who has broad interests, and he told me he had already heard of it (before the book is even at the printer)! Someone had telephoned him to recommend it.

No doubt you will be in the Bay Area some time soon to publicize the book, and I hope to have a chance to meet you personally.

4.9. Ron Calais

Ron Calais is an archivist specializing in reports of archeological anomalies. Material provided by him is featured in Appendix 2 of Forbidden Archeology.

4.9.1. Letter to Bhaktivedanta Institute from Ron Calais, October 20, 1993.

I have recently completed reading a copy of Michael Cremo & Richard Thompson's fascinating volume: *Forbidden Archeology*, and found it to be both comprehensive and above all, intellectually stimulating. Their in-depth research efforts are impressive.

Having conducted nearly 30 years of intensive research work on this very subject myself, I can say without hesitation that this encyclopedic collection of "misfit" anthropologic discoveries is about the most convincing I have ever digested. Which brings me to the purpose of this communication.

As I said previously, I have a mass of similar data on file, some of which the authors have not mentioned in their book. Specifically, there are several interesting archeological finds covered by my material which would fit nicely into their "Appendix 2" section.

The only demand I make in order to use this data in a future printing of their volume is simply that I receive proper credit in print.

If either the editor or the authors are interested in reviewing this material, please inform me as soon as you are able. I will then proceed to photocopy relevant material for your perusal.

In the meantime, I shall eagerly anticipate your reply.

4.9.2. Letter from Ron Calais, July 10, 1994.

I hope you received the data I sent you for the new edition of your fine volume *Forbidden Archeology.* I have more material I uncovered recently and should be getting more in the near future, most of which you[r] book does not cover. Are you interested?

I need a favor from you. Can you send me a good copy of the Ameghino article—both the original Spanish and English translation—"El Femur de Miramar". I'm doing an article on related discoveries and I need a copy of this one for comparison, etc.

Please let me hear from you at your convenience. Thanks again!

4.9.3. Letter to Ron Calais, July 28, 1994.

I am enclosing a copy of "El Femur de Miramar" by Carlos Ameghino. I do not have a complete English translation. I translated the most relevant passages for inclusion in *Forbidden Archeology*. So whatever is translated into English you will find in the book, which I think you have.

Yes, Richard and I would like to see the new material that you have uncovered.

4.9.4. Letter to Ron Calais, May 10, 1995.

Thank you for the pages from the Gideon Mantell book, with the footnotes about objects discovered in flints. My working method would be to go back to the sources cited by Mantell, and even further back if possible. It would be good to know exactly what kind of coins were found in the pieces of Danish flint. Please do send more. Eventually I may do another book, when I have accumulated enough material.

4.10. Dr. George F. Carter

Dr. George F. Carter, of Texas A&M University, is an expert in North American archeology. He is the author of the influential book Earlier Than You Think, *which pushes back the time of human occupation of the Americas to over 100,000 years. Dr. Carter and I have had an extensive and very useful correspondence, but he decided he would rather not have his letters to me published. He expressed himself rather freely on many topics and personalities, and I can understand his reluctance to have such communications made public. Some of my letters to him follow.*

4.10.1. Letter to Dr. George F. Carter, October 9, 1993.

Knowing your interest in archeological evidence that challenges currently held ideas, I am sending you a copy of my book *Forbidden Archeology: The Hidden History of the Human Race.* You'll find several references to you and your work in the index. Of course, I would be very interested in hearing your comments about the book.

4.10.2. Letter to Dr. George F. Carter, December 31, 1993.

Richard Thompson and I have received with much pleasure your letters of November 2 and November 4, in which you give your impressions of our book *Forbidden Archeology.*

One of our hopes when we published the book was that we would get some feedback from insiders like yourself. So now that hope has been realized.

If you ever come to San Diego, Richard and I would be honored to accompany you to Texas Street. And if we could do a search with you along Pliocene shorelines in this area, that would be an even greater treat. As you can gather from our writings, Richard and I are not field archeologists. We have approached the subject from an epistemological standpoint, analyzing how archeological truths are generated in written reports and how these reports are related to the work at archeological sites. So it would be good to get out there in the field with you.

Lewisville. Why is it that charcoal makes the carbon dates bad? I thought charcoal was one of the most reliable materials for carbon dating, with bone being among the worst. Timlin. Does Raemsch still stand by his age for his site? Sheguiandah. Some of the published reports by Tom Lee and Sanford seem fairly detailed, especially the geological report by Sanford.

In general, you seem to have some resistance to finds of advanced stone implements in very old geological contexts, as at Hueyatlaco and Sandia Cave, for example. One can always say that a worker misread the geology. I am sure many professionals have reacted to your own work in that same way.

Concerning the auriferous gravel evidence, we are prepared to find that the geology of the entire area has been misread. But that would be a fairly large admission for geologists to make. Up to now, the age of the gravels in the areas where implements have been found seems to be pretty well established. We took the trouble to get opinions from the state geology department about the situation at the exact sites. So our point would be that given the currently accepted geology, the archeological evidence would have to be very, very old. That would go against the general archeological scheme that you outline—blades and mortars very late, not more than 10,000 years, cruder implements up to 200,000 to 300,000 years old. But until the geology of the auriferous gravel region actually does change, then we have to say, using our analytical method, that it appears the archeological evidence as reported is quite old. You say that the geology of the area may be more complex than anyone has imagined, and it may be like that. But it is also possible that the history of human beings in the same area might be more complex than anyone can imagine, and cannot be encompassed in some linear progression of cultural materials, going from simple to complex over a fairly short span of time. Richard and I tend to see complexity in the archeological record, which leads us to posit the coexistence of beings of various levels of cultural advancement over very long time frames, that may encompass cyclical rather than linear patterns of development.

In your last letter, you promise more comments as you get on to the remainder of the book. I am looking forward to receiving them!

4.10.3. Letter to Dr. George F. Carter, April 19, 1994.

I have been traveling a lot lately—India, New Zealand, and Colorado, hence the delay in replying to your letters.

Forbidden Archeology was the subject of an extremely negative review by Jonathan Marks in the January issue of the *American Journal of Physical Anthropology*. I kind of expected that. He really misrepresented the book. But there was a brief notice of the book in J*ournal of Field Archeology* that accurately portrayed what it is really about.

In one of your letters, you asked about my thoughts on Table Mountain—did people live there millions of years ago and did another group come in millions of years later, making the same kind of implements? Given the human factors and the natural incompleteness and complexity of the entire archeological record, I am not optimistic about ever being able to conclusively demonstrate any idea about the past, other than it is quite mysterious and does not lend itself to linear, simple-to-complex kinds of explanations. I am convinced, from other sources, that time and hence history are cyclical in nature, that advanced cultures have come and gone many, many times, over millions of years, while simultaneously there have existed many more primitive cultures, and varieties of apemanlike creatures. I don't think I can conclusively demonstrate that from the archeological or paleontological record, but some of the things that are there do seem to be consistent with such a view and inconsistent with most current views.

About Mochanov, I have always been very interested in his work, and included it in *Forbidden Archeology*, with dates of around 2 million years. Three million years would be quite extraordinary and, if accepted, would demolish the whole current scenario of hominid progression. So would 2 myr, for that matter.

When I have time, I shall look up the BAE [*Bureau of American Ethnography*] report on Miramar, where Ameghino made some finds. It is very interesting about the 2 teeth found there.

Want to know something else interesting? When I was putting together *Forbidden Archeology,* I totally came under the spell of the idea that the Moulin Quignon jaw was a hoax. The secondary reports I read were so negative that I ignored my own maxim of going back and checking the original sources. Later on, just on a whim, I decided to look more deeply into Boucher des Perthes.

What I learned was surprising. He was so stung by the "bad rap" he got on the Moulin Quignon jaw that he organized another series of excavations in such a manner as to rule out any chance of cheating by workmen. He would proceed without warning to certain locations, begin digging, and, often in the presence of very respectable witnesses, came up with about 200 human teeth and bone fragments. These could be up to 700,000 years old. The descriptions of the stratigraphic positions are very exact, and can be checked against modern geological work in that region. I am going to put this out in a small book. It is a very interesting case. Considering the very good quality of the later work, I am inclined to take a second look at the notorious Moulin Quignon jaw itself.

On bolas, I have heard suggestions that some of the round stones in Africa are not bolas but are the natural product of using one stone to chip another. The chipping stone, or hammerstone, it is said, naturally becomes rounded. Any comments?

I can't reply now to all of your comments, but I have a file of your letters that I shall refer to when it comes time to do a second edition of the unabridged version of *Forbidden Archeology*.

You have also provided me many leads for further research.

Thank you for mentioning Richard and me to Dee Simpson. A couple of years ago, we did go to visit her. We showed her some of the illustrations we included in the book, asked her lots of questions about how to recognize artificial work on stone, and examined a lot of the Calico implements in her workshop in the museum. I previously had been to the Calico site once for a quick look around. I sent Dee a copy of *Forbidden Archeology* and asked for her comments but have not heard anything from her. I would, however, like to have another look at Calico. What I would really like is some sample implements and maybe a cast of that beaked graver that is so often pictured.

Do you have any offprints of your work on transoceanic diffusion of languages and plants into the Americas?

Oh, something else—a few years ago, I talked to Julie Parks, a nice lady who had worked with George Miller on the Anza-Borrego incised bones. This was just after Miller passed away. She was really pushing the matter. Late last year, I called her again, and she related how a new director had come to the museum (formerly at the La Brea museum in L.A.), and he had told her that investigating those bones should not be such a big priority. There were lots of other things to be catalogued and displayed. And so that was that. It is amazing how things work.

Thanks for the Quecha blessing.

4.11. John L. Cavanagh

John L. Cavanagh was formerly a graduate student in sociology of knowledge at the University of Wisconsin.

4.11.1. Letter from John L. Cavanagh, March 26, 1996.

I've just spent three hours with *Forbidden Archeology* and memory insistently draws me back to Madison, Wisconsin, in the Spring of 1970. I was a graduate student in the sociology of knowledge and a teaching assistant. We TA's were out on strike, ostensibly over labor grievances, but at least in our department everyone knew the real stakes: The previous year, 26 out of 41 TA's had been fired, not for any job deficiency, but for failing to major in methodology and statistics. The chairman pleaded that all Ford Foundation monies would be cut off unless a quota of 80% of us were pursuing that major. I'll pass along two incidents without further comment.

The strike suspended university activity for six weeks. We TA's felt obliged to our students, so we took turns answering the phone in our office. There were few calls, and most of them were wrong numbers. To relieve boredom, I would pick up the phone and say, "University Thought police, Lieutenant Pinwheel speaking. Can I help you?" But for the lack of tape recorder and a little imagination, there was meat for a doctorate in the sociology of knowledge. Some of the responses were, "Oh, my God." and a soft hang-up; "Who is this? Who do you think you are?" in high tremolo; and, my favorite, "Look, this really is a mistake. Don't tell anyone I called, OK? Please..." click...

The floor below ours was Economics and about a quarter of it was behind a glass door inscribed, "Institute for Research on Poverty—A Project of the Ford Foundation." A few of us had a running gag that we were going to start an institute for research on the rich, rent tuxes and crash debutante parties in Grosse Pointe. The chairman got wind of it and it took us half an hour to convince him it was a joke.

Thanks for a wonderful, brave book.

4.11.2. Letter to John L. Cavanagh, May 10, 1996.

I really enjoyed your letter, in which you described your experiences as a graduate student in sociology of knowledge at the University of Wisconsin in Madison. You really gave a good snapshot impression of the pressured intellectual trackways. Of course, I am very curious to know what happened. For example, did you eventually finish your graduate work? Did you go into sociology of knowledge? Also, how did you happen to encounter *Forbidden Archeology*? It's interesting and helpful for me to see by what pathways the book is getting around (from the sociology of knowledge standpoint!). A book really is a social document, and the documentation is ongoing. I like documenting it.

4.12. Dr. Bruce Cornet

Dr. Bruce Cornet is a paleobotanist at the Lamont-Doherty Earth Observatory, Columbia University.

4.12.1. Letter to Dr. Bob and Zoh Hieronimus from Dr. Bruce Cornet, August 4, 1994.

Dr. Bob Hieronimus forwarded me this letter, as part of his very much appreciated network-building efforts.

Thank you for having me on your radio program on 24 July, and for sending me the box of books, papers, and news articles. I am enjoying reading them immensely, especially *The Hidden History of the Human Race,* by Cremo and Thompson. I can identify with their pursuit of the truth despite the paradigm that has been politically accepted as conventional wisdom (the ordained "truth"). As a paleontologist, I can be critical of some of their evidence, particularly the shoelike "print" discovered in Cambrian shale, Utah, and the shoe sole "imprint" from Triassic rocks of Nevada.... There is other evidence presented by Cremo and Thompson that is credibly documented and validly interpreted, and is therefore highly relevant to the issue of when *Homo sapiens* first walked this earth. I agree wholeheartedly with the authors that the building of a paradigm requires (by definition) model-dependent thinking and knowledge filters in order to make the paradigm strong, supportable, and defensible. I am also very familiar with morphological dating, and how circular any arguments can become that deal with conflicting evidence. It is unfortunate that scientists such as Richard Leakey call the book pure humbug, and that it does not deserve to be taken seriously by anyone but a fool. It does not take a genius to recognize that time after time old ideas and paradigms have fallen out of favour as have their supporters disappeared (died). Why is that so? Could it have anything to do with the majority of scientific ideas being wrong, simply because those ideas have been elevated to a belief with the development of tactics for minimizing and eliminating tests that could challenge or invalidate that belief?

One common statement made by the defenders of paradigms in the field of paleobotany has been, "It is not up to us to prove our theory wrong. You must provide convincing evidence that will compel us to rethink all our ideas on the subject." In other words, the burden of proof is on the challenger. No one who supports the established paradigm will look for contradictory evidence. That is one reason why it has become a joke among the scientists that progress in science can be measured by the lengths of lives (longevity) of those who are regarded as authorities on any given subject. Kingdoms in science can be overthrown, but it requires a revolution. If one is patient, death will cause paradigms to die; good ideas and theories will be recycled, perhaps to be reborn anew with different labels and supporters. Rephrasing comments in Christopher Bird's book on Gaston Naessens, "It is an unfortunate fact that [science] is not motivated to search for truth. The forces behind it are those that promote...money and profit [and power and control])." (p. 287).

I consider myself to be a 21st Century Dragon Slayer. I enjoy testing paradigms, because I know, given human nature and the temptation for emotional/material reward, that it will not be hard for me to find controversial or contradictory data that has been overlooked or intentionally ignored and forgotten.

After discovering that an area near Middletown, NY, where I used to live (1986-1992), was considered to be a UFO hotspot (*Silent Invasion*, by Ellen Crystall), I wanted to see for myself if it was true. I had never seen a UFO, I thought, and after seeing "intruders" on television, I was curious enough to explore the unknown. After three evenings in the field with Ellen Crystall, I was convinced that there was something going on there that could not be explained away as misidentification of conventional aircraft. My involvement to date includes doing a 20 square mile magnetic and geological survey of the area (with spectacular results), as well as the photographic documentation of more than 80 UFO sightings. I am currently writing a book on my findings, and hope to be able to present a scientific format as well as a popular account of my experiences. I have experienced many things since my project began, including intervals of missing time and abduction, with memory of craft landing next to my van! The photographic evidence I have accumulated now fills three albums, and has blown the minds of those scientists and physicists to whom I have shown the data. They tell me that they have never seen a more thorough and compelling investigation into the UFO phenomenon, proving beyond question that the phenomenon exists and that it involves intelligently controlled craft with technological capabilities far beyond our own. For example, I have the first photographic proof that UFO lights are produced in an unconventional way—via ionized plasmas that can, when intensified, ionize the atmosphere, producing plumes of nitrous oxide gases (ironically, laughing gas!—perhaps another reason for all the ridicule?). This halo of amber-brown gas around the light source is what many witnesses describe as the "Fire in the Sky" effect (I will send you some of my business cards, which show the photographic proof of an ionized plasma). I have experienced so many bizarre phenomena associated with UFOs (with photographic documentation) that I have no doubt that I can produce a best seller.

Why am I telling you all this? I am a conventional mainstream scientist who has had his eyelids peeled back and his mind expanded by the awareness of other realities. I love complex intellectual puzzles, and the UFO phenomenon, along with the artifacts on Mars and the Moon, represent, in my opinion, the most important last frontier for mankind, next to knowing where we fit into the overall plan/architecture of reality and spirituality. My education has been not from reading about what others have interpreted and believe, but from experiencing for myself those realities. In a way, I am a late bloomer, a "flower child" (expressed literally as my botanical interests?) from the 60's who has quietly gathered data and evidence for decades, and who is now exploding at the seams wanting to share what I have discovered.

Perched on the horizon is perhaps the biggest discovery of my life. After having some close encounters with UFOs in Pine Bush in May of this year, I began having visions and awareness of knowledge that I cannot explain. That knowledge led me to investigate various unusual morphological features in the Wallkill River Valley, NY, where all the UFO activity is concentrated. Hoagland believes that this activity is concentrated as a sign indicating where we should look. Without going into the details, I will say that I may have discovered what may be Cydonia II on earth, complete with eroded remnants of the Face, City, D&M pyramid, Tholus, Crater, and Cliff! Even the latitude of the Crater is the same as on Mars! In addition, there are other morphological and geometric features carved out of granite mountains and plateaus that indicate tetrahedral and Sacred geometry. Although it is too early for me to go public with this discovery, I am hoping that you and Zoh may be able to put me in contact with people who might want to help or even fund further research, which will require archeological excavations. For example, I have discovered in the middle of all these monuments two prominent granite mountains rising hundreds of feet above the valley floor. One is carved in the form of a recent key crop glyph (surrounded by a plateau with raised hills in the form of a golden mean rectangle and fibonacci spiral), and the other in the form of a four sided pyramid complete with stepped sides on one side like the Mayan pyramids. I cannot give out the names of these mountains yet, but they have Biblical significance! If we can prove with artifacts that these features are carved and/or constructed, the age will push civilization back at least 100,000 years (prior to the last glaciation)!

Thank you again for allowing me to help Richard Hoagland with his discoveries, which have incredible implications for mankind's future. As I publish, you will get copies of my work for your library and show. I hope to meet you some day. From reading the articles about you, you both are people worth knowing and supporting!

4.12.2. Letter to Dr. Bruce Cornet, August 18, 1994.

Bob Hieronimus recently forwarded to me a copy of your recent letter to him, in which you have some things to say about my book *The Hidden History of the Human Race,* coauthored with Richard Thompson. Thank you for your kind words.

I agree that the footprint evidence would be stronger if there were additional prints to make trackways. Of course, the brief report about a print in the Jurassic of Central Asia is not very strong in any case. But if the print from the Triassic of Nevada really does have marks of stitching around the sole, that seems to put it in another category, even if it is just a single impression. Also, Richard Thompson personally inspected the Meister print, and although he is quite a skeptical person found it convincing.

I read your accounts of your UFO experiences with some interest. Richard Thompson has recently come out with a book on the UFO topic. It is titled *Alien Identities* and compares modern UFO and UFO entity experiences with reports of similar experiences recorded in the ancient Sanskrit texts of India. Richard and I plan to use some of this material in our next book, *Human Devolution: An Alternative to Darwin's Theory.* Therein we will give an account of human origins that places our species within a cosmic hierarchy of beings.

Amazingly enough, I am also working on a book of paleobotanical mysteries, specifically occurrences of angiosperms and gymnosperm fossils and microfossils in the Cambrian. Do you have any insights on this topic? If you like, I can keep you informed as the work progresses.

Dr. Cornet supplied me with copies of his many papers demonstrating the existence of angiosperms (flowering plants) as early as the Jurassic. The conventional view is that the angiosperms originated much later, in the Cretaceous.

4.13. Laura E. Cortner

Laura E. Cortner is executive producer of the Bob Hieronimus radio show, which features alternative science topics. I have twice been a guest on the show.

4.13.1. Letter from Laura E. Cortner, December 18, 1993.

We are very interested in reviewing a copy of your book *Forbidden Archeology*, which we read about in an article by Ed Conrad in the

Hazelton (PA) *Standard Speaker*, 11/17/93. We have long been interested in doing a special program on the subject of archeological finds that challenge the earliest recorded history of humans for quite some time, and we are encouraged to learn of an academic book with authors we could interview on the radio.

Our programs are designed to educate our listeners on a wide variety of subjects that are usually not covered in the major media, and we know your book will be of interest to them.

Please send us a review set of *Forbidden Archeology* as soon as you can. If for some reason you are unable to do so, we would appreciate it if you could drop us a line or give us a call to let us know.

4.13.2. Letter from Laura E. Cortner, February 7, 1995.

Enclosed is the latest letter of praise we have received about your book from one of our recent guests on 21st Century Radio to whom we have presented a copy. Thank you very much for supplying us with the extra copies to continue this type of promotion. In the last month we have also sent copies to two of the creators of Howdy Doody, and to Peter Occhiogrosso, author of *The Joy of Sects: A Spiritual Guide to the World's Religions* (Doubleday).

We want to keep you appraised of the thank you letters we receive with your courtesy.

4.14. Jim Deardorff

Jim Deardorff is a member of the Ancient Astronaut Society and has published writings in the UFO field.

4.14.1. Letter from Jim Deardorff, October 4, 1995.

I read your article on *Forbidden Archeology* in the latest *Ancient Skies [newsletter of the Ancient Astronaut Society*] and was very impressed. It seems to be a significant cut above their usual article, in that yours was carefully researched, carefully documented, well presented, and nicely augmented with plausible explanations why orthodox archaeologists have failed to acknowledge the data that lie outside the perimeters of what they've been taught.

Congratulations on a job well done. I'll be saving your article in case I need information on the topic.

By the way, I was once in correspondence with a Mr. Ron Calais (in 1988) who had his own substantial collection of documentation of ancient artifacts of the type you presented.

4.14.2. Letter to Jim Deardorff, October 18, 1995.

Thanks for your kind words about my article on *Forbidden Archeology* in *Ancient Skies.*

I do already know Ron Calais. He supplied me with some of the cases that went into my book *Forbidden Archeology*, and we have stayed in touch over the years. He sends me leads from time to time.

By the way, your name sounds very familiar to me. Have we met? Or have you perhaps written something I might have read or read about—maybe in the UFO field?

4.14.3. Letter from Jim Deardorff, November 3, 1995.

Thanks for your letter of 18 Oct; if you already know Ron Calais you're then already well aware of my only source of knowledge concerning *Forbidden Archeology.*

I don't know if we've met before or not. I've written a couple of articles in UFO journals quite some time ago, mainly about the Meier contactee case, and have given a few talks at UFO meetings. And wrote one book about a document that is connected to the Meier case, and then two later books about topics of concern to New Testament scholars, which again have connections to information supplied by this document (related to the Jesus-in-India tradition).

Possibly we might have met at the AA [*Ancient Astronaut Society*] meeting in Chicago in 1989, where I gave a paper.

I just now noticed your e-mail address. If I had seen it sooner, I would have responded the quick way!

4.15. Dr. Thomas A. Dorman, M.D.

Dr. Thomas A. Dorman is a physician residing in California.

4.15.1. Letter from Dr. Thomas A. Dorman, August 18, 1993.

Thank you for writing *Forbidden Archeology*. I recently finished reading the book and would like to congratulate you and thank you for writing it.

I am a practicing clinician here in California and have a special interest in back pain which you probably know is becoming more and more frequent in modern civilized man. The reason for the increase is unknown. Theories include a deterioration in the quality of modern man's ligaments and it is the ligaments which regulate the function of the iliac bones and the pelvis. As you rightly point out, *Homo sapiens* is the only true biped. My own research in connection with function and dysfunction in the human pelvis leads me to believe that *Homo sapiens* is unique and almost certainly did not have a common ancestor with simians.

I was particularly interested in the section in the book you deal with Lucy as I know Professor C. Owen Lovejoy and have debated this issue with him on more than one occasion and I am particularly grateful to you and to Dr. Richard Thompson for bringing out the disinformation which comes from the establishment.

The appendix about carbon dating and other methods for archeological dating was also extremely timely. Are you aware of Donald Patten's work on catastrophism? Are you aware that he has recently adduced evidence that complex life on earth might be no more than 100,000 years old? You drew the conclusion that the time scale ascribed by official science is unreliable. Do you and Dr. Thompson have an opinion whether it is compatible with 100,000 years? If you are not familiar with Donald Patten's work, you may wish to contact him directly ...

I am intrigued by the Bhaktivedanta Institute. Do you have any literature about it?

I hope you will find a few minutes to write to me.

4.15.2. Letter to Dr. Thomas A. Dorman, December 30, 1993.

Please excuse my delay in answering your letter of August 18, 1993. The last six months have been a very hectic time for me. I have never before been in a position where I have not been able to keep up with my correspondence, but with the publication of *Forbidden Archeology*, I suddenly found myself a bit overwhelmed. Also, I recently made an extended trip to India and Pakistan.

Thank you for the papers that you enclosed with your letter. I already have the one by Latimer and Lovejoy on the calcaneus of *Australopithecus afarensis*, but the others are new to me. You mention that you are personally acquainted with C. Owen Lovejoy. I wonder what he would think of *Forbidden Archeology.* If you happen to see him, perhaps you could ask him if he would like a copy.

I would be interested in hearing more about your research on the human pelvis, which leads you to believe that we "almost certainly did not have a common ancestor with simians."

I am aware of Donald Patten's work on catastrophism. I heard him speak at the Ancient Astronaut Society convention in Las Vegas a few months ago. He has a lot of interesting insights. Richard Thompson and I eventually want to write a book on the shortcomings of the geological record. Our basic message would be that it is exceedingly mysterious. From various sources, Richard and I believe that the most reasonable picture is that complex life has been present on this planet for hundreds of millions of years, rather than just 100,000. That does not, however, diminish our interest in Donald Patten's work.

You asked about additional Bhaktivedanta Institute literature, but I think that since you first wrote to me you have received some information from our secretary.

If I can be of any further help to you, please let me know. As I am now fairly well caught up on my correspondence, you can expect a more timely reply.

P.S. If you were to write a brief review of *Forbidden Archeology,* I would find it a nice addition to our collection.

4.15.3. Letter to Dr. Thomas A. Dorman, April 21, 1994.

Forgive me for the late reply to your letter of January 20. For the past few months I have been traveling in India, New Zealand, and Denver, and I have gotten behind in my correspondence, as I always seem to be.

I have also seen the brief reports redating Lucy to 3.2 million years from about 3 million years. In terms of my interests, that is not very significant. But I suspect there must be some reason for the redating—keeping all the Hadar *A. afarensis* material in proper sequence, for example. To get the real picture, I would have to get a hold of the detailed reports and read them carefully.

By the way, what intellectual path took you to your present outlook on human evolution?

4.15.4. Letter from Dr. Thomas A. Dorman, April 26, 1994.

Thank you for your letter of April 21st. Don't worry about the delays in writing to me. We have waited 3.2 million years and haven't solved the evolutionary problem, so a few weeks...!

Did I send you my newsletters on the origin of man? As I can't remember, I am enclosing these issues.

I was raised in what was the modern Marxist/Darwinism intellectual environment and as I started reading early zoological writings and being very much of an outdoors boy had a lot of hands-on contact with creatures and plants. I developed my first phase of skepticism in my early 20's in the archives of the Liverpool public library which had an extremely fine collection of books on early zoology and botany. And the interest has stayed with me since. Having encountered a (cryptic) alternative agenda in the medical political arena as a practicing physician, I realized that science, politics and philosophy are woven together in the history of our civilization, at least since the dark ages, not entirely by chance. It is not possible to view scientific questions in isolation from the others.

If you have time to share your answer to the same question (What intellectual path took you to your present outlook on human evolution?), I would be very interested to hear.

I can't remember if I have brought DeSarre's work (in France) to your attention.

4.15.5. Letter to Dr. Thomas A. Dorman, June 27, 1994.

I have recently relocated from southern California to northern Florida. You did previously send me your newsletters on the origin of man. I think you also mentioned the work of DeSarre in France. Thank you for sharing some of the history of your intellectual development. As far as my own history goes, I shall start with Thomas Huxley's maxim that "autobiography is fiction." Much of my education has taken place outside the normal institutional channels. My early years were spent traveling with my military family to various places around the world. My tendencies were toward mysticism, literature, history, and political intrigue. At one time I was enrolled in the school of international relations at The George Washington University, preparing for a career in one of the intelligence services. But I gave that up after a short time and pursued my very individualistic mystical bent, which led me in time to the International Society for Krishna Consciousness. I then began to put my literary talents to use, writing articles and coauthoring several books. Along the way, I absorbed a lot of the

Vedic ideas about life and the universe. In 1984, I met Richard Thompson, and we began collaborating on science topics. Initially, I was just helping him get some of his ideas into a more popularly accessible form, but I gradually became more directly involved in research and writing. It was at that time that I began to really examine the shortcomings of modern evolutionary ideas and seriously think about how to convince others of those shortcomings.

4.16. Deanna Emerson

Deanna Emerson is author of the book Mars/Earth Enigma.

4.16.1. Letter from Deanna Emerson, June 1, 1996.

I'm the author of a new book that just came out, and I have some very important information that you need to know about. Please call me. This information pertains to the Letters on the Block of Marble found in Philadelphia on page 105 of *The Hidden History of the Human Race.*

Six months ago, I read *Forbidden Archeology* and recognized that the two signs on this slab of marble were in a Sacred Script "language" used by a mysterious Goddess worshipping civilization in Old Europe, symbols that no one understood the meaning of. That is, until I discovered that all the Crop Circles are "written" in this 8,000-year-old script, and most amazingly—at least four of these signs can be seen etched onto the Mars Face, according to computer enhanced photographs of the face.

To discover two of these symbols on a piece of marble from Philadelphia is almost too funny. However, if this slab of stone is as old as I think it is, this is a very exciting discovery.

Looking forward to hearing from you.

P.S. This script was also inscribed on ET-like almond-eyed masks worn by worshippers of the female deity—as a form of communication between humans and "deities."

I visited with Deanna during a visit to Washington State in 1996, and we had a very good conversation about my book, her book, and her interest in UFOs.

4.17. Jim Erjavec

Jim Erjavec is a professional geologist with an interest in scientific anomalies.

4.17.1. Letter to Dr. Bob Hieronimus from Jim Erjavec, July 16, 1996.

After appearing on the Bob Hieronimus radio show, Jim Erjavec received a copy of The Hidden History of the Human Race *from Hieronimus, and wrote back to him. Laura Cortner, Hieronimus's producer, then forwarded me a copy of Erjavec's letter.*

Thanks for allowing me to say a few words about the Cydonia area on 21st Century Radio. It was a new experience for me and fortunately, after listening to the tape of the show, I didn't come off too badly. Your questions about the study were both excellent and necessary. Though I know the article was written in geo-lingo, I appreciate your effort to cut through it and make the interview understandable and meaningful to the listeners.

As well, I would also like to thank you for the copies of *Hidden History* and the *Message of the Sphinx*. I have been able to read both books during the past few weeks and will admit I'm certainly impressed. Please do inform the authors/publishers, if the occasion arises, that as a geologist, I believe both books have provided much credible evidence and rational analyses of the evidence to support their findings.

In *Hidden History,* despite what one wishes to believe about the validity of the available anthropological evidence, the authors have clearly shown that there is a strong bias in the reporting of such data. If anything, their arguments enforce my conviction that some (if not much) of the evidence for the evolution of humans has been channeled down a "mainstream path" to support the theory, rather than considering the data on a case-by-case basis. Sadly, I feel that such tactics have overwhelmed almost all areas of science in this age and it is unfortunate that truly creative thought often appears to be shackled by underlying political and social venues. It's also amusing to see the way in which mainstream science uses the double-edged argument to both support one theory and ridicule another. At best, that is pseudoscientific approach and most certainly it is not a cognitive argument nor is it beneficial to the search for the truth. Most distressing, though, is that this (pseudo) science has become such an ingrained entity within today's research that it is extremely difficult to see, not only by the public but often by the scientists as well.

I cannot tell you the number of times I have either laughed or been appalled over so-called sanctioned findings by one author or another that have been printed in prestigious bulletins and journals. I believe that many such authors have noble intentions, but are stretching the evidence to cover the holes in their theories to a degree, that to me, is often unpalatable.

As for the "*Sphinx*," it is another well-written work and reestablishes many inconsistencies with previous studies and highlights some new ones. What amazes me (though it shouldn't) is that many experts are close-minded to even a discussion of these findings. Without a written history to document Egyptian chronologies, the Egyptologists are basing their time-lines of that vast empire on little but speculation. I know as that is my field of expertise—speculation of the past. And speculation is proper (and required) as long as it doesn't become a history written in stone. The question is—Can Egyptologists make such a claim?

Anyway, I'm supportive of your efforts to bring to light some of the mysteries and inconsistencies which continue to plague us concerning these "controversial" issues. Only through continued research-driven pressure and a rational discussion of the evidence will meaningful results be obtained.

Enclosed is the final version of the Cydonia paper (just completed last week) along with a number of B/W copies of the map, which I believe are essential for an understanding of what is described in the paper. Feel free to copy the paper/maps and distribute them to whomever is interested. Sorry I can't provide you with some published materials (i.e. books) but as of this date, I'm still a small-time operation. Perhaps down the road that will change.

4.17.2. Letter to Jim Erjavec, August 4, 1996.

Laura Cortner of 21st Century Radio forwarded me a copy of a letter you sent to Bob Hieronimus. In that letter you had some nice things to say about my book *Hidden History of the Human Race*. Thank you. You mentioned you are a geologist. What kind of things are you working on?

4.17.3. Letter from Jim Erjavec, August 21, 1996.

I'm glad that Laura Cortner forwarded my letter to you concerning *Hidden History*. I think, above all, you clearly exposed the bias that permeates much of academia in regard to origins of humans on earth. In itself that approach is enough to spark rebuttals like the one from Richard Leakey. That isn't to say that outright deceit and dishonesty are prevalent in research, but as you and Thompson stated both strongly and succinctly in *Hidden History*, evidence appears to have an unequal weighting in relation to the theories.

I have yet to purchase the unabridged version, *Forbidden Archeology,* but hopefully will do so soon. I'm very much interested in going over specifics which most readers are unlikely to entertain. After working nearly 20 years in the geological sciences, I have found that the details are either the weakest or strongest arguments of any research work and are the most likely to be overlooked or taken on assumption.

As for myself, I'm currently employed as a geologist/Geographic Information Systems Specialist with Parsons Engineering Science, Inc. The cleanup of a Superfund site is not the most enlightening work, thus I've spent much time involved with Professor Stan McDaniel and the Cydonia Mars research. I wrote a paper last year (coauthored with Ron Nicks, another geologist) and spent six months developing a CAD map of the Cydonia region in question. The brunt of my work was not to prove or disprove any extraterrestrial arguments concerning Cydonia, but to establish some basic facts and possible geologic scenarios for that area of Mars. Not surprisingly, it was quite easy to refute arguments by NASA concerning the geology and development of the landforms in that area—using both visual interpretation and published references which NASA should have been privy to. As I learned, arguments which do not support NASA's generalized interpretation have been for the most part ignored. Just one more encounter with "selective" referencing.

That brings me to another point. I understand how social and political undertones can essentially influence sanctioned research works, but I'm still not sure—why. Case-in-point: the extinctions at the end of the Cretaceous period. The assumptions and conjectures that have been conceived from the evidence for an alleged asteroid impact are, to put it mildly, mind boggling. Anyone who researches the complexity of the Cretaceous fauna and flora, the environmental changes that were taking place prior to the "big event," the animals which did not go extinct, animals which went extinct before the boundary, the fragmentary nature of the fossil record itself, the fragmentary nature of the sedimentary record (much, much overlooked), the lack of understanding of sedimentation rates, etc. will undoubtedly have misgivings with the asteroid impact theory as currently stated, and most certainly would not take it as the absolute truth.

Yet, despite the impact theory being just that—theory, many scientists (even some I personally know) will speak of it as a totally proven fact. Add to that the alleged Martian meteorite and its alleged biological products and one begins to wonder if anyone actually reviews the evidence and assumptions behind these "theories." The complacency of the lay public with such theories is understandable, but that of the scientific—not acceptable.

So, why has the asteroid impact theory and Martian "bio"-meteorites become mainstream science? It certainly isn't because of hard science and hard evidence...

Anyway, I appreciated your correspondence. I understand that you are currently working on a follow-up book to *Forbidden Archeology*. Since my background is quite strong in paleontology, if I can be of any assistance to you in any way, please let me know. In addition, I've been collecting fossils for over 20 years, so I am quite at home with the geologic record and the fauna and flora that are often associated with it. Take care.

4.18. Dr. Paul Feyerabend

Dr. Paul Feyerabend, now deceased, was a leading philosopher of science.

4.18.1. Letter to Dr. Paul Feyerabend, February 12, 1993.

I have a great deal of respect for your books such as *Science in a Free Society*. I especially liked the section in *Science in a Free Society* titled "Laymen can and must supervise Science." Therein you say: "Duly elected committees of laymen must examine whether the theory of evolution is really so well established as biologists want us to believe...."

In my book *Forbidden Archeology* I have undertaken such an examination, as a layman and from a standpoint outside that of the modern university and its knowledge tradition. I am sending you an advance, prepress copy of *Forbidden Archeology*, and if your time permits, I hope you will look through it. I am gathering some prepublication reviews, and I would welcome any comments you might care to make.

4.18.2. Letter from Dr. Paul Feyerabend, April 26, 1993.

Thank you very much for sending me a preprint of your opus. I have no time to look at it right now but I shall certainly study it with care. I know very little of the field and therefore can't provide a review, either pre- or post publication. Still I hope to profit from reading the book.

I have not sought any permission to publish this brief note from Dr. Feyerabend. I include it as an example of a polite noncommittal response from an important scholar.

4.19. Duane Franklet

4.19.1. Letter to Duane Franklet, May 15, 1996.

Yes, you are correct that the dates geologists give to coal formations could be wrong, as could many of their other dates. But for the sake of discussion, my coauthor and I decided to accept the standard dates as given, and then document the existence of human artifacts and skeletal remains in those very ancient deposits. So there is a problem. That was the main point of *Forbidden Archeology*. There is a problem, and it has to be dealt with.

One can deal with the problem in different ways. Personally, I get a lot of my insights from the ancient Sanskrit writings of India. These books tell of human civilizations going back hundreds of millions of years on this planet. They also talk about humans coexisting with more apelike but still very intelligent beings. So the evidence reported in *Forbidden Archeology* is consistent with that version of things. Also, the ancient Sanskrit writings talk about periodic catastrophes that devastate the earth over the course of hundreds of millions of years. The evidence for catastrophism is consistent with that. It thus appears that catastrophism is compatible with long time frames as well as short ones.

Perhaps you could tell me how you encountered *Forbidden Archeology*. It is helpful for me to know by what paths the book is getting around.

4.19.2. Letter from Duane Franklet, August 9, 1996.

A belated response to your letter of May 15, as I have been traveling in the NW—Washington to Alaska.

A friend has similar interests and we share reading material as well as thoughts. He spotted your book as it was added to our library and we were the first to read it. Credit must go to one of our librarians—an intelligent woman ...with an inquiring mind. Why your book was selected I can not tell you at this time, but when I return home I will try to get you an answer.

4.20. Dr. Horst Friedrich

Dr. Horst Friedrich is a retired German businessman with a doctorate in the history of science. He publishes articles on archeological and historical anomalies.

4.20.1. Letter from Dr. Horst Friedrich, December 18, 1993.

In possession, since yesterday, of your excellent and most meritorious book *Forbidden Archeology*, I can only congratulate you on it. Of course I will have to study it in detail, but I already greatly enjoyed your sub-chapters on [*he lists the sections of the book he has read*]. I have no knowledge what the Bhaktivedanta Institute is all about, but without doubt—apart from the connection with the life-giving stress of the religion-philosophy of India, which we need so desperately today—you have done Western culture a great service by clearly showing how extremely "windy" is the scenario, which Western Neo-Scholasticism is presenting (in reality nothing more than ideologically-based "dogmas"), and by showing that other views about our ancient past, especially that of ancient India, could come much more nearer to the truth. Again, my heartfelt congratulations!

The only two points about which I would like to ask a question at this moment, are: 1) Why didn't you mention, in your chapter A.2, the traditions of India, Ireland, etc. about air/space travel ("*vimanas*") and super-weapons "before the Flood", with the sub-question re "Flood", why didn't you mention the pre-/protohistoric annihilating cataclysms?, and: 2) is it unknown to you that a certain nonconformist "school" in Germany—especially Prof. Heinsohnn of Bremen University and Dr. Heribert Illig—are proposing that there existed no higher civilization at all on this planet before ca. 1100 BC!?

4.20.2. Letter to Dr. Horst Friedrich, April 21, 1994.

I am pleased that you found *Forbidden Archeology* so interesting and useful. I quite agree with your characterization of much of Western intellectual life as "Neoscholasticism" and "ideologically-based dogma."

Regarding your question about *vimanas,* these are discussed at length in Richard Thompson's book *Alien Identities*, which examines the modern UFO phenomenon in light of the Vedic accounts of *vimanas* and superhuman entities.

Regarding prehistoric cataclysms, Richard and I will discuss them in *Descent of Man Revisited.* This is the companion volume to *Forbidden Archeology*, and in it we will present the Vedic account of human origins and antiquity, which, as you know, involves a cyclical version of history, including periodic catastrophes. In *Forbidden Archeology*, we simply wanted to catalog the physical evidence that contradicts the current view of human origins. In *Descent of Man Revisited* we shall present our positive alternative view.

I was not aware of the Heinsohn-Illig proposal that there was no high civilization on this planet before 1100 BC. Can you tell me any more about what they are saying and why?

Also, I would like to know more about you and your special interests.

The current working title of Descent of Man Revisited *is* Human Devolution, *of which I am the sole author.*

4.20.3. Letter from Dr. Horst Friedrich, April 28, 1994.

Many thanks for your kind letter! I will try to find Richard Thompson's book about the *vimanas* and, of course also, your co-authored *Descent of Man Revisited*, as soon as it is published.

My interest, I suspect, will be relatively similar to yours. I have earned my money, until my retirement, in business and industry but, between times, have made my doctorate in the history and philosophy of the sciences. I have studied many examples of scholasticism versus nonconformists controversies over the decades. Also, I am deeply interested in India's spiritual heritage to the world, especially in its practical (!) aspect.

I enclose copies of two articles of mine, where your book is mentioned. Hopefully somebody will be at the Bhaktivedanta Institute who can translate the German text.

NEARA Journal in some issue had an article by Dr. Illig (if I remember correctly, I even translated it into English), but I have no copy at this moment. Please write to *NEARA Journal*'s Editor: Mrs. Katherine Stannard, 3 Whitney Drive, Paxton, MA 01612 and ask her for a copy. There are, I think, English versions of Heinsohn's books around: please ask the editor of *The Velikovskian:* Mr. Charles Ginenthal, 65-35 108th Street, Suite D15, Forest Hills, New York 11375. In my opinion Heinsohn has adduced clear proof that the alleged "Sumerians" of the 3rd millennium are none other than the Chaldaeans of the 1st, with the consequence that whole "ghost civilizations" will have to be eliminated from our historical atlases. But this, of course, you will have to decide for yourself.

I would be interested to learn more about your *guru* (e.g. do you regard him as an "avatarian" manifestation?) and the activities of the Bhaktivedanta Institute. Is there also a branch in Germany?

4.20.4. Letter to Dr. Horst Friedrich, June 27, 1994.

Thank you for sending me a copy of your article from the *NEARA Journal.* I am in full agreement with your application of the word "sectarianism" to the modern scientific establishment. It really does seem to operate in the manner of, to use your very apt terminology, "ideological scholasticism." Despite my somewhat imperfect knowledge of German, I was nevertheless able to read through your "Hochkulturen im Tertiär" with some degree of comprehension. In your review, you note that Richard Thompson and I did not discuss the idea of recurring catastrophes or the evidence for advanced civilization mentioned in the Vedic literatures of India. That was deliberate on our part. In *Forbidden Archeology* we wanted first of all to demonstrate the need for an alternative view of human origins. In our next book, tentatively titled The *Descent of Man Revisited*, we shall outline that alternative, drawing extensively upon Vedic source material. This will include, of course, the recurring cataclysms of the *yuga* cycles and *manvantara* periods, as well as discussion of Vedic descriptions of advanced civilization in ancient times, and in an interplanetary context as well. I hope that will satisfy you! A new picture of human origins will have to be comprehensive, in the manner you suggest in your *NEARA Journal* article, incorporating evidence not only for archeological and geological anomalies, but also paranormal phenomena of all types, including evidence for extraterrestrial civilization.

You ask if I regard my *guru* as an avatarian manifestation. According to the Vedic texts, *guru* is an outward manifestation of Paramatma, or Supersoul. The idea is that within each person, there is the individual soul and the Supersoul. The Supersoul gives guidance from within, but also appears outside in the form of *guru* to give external direction, which is also necessary. The *guru* is an individual soul, a distinct person, who is empowered to act as a representative of the Supersoul. The extent of empowerment may vary. A *guru* manifesting a certain magnitude of divine empowerment may be termed an *avatara* of a specific type. *Avataras* are of three varieties. First is a descent into this world of the Supreme Personality of Godhead in His original form. Second, the Godhead may also expand into other equally powerful forms of Godhead, and these may also descend into this world as *avataras.* But then there is a third category of *avatara,* whereby an individual soul is specifically empowered with the personal potencies (*shaktis*) of the Godhead to manifest some extraordinary spiritual accomplishments. Such a soul is called a *shaktyavesha avatara*. Because my *guru* His Divine Grace A. C. Bhaktivedanta Swami Prabhupada (Shrila Prabhupada for short) performed the remarkable feat of spreading Krishna consciousness around the world in a very few years, he is sometimes called a *shaktyavesha avatara.* But keep in mind that this does not mean he is an *avatara* of the God type. Rather he is a living entity who has received some extraordinary empowerment from God. In general, the *guru* is seen by the disciple as on the same level as God, not because he is God but because he is a confidential representative of God.

There is not an official office of the Bhaktivedanta Institute in Germany, but Steven Bernath, who did much of the library research for *Forbidden Archeology* is in Germany. He would be a good source of information for you.

I should mention that the abridged German edition of *Verbotene Archäologie* has been brought out by the Bettendorf company.

4.20.5. Letter from Dr. Horst Friedrich, July 5, 1994.

Thank you most warmly for your kind letter of June 27th and your kind words about my article "Sectarianism versus comprehensiveness!" The scenario you will present in *The Descent of Man Revisited* [the previous working title for my forthcoming book *Human Devolution*] will obviously satisfy me. In its inherent value this new book will certainly become an apt companion to your most meritorious *Forbidden Archeology.* I have indeed found out that the German version is already available (at DM 44,—) and hope to possess that, too, soon.

Also many thanks for your kind information concerning Godhead/*Guru* etc. and your *Guru,* Shri Prabhupada! I'm very grateful that he is promoting such work, helping to free Western culture from ideological distortions of truth, truly the curse of....Let's hope that we will arrive as soon as possible at this stage of consciousness where we can truly understand and appreciate the difference between those three (or rather 3M?) kinds or levels etc. of *avataras* you mention!

4.21. Dr. Duane T. Gish

Dr. Duane Gish is a prominent Christian creationist scientist with the Institute for Creation Research. He is the author of several books against Darwinian evolution and is an accomplished debater and speaker.

4.21.1. Letter from Dr. Duane T. Gish, June 28, 1993.

In 1993, I sent Dr. Gish a copy of the bound galleys of Forbidden Archeology *and later received the following reply.*

I regret to report that I have not yet had time to read your book *Forbidden Archeology.* I did glance through part of it and it did seem to be quite interesting and perhaps useful to us. I will let you know as soon as I have had time to carefully examine the book.

4.21.2. Letter to Dr. Duane T. Gish, July 7, 1993.

Thank you for your recent letter, in which you say *Forbidden Archeology* seems "quite interesting and perhaps useful." I am looking forward to receiving some more detailed impressions from you, after you have had time to more carefully go over the book—which is, I admit, somewhat forbidding in its length.

Since I last wrote you, I have received some printed copies of the book, and I am enclosing a copy for you, along with some excerpts from the reviews received thus far.

We are now in the process of sending review copies to the mainline evolutionist scientists and journals. It will probably take some time to get their responses, but, if you like, I can keep you informed.

Thank you for your consideration, and best wishes in your service to Lord Jesus Christ.

I later visited Dr. Gish at the Institute for Creation Research in San Diego, and at that time he told me he would order one copy of Forbidden Archeology *for the research staff and another for the library.*

4.22. Dr. Rupert Holms

Dr. Rupert Holms is a microbiologist in England.

4.22.1. Letter to Dr. Rupert Holms, January 3, 1995.

For the past several months I have been traveling in the United States and India (with stops of a few days in London coming and going) and have therefore only just now read your letter of November 1, 1994.

Thank you for your kind words about *Forbidden Archeology*. If I may ask, how did you learn of the book? As for my next book, also coauthored with Richard Thompson, it is still in the research and writing stage. I am hopeful it will come out in 1996, but that might be too optimistic. We are very committed to thorough research and careful presentation in our work, and that does mean taking enough time to do the job well.

I did, however, recently present a paper at the World Archeological Congress 3, held in New Delhi, December 4-11. It was titled "Puranic Time and the Archeological Record," and briefly treats one of the themes of our forthcoming book, namely the Vedic concept of cyclical time. If you would like to see a copy of the paper, please let me know and I will send one to you.

4.22.2. Letter from Dr. Rupert Holms, January 22, 1995.

Thank you for your letter of 3rd January. I learned of your book from a review in the journal of The Scientific and Medical Network (an international organization of scientists and doctors aimed at advancing knowledge through both rational and intuitive insights). Although my Ph.D. is in molecular biology and my principal research focus has been on studying AIDS as an autoimmune disease, I am now very actively interested in the origins of civilization.

I am currently researching my hypothesis that the burnt-out core of a star (a black dwarf) is orbiting the sun in a highly elliptical orbit with a periodicity around 3500 years which was responsible for the destruction of an early civilization on earth. The idea that the sun was originally part of a binary star system, and that the matter of the planets came from the debris of a nova explosion of the sun's partner was originally proposed by Fred Hoyle in the 1930's (the hypothesis was quickly buried without being refuted). In the 1950's, Immanuel Velikovsky proposed that there was a planetary "near miss" in about 1500 B.C. but I think he incorrectly identified the object responsible as Venus. Greenland ice cores, pollen data and tree ring analysis all seem to suggest that there have been a series of environmental catastrophes with approximately a 3500 year periodicity in the last 12,000 years.

I would be very interested to read your paper titled "Puranic Time and the Archeological Record."

4.22.3. Letter to Dr. Rupert Holms, February 10, 1995.

I am enclosing a copy of my paper "Puranic Time and the Archeological Record," delivered at the World Archeological Congress 3, New Delhi, December 4-11, 1994.

I've just returned from the West Coast of the United States, where I appeared on some television and radio shows in Los Angeles and Seattle, and gave in Seattle a 4-hour auditorium lecture on *The Hidden History of the Human Race*, accompanied by slides and video clips. I have also recently received an invitation to address the Ancient Astronaut Society meeting in Bern, Switzerland in August. On the mainstream academic front, I am awaiting additional reviews of *Forbidden Archeology* in *British Journal for the History of Science* and *Social Studies of Science*, among others. I also have some opportunities to address faculty seminars in some American universities.

4.23. Dr. William W. Howells

Dr. William W. Howells of Harvard University is one of the most influential physical anthropologists of recent times.

4.23.1. Letter from Dr. William W. Howells, August 10, 1993.

Thank you for sending me a copy of *Forbidden Archeology,* which represents much careful effort in critically assembling published materials. I have given it a good examination; you are doubtless aware that specialists do not sit down and read books in their field from the first to the last word, but rather go for what seem like the key points. I have to say I remain a crusted skeptic.

Most of us, mistakenly or not, see human evolution as a succession of branchings from earlier to more advanced forms of primate, with man emerging rather late, having feet developed from those of apes, and a brain enlarged from the same kind. Teeth in particular strongly indicate close relations with Miocene hominoids. This makes a clear phylogenetic pattern.

To have modern human beings, with their expression of this pattern, appearing a great deal earlier, in fact at a time when even simple primates did not exist as possible ancestors, would be devastating not only to the accepted pattern—it would be devastating to the whole theory of evolution, which has been pretty robust up to now. An essential mark of a good theory is its disprovability, as you know. The suggested hypothesis would demand a kind of process which could not possibly be accommodated by evolutionary theory as we know it, and I should think it requires an explanation of that aspect It also would give the Scientific Creationists some problems as well.

Thank you again for letting me see the book. I look forward to viewing its impact.

The last sentence of Howells's letter suggested the title of this book.

4.23.2. Letter to Dr. William W. Howells, September 25, 1993.

Thank you for your comments about my book *Forbidden Archeology.* You say you "remain a crusted skeptic." I'm also a skeptic, although at the tender age of 45, my crust is probably not as crispy as yours.

You raise interesting questions, and I want to answer as fully and straightforwardly as possible. In fact, I'd welcome the chance to enter into a dialogue with you on the topics you brought up. I wonder how you feel about that. I'm sure you wouldn't want to get entangled in some lengthy and tiresome correspondence. However, if you think there might be some small profit in us two skeptics exchanging views, as time permits, on a topic of mutual interest, I'd like that very much.

That said, I'll respond to some of your points....

I agree with you that the theory of evolution has been pretty robust up to now—in terms of its survivability. Of course, there are other explanations for the origin of life floating around the planet, as you well know. Probably the vast majority of people believe in something else. That is apparently true even in the United States, according to surveys I have seen. But within certain boundaries, the theory of evolution reigns supreme. The principal areas of nearly absolute supremacy are the modern educational institutions and the minds of most university-educated people, who now hold the most important and influential posts in modern technological society, especially in government, the judicial system, and the media. So that is certainly a tribute to the genius of Darwin and his modern disciples. The ascendance of Darwinian evolution was in some ways a genuine intellectual triumph. Nevertheless, I think the theory is open to challenge and critique on many levels, including the most fundamental.

Regarding teeth, Charles Oxnard recently sent me in exchange for *Forbidden Archeology* a copy of his book *Fossils, Teeth, and Sex,* which I found quite intriguing. Even confining ourselves to currently accepted fossil evidence, it seems there are radically different ways of outlining the branching relations among the hominoids. Dr. Oxnard, as you doubtlessly know, regards the robust australopithecines, the graciles, the African apes, *Homo,* and *Pongo* as separate lines of a greater hominoid radiation.

But getting back to the main point you raise—the necessity of coming up with a general counter explanation. In one sense, I feel that it is just as important to establish the real need for a counter explanation as to give one. Otherwise, the proposed counter explanation, no matter how elegantly and thoughtfully developed, will, I suspect, seem trivial to those now most strongly committed to evolutionary theory on genuinely intellectual grounds (as opposed to those who accept it on the basis of authority or for sectarian reasons).

I selected human evolution as a specific case to test the general ideas on evolution because many workers claim that human evolution is one of the best documented instances of the evolutionary process. In this specific test case, I found, however, that there are grounds to suppose that there might be something wrong not only with human evolution but evolution in general. What's wrong is something twofold. First, there appears to be quite a mass of evidence contradicting current views, and second, there appears to be a general process of "knowledge filtration" that suppresses such evidence or renders it harmless.

It is the latter factor that is of most interest to me. The filtration of evidence in human evolution is a continuing process, as far as I can see. I try to stay current with the literature, and I see signs of such filtration everywhere. For example, Glenn Conroy says in a review of *Other Origins: The Search for the Giant Ape in Human Prehistory* by Russell Ciochon et al. (*J. Hum. Ev*. 1991, 21:233-35): "the authors themselves did not seem too excited about the discovery of five isolated Homo molars until after they learned the fossils were apparently older than they originally thought (about 500,000 years B.P. as determined by Electron Spin Resonance dating of a single pig tooth undertaken by Henry Schwarcz of McMaster University). In fact, their initial attribution of the teeth to *Homo sapiens* was altered to *Homo erectus* solely because of the dating results. This may raise some eyebrows since most paleoanthropologists realize that if taxonomy is based solely on geologic age, phylogenetic reconstruction simply becomes an exercise in circular reasoning." Now I have not looked into these teeth beyond this, but given my experience in compiling *Forbidden Archeology*, I have to say that my suspicions are aroused. Is this one more example of anomalously old *Homo sapiens*?

Another example, this time from *L'Anthropologie* (1990, 94:321- 34), is a report by G. Onoratini et al. on Acheulean stone tools being found in a Moroccan formation (the Saissian) previously regarded as Pliocene (cited are Choubert 1965, Taltasse 1953, Choubert and Faure-Muret 1965, Martin 1981). The discovery of advanced stone tools in one of the units of this formation caused the workers to redesignate it as Middle Pleistocene. They said about the tools: "Ceci constitue le premier element de datation anthropique d'une formation qui n'a pu etre datee par aucune autre methode." In other words, it was first and foremost the presence of the tools that caused them to redate the unit from Pliocene to Middle Pleistocene. It is taken for granted that advanced Acheulean tools cannot be earlier than Middle Pleistocene. Again, I have not looked deeply into this case but, being a skeptic, am somewhat suspicious.

In your letter, you mention the incongruity of having anatomically modern humans before their primate ancestors. But I really do not know what would happen if one were to do a thorough search into the history of fossil primate discoveries. When I first started looking into hominids, I only knew the current hypotheses and their supporting evidence. Concerning this supporting evidence, nicely laid out in the textbooks, I thought, well, this is the evidence. That's all there is. I was genuinely surprised to find that was not the case, and that a process of knowledge filtration had been going on. So maybe the primates go back further than many now think possible. If I looked, perhaps I would find numerous cases of the ages of geological formations being adjusted to fit the character of primate remains found in them. Or maybe not, but given my experience with hominid evidence, I cannot say for sure without first looking into the matter quite thoroughly.

At the moment, I am investigating a case of advanced plant and insect remains being found in a Cambrian formation. Of course, the obvious explanation is that they are intrusive. But my study of the rather extensive original reports tends to discount that possibility. I will document this in a forthcoming book.

Well, more of the same, you are bound to say. What about the counter explanation? Assuming just for the sake of argument that you genuinely see the need for one, I shall give you a hint of the beginnings of one.

The current theory of evolution takes its place within a worldview that was built up in Europe, principally, over the past three or four centuries. We might call it a mechanistic, materialistic worldview. Even if God is involved, it is only as a distant, transcendental guarantor of the laws governing the mechanism. This view, which a friend has called "the Enlightenment consensus," differs somewhat from a previous European view of the world as an organic whole, with its material substance pervaded and governed not only by a Supreme God, but a multiplicity of demigods, demigoddesses, nature spirits and so forth. In this view of the world, the ordinary laws of physics existed alongside other laws that went beyond them. Historically, I would say that the Judaeo-Christian tradition helped prepare the way for the mechanistic worldview by depopulating the universe of its demigods and spirits and discrediting most paranormal occurrences, with the exception of a few miracles mentioned in the Bible. Science took the further step of discrediting the few remaining kinds of acceptable miracles, especially after David Hume's attack upon them. Essentially, Hume said if it comes down to a choice between believing reports of paranormal occurrences, even by reputable witnesses, or rejecting the laws of physics, it is more reasonable to reject the testimony of the witnesses to paranormal occurrences, no matter how voluminous and well attested. Better to believe the witnesses were mistaken or lying. In my opinion, there is even today quite a lot of evidence for paranormal phenomena. Unfortunately, this evidence tends to be suppressed in the intellectual centers of society by the same process of knowledge filtration that tends to suppress physical evidence that contradicts general evolutionary ideas.

In one sense, I hesitate to present all this in the form of a letter. It demands a book, perhaps several. But much depends upon your willingness to enter into a discussion of these things. If you are interested, I could fill in the outline a bit more. I will just say for now that the presentation of an alternative to Darwinian evolution depends upon altering the whole view of reality underlying it. If one accepts that reality means only atoms and the void, Darwinian evolution makes perfect sense as the only explanation worth pursuing. Fortunately, there are already some holes in the Enlightenment consensus, and even in the intellectual centers of modern society there are many scholars who are now actively pursuing investigations of paranormal phenomena, ranging from UFO abduction cases, to reincarnation memories, to psychic healing. For example, there is an International Society for the Study of Subtle Energies and Energy Medicine, with 1,900 members, 17% of whom are physicians and 28% of whom hold doctorates. The primary areas of study for this group is physiological transformations apparently governed by paranormal processes. I could give other examples. The relevance is this: if it can be demonstrated that organic tissues may transform by the influence of "subtle energies," that opens up some possibilities for an alternative to Darwinian ideas of evolution.

In terms of the intellectual drift in the elite centers of modern society, I think we are heading back to the situation that existed at the time Darwin presented his theory. At that time, a multiplicity of explanations for the origin of living things, founded upon different perceptions of the ultimate nature of reality, were competing in the elite centers. Not only were there Biblical creationists and Darwinian materialists, but a variety of alternate views. Wallace, for example, posited some kind of spiritually guided evolution. Others spoke of vital principles. Still others, such as Richard Owen, had ideas resembling Platonic archetypes.

As you say, all this should "give the Scientific Creationists some problems as well!" I agree. But some of them are a little more openminded than I thought. For example, Siegfried Scherer, who gave a nice review of *Forbidden Archeology*, is a German young-earth Biblical Creationist.

You say you look forward to seeing the impact of *Forbidden Archeology*. The impact, although as yet small, is as I desired. A small but growing number of scholars—in the fields of the history and sociology of scientific knowledge, philosophy, religion, and even physical anthropology—seem to be taking it politely, even seriously. Some Christian creationists, including Duane Gish, have responded nicely to it. And, as you might expect, there is interest among what some would call the "fringe" science crowd. That the book has appeal across these boundaries is satisfying to me.

Hope this finds you well up there in the southernmost bit of Maine.

4.23.3. Letter to Dr. William W. Howells, August 16, 1994.

When I sent you a copy of my book *Forbidden Archeology* last year, you wrote that you would "look forward to viewing its impact." You doubtlessly saw Jonathan Marks's bad-tempered, name-calling "review" in the

American Journal of Physical Anthropology (93,1:140-141). You may, however, have missed a recent review by Kenneth Feder in *Geoarchaeology,* so I am sending you a copy. I am satisfied with Feder's review, in the sense that he does not grossly misrepresent the substance of the book, as did Marks ("drivel," "dreck," "goofy popular anthropology"). Of course, I do not agree with everything Feder said (see my enclosed reply). I am also going back and forth with Matt Cartmill about a rejoinder to Marks in the *AJPA.* I think it would be in his best interest, and that of the journal, to allow one. In any case, I have learned that additional reviews of *Forbidden Archeology* are forthcoming in other anthropology journals as well as journals of other disciplines (science history). So it looks like there has been some little impact. How is your new book doing?

4.24. Dr. Alexander Imich

Dr. Alexander Imich is a researcher and author in the field of parapsychology.

4.24.1. Letter to Dr. Alexander Imich, May 20, 1995.

I've recently been corresponding with Colin Wilson. He mentioned that you had enthusiastically recommended my book *Forbidden Archeology* to him. Colin has told me he liked the book very much, and is also recommending it to friends. You cannot imagine how much this pleases me. So I am grateful to you. I am curious to know how you yourself came to know of *Forbidden Archeology.*

Forbidden Archeology, as you know, is mostly about stones and bones. But in my next book, tentatively titled *Human Devolution*, I am going to offer an alternative view of human origins based on the Vedic texts of ancient India. That such an alternative is genuinely needed is demonstrated by *Forbidden Archeology*. The alternative I propose takes into account not only the suppressed archeological data but also the suppressed psychical data. In that regard, I am interested in your work. Colin tells me that you have a book coming out that features new psychical research from Russia and China. I have heard that in China there has been quite a bit of research focusing on the *qi qong* related phenomena.

The literature on psychical phenomena is vast. I am going through as much of it as I can—the early mesmerist reports, the nineteenth century mediumistic phenomena, the experimental work of this century. Of course, I am looking for the best and most convincing cases. If you have any thoughts on this, I would like to hear them.

4.24.2. Letter from Dr. Alexander Imich, May 29, 1995.

Jean Hunt told my friend, Ingo Swann, of *Forbidden Archeology*, and Ingo told me about it. I realized the importance of the book, and recommended it to Colin, who, as he wrote, has now two copies of the book.

Findings you describe seemed to undermine the fundamental notions of the biological evolution theory. After getting acquainted with the content of your book, I came to the conclusion that the artifacts from the Paleozoic through Cenozoic era are not products of *Homo sapiens.* If technologically advanced societies did exist at that time, we would find many more remnants than the ones that have been unearthed. The only plausible explanation is that what we find in these archaic times are items left by extraterrestrial visitors of our globe. Are you familiar with the works of Zacharia Sitchin? They should be useful for your work on human origins. I am expecting this future work of yours with true anticipation.

Incredible Tales of the Paranormal, the book I edited, should be on the market this fall. *Matylde,* written by myself, and *Kluski,* written by Dr. Bugaj, are two strongest and, for the parapsychologists, most convincing stories. Russia and China are also represented in the book. I believe that Brian Inglis's historical works and Richard S. Broughton's *Parapsychology The Controversial Science* (Ballantine Books 1991) would be useful for your work.

4.24.3. Letter to Dr. Alexander Imich, June 25, 1995.

One reason I asked about how you learned of *Forbidden Archeology* is that I find it interesting to trace out the psychic pathways by which the book circulates. There is some value to that, as a kind of research into subtle phenomena, because, after all, a book is simply an encoding of a mental process.

Jean Hunt was someone very dear to me, although my contact with her was only through extensive correspondence and several telephone conversations, for about a year, before her departure from her last embodiment on this physical plane. She took a liking to me and was quite generous in introducing *Forbidden Archeology* to her contacts. Reflecting back on things, I can see that her range of contacts extended in two directions (at least). One was through the alternative archeology network. And the other, I now see, was through the psychical network. Most of my dealings with Jean were connected with her archeological aspect. I do recall that she told me she was going to give *Forbidden Archeology* to Ingo Swann. She asked him to review it for *FATE,* which he did. But I never did explore with Jean her personal interests, if any, in the paranormal. I do know that she had studied psychology. If you have an address for Ingo Swann, I would like to write him with a belated thanks for his *FATE* review.

In your letter you said, "I came to the conclusion that the artifacts from the Paleozoic through the Cenozoic era are not products of *Homo sapiens.* If technologically advanced societies did exist at that time, we would find many more remnants than the ones that have been unearthed. The only plausible explanation is that what we find in these archaic times are items left by extraterrestrial visitors of our globe."

Of course, *Forbidden Archeology* is basically a sourcebook, and although I did introduce a subtext of extreme terrestrial human antiquity and coexistence with more primitive hominids, I did so only lightly. As such, anyone is free to offer any interpretation they wish to the reported facts. In my next book, I shall offer my own interpretation, in the context of other evidence from cross-cultural studies and psychic research. There is definitely going to be a thread of extraterrestrial contact running through the tapestry, and one might certainly expect to find some physical relics of these contacts.

But as for the exclusivity of your own suggestion, I offer some considerations:

1. High technology metal artifacts are indeed rare, but you will see that there are lots of stone tools, and stone tools of a certain level of sophistication, such as the many mortars and pestles found in the California mines (up to 55 million years old) are just as good evidence for a human presence as a computer. Also, it isn't likely that space traveling extraterrestrials would need to make stone mortars and pestles from the local rock, or use obsidian spearpoints like those found in the California mines. So it seems there is no reason not to attribute them to terrestrial humans, especially since human skeletal remains were also found.

2. The few high tech artifacts in *Forbidden Archeology* are doubtlessly just the tip of the iceberg. I know of several others, and I have heard that coal miners have often found such things that never get widely reported. So probably a lot of things have been unearthed that we have not heard of yet.

3. There is an interesting discussion now taking place on the internet, in a scientific discussion group on paleoanthropology. The premise of the discussion is to assume a high tech dinosaur civilization that existed 100 or more million years ago. How would one be able to recognize it, archeologically? (Keep in mind these are professional archeologists and anthropologists discussing this.) Apparently, it would not be so easy. Metals tend to disintegrate, what to speak of other materials. The consensus seems to be that signs would be hard to recognize.

4. Another thing—even looking at just the past ten thousand years, the kind of high tech civilization we know today is an anomaly. It has only

existed for about two centuries, and may not continue more than another two. Before that, there were other advanced urban civilizations that did not make use of our present type and level of industrial manufacturing technology. If you look at the other civilizations in China, Rome, Greece, Egypt, Mesopotamia, India, the Americas, etc., they were quite advanced in many ways but used relatively simple technologies. So it may be incorrect to suppose that we should be finding lots of computers and cars in Mesozoic formations (and re point 3, even if they were there they might now be, as one of the discussants said on the internet, just a multicolored layer of rock).

5. High tech in ages gone by may have been psy-tech. That is what the records of ancient cultures tend to indicate. Warriors would get subtle weapons and chariots and even flying craft from subtle beings (demigods) higher in the cosmic hierarchy. Psy-tech might not preserve in the record of the rocks.

I will look for your book *Incredible Tales of the Paranormal* this fall. And thank you for the other references. Another question—what are the possibilities for someone like myself to actually witness some of the stronger paranormal phenomena, such as telekinesis, materializations, or levitations?

To my last question, Imich responded that he was organizing a "definitive demonstration," to be witnessed by scientists.

4.25. R. Wayne James

R. Wayne James is a retired army intelligence officer with a degree in anthropology.

4.25.1. Letter from R. Wayne James, June 3, 1996.

I recently learned of your book, *Forbidden Archeology*, while watching a program on The Learning Channel, and managed to pick up a copy at the local Barnes & Noble bookstore. I haven't been able to put it down! Even though I'm not finished yet, I decided to pen a quick letter (or maybe that should be keystroke) to congratulate and thank you for your effort.

I am a recently retired Army officer who picked up a BS in Anthropology along the way. I have tried to keep abreast of the latest developments/theories for the last few years. I was in Military Intelligence and did a great deal of research and analysis on several military topics, so

I have some experience in ferreting out facts and conjectures. As such, I really wish to applaud your efforts in finding all these obscure, and probably multi-lingual, monographs and conference summaries, etc. Great job!

It is truly amazing how much material you have published! Your premises about the double standard for the last 100 years or so has been proven without a shadow of a doubt. Thank you.

I have greatly enjoyed reading books/articles by Leakey, Johanson, Boaz, et al, but I have also held on to a little skepticism, because every new find seems to dictate a redrawing of the tree, or a naming of a new species. It's been very thought provoking, but I still have the feeling we're missing something big. After getting a few hundred pages into your volume I can see why. Please keep up the good work. Are there plans for a companion volume or a sequel? I hope so.

4.26. Dr. Lonna Johnson

Dr. Lonna Johnson is an archeologist.

4.26.1. Letter from Lonna Johnson, 1996 undated.

I enjoyed your new book very much. I worked professionally for environmental companies as an archaeologist for 13 years. For much of that time I held the title of lithic analyst and recall numerous anomalies that caused me to know some of the precepts put down in your book are true.

I spent those years as an atheist but since then had a profound spiritual awakening—I had no idea on the day that your book caught my eye that it was written in the interest of a spiritual origin—what a delightful "coincidence."

4.26.2. Letter to Dr. Lonna Johnson, May 23, 1996.

My publisher passed on to me your note about my book *Forbidden Archeology.* You mentioned that in your work as an archeologist (lithic analyst) for construction companies, you often encountered anomalies. I would really like to know more about that—whatever you feel you can tell. I am also grateful for your recognition of the spiritual inspiration for *Forbidden Archeology.* In forthcoming books, I want to make the spiritual content more explicit, but it requires very careful presentation.

Dr. Lonna Johnson replied to this letter with an account of a particular case involving an archeological anomaly.

4.27. Dr. Phillip E. Johnson

Dr. Phillip E. Johnson is a professor of law at the University of California at Berkeley. He is author of Darwin on Trial. *He contributed an endorsement for the back cover of* Forbidden Archeology *and a foreword to* The Hidden History of the Human Race.

4.27.1. Letter from Dr. Phillip E. Johnson, November 30, 1992.

Here is a prepublication review which you may use:

"*Forbidden Archeology* is a remarkably complete review of the scientific evidence concerning human origins. It carefully evaluates all the evidence, including the evidence that has been ignored because it does not fit the dominant paradigm. Anyone can learn a great deal from the authors' meticulous research and analysis, whatever one concludes about their thesis regarding the antiquity of human beings."

As you can see, I think highly of your book. As I said on the telephone, I knew nothing about the research activities of your organization and was stunned (and delighted) by its quality. I would now like to learn more about your opinions on the geological column, radiometric dating, molecular classification, and whatever else you are researching. I will look forward to an opportunity to meet you and Richard Thompson in San Diego, or perhaps in Northern California if you ever come this way. In the meantime, feel free to write or call on any matters you would like to discuss.

Richard and I did visit Dr. Johnson at his Berkeley home. We had some good talks during a hike along a trail on the hills overlooking San Francisco Bay.

4.27.2. Letter to Dr. Phillip E. Johnson, April 21, 1994.

The American Journal of Physical Anthropology recently delivered a blast at *Forbidden Archeology* (see enclosed). We also got a brief notice in

Journal of Field Archeology that gave a fair picture of what the book is about and another briefer notice in *Antiquity.* More full journal reviews are coming, two that I know of so far. I have written a reply to Jonathan Marks, who wrote the review in the *AJPA.* Let's see if they print it. By the way, I went to the annual meeting of the American Association of Physical Anthropologists in Denver recently (we had a book display) and despite the caustic review, a good number of members purchased the book on the spot, and hundreds took order forms.

Printed copies of the abridged version should be available in a month or so. I will be sure and send you some copies. The actual launch of the book is scheduled for the fall, however. Meanwhile, we will be soliciting reviews, going to publishing industry trade shows, etc. I have also been doing some radio interviews.

A favorable review of *Forbidden Archeology* appeared in the science section of *Politiken,* Denmark's biggest paper. As a result, we have an offer from a Danish publisher to bring out a Danish edition, and Danish national television is interested in interviewing Richard and myself.

I also have an opportunity to present a paper at the World Archeological Congress in New Delhi this December.

I am working on another book on an anomaly in plant evolution. During the 1940s and 1950s, remains of advanced land plants were found in the Salt Range Formation, in what is now Pakistan. The formations are now considered to be Cambrian. And according to standard views there were no land plants in the Cambrian period.

How are things going on your end? Are you working on any new writing projects in the evolution field?

How is the Dean Kenyon case going? A few months ago, I saw a lot of articles and editorials about it, but not much since. Also, does he have a copy of *Forbidden Archeology*? If not, give me his address and I would be happy to send him a copy.

Dr. Dean Kenyon, a professor at San Francisco State University, was disciplined for teaching intelligent design as an alternative to Darwinism. Dr. Johnson was involved in his legal defense.

4.27.3. Letter from Dr. Phillip E. Johnson, May 2, 1994.

It is good to hear from you. Jonathan Marks is an old adversary of mine, and a thoroughly unreasonable and unpleasant person. An obnoxious review from him is an honor. I am delighted to hear about your success in Denmark, and am very interested in your work on plant evolution.

I am enclosing a couple of my own recent publications. Dean Kenyon's address is at the bottom of this letter; I am sure he would enjoy receiving a copy of your book. Keep in touch.

4.27.4. Letter to Dr. Phillip E. Johnson, September 6, 1994.

I am enclosing page 340, the missing page of the *Geoarchaeology* review [*by Dr. Ken Feder*] of *Forbidden Archeology*. I am also sending you, because you asked, a copy of the paper I am going to deliver at the World Archaeological Congress in New Delhi this December. I recently received word that it has definitely been accepted and scheduled for one of the sections (Time and Archeology). As you will see, this paper does not constitute a full development of an alternative to Darwin's theory of human origins. That will come out in the book I am now working on with Richard, which we are tentatively calling *Human Devolution*. The reason I brought the paper up in my letter to Feder was to show him that I am not shy about discussing things related to my spiritual outlook. In his review, Feder was trying to use the rhetorical stratagems commonly employed against his usual adversaries. But I think he is going to have to come up with some new ones for me. The rhetorical strategy behind my paper is to effectively turn the "no religion in science" thing back against the Darwinians. I think they are going to find it quite shocking.

4.28. Dr. Bennetta Jules-Rosette

Dr. Bennetta Jules-Rosette is a professor of sociology, specializing in sociology of religion, at the University of California at San Diego.

4.28.1. Letter to Dr. Bennetta Jules-Rosette, February 7, 1993.

I have some advance, prepress copies of my book *Forbidden Archeology*, so I thought I would send one to you. I am gathering some prepress reviews and would welcome any comments you might have.

Of course, I would not expect you to comment on the archeological evidence as such, but I thought you might have something to say from your perspective as a student of the sociology of religion.

Really, the book is part of a centuries-old dialogue, between the representatives of the Vedic knowledge-tradition and the Western knowledge-tradition, on human origins, identity, and destiny. For a long time, the "conversation" between the traditions has been pretty one-sided, with the Western view prevailing, even in the homeland of the Vedic tradition. But Richard Thompson and I are raising our voices in a response calculated to make the monologue once more a dialogue.

4.28.2. Letter from Dr. Bennetta Jules-Rosette, April 14, 1993.

Thank you so much for the two volumes of *Forbidden Archeology* that you recently sent me. They are truly wonderful and very provocative. Congratulations on such an excellent piece of work!

Once again, I am teaching Sociology 157 Religion in Contemporary Society. The students would enjoy a presentation by you at your convenience, if you are available during this academic quarter. The class meets on Tuesdays and Thursdays. Tuesdays are the best days for a guest presentation. Tuesday, April 27 would be the ideal day for a presentation. Tuesday, May 11 would also be good. Please let me know at your earliest possible convenience whether either of these days fits into your schedule, or whether another Tuesday would be possible if these dates are not convenient.

I may best be reached by telephone ... or you can send me a note. I look forward to hearing from you soon.

I made a presentation on Forbidden Archeology *to Dr. Jules-Rosette's class on Religion in Contemporary Society.*

4.29. Dr. P. C. Kashyap

Dr. P. C. Kashyap is an Indian anthropologist. I met him in 1994 at the World Archeological Congress 3 in New Delhi. He was favorably impressed with the paper I presented there ("Puranic Time and the Archeological Record"). At that time I presented him with a review copy of Forbidden Archeology.

4.29.1. Letter to Dr. P. C. Kashyap, February 14, 1995.

Namaskar. It was a pleasure to meet you at the World Archeological Congress in New Delhi. I hope by now you have had a chance to go through some of the pages of *Forbidden Archeology*. I would very much like to hear your thoughts about the book. If you are favorably impressed, perhaps you might also consider writing a brief review, which my publisher could use to introduce the abridged version of the book to Indian audiences.

4.29.2. Letter from Dr. P. C. Kashyap, February 13, 1995.

Your book has proved to be most useful to me in the field of anthropology and the beginning of the human race. I enclose a copy of my paper titled "Pre-Rigvedic and Early Rigvedic History—Western Himalayas, the Workshop." Western Himalaya in this sense extends up to Pamir and covers the northern and eastern parts of Afghanistan as well.... Your paper gave me a lot of support.

4.29.3. Letter to Dr. P. C. Kashyap, March 17, 1995.

I have read your paper "Pre-Rigvedic and Early Rigvedic History—Western Himalayas, the Workshop." Your ethnoarcheological method has wide application in unraveling the early history of India, and many other questions of human history throughout the world.

My publisher Alister Taylor has recently been in India, and he informs me that he has been having fruitful discussions with publishers there who have expressed interest in bringing out an Indian edition of The *Hidden History of the Human Race,* which is the abridged version of *Forbidden Archeology.*

Dr. Kashyap finds in the current customs and rituals of the Himalayan peoples evidence that is helpful in unfolding their history—hence my use of the term "ethnoarcheology" to describe his work.

4.29.4. Letter from Dr. P. C. Kashyap, February 15, 1997.

I am glad to hear that your new book, *Forbidden Archeology's Impact,* is scheduled to come out in September. I have little doubt that it will be as outstanding as the earlier one.

Your deeply researched *Forbidden Archeology* is the most precious parting gift of the 20th century to the human race. It is a vast storehouse of analytical and factual data. It charts a new path for man to study his origins.

4.30. Richard E. Leakey

Richard E. Leakey is well known for his discoveries of fossil hominids in East Africa. He did not respond to my formal request for permission to publish this letter.

4.30.1. Letter from Richard E. Leakey, November 8, 1993.

I received your letter and your book. As you may know, I am no longer actively involved in human palaeontology due to other public duties but a quick glance at some pages suggests to me that your book is pure humbug and does not deserve to be taken seriously by any one but a fool. Sadly there are some, but that's part of selection and there is nothing that can be done.

4.31. Marshall Lee

Marshall Lee is president of Interworld Press in New York City.

4.31.1. Letter from Marshall Lee, August 26, 1996.

I have been reading your very interesting book, *Forbidden Archeology,* and find myself agreeing with your thoughts about the antics of the science establishment. I agree also that it is time for an overhaul of the "scientific method" to make it more open to anomalous discoveries and phenomena.

I wonder why you make no reference to the engraved stones of Ica, Peru, which seem to be artifacts of a very early human society. Following your reasoning, there is as much reason to believe that these stones are authentic evidence as there is to reject them because they contradict the established chronology of human evolution. It seems quite illogical that tens of thousands of these stones would be created by someone with such sophisticated scientific knowledge, only to bury them in the desert. The only alternative is that they were made at some time in the distant past of which we have no knowledge, as proposed by Dr. Cabrera. Whether the engravings were made one million or a hundred million years ago may be disputed, but I should think that you would find some interest in a large body of evidence that supports your belief that man existed a million or more years ago.

4.31.2. Letter to Marshall Lee, October 23, 1996.

Thanks for your kind words about *Forbidden Archeology.* Regarding the Ica Stones of Dr. Cabrera, I am aware of them. I did not discuss them in my book because the discoverer had not revealed the actual place of discovery. This is crucial for dating purposes. Without any verifiable geological context for the discoveries, it was not possible for me to assign them any particular age. They could, perhaps, if genuine, still be only a few thousand years old, which would not have helped me make my case for extreme human antiquity. That the stones show humans and dinosaurs together does not necessarily mean they are extremely old, as there are reports of living dinosaurs even today. I do, however, have an open mind about the Ica stones, and if some more definite information about the locales of their discovery is forthcoming, I am ready to consider this.

4.32. Dr. Joseph B. Mahan

Dr. Joseph B. Mahan, now deceased, was President of The Institute for the Study of American Cultures.

4.32.1. Letter to Dr. Joseph B. Mahan, October 6, 1994.

I've heard quite a bit about yourself and ISAC from my late friend Jean Hunt and from Vincent J. Mooney, Jr. Jean once advised me to send you a review copy of my book *Forbidden Archeology: The Hidden History of the Human Race*, coauthored with Richard Thompson. Because of a heavy schedule of traveling, I never got around to it. But a couple of days ago, I did a call-in radio talk show about the book on WRCG radio in Columbus [*Georgia, where ISAC is located*]. I think the host's name was Doug Kellette. In any case, that got me thinking again about ISAC, and I would like to offer you a review copy of *Forbidden Archeology,* if you have not already received one. Please let me know, and I will send one to you.

4.32.2. Letter from Dr. Joseph B. Mahan, October 11, 1994.

Thank you for your letter of October 6 with the kind offer to send me a review copy of your book *Forbidden Archeology: The Hidden History of the Human Race.* I do not have a copy, although I know something of its scope and competence from reviews and comments I have read, especially those of our mutual friend Jean Hunt. I will appreciate and treasure a copy I may receive from yourself.

As you perhaps know, I have spent the past several years in the effort to establish this institute, including a research center containing as much specialized material as I can assemble relating to the subject of America's role in the Ancient World and the continuing effort on the part of establishment "scholars" to ignore this information when it finally comes to their attention. Your book will add materially to this facility.

Enclosed is a copy of the program for our annual research conference, which begins Friday. I am going to display our rather sizable library collection during the conference. We intend to make it available at a formal opening next April, along with other collections of manuscripts, photographs, and videotapes. I am happy a copy of your book will be included.

I wish you could attend the conference.

4.32.3. Letter to Dr. Joseph R. Mahan, November 3, 1994.

I am sending along with this letter a copy of my book *Forbidden Archeology*. I would be interested in hearing your impressions. I am honored that it will become part of the ISAC research library collection. I regret that I was not able to attend this year's conference. At the time, I was in the Midwest, on an author's tour promoting *The Hidden History of the Human Race*, the abridged edition of *Forbidden Archeology*. In a few days I will be departing for the West Coast leg of the tour, and then in December I will be going to the World Archeological Congress in New Delhi, where I will be presenting a paper based on the material in *Forbidden Archeology.* Now that should be interesting!

4.32.4. Letter from Dr. Joseph B. Mahan, November 17, 1994.

Your letter with the copy of *Forbidden Archeology* came today as a most welcome surprise. The book, which truly, I believe, is one of the greatest contributions that has been made to the study of human history, will certainly be a most welcome addition to the ISAC research library. I am most appreciative for it.

I have only read the Introduction and scanned other parts, which I expect to do more of as soon as time and my ability to absorb the weighty contents permit. I can tell you, however, that I am tremendously impressed by both the volume and quality of your work.

You have documented and most effectively called attention to the stubborn stonewalling by students of ancient man in the manner I have tried to do to American archaeologists and ethnologists for the past thirty years. Congratulations!

I am sorry you were not able to attend the research conference last month. Maybe you will be interested in coming to a one day symposium and news conference we are planning for April 8. At that time we will present papers by persons who have deciphered many of the inscriptions from the Burrows cave in Illinois. They will announce some extraordinary information they have learned from the inscriptions. This should end any reasonable doubt about ancient America's transoceanic contacts. We are challenging certain conservative scholars to come and prove the information is not true. The information is of such general interest we are inviting news media to help themselves to the expected heated exchanges between the intellectual fundamentalists and the seekers after truth.

I hope your paper will be well received at the World Archeological Conference in New Delhi. I hope that at the next conference of this august group someone will report that the tomb of Alexander the Great has been found intact, complete with gold coffin, Trojan shield, and gold embroidery purple robe in hinterland America where it was hidden away from seizure by the Roman forces under Octavian in 30 B.C. This is the announcement I expect to make at the news conference in April. The papers that are to be presented will prove it.

This will provide the Establishment a challenging opportunity to use their "presumption of fraud" defense. It should create interest elsewhere as well.

Thanks again for the magnificent book.

4.32.5. Letter to Dr. Joseph B. Mahan, January 3, 1995.

Thank you for the kind words about *Forbidden Archeology*. My trip to India went well. The paper I gave at the World Archeological Congress was well received, despite its controversial nature. There is a chance it will be published in the conference proceedings, which are brought out by the British publisher Routledge.

I may be returning to India in February, but I think I will be back in Florida by April. So please send me the details of your April 8 news conference and seminar—i.e. exact time and place. I really would like to attend.

4.33. Madan M. Mathrani

Madan M. Mathrani is a retired aerospace engineer living in the United States. He is also a former vice president of the Vishwa Hindu Parishad in the United States.

4.33.1. Letter from Madan M. Mathrani, July 8, 1995.

I read your scholarly excellent article, "Puranic Time and the Archeological Record," in the *Back to Godhead.* I had earlier also read about the Archeological Congress in the Organizer weekly. I pray to Lord Krishna that your efforts to get the Puranic Time accepted by the establishment Scholars, and specially the Indian Government, are successful. I have also been trying to get the Puranic History recognized in my very modest way (I am a retired Systems Engineer from Aerospace Industry. I am also a retired V.P. of VHP of America). I have written a couple articles and given a few talks on the subject to Hindu audience.... Incidentally, I hope you know the Apte Samarak Samiti, their Director, Shri Sriram Sathe, and some of their publications. They are also working to get the Puranic History recognized. If not, I will be glad to introduce you. Thank you and Jai Shri Krishna.

4.34. Christopher Meindl

Christopher Meindl is a producer for the nationally syndicated American television program Sightings. *An episode of the program broadcast in early 1995 featured a segment that included material from* Forbidden Archeology.

4.34.1. Letter to Michael Cremo and Richard Thompson from Christopher Meindl, September 21, 1994.

I'm writing to thank you both for all of your help with our recent taping.

Our interviews with you were terrific and will make a great contribution to the finished piece.

We appreciate the time you took from your schedules to be with us.

Once again, on behalf of director Tod Mesirow and myself, thank you for all of your help and cooperation. It was a pleasure working with you.

4.35. Ivan Mohoric

Ivan Mohoric is a television and newspaper journalist from Slovenia. I met him at the Ancient Astronaut Society world conference in Bern, Switzerland, in 1995. At that time, he filmed an interview with me that later aired on Slovenian television.

4.35.1. Letter from Ivan Mohoric, November 20, 1995.

Enclosed, I am sending you a copy of the article related to our conversation in Bern, Switzerland on August, 19th, 1995. I put almost all of it in this article. Therefore you know what it is about in spite of the "exotic" language in which it is written. Perhaps for this reason it will find "a place of honour" in your personal files! ....

Besides, I'd like to ask both of you another favour. Please send me a copy of your book *Forbidden Archeology* via air mail including the bill (price of the book + postage) that I shall be able to pay all the costs immediately. I hope you will permit me to use some material from the book for my articles, provided I shall mention your names as original authors. If there is any additional request (address of the publisher?), please let me know. People in our country will be pleased to read this material. Thank you in advance.

4.35.2. Letter from Ivan Mohoric, December 1995.

I hope you received the copy of the article I sent you.

Recently I had a TV show in our country where I mentioned your latest book, *Forbidden Archeology*. Since there was an interest shown I gave your address to some people as possible option to get the book.

I hope you would not mind if you will get some mail from our country. Hope to hear from you in a short time.

4.35.3. Letter to Ivan Mohoric, January 10, 1996.

I was very pleased to see the article in the Slovenian newspaper, featuring your interview with me in Bern. I know a little Russian, so I can follow the Slovenian language somewhat. But I shall look for someone here in Los Angeles to translate it for me.

I am asking my publisher Alister Taylor to send you a copy of *Forbidden Archeology*, along with a bill for the cost of the book plus airmail postage.

You may be interested to know that we have brought out an abridged edition of the same book.

If you do publish any more articles based on the book *Forbidden Archeology,* please send me copies for my files. They may be helpful for me. Perhaps I may visit Slovenia someday, and I can show people the articles.

Mohoric replied that he would search for a Slovenian publisher for the abridged version of Forbidden Archeology.

4.36. Vincent J. Mooney, Jr.

Vincent J. Mooney, Jr., is a mathematician employed in the computer industry.

4.36.1. Letter from Vincent J. Mooney, Jr., March 20, 1994.

I have a copy of your book *Forbidden Archeology* and have been reading it. I have several points to make in this letter.

First, in order to get more copies into the hands of people that will appreciate it, I need to know how to get a volume buy discount. There is a group here in Frederick County MD that has been meeting on the subject of evolution and its history. Some are Ph.D.'s including a Medical Doctor. We get books in 10 or more (sometimes 20+) depending on how many orders are taken. For example, Phillip Johnson's book *Darwin on Trial*, Michael Denton's *Evolution: A Theory in Crisis*, and Francis Hitching's *The Neck of the Giraffe* have been purchased in quantity (one book, I forget which, was purchased in quantity twice to meet the demand). These were all covered in 1992 and 1993; there were other books as well.

I function as the unofficial buyer (If you *think* of an idea, you are *sure* to be asked, i.e. *told,* to carry it out!) so I am writing.

A second sales source will be the ISAC (Institute for American Studies) meeting next month in Columbus, Georgia (late April 1994). A recently deceased member, Jean Hunt, wrote a highly favorable review of your book (which is how it came to my attention). The review will be known to many who attend. Jean Hunt believed that there was a long history of humanity, maybe 2,000,000 years. The ISAC group is primarily one that examines America Before Columbus (ABC) but also gets into your area. ISAC sells books at its meetings and would probably only do so on consignment; the Frederick group would collect money first and send an order with payment enclosed. ISAC sales cannot be predicted; attendance may be middling (75-100) or good (150+) and if, like last year, they did not have too many books for sale, that can affect the volume. In 1992 there were over 50 titles for sale.

The book by (the late) Jean Hunt was called *Hunting for the Flood Survivors* (I think I have it right) and is still available from The Louisiana Mounds Society at their new address, C/O Mr. Bill Rudersdorf, 215 Hawthorne, Houston, Texas 77006 (713-522-6220), FAX:713-522-5256.

Second, I'd like to compliment you on the effort and the problems. Central to all of this is the process of knowledge filtration, of a radically incomplete collection of facts. (Both phrases are from your book, page xxv.) It is such a common problem in these days that I and many others see it in many ways. I do some writing; I enclose an example of a "fixed idea" that can be debated but isn't, as a brief article I did on Columbus.

Concerning the dispute over Columbus as a Spanish Jew versus Columbus as a Genoese Italian, I am fortunate to have in town here Dr. Charles Merrill, Ph.D., who is fluent in both Spanish and Catalan, and I listen to him and the problem of getting Columbus recognized as Colon, the Spaniard. Dr. Merrill is a professor of history at Mount St. Mary's College in Emmittsburgh, MD; his wife is a Ph.D. in Mathematics and teaches at the same college. Madraiga's book *Christopher Columbus* was published in 1940 in English (it was written in English; Salvador de Madraiga had been his country's—Spain's—ambassador to England, to France and to the League of Nations). Then Samuel E. Morison wrote his book on Columbus just two years later and does not even deal with Madraiga other than to dismiss the "Colon Espanol" theory in a few pages. But Morison does say that there are no record[s] that show Columbus ever wrote in Italian.

I am mildly involved in another area of academic quarrel. It is on the question "Who wrote Shakespeare" (it wasn't the William Shaksper of Stratford on Avon): The debate has been going on for over 200 years and academia simply does not acknowledge the opposition. The opposition is usually just ridiculed. See, for example, the March 1994 *Dartmouth Graduates* magazine and their treatment of William Fowler, Class of 1921, who died January, 1993 (at 93 years old). I have been to his house, spoke to him on the phone, written back and forth, and the article is just junk. Even if you did not know Fowler, I suspect you'd say it was junk. I have given a speech on this subject in January, 1994 to the Baltimore MENSA. I can send a copy of that material (about 40 pages) if you want it. I suggest this to help convince you that there are other areas where legitimate debate is not permitted. (And if you become believers in the idea that not Shaksper but someone else wrote the plays, all the better).

(In order to not confuse you, it is my son who graduated from Dartmouth in 1993, not me.)

Third, I'd urge you to collect stories from those who have been excluded by preconceptions. I have listed three possible persons above; you mention Virginia Steen-McIntyre and Thomas E. Lee; there are sure to be many more. As a guess, ISAC can help (ISAC members are familiar with Dr. George Carter, whom you mention). Just concentrating in your own area should be enough; but as I've noted, there are other areas as well. I suggest this because the lack of openness is wide but not permanent. I know that there is idiocy knocking on academic doors as well; one just has to be discriminating. Part of the story is the publication of bogus science as noted by the National Institutes of Health's Office of Scientific Inquiry. This OSI group in NIH is a recognition that even referreed paper can be false (ask Dr. David Baltimore). I also will mention that the fact that the book is published by your institute will work against you; but it does not affect me.

Fourth, for your information, I am employed full time in the computer industry. I have a BA in math, training in linguistics, and a penchant for inquiry but not enough free time to do all that I would do.

4.37. Dr. Michael Mulkay

Dr. Michael Mulkay is a prominent sociologist of scientific knowledge at the University of York in England.

4.37.1. Letter to Dr. Michael Mulkay, February 7, 1993.

Because you are one of the core set of the sociology of scientific knowledge community, I am sending you an advance, prepress copy of my book *Forbidden Archeology: The Hidden History of the Human Race.* I am gathering some prepublication reviews and would welcome your comments....

I have recently read your book *Sociology of Science: A Sociological Pilgrimage*, in which I found ideas that were very close to my own—not in some formal scholarly sense, but in terms of personal realizations I have come to in my own pilgrimage, which has taken me into a non-Western knowledge tradition. Some sentences that I particularly liked from your preface were: "It may be that, despite its achievements, or perhaps partly because of its achievements, scientific knowledge engenders a corresponding ignorance which hides from view the more important dimensions of our world" and "I have come to see sociology's ultimate task, not as that of reporting neutrally the facts about an objective social world, but as that of engaging actively in the world in order to create the possibility of alternative forms of life."

My own interests are fairly wide ranging, extending from archeology and anthropology to cosmology and particle physics. And my existence also incorporates one of those "alternative forms of life" and other ways of knowing.

In any case, I hope you will be able to find some time to have a look from *Forbidden Archeology* and let me know what you think.

4.37.2. Letter from Dr. Michael Mulkay, May 18, 1993.

Dr. Mulkay, for unstated reasons, has asked that I not print this extremely interesting letter. In it he says that he does not, and will not, have time to read Forbidden Archeology. *He consoled me that his own best book met with silence from reviewers. I attribute Dr. Mulkay's reticence to the natural reluctance of a professionally accredited scholar to engage in privileged discourse with someone perceived as an intruder in his disciplinary sanctum. To engage in such discourse would be to devalue one's own status.*

4.37.3. Letter to Dr. Michael Mulkay, May 27, 1993.

Thank you for your courteous response to my letter inquiring about your opinion of my book *Forbidden Archeology*. I did not find your plea of no time to be "irritating." My inner cultivation of consciousness has engendered in me a sense of detachment from, as well as commitment to, my work. The Bhagavad-gita recommends: "Without being attached to the fruits of activities, one should act as a matter of duty, for by working without attachment one attains the Supreme."

Also, I am tracking the reception of *Forbidden Archeology* in different knowledge communities. So your response of any kind is a useful datum (as would have been your nonresponse). Among the response categories thus far perceived are:

(1) no response;
(2) positive responses from
  a. standard scholars personally acquainted with me
  b. standard scholars not personally acquainted with me
  c. nonstandard scholars (Christian creationists) not personally acquainted with me.
(3) promises to read and comment from standard scholars not personally acquainted with me.
(4) no time, will neither read nor comment responses from standard scholars not personally acquainted with me.

Your response fits rather nicely into category (4,) although I confess I had hoped for a (2)b and rather expected a (3). I have not yet circulated the book among those most likely to give negative comments. When I do, I shall be able to add some more categories. And of course the categories could be divided according to academic discipline (i.e., history of science, anthropology, archeology, SSK [*sociology of scientific knowledge*], etc.).

Thank you for sharing your experiences with your own books. I shall take your counsel of patience to heart in all matters, not excepting my hope of your eventually finding some time to read *Forbidden Archeology*.

4.38. Rev. Father Donald E. Nist, J.D.

Donald E. Nist is an Anglican priest, as well as a patent attorney, chemist, and biologist.

4.38.1. Letter from Rev. Father Donald E. Nist, J.D., July 3, 1996.

I have just finished reading *Forbidden Archeology* and have found it to be fascinating and important. It confirms my own views concerning the tendency of "The Establishment" to use its influence in science and in other fields to control thought by exclusion, that is, by forbidding other views to reach the general scientific community and the general public.

I am most anxious to have you publish your sequel in which you summarize your views and evidence concerning the remote antiquity of modern man. That new book will be most welcome.

If you will permit me a few observations on *Forbidden Archeology*, I found the book was difficult to handle due to its massive size and weight. It would be better to publish it in two volumes, even though that would be more expensive. Moreover, a more concise treatment in the main volume of each subject would help the general reader, with an end-of-chapter summary of the most salient points of the chapter. The Appendix could then be reserved for a more detailed treatment of the subjects of the main volume.

One thing I had trouble with was the use of the technical terms for the various epochs, without in all cases listing the absolute times involved. I had to keep referring back to your Table 1.1 of the various eras to have a clear picture of the antiquity of each item being discussed.

All in all, however, the quality of your work and your attempt to summarize all available information, both pro and con, was admirable and astonishing in its scope. All who read this book will be impressed by it. My own background as a chemist, biologist, professional gemologist, patent lawyer and Anglican Catholic priest made it easier to follow your detailed analyses and explanations of test procedures, etc. But I am sure the general public will also benefit greatly from this book.

Congratulations on a job well done and long overdue.

4.38.2. Letter to Reverend Father Donald E. Nist, J. D., August 4, 1996.

Thank you for your words of appreciation about *Forbidden Archeology*. The large size of the book was a deliberate choice. Psychologically, it has an impact. It says there really is a lot of evidence that goes against the current ideas of human origins—enough to fill a big book. But it is available in an abridged volume of some 350 pages called *The Hidden History of the Human Race*. Could you tell me how you learned about *Forbidden Archeology*? I am interested in seeing by what channels the book is circulating.

4.38.3. Letter from Reverend Father Donald E. Nist, J.D., August 8, 1996.

In answer to your question, I subscribe to *Science News, Archeology* magazine, *Bible Review, Biblical Archeology Review* and *Discover* magazine and cannot recall in which publication a short review of *Forbidden Archeology* was presented, which enabled me to learn about your interesting book.

Forbidden Archeology mentioned that there would be a new book in which you and your co-author would present your explanation of the presence of the numerous so called anomalies, those artifacts and skeletal remains which modern science cannot explain and which do not fit their evolutionary theory. I believe that the new book will be even more valuable and interesting than *Forbidden Archeology* and I look forward to reading it. I hope that when it is published you will let me know so that I can order it. I do not trust my current sources for new books to alert me to it.

I wish you the best with your continuing investigations. Your work has blown huge holes in the neatly packaged evolutionary theories we, the public, are supposed to swallow whole without questioning their shaky basis.

4.39. P. N. Oak

P. N. Oak is an Indian historian and author. I find his work of uneven quality. For example, although I share Oak's opinions about the antiquity of worldwide Vedic culture, I find unconvincing and superficial his assertions that European place names are equivalent to Sanskrit words that happen to have vaguely similar spellings (e.g. Russia is "Rishiya," the land of the Vedic rishis, *or sages).*

4.39.1. Letter from P. N. Oak, May 31, 1996.

Thank you for your letter dated May 15 received two days back saying my letter to you having reached in a torn condition was not fully readable. So I am writing again in some detail.

Vedic culture is the only one which has retained human history from the very first generation almost 2,000 million years ago (as per the tally summary given in the introductory portion of the annual Vedic almanac in an unbroken year-to-year tradition) down the Krita, Treta, Dwapar and Kali eras.... Since about 2/3rds of the world is ruled by Muslims and Christians, all pre-Mohammad history has been wiped out from Islamic countries and all pre-Christian history has been destroyed in Christian regions....

Though Buddhism was not spread by cruel compulsion (because the Buddha was a Vedic saint and not the founder of any religion), all history in Buddhist regions, such as China, Japan and Mongolia, doesn't go beyond the Buddha. That means to say that out of the multi-million-year history of humanity, Muslims learn a history only of the last part of 1,374 years, Christians, of about 1,996 years and Buddhists only of, say, 2,500 years. This is a very sad state of affairs.

Considering all that I feel very keenly the need for founding a World Vedic Heritage Academy which will train scholars fluent in foreign languages to awaken all Buddhists, Christians, and Muslims to their primordial Vedic, Sanskrit Unity.

That task has been partly begun by the St. James Schools in Britain which have separate institutions for boys and girls as per Vedic culture and where every standard has a Sanskrit period compulsory...from the 4 1/2-year-old entrant to 'A' level.

U.S.A. too has Lincoln school working on the same lines, I am told.

They are likely to be our future collaborators in the task of reviving Vedic culture throughout the world.

I, therefore, feel the ISKCON [*International Society for Krishna Consciousness*] with its numerous branches and worldwide centres should set up World Vedic Heritage Academies or Universities in different countries to train scholars, fluent in world languages, to fan out into the world to inform people of their primordial Vedic and Sanskrit unity....

If, therefore, the above brief analysis persuades you to feel that setting up of World Vedic Heritage Academies (or Universities) is an important and noble task, you may take up the matter with ISKCON leaders.

4.39.2. Letter to P. N. Oak, August 4, 1996.

Namaste. Thank you very much for your letter of May 31. I am pleased to learn that you accept the entire Vedic chronology, with human civilization existing for millions of years on this planet. There is some evidence in favor of human beings having been present on earth for hundreds of millions of years, and I have documented it in my book *Forbidden Archeology*. I have asked my publisher to send you a complimentary copy. After you receive the book, perhaps you will consider sending me a copy of your book *World Vedic Heritage* in exchange.

ISKCON is presently attempting to do something in the educational arena. The Bhaktivedanta Institute at the Bombay ISKCON center in Juhu Beach has started a graduate school in connection with the Birla Institute of Technology. It will give masters degrees in certain subjects. There is already existing in the United States an institution called Florida Vedic College, founded by ISKCON members. It is licensed to give degrees in various subjects. The college gives courses through correspondence and by computer. The director is Maha Buddhi Dasa, 10343 Royal Palm Road, Suite 207, Coral Springs, Florida 33065, USA. In the Vrindaban [*India*] ISKCON center you can find the Vrindaban Institute for Higher Education. Other institutions are in the planning stages.

4.39.3. Letter from P. N. Oak, February 17, 1997.

About two months back I received a copy for review of the 914-page volume published by you, titled *Forbidden Archeology: The Hidden History of the Human Race,* authored jointly by Michael A. Cremo and Richard L. Thompson.

The volume is welcome as a first step in exposing the fraud, forgery, and hypocrisy rampant in archeological studies of Western Christian scholars, as is apparent from observations such as "The Problem of Forgery" ...[*he lists several other subchapters of Forbidden Archeology*]. In that context, I would like to bring some fundamental facts to the notice of the ISKCON [*International Society for Krishna Consciousness*] scholars.

(1) The entire present tradition of Western archaeological studies needs to be totally scrapped *ab initio* because it concerns itself with very flimsy and inconsequential data such as some rocks, skulls, jawbones, and stone implements. The ponderous conclusions they seem to reach make much ado about nothing.

(2) I myself participated in the weeklong World Archaeological Congress hosted by Southampton University, U.K., September 1 to 7, 1986 ...scholar after European scholar presented some hair-splitting arguments over some miniscule data....

(3) Since European heads dominate the present educational set-up, Africans, Asians, etc., have to willy-nilly toe the Western-Christian line and appear equally serious about the issues sketched out by the Western-Christian lobby.

(4) The whole basis of Christian archaeological studies was artificially rigged up after a well-entrenched Christianity had totally destroyed all memories and record of its Vedic Sanskrit past from the 4th to the 14th century ...

(5) Vedic archaeology begins with a Sanskrit treatise known as *Brahmanda Purana*, which describes conditions before the creation and how the cosmos was created in a very orderly manner, as an ongoing concern, with the inaugural divine sound "aum" ringing through the firmament, the creation of the first generation of all life including godly humans, the handing over to them of the divine Vedas (in Sanskrit), which therefore became the God-given universal language. The Bible exactly records that same Vedic version by stating "at first there was the word" (i.e. aum) and "the spirit of God was seen floating on the water" (i.e. Lord Vishnu was seen reclining on the Milky Ocean) and "at first the world was of one speech" (i.e. Sanskrit was spoken throughout the world).

(6) Vedic culture, Sanskrit language, hermitage schools of Vedic gurus known as *gurukulam* ...pervaded the world from the 1st generation of humanity for millions of years down the Krita, Treta, Dwapar, and Kali Yugas. That history is recorded in Sanskrit volumes known as the Puranas for anyone to read ...Christianity wiped out all pre-Christian history from the regions it dominated. Similarly, Islam, too, destroyed all pre-Mohammed history.

(7) At the end of that Dwapar period came the Mahabharat War, beginning November 15, 5561 B.C. It lasted 18 days, being a great internecine war in which highly destructive armaments were used bilaterally, it reduced the world polity to a shambles....

(8) It was the total Christian destruction of every vestige of Vedic, Sanskrit erudition, which pervaded the ancient world, that ushered in the Dark Ages in Europe, lasting for nearly a thousand years. The Renaissance assumes importance in European history precisely because it made

Europeans wake up to realize that they knew nothing of their pre-Christian forefathers.

(9) .... With the uniform pre-war worldwide Vedic theology propagated through Vedic hermitages ...throughout the world, having been reduced to a shambles, humanity got divided into various cults....

(10) ...The lack of any authentic history made the Western intelligentsia clutch at the speculation of the physicists' Big Bang theory and the Darwinian evolution of the species theory....

(11) .... [*section about early history of Christianity deleted*]

(12) That multi-million-year period conjured up by evolutionists and clutched at by so-called archaeologists and historians has led to bizarre speculation ...of an Ice Age, a Pleistocene Age, etc., and hairsplitting arguments about miniscule finds of a jaw-bone here or a chiseled rock there.

(13) ...I suggest that all Western scholars abandon their present contrived archaeological studies and adopt the Vedic account of creation and subsequent developments as recorded in the Puranas and other Sanskrit literature....The Bhaktivedanta Institute with its network of numerous centers throughout the world has the vision, the inspiration, and drive to restore that precious, pristine, pious Vedic Sanskrit culture to all humanity.

(14) ...In the 1,000 years (4th to 14th centuries) of wanton destruction of Vedic culture in Europe, all ancient Vedic temples were seized and made to masquerade as churches. This is apparent from the *Encyclopedia Britannica* observation that most churches in ancient Europe were astronomically oriented. Such astronomical orientation is a Vedic penchant. Secondly, Vedic idols consecrated in those temples...have been discovered in pits and bogs in Europe. For instance, a solar disc drawn by a horse, an idol of the Vedic lion-god (Narasimha), goddess Lakshmi flanked by two elephants, a Ganesh idol discovered in Bulgaria (presented to President Shankar Dayal Sharma of India during his official visit to Bulgaria in 1996), the numerous emblems dug up all over Italy from time to time, Krishna idols discovered in Greece, Ramayanic episodes found painted in ancient Etruscan sites discovered in Italy...may be mentioned as some glaring instances of how Western Christian archaeologists have conspiratorially hidden substantial Vedic, Sanskrit evidence, and deliberately channeled and confined archaeological studies to paltry jaws, bones, and crude aboriginal stone implements. [*A note at the bottom of this page of the letter reads*: A king of Greece in the 2nd century had figures of the divine brothers Krishna and Balarama etched on his coins.]....

4.39.4. Letter to P. N. Oak, March 15, 1997.

Namaste. I am pleased to hear that your review copy of my book *Forbidden Archeology* arrived safely in India and that you were able to look it over. Of course, it is true that the Western and Western-influenced archeologists engage in much hair-splitting debate over bits of bone and stone. But even taking things in this way, we find that their picture of human origins and antiquity is very wrong, and that even this kind of evidence is consistent with the Puranic view of extreme human antiquity. At least this is a good starting point for debate with the orthodox archeologists and their followers.

From time to time, evidence of advanced culture is discovered, as you will see in Appendix 2. Even more important is evidence for Vedic culture in the distant past, in India or in other parts of the world.

I would someday like to make a catalog of monuments in India that have been standing for tens of thousands or even millions of years. My spiritual master His Divine Grace A. C. Bhaktivedanta Swami Prabhupada, said, for example, that the Hanuman temple at Allahabad has been standing since the time of Ramacandra, which would be at least one million years. And I have heard of other such ancient monuments. I have also heard that some of the temple Deities, such as the form of Lord Ramacandra at Udipi, are millions of years old.

In your letter to me, I was especially intrigued by your listing of Vedic idols found in various parts of the world, such as 1. a solar disc drawn by a horse, 2. an idol of Narasimha, 3. goddess Lakshmi flanked by two elephants, 4. a Ganesh idol from Bulgaria, 5. Krishna idols from Greece, 6. Ramayana episodes in Etruscan paintings, etc. Do you have documentation of those discoveries, including estimated ages of the objects. That would be quite valuable for my work.

4.40. Geoff Olson

Geoff Olson is a Canadian journalist with an interest in scientific anomalies.

4.40.1. Letter from Geoff Olson, January 12, 1995.

Congratulations on your fantastic book! I'm a journalist in Vancouver, and have long had an interest in scientific anomalies. I stopped reading paleontology some years ago, assuming that the remaining work in human evolution was just a matter of dotting a few more i's.

I must admit I felt some annoyance when I first read your book, since I had assumed that, at the very least, the broad outlines of human evolution had been fully demonstrated. But the great rigor in your investigations into primary sources makes it undeniable that there's something very wrong with the standard view. I think it was wise to admit your own metaphysical bias in your investigations, given your involvement with Eastern religion. It's going to be used against you both no matter what, so it's much better to announce it up front, and to ask that your work be judged on its correspondence with the truth, not on the author's motivations.

My main reason for writing is to ask if you have had any interest expressed by any film-makers to do a documentary on your book, or if you have the resources through the Bhaktivedanta Institute to hire some professionals to put something together that conceivably gets wide distribution, if only through video rentals. Alas, the knowledge filter is going to work overtime trying to strain this one out!

Also—I'm sure you both have heard of William Corliss and his sourcebooks on strange phenomenon. If I remember correctly, he has at least one on archaeological material. Have you consulted this, or have you already for a future book?

One further point. I taught astronomy for a time here at the local planetarium and in the school system, and have always had a fascination with the UFO phenomenon. My work in this area (I've had some articles published in *MUFON Journal, IUR,* and *UFO* magazine) has led me to appreciate how intensely political the scientific process is. Your book demonstrates that this holds for the field of physical anthropology, and I wonder how many other of our current "ologies" are consensus reality by decree!

Again, thanks to you both and all those who worked with you on the research for this excellent piece of science.

I wrote back asking Olson to consider writing a review of The Hidden History of the Human Race. *I also informed him that I was working with a television producer on a program featuring material from the book.*

4.40.2. Letter from Geoff Olson, June 4, 1995.

Hello, Michael. I'm the journalist from Vancouver who faxed you a while back—sorry for the delay responding. I don't have an excuse as good as holidays, unfortunately! Plain old sloth will have to do for me.

I'm glad that you've had good response to your being up front on your religious beliefs. As I said before, in no way do I think this detracts from the actual work, since the archaeological data can stand on its own, regardless of anyone's particular worldview.

You inquired about me doing some article on the book. Unfortunately, I don't think I could do justice to your book's findings in anything less than 1200 words, and my regular stint is a humour column of six hundred words. If I tried it in my usual forum, I think everyone would figure I'm pulling their collective legs.

I have also written on scientific topics for the *Globe and Mail*, Canada's national newspaper. However, I know with some certainty that an article on *Hidden History*, especially a sympathetic one, would never see the light of day. It's a very conservative paper, and the topic would undoubtedly be blown off by the editor there as some creationist propaganda. (I don't think I would have ever accepted your thesis myself unless I had actually read the book in its entirety.) My last proposal was too radical for them—an analysis of the varying concepts of God held by Stephen Hawking and contemporary cosmologists.

Here is something perhaps you and Richard Thompson may like to follow up. On page 13 of zoologist Lyall Watson's book *The Dreams of Dragons* (Morrow, 1987), he talks about skeletons found on Africa's west coast:

"The limbs were weak and spindly and the ribs no thicker than paper. But the head was incredible. Beneath the high arch of the forehead, the face was straight, and small with delicate jaws and tiny teeth. It was a child body, with childish face, driven by a gigantic brain.

"I learned later that his people are known to science as Boskopoid, after the site in South Africa where they were first discovered, and are regarded as interesting if somewhat meaningless freaks, having a cranial capacity 30 percent larger than ours. They are also thought to be ancestral to the Bushmen who still eke out a precarious existence in the Kalahari."

For his sources for this chapter, Watson lists the following:

Eisely, L. *The Immense Journey.* New York: Vintage, 1959.

Holloway, R. L. "Endocranial Capacities of the Early African Hominids" *Journal of Human Evolution* 2:4449, 1973.

Krantz, G. S. "Brain Size and Hunting Abilities in Earliest Man" *Current Anthropology* 9 :450, 1968.

Tobias, P. V. *The Brain in Hominid Evolution*. New York: Columbia University Press, 1971.

(Hopefully, one of these will back up Watson's slightly dubious tale. The man has an unfortunate knack for apocrypha — he's the one responsible for the "Hundredth Monkey" tale, if you've heard that.)

You mentioned Richard Thompson's book *Alien Identities.* I haven't yet read it, but I've heard some praise for it. This is an area of interest to me personally, and I've written a few articles for some American magazines on the UFO phenomenon. While I think there is a lot of wishful thinking and borderline personalities in this area, there is also a core phenomenon that I don't think is reducible to any mundane explanations. I have found some people's entity reports seem to have more in common with spiritual or daemonic lore than any standard sci-fi narrative. Not that this reduces it to a Jungian, attenuated mythological experience — there also seems to be some external physical phenomenon associated with abduction reports, in addition to multiple witness experiences.

What is your and Richard Thompson's feeling on the genesis of humanity through the aegis of spiritual/dimensional beings? Are you more literalist in your reading of the Vedic literature, or do you feel the extraterrestrial/religious dichotomy reduces to a unitary intelligence?

Good luck with your video — is it out yet, and what is it called? Look forward to your next book....

4.40.3. Letter to Geoff Olson, June 1995.

The video I have been consulting on is being produced and directed by Bill Cote of BC productions in New York for NBC. The tentative title is *Forbidden History of Man,* to be delivered to NBC by August for fall 1995 broadcast. [*It was broadcast as* Mysterious Origins of Man *in February 1996.*]

I take a literal reading of the Vedic texts, but a literal reading of those texts can be quite deep and intricate. Generally, when we think of a literal reading of a sacred text, we imagine something inanely simplistic. Not necessarily so, especially in the case of the Vedas, which are vast and treat the subject from many different angles. To put it simply, the Vedic view contains elements of the Darwinian view (descent with modification—but starting out with the more advanced forms instead of the least, and the process is directed rather than random), the creationist view (God is ultimately responsible, but there is a complicated process extended over time, and not a simple creation ex nihilo), and the extraterrestrial contact hypothesis (reproductive contact between earth residents and extraterrestrials many times throughout history). All will be explained, with much supporting evidence, in the next book on which I am now working.

It's interesting. *Forbidden Archeology* and *Hidden History of the Human Race* have generated considerable response in academic journals, "new science" and New Age publications, and in broadcast media. But the mainstream print media (book reviewers and science writers) have ignored them. I take it that these groups are much more conservative, in the sense of consciously or subconsciously toeing the line put out by the scientific establishment.

4.41. Dr. Charles E. Oxnard

Dr. Charles E. Oxnard is a distinguished physical anthropologist, now at the University of Western Australia. During his long career, he has often challenged consensus views on human evolution. For example, during the 1960s and 1970s, when most physical anthropologists promoted Australopithecus africanus *as a direct human ancestor, Oxnard offered convincing anatomical evidence to the contrary.*

4.41.1. Letter to Dr. Charles E. Oxnard, February 7, 1993.

Over the years I have been a great admirer of your books such as *Uniqueness and Diversity in Human Evolution* and *The Order of Man*. I am therefore sending you an advance prepress copy of my book *Forbidden Archeology: The Hidden History of the Human Race*. I am gathering some prepublication reviews and would welcome your comments.

4.41.2. Letter to Dr. Charles E. Oxnard, May 12, 1993.

A few months ago I sent you an advance review copy of my book *Forbidden Archeology.* I can well imagine that your academic duties limit your reading time, but if you have had a chance to go through any of *Forbidden Archeology,* I would welcome your comments. I am not (necessarily) looking for endorsements of the book's central thesis, which is somewhat radical. But I would be interested in your impressions of the book's general level of scholarship and whether or not you think the book would be of interest to members of your discipline's research community.

4.41.3. Letter from Dr. Charles E. Oxnard, May 19, 1993.

Thank you for your May 12 letter. In fact I have received your book but have only just started reading it because I have been in the US for anthropology meetings.

A quick skim indicate[s] to me that it is very interesting and needs a much deeper reading. I am particularly interested in your over theme—forbidden anthropology (archeology) as you will realize from my writings. I am sending you my two latest books under separate cover. One—*Fossils, Teeth and Sex*—relates strongly to your thesis.

4.41.4. Letter to Dr. Charles E. Oxnard, September 18, 1993.

Thank you for your comments about my book *Forbidden Archeology*, which I sent to you in bound galley form. I am now sending you separately a copy of the published version. You should receive it in a couple of weeks.

I also thank you for sending me copies of your two latest books, *Animal Lifestyles and Anatomies* and *Fossils, Teeth and Sex*. I delayed writing you until I had time to read both.

Animal Lifestyles and Anatomies puts the brakes on the often speculative recreations of fossil hominid, particularly australopithecine, lifestyles. Your work shows that much careful study and analysis is required to form hypotheses regarding the relation of anatomy to lifestyle even among living species. So what to speak of nonliving species known only through rather incomplete fossil evidence. Also the past environmental surroundings are not always easy to recreate. But I see that you do find some correlation between extant primate lifestyle groupings and previously determined anatomical groupings. Do you have at the moment any tentative suggestions about australopithecine lifestyles and environment, as deduced from their anatomical groupings relative to living primates?

Fossils, Teeth, and Sex was, as you predicted, of even more direct interest to me. Operating within the bounds of currently accepted fossil evidence, your first, second, and third order conclusions about the australopithecines are the most reasonable and convincing evolutionary interpretations that I have seen. But, as you might anticipate, I very much agree with your statement in Chapter 1: "...the possibility of a totally new and unexpected fossil find is always present. Complete denial of hypotheses is not, and must not be, excluded. It must be possible for occasional fossils to suggest new evolutionary vistas to us." One might even add nonevolutionary vistas. [*This hint at intelligent design or creationism is entirely my own suggestion, and was not implied by Dr. Oxnard, who regards such proposals as outside the scope of scientific investigation.*] Your final picture of the relationships of the extant and fossil hominoids resembles somewhat the coexistence of the various forms suggested in *Forbidden Archeology*. You have the gorillas and chimpanzees, the gracile australopithicenes, the robust australopithecines, the *Homo* line, and the orangs as distinct elements of a radiation, but you were careful not to link the lines in your figure.

I've recently received a letter from W. W. Howells. He writes:

> Most of us, mistakenly or not, see human evolution as a succession of branchings from earlier to more advanced forms of primate, with man emerging rather late, having feet developed from those of apes and a brain enlarged from the same kind. Teeth

> in particular strongly indicate close relations with Miocene hominoids [*having read* Fossils, Teeth, and Sex, *I am now equipped to pursue that topic with him in some detail*]. This makes a clear phylogenetic pattern. To have modern human beings, with their expression of this pattern, appearing a great deal earlier, in fact at a time when even simple primates did not exist as possible ancestors, would be devastating not only to the accepted pattern—it would be devastating to the whole theory of evolution, which has been pretty robust up to now. An essential mark of a good theory is its disprovability, as you know. The suggested hypothesis [*coexistence through geological time*] would demand a kind of process which could not possibly be accommodated by evolutionary theory as we know it, and I should think it requires an explanation of that aspect. It also would give Scientific Creationists some problems as well!

In fact, I did look into human evolution to test the current evolutionary consensus. And when I collected all the relevant data, I found much well-documented archeological and anatomical evidence suggesting humans of our type have been around for a very, very long time. This was surprising to me. The key is one of your own statements in *The Order of Man*, on the history of the debate about the australopithecines: "In the uproar, at the time, as to whether or not these creatures were near ape or human, the opinion that they were human won the day. This may well have resulted not only in the defeat of the contrary opinion but also the burying of that part of the evidence upon which the contrary opinion was based. If this is so, it should be possible to unearth this other part of the evidence."

This burial of evidence in human evolution is a continuing process, as far as I can see. I try to stay current with the literature, and I see signs of such burial everywhere. For example, Glenn Conroy says in a review of *Other Origins: The Search for the Giant Ape in Human Prehistory* by Russell Ciochon et al. (*J. Hum. Ev.* 1991, 21:233-35): "the authors themselves did not seem too excited about the discovery of five isolated *Homo* molars until after they learned the fossils were apparently older than they originally thought (about 500,000 years B.P. as determined by Electron Spin Resonance dating of a single pig tooth undertaken by Henry Schwarcz of McMaster University). In fact, their initial attribution of the teeth to *Homo sapiens* was altered to *Homo erectus* solely because of the dating results. This may raise some eyebrows since most paleoanthropologists realize that if taxonomy is based solely on geologic age, phylogenetic reconstruction simply becomes an exercise in circular reasoning." Now I have not looked into these teeth beyond this, but given my experience in compiling

Forbidden Archeology, I have to say that my suspicions are aroused. Is this one more example of anomalously old *Homo sapiens*?

Another example, this time from *L'Anthropologie* (1990, 94:321-34), is a report by G. Onoratini et al. on Acheulean stone tools being found in a Moroccan formation (the Saissian) previously regarded as Pliocene (cited are Choubert 1965, Taltasse 1953, Choubert and Faure-Muret 1965, Martin 1981). The discovery of advanced stone tools in one of the units of this formation caused the workers to redesignate it as Middle Pleistocene. They said about the tools: "Ceci constitue le premier element de datation anthropique d'une formation qui n'a pu etre datee par aucune autre methode." In other words, it was first and foremost the presence of the tools that caused them to redate the unit from Pliocene to Middle Pleistocene. It is taken for granted that advanced Acheulean tools cannot be earlier than Middle Pleistocene. Again, I have not looked deeply into this case but am somewhat suspicious.

So, getting back to W. W. Howells, I really do not know what would happen if one were to do a thorough search into the history of fossil primate discoveries. When I first started looking into hominids, I only knew the current hypotheses and their supporting evidence. I thought, well, this is the evidence. That's all there is. I was genuinely surprised to find that was not the case, and that a process of knowledge filtration had been going on. Over the course of 150 years, the triumph of certain opinions had led to the sinking of whole categories of evidence, somewhat unfairly. So maybe the primates go back further than Howells now thinks possible. Perhaps I would find numerous cases of the ages of geological formations being adjusted to fit the character of primate remains found in them. Or maybe not, but given my experience with hominid evidence, I cannot say for sure without first looking into the matter quite thoroughly.

I do intend to satisfy Howells's request for a counterexplanation, but I'd like to go step by step. And the first step is to convince him and others that there is a genuine need for a counterexplanation.

Perhaps you could satisfy an element of my curiosity about the Lufeng finds. You mention they are from Chinese coal fields, which you characterize as late Miocene. Generally, coal fields are Carboniferous or Permian. I know that there are coals that might be more recent than that, but, given my natural suspicions, I wonder just how the age determination for the Chinese fields was made. I hope it was not simply on the basis of the presence of advanced hominoids. But if you look at Chapter 9 in *Forbidden Archeology*, you will see that, in China, dating of geological formations simply on the basis of hominid morphology is quite common. Of course, I will eventually try to obtain the citations you gave, and track down other geological studies, but for starters I thought I would ask you.

Finally, I really liked your negative evaluation of cladistics and your critique of biomolecular studies.

P. S. If you think Lord Zuckerman would appreciate a copy of *Forbidden Archeology* (I trust he is still with us), please supply me with his address and I will send him one.

4.41.5. Letter to Dr. Charles E. Oxnard, July 18, 1994.

I have just finished reading *The Neandertals,* by Erik Trinkaus and Pat Shipton. In one of the chapters, I encountered a fleeting reference to W. E. Le Gros Clark meeting with "bitter opposition by one of his former junior colleagues, Solly Zuckerman" on the *Australopithecus* question. I am also reading a biography of Bishop Wilberforce, which has material about Huxley and Owen. All of this has reminded me of our conversation in Denver, after one of the sessions at the AAPA [*American Association of Physical Anthropology*] annual meeting. At that time, you alluded to some of the personal, financial, academic, and religious factors bearing upon the anatomical debate about the interpretation of the australopithecine fossils. You also suggested that this complex of factors extended back to the time of Huxley and Owen, and even beyond. If you have written anything on this, I should very much like to see it. And if you have nothing in print, I would be grateful for any clues you could give me, clues that would help me pick up the trail in the course of my own investigations. For example, any light you could shed on Lord Zuckerman's mentors and their rivals, and likewise the previous generation, going back to the time of Huxley and Owen, would be useful for my understanding.

4.42. Leland W. Patterson

Leland W. Patterson is a chemical engineer. He is also a recognized expert in lithic technology (stone tools).

4.42.1. Letter from L.W. Patterson, August 10, 1993.

In reference to your letter to me of July 27, 1993, thank you for sending me a copy of your book *Forbidden Archeology.* You have reviewed a large amount of literature on archeology and physical anthropology. I am afraid that I can only give you mainly negative comments, however. It appears that data have been selectively used in an unjustified manner to refute the theory of evolution. I am especially unimpressed by your reasons why anatomically modern man could have existed for millions of years together with other types of hominids.

You have quoted profusely from my publications on the criteria for identifying man-made lithics. However, you then selectively use a few of these criteria to support your conclusions. It is not enough to have a few flakes with bulbs of percussion or other attributes of man-made lithics. Multiple criteria must be applied to significant size samples. In particular, all accepted lithic industries by hominids have examples of manufacturing activities with many byproduct flakes.

In my opinion, failure to verify that stone tools were made and used by hominids in Europe during the Tertiary was because: (1) only a few of the proposed Eolith specimens resembled tools made by hominids, (2) the geological contexts of specimens were not always well-defined, (3) no definite sites were found where concentrated groups of flakes would demonstrate repetitive manufacturing patterns for stone tools, and (4) edge damage patterns on Eolith specimens only occasionally resembled patterns of retouch on stone tools made by hominids. Far from being a scientific cover-up, the Eolith issue was given full public debate.

Unlike many of the examples of Eoliths that you have noted, the Calico site in California has many attributes that indicate human lithic manufacturing, including thousands of percussion flakes with a flake size distribution pattern that has been replicated experimentally. Proposed very early sites in the New World, such as Calico and Toca da Esperanca, would require no change in the theory of human evolution, but would indicate that late *Homo erectus* or early *Homo sapiens* had entered the New World. The pre-Clovis time barrier in the New World has definitely been broken, but it will take many years of further data collection to obtain an interpretation of a Pre-Clovis cultural sequence in the New World.

Your book reminds me somewhat of Goodman's *American Genesis,* where much conventional archeological literature was used to give the superficial appearance of a scientific study. My background is training in the physical sciences, as a chemical engineer, and my response to your book is from that background. You are correct that there is some questionable dogma in many scientific fields, but there is also a clear trend over time toward refinement of theories.

My comments on your book may seem somewhat harsh to you, but your main conclusions that the theory of evolution is not correct and that anatomically modern man has existed for millions of years do not seem to be well-supported by the examples in the literature that you have cited.

4.42.2. Letter to Leland W. Patterson, October 1, 1993.

I genuinely thank you for your comments on *Forbidden Archeology.* I don't take them as harsh or negative, but as your honest reaction to the presentation made in the book. When a second edition of the unabridged book comes out (I don't know for sure when that will be), I will register your views and make clear that you do not support the conclusions that I myself have reached. [*I have not yet had time to revise and update* Forbidden Archeology, *but I regard the publication of Patterson's letter above as fulfillment of the promise I made in this letter.*]

I do have some questions for you. Is it possible that certain European eolith sites might have to be reinvestigated to determine whether or not they meet your criteria? It might be that neither the supporters nor the opponents of certain discoveries applied the tests that you deem advisable. For example, I doubt whether any careful studies of flake size distribution patterns have been performed.

Let's evaluate the late Pliocene Foxhall site in England (see starting p. 124), in light of your list of reasons for the rejection of European eolith sites.

1. *no resemblance of tools to those made by hominids*—according to Henri Breuil and Henry Fairfield Osborn and many other experts, the tools did resemble those made by humans.

2. *no well defined geological context*—see p. 142, where Coles, a modern worker, says that it is likely the layers in which the tools were found represent habitable land surfaces and that natural flaking forces were probably not operative. The reports and photographs I have seen show that the Foxhall finds at the 16-foot level in the Red Crag do have a very well-defined Late Pliocene geological context.

3. *no definite manufacturing sites*—Moir said: "The finds consisted of the debris of a flint workshop, and included hammer-stones, cores from which flakes had been struck, finished implements, numerous flakes, and several calcined stones showing that fires had been lighted at this spot." There are also reports of a human jaw having been found at the same level.

4. *edge damage patterns did not resemble retouch*—even strong critics of eoliths such as Hugo Obermaier were willing to accept the Foxhall finds. Louis Leakey, who studied the flakes at Calico and the Oldowan tools in Africa, also accepted Moir's discoveries in England, presumably including the Foxhall discoveries, which were among the most significant. I think your points 4 and 1 go together. The many experts who studied the tools definitely considered the nature of the retouch and found it artificial rather than natural.

I could perform this exercise with some other sites, showing they actually meet your criteria. But just sticking to this one, given the reported evidence, especially as detailed in #3, can you say that you are absolutely convinced there is no possibility Moir might have been correct?

I never said that the European eoliths were not publicly discussed. Obviously, I cited in *Forbidden Archeology* a lot of published material. But I think what happened to the eoliths is something like what might happen to Calico. I do not see that Calico has as yet won wide acceptance in anthropological and archeological circles. You and others have made a good case, but, as far as I can see, you have not prevailed. When pre-Clovis evidence is mentioned in authoritative studies and in NOVA type TV shows, it is the evidence that falls in the 25,000-year range, such as Meadowcroft and Monte Verde. Now what might happen is that the principal supporters of Calico will someday pass from the scene (as we all will). And if nobody picks up the ball and runs with it, that might be it. With no central figure to motivate them, the helpers and workers might lose interest. A new museum director might find that there are other priorities for volunteer work. Funding to keep the site open might dry up. No more articles will be written. And after 50 years, nobody at all will be paying much attention at all to Calico. That does not mean that the establishment won the debate. It just means they outlasted their opponents. And if somebody like me digs back into the history of the debate, many will say, "Why dredge that up? It was already publicly discussed and dealt with." Now, I hope that does not happen, but it could. And that is, in my opinion, what happened to a lot of the European eolith sites.

But, anyways, I really would like to hear what you specifically have to say about Foxhall. What if somebody told you that there was another Middle Pleistocene site like Calico in southern California, with hammerstones, cores, finished implements, numerous byproduct flakes, and signs of fire, along with a human fossil bone? Even if you heard through the grapevine that the New World archeology establishment had dismissed all this, wouldn't you at least want to take a look?

Again, thank you for taking the time to send me your comments about *Forbidden Archeology*.

Patterson did not respond to this letter.

4.43. Gene M. Phillips

Gene M. Phillips, an attorney, is founder and president of the Ancient Astronaut Society.

4.43.1. Letter to Gene M. Phillips, June 21, 1995.

In a letter you wrote to my colleague Richard Thompson regarding the world conference, you mentioned that you would like to have drafts of papers by June 15, for translation purposes and for possible publication in *Ancient Skies.* I am therefore enclosing a copy of my paper, "Forbidden Archeology: Evidence for Extreme Human Antiquity and the Ancient Astronaut Hypothesis." I apologize for sending it a little late and hope I have not caused any great inconvenience. In my slide presentation I will not be reading the text verbatim, but the slide presentation follows the paper very closely. I am looking forward to seeing you in Bern.

4.43.2. Letter to Gene M. Phillips, May 15, 1996.

Thank you for sending me the copies of *Ancient Skies* that featured my lecture at the Bern conference. I've received a number of letters in response. Since we last met, I've separated my activities from those of Richard Thompson. Although we had worked together productively for a number of years, of late there had been some tension, especially concerning authorship of *Forbidden Archeology* and the disposition of sales proceeds. As part of our separation agreement, Richard has surrendered all financial and copyright interest in the book, although his name will remain as coauthor. I am presently working out of Los Angeles. This past February, *Forbidden Archeology* was featured in an NBC television special titled *Mysterious Origins of Man*, hosted by Charlton Heston. Following up on that, I will be going on a university speaking tour this coming fall. I am also working on new book projects, but it's going slowly at the moment. I've also presented some papers at scientific conferences. Please keep me informed of news about the Orlando conference. With my best wishes to you and your wife, I remain,

4.43.3. Letter from Gene M. Phillips, June 21, 1996.

Thank you for your letter of May 15 and please excuse my delay in responding. Doris and I were in Peru from May 14 to June 1 on our Member Expedition. It was a fantastic trip. Thanks for informing me that you and Richard Thompson have parted ways. I am neither surprised nor sorry....

Speaking of Orlando, your invitation to deliver a lecture still stands—in fact, we expect you to be one of the principal speakers. Soon I will send you a letter with all the details of the Conference, but in the meantime, just know that it will be held on August 3 to 8, 1997, at the Sheraton Plaza Hotel at the Florida Mall in Orlando. Arrival day is Sunday, August 3, and departure day is Friday, August 8. If you have any questions about it at this time, let me know.

We saw your segment on the "*Mysterious Origins of Man*." You were great! I am pleased that you are getting the attention and coverage that your work deserves.

I am currently negotiating with a producer for a TV series on our subject and if it materializes, you will be contacted for one of the segments.

If you wish, I could give *Forbidden Archeology* another "plug" in *Ancient Skies*. Let me know if our members should still order the book from Torchlight Publishing.

Well, Michael, it was great to hear from you again and Doris and I look forward to seeing you in Orlando in 1997.

4.44. Dr. K. N. Prasad

Dr. K. N. Prasad is a vertebrate paleontologist and former director of the Geological Survey of India.

4.44.1. Letter from Dr. K. N. Prasad, November 26, 1993.

I received a copy of your book entitled *Forbidden Archeology*. Thank you so much. I find the entire gamut of human origins and prehistory has been brought out, in one single comprehensible volume, a task few people can achieve. I congratulate you and your colleague Dr. Thompson for writing this excellent reference book, which will act as a catalyst for further research on a subject of immense interest, not only to scholars and students but also laymen.

Several human episodes have originated in the Himalayan region for the past 10-15 million years. Valuable data on human origins in the form of dentition, skull and post-cranial skeletons have been lost or buried in the sediments, due to several tectonic episodes in the Himalayan orogeny.... I wish you and your colleague Dr. Thompson all the best.

4.45. R. René

R. René is the author of books on extreme scientific anomalies and conspiracy theories. He is perhaps best known for a book alleging that the manned moon landings were faked.

4.45.1. Letter from R. René, November 26, 1995.

[*Alternative science radio talk show host*] Laura Lee has just sent me a copy of *Hidden History* which I have read half of as of this minute. In appreciation I am sending you a copy of my book on scientific anomalies. Even if you are too busy to read it or totally unimpressed with my work, you should read the "Roll-Over" chapter. If this periodically happens, it would go far to explain why very few technical artifacts are found, despite the mass of proof that modern man has been prowling the Earth for millions of years.

When whole oceans scour their adjacent continents, not much can remain. This would also tend to confuse stratification levels. The argument used against anomalous human skeletons about evolution standing still is specious, at best. Do they say that about sharks, skates and the coelacanth?

Specifically, what about the *Megalodon,* whose teeth and jaw are identical with the white shark? Did you know that Australian fisherman periodically report 50-foot long white sharks? About 25 years ago a man named Pete, who owned the marina where I bought my boat in the winter, told me of inadvertently cruising up alongside of a hammerhead one day to discover that it was as long as his 40-foot Sportfisherman. He didn't drink.

The loss of anomalous bones, relics and artifacts is not accidental. There is a Masonic Lodge in London whose mandate is just such an activity. They use force, theft, persuasion and purchase to acquire to accomplish their goal. A man named John Daniels, author of *Scarlet & The Beast*, can be contacted because I believe he would be able to give you precise info on this lodge. I believe that David Childress, the author of many books on archaeology, may also have some info about this.

4.45.2. Letter to R. René, January 10, 1996.

I was very happy to receive the copy of your book *The Last Skeptic of Science.* I am very interested in all the topics that you discussed. I really do think that a lot of physics and cosmology current today is going to have to be drastically revised, and it looks like you are taking steps that make this clear. I am not well enough informed about mathematics and physics to judge myself whether all your calculations are correct, but over the next few years I would like to write something on those topics, and when the time comes, your work will certainly prove helpful. I've read other works about anomalous red shifts, gravity anomalies, orbital anomalies, etc., but you seem to have gathered a lot of that material into one volume, which is helpful.

I agree that catastrophic events in the past have had an effect on the fossil record. Recently, however, I read a book written by a diffusionist scientist back in the 1920s (*Children of the Sun* was the title, I forget the author's name). In that book he gave a lot of evidence of mining, foundations dug into solid rock, remains of old stone work, etc. in various parts of the world. How can you tell how old a foundation or mine dug into solid rock is? Could be quite, quite old. Interesting.

I also liked your reports of oversized sea creatures. There may be even stranger things lurking in the depths, according to some reports (sea monsters). Thanks for the connection to John Daniels. I will write him soon. I'm already familiar with David Childress. He sells my books in his catalog, *Adventures Unlimited.*

By the way, I am scheduled to be on an NBC television special, *The Mysterious Origins of Man*, which is supposed to air February 25 at 7 pm. The air date is subject to change, but you might want to watch out for it.

4.46. Jeffrey Rense

Jeffrey Rense is host of a radio talk show called The End of the Line, *in Santa Barbara, California.*

4.46.1. Letter from Jeffrey Rense, March 1, 1996.

Thanks very much for doing a superb program. Here is a tape. I had many responses ...all of them raves and several from devout Christians.

The NBC show was done very well. I think your next book will be extremely timely, given the way things are unfolding.

I hope to have you on as a guest again in the future.

4.47. Dr. Beverly Rubik

Dr. Beverly Rubik is a biophysicist and executive director of the Institute for Frontier Sciences. My correspondence with her is a good case study in networking.

4.47.1. Letter from Dr. Beverly Rubik, June 26, 1996.

Thank you for inquiring about IFS. The Institute for Frontier Science is a new nonprofit corporation that grew out of my work in academia. Although we have not yet printed a brochure, enclosed is a page describing the goals of IFS for your information. I've also enclosed a brochure detailing the first publication of IFS, an anthology of 14 of my papers, and a short biography that details my background. An electronic journal is underway.

One of the services IFS provides is a nonprofit umbrella organization for frontier scientists and scholars working in isolation. We are available to help them procure funding through our organization and manage their funds for a small fee.

Thanks for your interest in our work. I am unfamiliar with your book but interested in hearing about it, especially since I have recently applied for funding to study the Mayan pyramids in relation to subtle energies and parapsychology. Perhaps we can continue this conversation over e-mail.

4.47.2. Letter to Dr. Beverly Rubik, August 1996.

I am having my publisher send you a complimentary review copy of my book *Forbidden Archeology*. It has received a significant response from mainstream scholars, with major reviews in *British Journal for the History of Science, Social Studies of Science, American Journal of Physical Anthropology, Geoarchaeology, L'Homme, L'Anthropologie*, etc. It has also gotten a good response from the alternative science community, new age, anomalists, ufo/alien researchers (even though it says nothing about such things!), and others. *Forbidden Archeology* documents the physical evidence contradicting current ideas of human origins. This establishes the need for an alternative idea, which I will be presenting in my next book, *Human Devolution*. This alternative idea will be founded upon evidence from the paranormal and subtle energy fields of knowledge.

Anyway, I hope you will find *Forbidden Archeology* interesting. Perhaps you will consider sending me a copy of your latest book (the anthology of your papers) in exchange?

4.47.3. Letter from Dr. Beverly Rubik, August 5, 1996.

Thanks, I would be very much interested in your book. I believe you have my current address on McCallum St. in Philadelphia, because I remember a recent letter exchange with you. If you have any doubts about my address, please let me know.

When I get my electronic journal up on the Web, I will mention your book and give a short description of it. I will also send you a copy of *Life at the Edge of Science.*

I leave for Italy tomorrow and won't have time to mail it today, but I will get it out to you by the end of August.

4.47.4. Letter to Dr. Beverly Rubik, August 1996.

Have a nice trip to Italy. I myself am off to Australia, returning to the U.S. on September 7. I'll be looking for your book when I return. My publishing assistant says your copy of *Forbidden Archeology* is on the way to your McCallum Street address. Please let me know if it doesn't arrive.

4.47.5. Letter from Dr. Beverly Rubik, August 25, 1996.

Just returned from Italy and now I leave for the Bay Area. However, I will mail you a copy of my book on Monday before I leave, to the Bhaktivedanta Inst. on Venice Blvd. in L.A. Have a nice trip to Australia. As I recall, it's 14 hours from LAX to Sydney. I look forward to a discussion over the net on our books sometime this fall.

4.47.6. Letter to Dr. Beverly Rubik, September 1996.

Did you get your copy of *Forbidden Archeology*? You didn't say in your last message. I did receive your book, and it looks intriguing. I am traveling now, however, and don't have it with me. I meant to take it, but somehow forgot to pack it in the last minute rush. Yes, you are right. The flight from L.A. to Sydney is 14 hours. Makes for a long night. I spent most of my time in Australia on a working vacation—staying at a house owned by an English pop star on an island near the Great Barrier Reef. Encountered lots of exotic wild life right in the gardens of the home—giant fruit bats that explode out of the trees as you walk by in the early morning, wallabies that go crashing away through the underbrush, and those kookaburra birds with their wild, hooting calls, screeching parrots, etc. I guess you know all about those things, but it was my first trip down under. I am in San Diego now, but am going to be in Potomac, MD Sept. 23 to October 8. Actually, I'm very likely going with a friend to Philadelphia on some business on September 28, Saturday. It might be possible for me to get away for a short meeting with you, if you are going to be there and it fits with your schedule.

I don't have your address and phone number with me, so please include them when you reply.

4.47.7. Letter from Dr. Beverly Rubik, September 25, 1996.

No, I never received a copy of your book. Sounds like a nice vacation you had. I never saw the fruit tree bats. Sept. 28 I will probably be in New York City conducting an experiment on a Tai Kwan Do master who claims to do extraordinary paranormal feats. This time, we plan to have a magician present as well as several video cameras placed around filming him. However, this event may be canceled. As of now, there are some extenuating circumstances that make it unprobable. Therefore, give me a call.

4.47.8. Letter to Dr. Beverly Rubik, September 1996.

I apologize for the delay in getting my book to you. Upon inquiring, I found my publisher shipped it out weeks ago by fourth class book rate. It normally takes four weeks for things like that to get from California to the East Coast. I should have told them to ship it first class. Mea maxima culpa. Anyway, within the past two days, another copy was shipped to you by UPS. If the other copy eventually does arrive, please keep it or give it to a friend. And if you don't get the UPS copy in the next couple of days, please let me know. I hope your New York experiment went well.

4.47.9. Letter from Dr. Beverly Rubik, September 28, 1996.

Thanks for your response regarding your book. I will also post a review or summary of it in my electronic journal, *Frontier Perspectives in Science and Medicine.* The NYC experiment was cancelled due to personnel problems between the magician we hired and the psychic. Too bad! Anyway, I am in Phila. this weekend, so if you are traveling through, do give me a call.

4.47.10. Letter to Dr. Beverly Rubik, September 1996.

Too bad about the NYC experiment. My travel plans changed, and I wound up not coming to Philadelphia. I am going from here (Potomac, MD) to Atlanta for a conference on Vedic history. Then to Dallas. Then to Los Angeles, my base place, with no fixed plans for travel. But there are some agencies booking me for university speaking engagements. We shall see what happens with those. I hope you finally got my book. Please confirm that you got it (or still didn't get it).

4.47.11. Letter from Dr. Beverly Rubik, October 3, 1996.

As of Oct. 2, I have not yet received your book. Best wishes in your travels.

4.47.12. Letter from Dr. Beverly Rubik, October 15, 1996.

I got your book, which is quite a long, scholarly, and impressive amount of work with a very interesting theme. Didn't read it yet, of course, but wonder if your thesis actually conflicts with conventional science, in the sense that humans and other creatures might well have been here millions of years ago, but then there was a holocaust that wiped out all life. Then life could have started up again, once conditions were optimum, and evolution occurred all over.

So the origin and evolution of life punctuated by world holocausts that destroy all life is one possibility that is consistent with both your evidence and that of conventional science, although conventional science does not yet admit to such cycles of life on earth.

Anyway, I appreciate receiving your book and will mention it in my electronic journal that I am still working on these days. Should go on the web this winter.

4.47.13. Letter to Dr. Beverly Rubik, October 1996.

You got the book. Great. So now we can actually talk substance.

You "wonder if ...[the] thesis actually conflicts with conventional science, in the sense that humans and other creatures might well have been here millions of years ago, but then there was a holocaust that wiped out all life. Then life could have started up again, once conditions were optimum, and evolution occurred all over."

According to conventional science, life arose according to some random chemical interactions that occurred in a very specific environment. And then further evolution occurred according to random genetic mutations, with no teleological impetus. The odds that in a rerun you would get the same series of genetic mutations (millions of them) that yielded humans the first time would be infinitely small. So the evidence in *Forbidden Archeology* does pose some problems for conventional science.

If you don't feel inclined to leap straight away into "*Forbidding*" *Archeology*, you can try the Introduction, which I wrote to give a solid overview of the material for those who might not get around to reading the whole tome right away. Also, you might find the case of Virginia Steen-McIntyre in Chapter 5 interesting.

But getting back to conventional science, I admit that the evidence drawn from stones and bones can probably be accommodated by conventional science, if it is stretched far enough (and that would be pretty far from what it is today). But my interest in human origins goes beyond stones and bones.

If we really want to understand human origins, we first have to understand what the human organism is. If we leave it as it is—a simple mechanism—then conventional science can always offer some mechanistic account for its origin. But if we take into account the kinds of things you talk about in your book *Life at the Edge of Science*, then that opens up other possible accounts of human origins. That is the subject of my next book, which I am calling *Human Devolution.*

I will be proposing that the human organism includes physical, subtle material, and spiritual components, and these all need to be taken into account in any explanation of origins. In establishing what the human organism is, I will be reviewing evidence of the kind you are most interested in. As you can see from *Forbidden Archeology*, I like to be comprehensive and thorough. And I like to present things in such a way that they cannot be ignored.

Forbidden Archeology recently received attention in a major review article in *Social Studies of Science*. The authors of the article said the book represents a genuine contribution to the literature on paleoanthropology for two reasons—first it examines the historical material in unprecedented detail and, second, it raises important problems concerning scientific truth claims. I've also presented papers on the book before mainstream audiences, such as the World Archeological Congress, Russian Academy of Sciences, etc. A paper about a specific case in the book has also been accepted for presentation at the meeting of the International Union of Historians of Science, which will be held in Liege, Belgium, this coming summer.

But I also like to communicate with the alternative science community and the general public (I have appeared on about 100 radio and television shows, including *Sightings,* the PBS show *Thinking Allowed*, and an NBC television special called *Mysterious Origins of Man*).

I read your book *Life at the Edge of Science* today. I did not find much to disagree with, hardly anything at all. Or to put it positively, I thought it was a fantastic book, a great survey of the anomalous evidence that challenges current mechanistic paradigms in the fields of consciousness studies, bioelectric phenomena, and alternative medicine.

You didn't say exactly why Temple University terminated your position. Is that something you can share with me? Also, on page 14 you said that up to 1993 yours was the only such university-based center for study of frontier science. Are there now others?

Anyway, I feel fortunate to be in communication with you. At some point, I hope our paths will cross and we'll get a chance to talk. It looks like I will be going to Budapest for a small conference at the end of November. Otherwise, I will be in LA until January.

4.48. Vash A. Rumph

4.48.1. Letter from Vash A. Rumph, July 17, 1996.

After hearing your book *Forbidden Archeology* referenced on the TLC [*The Learning Channel*] program *Paleo,* I obtained a copy from the library and have been reading it. I must admit that I skim parts of it as I am not a scientist. It has been very interesting and the things that you point out are very logical. At this point I agree with you that the current theory of evolution is seriously flawed and in my opinion should be discarded.

Of course this does not eliminate the possibility that man was created by the process of evolution. One could postulate that hundreds of millions of years ago there was a common ancestor to both men and apes. I am not at all convinced that just because man came to his present form via evolution it means that he has to continually change. There are examples of animals that have existed for millions of years in basically the same form. Unfortunately, if this is in fact how man came to the physical form that he has today, I suspect that it will never be proven, for the records get much more fragmented as one goes back in time.

This possibility does not bother me and it is as good a theory as any other. I can also accept that man was created by God instantaneously. Either theory could be the work of God, as the Bible does not explain how man was created. In fact there is no conflict, in my opinion, between the theory of evolution and the Bible story of creation....

If your book stirs the interest of the general public and you organize them in any fashion, I would like to be included. I believe that it will take such a group to convince the defenders of the current theory to discard it and take a fresh look at all of the facts and start anew.

I would also like to know more about the Society for Krishna Consciousness if you would care to refer me to books that are about the Society.

Thank you for your time in reading this rather lengthy letter and I look forward to your next book.

4.48.2. Letter to Vash A. Rumph, August 4, 1996.

Thank you for your kind words about *Forbidden Archeology*. It is news to me that the book was referenced on the *Paleo* program on The Learning Channel. Can you recall any of the details?

Of course, you are correct that even if the current evolutionary account of human origins is incorrect, one could always propose that humans evolved earlier. But the evidence documented in *Forbidden Archeology* puts a human presence so far back, hundreds of millions of years, that it is hard to see how a viable evolutionary account could be reconstructed. In any case, the real point is that we need an alternative to the current version of human origins.

If you wish to help, one thing you could do is recommend *Forbidden Archeology* to others. It is available through most bookstores, and people can also get it by calling 1-800-443-3361.

I have asked the secretary of the International Society for Krishna Consciousness to send you information about the Society and its publications.

4.49. Kishor Ruparelia

Kishor Ruparelia was at the time of this correspondence General Secretary of the Vishwa Hindu Parishad, a worldwide Hindu cultural organization, in the United Kingdom.

4.49.1. Letter to *Back to Godhead* magazine from Kishor Ruparelia, August 2, 1993.

I have just received the May/June 1993 issue (Vol. 27, No. 3) of the magazine *Back to Godhead,* and I am writing with reference to the condensed form article on the book *Forbidden Archeology* written by ISKCON researchers Michael Cremo and Richard Thompson, respectively Drutakarma Dasa and Sadaputa Dasa.

Having read the article, I consider it very important that a meeting be held between the authors of the book and some of the Indian scholars who are in the USA at present to participate in a world conference organized by the VHP of America under style of 'Global Vision 2000' to take place in Washington DC on Aug. 6th, 7th & 8th.

I shall be very grateful if you will fax to me the contact telephone/fax numbers of the authors so that I can contact them directly to discuss possible arrangements of a meeting.

Alternatively, you may please pass on to the authors my humble and earnest request that they join us at the event in Washington where a meeting will be arranged without any great difficulty.

As you are perhaps aware, this event in Washington has been organised to commemorate the Centenary of the Vedantic Message Swami Vivekananda gave to the West at the World Parliament of Religions in 1893.

Inspired by the Vedic writings and encouraged by His Divine Grace A.C. Bhaktivedanta Swami Prabhupada, the scholarly authors have made a tremendous and painstaking effort to compile and compare umpteen evidences to make archeological scholars rethink about the predominant paradigm on human origin and antiquity.

We very much look forward to hearing from the authors either direct or through yourselves. A reply to this request may be sent direct to the undernamed persons in Washington to whom this letter is being copied.

4.50. George Sassoon

Author and computer expert George Sassoon is a graduate of Cambridge and son of renowned English poet Siegfried Sassoon.

4.50.1. Letter from George Sassoon, December 31, 1993.

I have obtained your book *Forbidden Archeology* from Arcturus Books Inc. of Port St. Lucie, FL and found it absolutely fascinating. I look forward to the companion volume which I understand is due next year (1994).

On p. 810, section A2.14.1, you ask for further cases of apparently high-tech objects found in ancient rocks—I believe that the archaeologists call these 'erratics', suggesting that there are enough around to merit a special technical term for them.

Many accounts of these have been published in the newsletter *Ancient Skies*, by the Ancient Astronaut Society, of which I am a member. I suggest that you contact the President, Gene Phillips, at 1921 St. John's Avenue, Highland Park IL 60035-3105.

There is also a German-language *Ancient Skies* with different content, published by the German branch of the A.A.S. at Baselstr. 10, CH-4532 Feldbrunnen/SO, Switzerland, edited by Johannes Fiebag.

I sometimes translate articles from the German edition which are published in the U.S. one. I remember doing one about an aluminum-alloy object found in a riverbed in Romania but cannot at the moment lay my hands on a copy or the computer file.

I am sure that Mr. Phillips and Dr. Fiebag would be glad to send you copies of items of interest to you.

With best wishes for success in your work.

4.51. Dr. Siegfried Scherer

At the time of my correspondence with Dr. Siegfried Scherer, he was director of the Institute for Microbiology at the Technical University of Munich in Germany. He is also a Christian creationist. Dr. Scherer did not give me permission to reprint any of his letters to me, but I will include here some of my letters to him, which will give some indication of the nature of our correspondence.

4.51.1. Letter to Dr. Siegfried Scherer, October 20, 1992.

Thank you for agreeing to review *Forbidden Archeology* in advance of publication, which is expected by January 1993. I am looking forward to receiving your impressions.

If you know anyone else who would be especially interested in receiving an advance copy for prepublication review, please inform me....

Yes, I would be very interested in seeing a copy of the book you edited, titled *Die Suche nach Eden*. (Ich kann Deutsch lesen, leider nicht sehr fliessend sprechen oder schreiben.)

4.51.2. Letter to Dr. Siegfried Scherer, December 12, 1992.

I hope you have by now received for prepublication review the complete typeset copy of my book *Forbidden Archeology*.

In addition to the review which you have kindly agreed to write for the German magazine *Faktum,* I wonder if you might also consider sending me a brief paragraph, summarizing your impressions of *Forbidden Archeology.* With your permission, I would very much like to use it on the jacket of the book.

Dr. Philip Johnson of the University of California at Berkeley, author of *Darwin on Trial,* has given a few sentences for this same purpose as follows: "*Forbidden Archeol*ogy is a remarkably complete review of the scientific evidence concerning human origins. It carefully evaluates all the evidence, including the evidence that has been ignored because it does not fit the dominant paradigm. Anyone can learn a great deal from the authors' meticulous research and analysis, whatever one concludes about their thesis regarding the antiquity of human beings." I am also expecting some brief notices from others.

The book is going to the printer at the end of December and the cover is now being designed. So if you are agreeable, please send a few suitable lines to me by FAX at your earliest convenience.

4.51.3. Letter to Dr. Siegfried Scherer, December 17, 1992.

Thank you for kindly sending a brief review for the jacket of *Forbidden Archeology*. Our editor/proofreader has made some small changes. Please let me know if they meet with your approval.

4.51.4. Letter to Dr. Siegfried Scherer, March 10, 1993.

After reading your book *Die Suche nach Eden*, I see in you a friend. I say this even though we represent worldviews that differ in some significant respects—the age of the earth, for example. But despite some great differences, we also have much in common. You and your colleagues, in your critique of modern evolutionary theory, are developing lines of argument and research quite similar to those with which I am familiar.

I found B. Steinebrunner's establishment of a connection between Darwin's evolutionary hypothesis and the evolutionary principles inherent in German idealism quite illuminating.

I was, of course, very impressed with the chapter by Sigrid Hartwig-Scherer on the paleontology and archeology of the paleolithic. Sigrid demonstrates that even if one confines oneself to currently accepted evidence, an evolutionary picture fails to emerge in any convincing fashion. This certainly opens the way for alternative explanations, such as the one you advocate (as well as the one I advocate).

In *Forbidden Archeology* my coauthor and I, in proposing an alternative, only go so far as to suggest the simultaneous coexistence of various hominid types, at various levels of culture, as a theory that accommodates the evidence better than the current human evolutionary theory. In a forthcoming book we will fully outline our alternative account of human origins in light of the Sanskrit Vedic texts.

The work we are carrying out in this regard involves, in part, studies very much like those of Werner Papke in "Ich suchte Gilgamesh." In examining the Sanskrit Vedic literature of India, particularly the cosmological sections, we have found some very good correspondences with modern astronomical data. This makes it possible to give some historical and scientific credibility to these literatures, which have been dismissed by Western scholars as simply mythological fantasies. We have not, however, confined our research to the Vedic literature, but have also looked into the Greek, Mesopotamian, Polynesian, and other cosmologies and have noted many correspondences.

Another area of research we find important in establishing an alternative view of human origins is the paranormal. You did not give any hint of this in *Die Suche nach Eden*, but I would think that in order to build a credible picture of the Biblical reality, including the angels, devils, heavens, hells, and all kinds of mystical occurrences, you would eventually have to offer a convincing case for the reality of such phenomena....

Not long ago, I happened to be at the library of the University of California at San Diego, doing some research, and I happened across an article by Sigrid in the *American Journal of Physical Anthropology*, the one about determining hominid weights from skeletal remains. It is very interesting to me that one can be a Biblical creationist and at the same time publish in the *AJPA*. In America, there is quite a sharp line drawn between the Biblical creationists and the mainstream anthropologists and archeologists. In Europe, is the situation somewhat different?

Forbidden Archeology is not on the same level as Sigrid's technical papers. But nevertheless, I would welcome receiving any comments Sigrid might care to make about *Forbidden Archeology*. If you could communicate this request to Sigrid, I would be grateful. Sigrid's comments, along with those of others, will be helpful in getting a wider reading for *Forbidden Archeology*. You may be interested to know that we are negotiating with a German publisher who has expressed a strong interest in the book. Of course, even if we come to an agreement, it will take some time for the book to be translated and published in German.

I was very interested to see in the bibliographies of the chapters of *Die Suche nach Eden* your own list of publications in the field of molecular biology. My coauthor and colleague Richard L. Thompson has published in some of the same journals and is very interested in this line of research. Some of his publications are:

A simulation of T4 bacteriophage assembly and operation; with N. S. Goel, *Biosystems*, vol. 18, pp. 23-45, 1985.
Simulation of cellular compaction and internalization in mammalian embryo development as driven by minimization of surface energy. With N. S. Goel and C. F. Doggonweiler, *J. of Theoretical Biology*. Vol. 48, no. 2, pp. 167-187, 1986.
Organization of biological systems—some principles and models, with N. S. Goel. *International Review of Cytology*. Vol. 103, pp. 1-88. 1986.
Moveable finite automata (MFA) models for biological systems I: bacteriophage assembly and operation. With N. S. Goel. *Journal of Theoretical Biology*, Vol. 131, pp. 351-385, 1988.
Moveable finite automata (MFA) models for biological systems II: protein biosynthesis. With N. S. Goel. *Journal of Theoretical Biology*, Vol. 134, pp. 9-49, 1988.
Biological automata models and evolution I: the role of computer modeling in theories of evolution and the origin of life; II: the evolution of macromolecular machinery. With N. S. Goel. Presented at the International Symposium on Organizational Constraints on the Dynamics of Evolution. Budapest, 1987. Published in *Organizational constraints on the dynamics of evolution*. G. Vida and J. Maynard-Smith, eds. Manchester University Press, 1989.

Richard has also coauthored a book with N. S. Goel, titled *Computer Simulations of Self Organization in Biological Systems* (London: Croome Helm, 1988). This book has a chapter showing the very severe problems in proposing evolution as an explanation for biological diversity. This point is also made explicitly and implicitly in the above articles.

I am now working on a book on biochemical evolution, along with a biochemist who became a member of our organization after reading some of our publications. The work on this book is only in the initial stages, so it will probably be about two years before it is printed.

I was thinking that it would be good for us all to get together sometime and compare notes. It is likely that Richard and I will be coming to Europe sometime this year, so if you are interested perhaps we could have some talks with you and your colleagues.

For your information, I am enclosing a copy of our publication *Origins*. This is not a periodical, but an attempt to communicate in summary form, for a popular audience, our entire program of research. *Forbidden Archeology* is an expansion of part of this program. With a small number of people, we are currently working in all the areas discussed in *Origins*.

Keep in mind that *Origins* was printed in 1984, so some of the material is a little dated. But it will give you some idea of the scope of our interests.

4.52. Dr. Robert M. Schoch

Dr. Robert M. Schoch, a geologist, is associate professor of science and mathematics at Boston University. His research indicates that the Sphinx is thousands of years older than its conventional age.

4.52.1. Letter to Dr. Robert M. Schoch, November 7, 1994.

For the past year or so, I have been working with Bill Cote on a proposal for a television special on human origins and antiquity, to be based in part on my book *Forbidden Archeology: The Hidden History of the Human Race*, coauthored by Richard L. Thompson. During our talks, Bill mentioned that you are also interested in this topic and might like to receive a review copy of the book. If you do not already have a copy, I would be happy to send you one. I am, of course, familiar with your work on the antiquity of the Sphinx. Have you done any more research into the question of the antiquity of civilization in Egypt?

4.52.2. Letter from Robert M. Schoch, November 21, 1994.

Thank you very much for your recent letter. Yes, I would most definitely appreciate receiving a review copy of your book, *Forbidden Archeology.* I am very interested in reading it.

Since you are in contact with Bill Cote, I suspect that he has kept you up to date on the progress of my Sphinx research. In a nutshell, not much has happened recently; I am still waiting for permission from the Egyptians to continue the work. Has Bill passed on copies of my articles (the major ones were published in *KMT, A Modern Journal of Ancient Egypt, Geoarchaeology* and *Kadath*)? If not, I will be happy to send you copies—just let me know.

I look forward to receiving the review copy of your book. Thank you.

4.52.3. Letter to Dr. Robert M. Schoch, January 3, 1995.

I am pleased to send you a review copy of *Forbidden Archeology* and would welcome your impressions of the book. It has been the subject of negative reviews in *American Journal of Physical Anthropology* and *Geoarchaeology.* Milder, yet still negative, reviews and brief notices have appeared in *Journal of Field Archeology, Antiquity, Journal of Unconventional History,* and *Ecology, Ethology & Evolution.* Additional, and, from what I hear, somewhat balanced, reviews are forthcoming in *Social Studies of Science, British Journal for the History of Science, L'Anthropologie,* and *Journal of Scientific Exploration.* A reviewer for *Isis,* journal of the History of Science Society, has also requested a review copy. And, as you might expect, the book has received many favorable reviews from nonmainstream, alternative science journals such as *Fortean Times.* The book is also out in an abridged popular edition called *Hidden History of the Human Race*, which has drawn considerable interest internationally. A German edition is already in print, and a number of other publishers have signed translation agreements or are in the process of negotiating them.

I have just returned from New Delhi, where I delivered a paper based on the material in *Forbidden Archeology* ("Puranic Time and the Archeological Record") at the World Archeological Congress 3. If you like, I can send you a copy.

I have not yet seen the papers of yours that you mentioned in your letter, and would very much appreciate receiving copies.

When I was on my way to India, and passing through London, I saw some letters in the *Times* of London responding to an article about the age of the Sphinx and the water erosion evidence. In case you have not seen the letters, I am enclosing a copy of them. I suspect that you have already answered in your formal papers some of the proposals made in these letters (such as the rainwater percolation idea). But if not, what do you say? Also, what do you know about the chlorine-36 rock exposure dating method? Does it require a particle accelerator? Do you have any opinion about its reliability? If it were somewhat reliable, it might be interesting to date some of the ancient temple structures in India, some of which are given traditional ages far in excess of modern historical ages.

4.53. Dr. R. Leo Sprinkle

Dr. R. Leo Sprinkle is a psychologist active in UFO and abductee research.

4.53.1. Letter from Dr. R. Leo Sprinkle, February 28, 1996.

Commendations to you for your work and wisdom, not only in the recent TV program but in the volumes *Forbidden Archeology* and *Hidden History*.

Last Sunday, my wife, Marilyn, & I watched the TV program & yesterday I received my copies of the two volumes. I look forward to reading your work, which—in my opinion—benefits both science and religion.

May we all experience more Love and Light.

4.54. Dr. Virginia Steen-McIntyre

Dr. Virginia Steen-McIntyre (known to her friends as Ginger) is a geologist. Her work dating the Hueyatlaco, Mexico, site at over 200,000 years old is documented in Forbidden Archeology. *Since the publication of* Forbidden Archeology, *Dr. Steen-McIntyre has revived her work on the controversial dating of Hueyatlaco. We have had an extensive correspondence, amounting to dozens of letters. I am including here only a few.*

4.54.1. Letter to Steve Bernath from Dr. Virginia Steen-McIntyre, October 30, 1993.

Steve Bernath was a research assistant for me and Richard Thompson. He made the initial contact with Dr. Steen-McIntyre.

I'm 4/5's through *Forbidden Archeology* and felt led to stop and write you and authors Cremo and Thompson, commending you for a difficult job extremely well done. What an eye-opener! I didn't realize how many sites and how much data are out there that "don't fit" modern concepts of human evolution. Somewhere down the line the god of the Vedas and the God of the Bible will clash head-on, and then there will be fireworks! Until then, the servants of both can agree on one thing—human evolution is for the birds!

On first reading, I found only one small error in the Hueyatlaco section. The publication *Science 80* (see p. 364) is not the AAAS *Science*, it was to be a popular science magazine, the name of which changed yearly—*Science 80, Science 81, Science 82*. It only lasted 3 years or so. You also should know that Cynthia Irwin-Williams died 6/91—prescription drug overdose ...a brilliant but unstable mind. Guess she just couldn't handle the stress of her own site questioning her own idol (human evolution).

I'm doing my bit getting the publicity out for your book. Have ordered a copy for the local library (wonder how long it will be before it, like critical skeletal material, "disappears"?). I'm also sending the book review that appeared in Sept./Oct. *Science Frontiers Book Supplement* to various friends and colleagues (almost 50 so far). I predict the book will become an underground classic. Whether it breaks into mainstream media is questionable—the Illuminati are tightly in control there....

How's this for an alternate hypothesis for modern man?

Modern man, wild man, apes, orangs, early hominids, monkeys, etc. *are* all related. We've all *de*-volved from something called "man" that used 100% of his brain instead of the less than 10% we now use. Instead of being the glorious end-product of natural selection, we are the dregs!

How ego shattering! But it does explain world myths, genetic and DNA studies, thermodynamics, occult (hidden residual knowledge), and it explains why the Illuminati must have human evolution—to convince mankind that fooling with the human genomes to make "super-man" is the next step in the evolution process towards godhood (they won't admit, of course, that they are trying to get *back* to the mind power of *Homo noeticus*—without God this time, of course). And I wonder how often this has been tried before—the tower of Babel (one magnetic reversal will confuse all computer language); mermaids, centaurs, satyres, wildmen, giants (experiments that didn't work out).

4.54.2. Letter to Steve Bernath from Dr. Virginia Steen-McIntyre, November 2, 1993.

Another point that needs clarification in the Hueyatlaco portion of the book *Forbidden Archeology*:

We did the '73 work under an NSF grant through the Laboratory of Anthropology, Washington State University. Roald Fryxell was on the staff of the lab, not an employee of the U.S. Geological Survey. Hal Malde was with the USGS (now retired), and came down on his vacation time. I was a part-time Physical Science Technician at the USGS at the time and also came down on my own time (I think).

We had some problems receiving payment for expenses we incurred, from WSU, after Fryx was killed in a car crash and we looked askew at each other. I now think that hotel employees helped themselves to cash we had deposited in the hotel safe to pay the workers (They would only accept cash.) Didn't dawn on us at the time that we couldn't trust the hotel staff—we were very naive.

You might see if you can get a set of reprints of the publications of

Jonathan Davis (late—died in a car crash). He was with the Desert Research Institute at the Univ. of Nevada(?) He was a tephrochronologist who took over as Fryxell's assistant after I married and moved to Puerto Rico in '67. He also was working on approximate dating methods and may have been on the lookout for early man beneath the dated tephra layer in Nevada. I wrote years ago asking for a set of reprints, but received none.

Received a call from Hal Malde yesterday (I had sent him notice of your book), telling me that we have to move the crates, Hueyatlaco monoliths (stabilized stratigraphic sections) and samples out of government storage. Haven't looked at them since they were collected 20 years ago: they spent many years languishing in a warehouse down in Mexico. Have no really safe place to keep them other than a rented storage shed. I'm talking 1/2 ton of material!

4.54.3. Letter to Dr. Virginia Steen-McIntyre, December 30, 1993.

Thank you for your letter about *Forbidden Archeology*. It does seem to be on its way to becoming an underground classic. William Corliss says it is selling well, and it is going out through other outlets to people interested in these things. Some standard academics in the fields of the history and sociology of science also seem to be taking it seriously. I am not going to mention any names, but one history of science scholar told me he is going to review it for a journal in his field and a major anthropology journal editor also wrote to me saying he was going to review it. I am very cautious about revealing any names at this point because of the danger of attempts to suppress the reviews if the word gets out prematurely.

It would be very nice if *F.A.* could get some attention in Christian anti-evolution circles.

As far as the magazine *Science 80*, etc., goes, I know it is not the same as *Science,* but if you look closely into the matter you will see that *Science 80* and the rest of the series was in fact brought out under the auspices of the American Association for the Advancement of Science (AAAS) in the 1980s as a popular science magazine to counter the Biblical creationist onslaught. After the pro-evolution "victories" in the courts the magazine was folded, but the AAAS then took a supervisory role over *Discover,* the popular science magazine put out by Time-Life, I think.

Thank you for clarifying the funding situation for the Hueyatlaco work and the affiliations of the researchers. When a second edition comes out, I will try to adjust the text accordingly.

George Carter has been writing to us about *Forbidden Archeology.* He had this to say about Hueyatlaco: "I am convinced that there is very old material there but it is not the bifacially flaked knives. The geomorphology is very tricky, and it seems to me that an erosion occurred that they missed in the field and that infilling put more advanced material at a lower level." George seems to have some resistance to having advanced implements at the 200,000 to 300,000 year level, but is prepared to find very crude implements at such levels. This conception of his tends to influence his reaction to finds at other sites as well. For example, he says "in terms of what we think we know of the lithic series in America, Clovis or Sandia points can not ... be of the 250,000 year time level." I wrote him back that if someone has such an idea, then they can always propose that the geology at a site has been misread. But I said that in the absence of any clear evidence that the geology has indeed been misread, then we should be prepared to adjust our idea of what the actual lithic series in America or anywhere else might be. So—just for the record—what do you have to say about the suggestion that you all might have missed an erosion and infilling at Hueyatlaco? I can imagine what your answer will be, but I would still like to have it in your own words.

In any case, I was happy to hear from George. He says he would like to show Richard and I around Texas Street sometime, and he even proposed looking for implements on the Pliocene shorelines around San Diego, if we can locate them.

I like your idea for an alternate theory of human origins. It is very close to what Richard and I will be putting out in our companion book to *Forbidden Archeology*. That reminds me: W. W. Howells wrote me a nice letter saying he was impressed with the book but that we should present our counter idea. Our counter idea is inverse evolution—i.e. that modern humans, hominids, apes, monkeys, etc., are the modified descendants of more advanced humanlike beings. That is the Vedic idea, as explained in various Puranas. I quite agree with you that Darwinian evolution is a big factor in the justification for population control efforts of various kinds.

Finally, I recently got a nasty note from Richard Leakey, who says *Forbidden Archeology* is pure humbug and that only fools would take it seriously. What an endorsement!

4.54.4. Letter from Dr. Virginia Steen-McIntyre, January 18, 1994.

Thank you for your informative letter of December 30. It caught me in the throes of year-end financial reports, a broken thumb for my elderly dad (88), and a new computer (my first) to learn how to use. I thought to wait to answer until I could dig through my files and give you some hard data, but decided that might be awhile!

Had to smile at George Carter's comment about the age of the bifacial tools at Hueyatlaco. Seems to me he's falling into the same pattern as his own critics. Why can't we just look at the data (*closely, critically*) then go on from there?

What I had hoped to do is find a field sketch the late Cynthia sent to Hal Malde way back in the early 60's. It showed the artifact-bearing beds as she mapped them. The stratigraphy of the beds is simple: the complexity comes in with the overlying, younger beds. We noted the same relationship in '73, once the trenches were cleared of water hyacinth. (The trenches were flooded by the Valsequillo Reservoir during the wet season.) The beds in the lower reaches were very durable, almost like adobe. Even after standing abandoned for 9 years, we could still see the lines marking Cynthia's stratigraphic units traced on the trench walls. When she was still speaking to me, she commented on the hardness of the sediment, even the layers that held the bifacial tools. As I recall, she said they had to remove the artifacts using chisels: no brush and trowel work here!

How much easier it would be for me if I could get a hold of copies of her data and trench diagrams! Fat chance for that. So, I'll have to work with what I have. Which is considerable!

Last month I "inherited" a ton of Hueyatlaco sediment samples and, most important, stratigraphic monoliths. Stratigraphic monoliths are stabilized columns of sediment taken directly from the trench walls. If they haven't been tampered with (they were crated and stored down in Mexico for several years), they represent the stratigraphy as it actually occurred at the site in 1973. There is the added possibility that I might recover an actual artifact from these monoliths if and when I work them up, especially the big one, weighing 500 lbs. or more, that Fryxell and I took from the artifact-bearing layers in the lower trench. Fryx thought he saw a worked edge protruding form the monolith face, which should now be nestled against a wooden backboard support, but left it in place because of our "spy". (We had found other interesting features, but when we came back to the trenches the next morning, they would have disappeared. Only in Mexico!)

Right now the cache of material is stored in an old chicken house. Will I ever be able to prepare those monoliths, look for debitage, soil and mineral weathering characteristics, fossils? Who knows. And would I be able to publish the findings if I did? There are some powerful folk out there that don't want this information to get out to the general public.

As an aside, have you read the original Sandia Cave report? The cave mouth is located a couple of hundred feet above the stream, I believe. A chore for the water carriers, unless at the time of occupation the stream was flowing at a *much* higher level!

Remember that bit about an editor editing my contribution to a tephrochronology symposium to delete the Hueyatlaco information? He backed down, for all except the table. In that table I gave the hydration date for a layer of volcanic ash from a sediment core taken in Mexico City. The layer was about 73 meters down in the hole, gave a hydration date that matched the Hueyatlaco ash, and was situated slightly higher than a grain of modern corn pollen. And no, it couldn't have fallen down the hole! I spoke with guy who collected it back in the 50's!

Any fellow who is snubbed by Richard Leakey is a friend of mine!

4.54.5. Letter to Dr. Virginia Steen-McIntyre, April 21, 1994.

Sorry for the delay in answering your last letter. For the past few months, I have been traveling in India and New Zealand. I also went to Denver, to attend the annual meeting of the American Association of Physical Anthropologists. I didn't realize until just now, when I decided to check a map of Colorado in my atlas, that you were so close by.

Let me know if you find any more stone tools or debitage in those stratigraphic monoliths from Hueyatlaco. That would be quite interesting.

Thanks for the information about the corn pollen found below the hydration dated layer matching the age of the Hueyatlaco ash.

Also thanks for your study on the grooved piece of tuff. It appears like more evidence for a human presence 250,000 years ago. From the report it is not clear where the specimen is from. What can you tell me about that?

Forbidden Archeology has been getting some attention from anthropology and archeology journals. See the enclosed. I have submitted a reply to Jonathan Marks, who wrote the review in the *American Journal of Physical Anthropology*. Let's see if they print it.

4.54.6. Letter to Dr. Virginia Steen-McIntyre, August 19, 1994.

I am sending you the latest review of *Forbidden Archeology*. Although the reviewer is unfavorable to our point of view, he is at least respectful of it. I am also including a copy of the letter I wrote to the reviewer.

As for that television show, I am still waiting to hear some final word on it.

The review was by Ken Feder in Geoarchaeology. *The television show was* Sightings. *Virginia Steen-McIntyre and I were to appear in a segment on* Forbidden Archeology.

4.54.7. Letter from Dr. Virginia Steen-McIntyre, September 9, 1994.

Dad sickened August 15th (day after 89th birthday) and died August 22. I'm in the final throes of a memorial Swedish smorgasbord for the folks' memory (expecting 100-150 people). So this is short.

On other side is what I sent the "*Sightings*" group. They will drop by for a live interview (after a long phone one) Monday afternoon. Never rains but it pours!

4.54.8. Letter from Dr. Virginia Steen-McIntyre, October 22, 1994.

Had planned a nice computer letter giving details on what's going on with me, but time is moving too fast, so decided on a "short version" while drinking yogi tea this early AM.

We're planning to attend the Calico site reunion the first week in November (the early man California site). George Carter will be there and Roy Schleman (soils specialist) and Lee Patterson (lithic specialist). Will bring the Hueyatlaco slides, photos and trench diagrams and maybe small samples of sediment (from samples I took back in my suitcase in '73).

Then, the second or third week, we'll be driving down to Florida to spread my folk's ashes at sea off the coast by St. Petersburg. Any chance of our visiting you either coming or going?

Learned only this month that Marie Wormington, "the name" in establishment early man views, is dead. Smothered in a fire in her house. I was just about to ask her for the Hueyatlaco data of the late Cynthia Irwin-Williams. I assume she inherited it at Irwin-Williams' death.

Of the seven people either intimately involved or potentially involved in Hueyatlaco, five are dead (two fatal car crashes, suicide?, one suffocation, one kidney disease). Of the two of us alive, Malde, a totally establishment type, has no interest in the site. Just leaves me! Getting a bit lonely out here!

Is there any way I could purchase copies of *Forbidden Archeology* at a discount (I *do* have a business license—for my bulk organic food business—very tiny). I want to send a copy or two (one for the University of Puebla) to Mexico to the daughter of the late Juan Armenta Camacho, who originally discovered Hueyatlaco and the other sites (and was put down by the Mexican establishment because of it) and to a young prisoner up in Montana.

I arranged for Dr. Steen-McIntyre to receive some complimentary copies of Forbidden Archeology *from the publisher, with the option of purchasing more at a big discount.*

4.54.9. Letter to Celine Armenta from Dr. Virginia Steen-McIntyre, November 22, 1994.

A note to tell you that I recently received three copies of the book *Forbidden Archeology.* I am sending you one by airmail today. It is a gift to you in memory of your esteemed father, Juan Armenta Camacho.

I wish to send another copy to the library at the University of Puebla, in his memory. Can you give me the name of a person to contact there?

The book has become very popular, especially in Europe. I believe it soon will be translated into Spanish. A smaller, less expensive edition of the book has also been printed. It is called *The Hidden History of the Human Race.* It is written for a popular audience. An advertisement for this book is included.

Your father's name is mentioned on page 354 of *Forbidden Archeology.* How I wish the authors had known of his great work, *Vestigios de Labor Humana en Huesos de Animales Extintos de Valsequillo, Puebla, Mexico!* I showed my copy to Dr. George Carter at a recent meeting at the Calico Early Man site in southern California. He was in awe. I believe he soon will write to you and request a copy. Dr. Carter is a very famous scientist who studies evidence for early man.

Is it possible to send a copy of this work of your father's to the authors of *Forbidden Archeology*? Those beautiful stone tools! The butcher marks on the bones! The green-bone fractures, and finally, those graved pictures and incised designs! The photographs do more justice than any written description can....

I believe the authors hope to bring out either a second volume or second edition of the work, in which they will incorporate additional scientific data.

Within the next week or two, the book will be discussed on a network TV show called "*Sightings*". It is a program that deals mainly with the psychic and paranormal. The media have discovered that much of the evidence for the age of mankind has been either ignored or covered-up by the scientific establishment. They "smell" a BIG story! Perhaps Juan and his work will finally receive the recognition they deserve!

4.54.10. Letter from Dr. Virginia Steen-McIntyre, November 23, 1994.

Dr. Steen-McIntyre enclosed a copy of her November 22 letter to Celine Armenta.

My thanks to you for arranging free and discounted copies of your book. I do suggest you follow up with a letter to Ms. Armenta. The press run for her father's monograph was small (1,000 copies) and it may be out of print. Some fantastic photographs in it. (If it is out-of-print, I'll bring my copy along on our trip to Florida so that you can at least make a xerox of it.)

We must postpone our Florida trip until after the first of the year. On our return trip from the Calico Early Man Site reunion, Dave found two complex computer jobs awaiting him: I found a jury summons for November 30. I'll contact you when we know more about our proposed schedule. Until then...

4.54.11. Letter from Dr. Virginia Steen-McIntyre, November 23, 1994.

This letter was addressed to Rob Bonnichsen, George Carter, Michael Cremo, Dave Love, John Montagne, Hal Malde, George Otto, Dee Simpson, and Henry W. Smith. All are researchers concerned with North American archeology.

Lots of water under the bridge since last I wrote to you collectively. My father died August 22, shortly after his 89th birthday party. The book closed on a life well lived.

September 12 I was interviewed by a TV team from "*Sightings*", a far-out network show on the psychic and paranormal. They are doing a segment on *Forbidden Archeology* and got my name through it. (The book is quickly becoming an underground classic, especially in Europe, and has been/is being translated into several languages.) The segment should air (finally!) late at night the first or second Saturday after Thanksgiving, at least in the Denver area.

Earlier this month Dave and I took a flying car trip to attend the 30th reunion at the Calico Early Man Site in the Mojave. A great time. Saw old colleagues, got caught up on what's new in the early man field (lots!) and collected a sample of ca 145,000 year-old Lake Manix ash to study "one of these days". (The ash is in the lake beds: lower, older layers contain stone tools.) Scotty Mac Neish gave an excellent presentation on his Pendejo Cave (New Mexico). Firm 40,000 year dates; an old fire hearth lined with mud that still bears the impress of human fingerprints; even human hair! If poor Vance Haynes doesn't recant his "12,000 years, no earlier" for human presence in the New World, he just may be laughed into retirement.

Learned only recently of Marie Wormington's death in a house fire. She was Cynthia Irwin-Williams's mentor, and I had planned to write to her and find out the whereabouts of Irwin-Williams's Hueyatlaco trench diagrams and field notes. Am also looking for something Irwin-Williams published on the site, after we did our work. She wrote it in conjunction with a French statigrapher, according to MacNeish. I gather she published a date of ca 25,000 years. (Wouldn't know from actual experience, as she never sent me a copy or let me know she was doing such a thing. Grrr!) There was a ca 25,000 year date she liked, using carbon from snail shells, the shells associated with a stone flake. Uranium series dates also gave a ca 25,000 year age. Only trouble is, the collecting site is in Barranca Caulapan, 5 km to the east of Hueyatlaco, and in sediments much younger (using as criteria the extent of weathering of associated ash layers). An awfully cold trail, but then I'm awfully stubborn. Must be all that Viking blood!

4.54.12. Letter to Dr. Virginia Steen-McIntyre, January 5, 1995.

I have just returned from India, where I delivered my paper "Puranic Time and the Archeological Record" at the World Archeological Congress 3 in New Delhi. It caused a minor sensation, especially among the Indian archeologists. There is a good chance the paper will go in the proceedings, which are published by Routledge, a big British firm. I also learned from an Australian archeologist named Tim Murray that he is writing a review of *Forbidden Archeology* for *British Journal for the History of Science.* Murray said he recommends the book to his graduate students, saying "if you want to make a case for extreme human antiquity, this is how to do it." So the reverberations are continuing to spread.

The results of my fall media tour for *Hidden History* were 38 radio shows and 12 television shows. Actually, I am still doing lots of radio shows by phone from here in Florida. January 5-13 I will be out of town. And then at the end of January I will be going back out West, to prepare a slide show for an auditorium lecture at the Seattle Center in Seattle. The producer of one of the radio shows I was on also promotes auditorium lectures. He liked how I did on the show and offered me a booking, so I took it. Then there might be some more media engagements. That will probably keep me busy through the first week of February. Then it looks like another trip to India, which might last through the first part of March. After that I plan to be in Florida fairly continuously for some concentrated research and writing.

On the foreign front, we just got an offer from a Mexican publisher to do a Spanish edition of *Hidden History* for Central and South America. *Hidden History* was also displayed at the Frankfurt Book Fair, and we got offers from Swedish, Russian, and Romanian publishers.

I have just written a letter to Celine Armenta, asking if she could send me a copy of her father's book. By the way, how old would the marked bones from Valsequillo be? As old as the Hueyatlaco tools?

Oh, if you would like to go on some radio shows, I think I could arrange it. Laura Lee in Seattle is very much interested in scientific anomalies. She has a nationally syndicated show. I have been on it a few times. I mentioned you as a possible guest, and she was very favorable to the idea. Let me know if you would like to try it, and I will advise Laura. Another possibility is the Bob Hieronimus show out of Baltimore. His show features alternative science ideas almost exclusively. I think you would like it. Both Laura and Bob are quite serious in their approach to these matters, and I think you would be very comfortable with them. Anyway, let me know, and I will do what I can to help you get on their shows.

4.54.13. Letter from Dr. Virginia Steen-McIntyre, January 19, 1995.

I recently received a copy of *Vestigios de Labor Humana en Huesos de Animales Extintos de Valsequillo, Puebla, Mexico* from Celine Armenta. Some of the scribed pictures of animals on bones of extinct animals are fantastic! If you've not seen the book (actually soft-cover, letter-size, 125 pp.) I can send it to you to copy (and send back. It's rare). Let me know.

Received a call from Bill Cote re a film while I was out. Who is he? He left a phone number, but I'm too low budget to place long distance calls!

Our *Sightings* interview certainly went fast! They started reruns...the week after, so I may see us again!

I called Dr. Steen-McIntyre to let her know that Bill Cote was a television producer who wanted to include her in his NBC television special called Mysterious Origins of Man. *Dr. Steen-McIntyre agreed to go to Mexico with Bill Cote to do some shooting at the Hueyatlaco site. I had asked Bill Cote to contact Dr. Steen-McIntyre.*

4.54.14. Letter to Dr. Virginia Steen-McIntyre, February 15, 1995.

I'm sorry I missed you and Dave when you stopped by in Alachua. Thanks for dropping off the copy of Juan Armenta Camacho's book. I have read through it carefully, and it seems quite good. From what I gather, all of the incised and carved mammal bones are from the Tetela formation, which he says is about 200,000 years old. Is that correct, according to your understanding?

I have made a xerox copy of the book and am returning it to you with this letter.

I really would like to document his whole story. If when you go down you could interview Celine and get all the details, plus any copies of relevant papers, photographs, etc., that would be great (I am assuming you would not mind sharing them with me). Especially important would be the record of how the scientific establishment in Mexico shut down the site, confiscated the fossils, and so on.

I agree with you about the Ica stones of Dr. Cabrera. He testified that the carved stones come from a cave in Peru, but the location of the cave has not been revealed. So I think one may be justly cautious about accepting the authenticity of the stones. But one might be equally cautious about rejecting them. Richard and I are at such a point. We find them interesting, and are willing to consider them as genuine, but we do not have enough information to warrant publishing anything about them—other than that they exist and are worthy of investigation. Same with those dinosaurlike figurines. Bill Cote seems to be attached to them.

I've learned that we have recently come to terms with a major Mexican publisher who wants Spanish translation rights to *The Hidden History of the Human Race* (the short version of *Forbidden Archeology*). The negotiations have been underway for some time, but it looks like it's been settled. We are just waiting for them to send back the signed contract and their check for a small advance.

When are you going to Mexico? I wish I could go. If it is all right with you and Bill Cote, perhaps I might go down there at my own expense, just to "see the sites" with you.

4.54.15. Letter from Dr. Virginia Steen-McIntyre, March 1, 1995.

Thank you for your letter of February 15 and for returning [*Juan*] Armenta's monograph. I sent it on to Bill Cote yesterday.... If you compare Juan's map (his Fig. 1) showing the location of his collecting sites with Fig. 2 in the Hueyatlaco paper which shows the distribution of the Tetela Brown Mud (shaded; dated by the pumice it contains at ca 250,000 years), you'll note that his sites are topographically lower. In most cases, since the area is one of active erosion and not deposition, and the artifact beds are primarily flat-lying stream gravels grading into fine-grained overbank deposits, lower means older. Of course, once the specimen is removed from the deposits which surround it, dates become suspect. Fortunately, the uranium-series method should be able to date the bone directly....

I'll be in Mexico March 18-24 according to present plans. Would love to have you come along if it's ok with Bill. Perhaps we could squeeze in a day when Bill is working on the equinox to drive down to Puebla and interview Celine Armenta. She has not yet replied to my letter of January 25. (Did I send you a copy of that one?)

4.54.16. Letter to Dr. Virginia Steen-McIntyre, March 15, 1995.

Although I was dreaming of perhaps coming down to Mexico, some commitments I had made are making that impossible. I am scheduled for an interdisciplinary conference on science and culture being put on by Kentucky State University. I had proposed a paper on "The Impact of *Forbidden Archeology*" in various knowledge communities. The conference starts March 30 and I haven't done anything about the paper, so I've got to get working on that. I also have to prepare abstracts for two other papers, due April 1 and April 3. Last week I gave three classroom lectures at Florida State University—in classes on human evolution (see enclosed article), epistemology (knowledge and belief), and anthropology. And last night I was on a nationwide radio program (150 stations) for three hours. Anyway, have a good trip to Mexico, and please write when you get back.

I liked your letter to Senator Campbell. If he expresses some interest in the book, let me know. If you are fairly certain he will actually get the book, I think I can convince Alister Taylor to send you a review copy for him.

4.54.17. Letter from Dr. Virginia Steen-McIntyre, April 11, 1995.

Thanks for the letter of March 15. (You forgot to include the article on human evolution given at Florida State University.) Good luck on all your coming talks!

My trip to Mexico last month went very well. The TV crew interviewed Celine Armenta for several hours on Sunday afternoon. She had kept a lot of her father's correspondence and the original photos of the Figures from his monograph. She couldn't remember how Armenta's finds got from Puebla to Mexico City. She *did* remember that men with guns came down to the site to frighten the workers. They wanted the workers to say that either Armenta or they had planted the artifacts. Out of *ca* 60 men, 3 signed (bribed?); the rest refused to do so. The government believed the 3 men, ignored the 57.

Celine still had one small bone in her possession .... and I held it in my hand. Hard as stone, and as dense. The bone with the scribed mastodon (Armenta's Figs. 64-74) was featured in a *National Geographic* article (Sept., 1979, Vol. 156, No. 3). I had forgotten. The *Geographic* photographed the artifact (p. 350), but was still quoting 20,000 year dates.

Celine's accent is quite heavy, but understandable. You may want to telephone and speak with her, or make a date to interview her.

Monday we drove down to the Valsequillo Reservoir area and found the Hueyatlaco Site on the Tetela Peninsula. The trench walls were still standing, vertical, after 22 years! Shows how indurated that sediment is. The upper walls were exposed, the ones containing the ash and pumice layers that were dated using the zircon fission-track method. The lower levels, the ones containing the artifacts, were under water at the time. (The reservoir is the open sewer for the City of Puebla. You wouldn't want to put your hands in it!)

We ran into the widow of Felix Morales, Cynthia's crew boss and devoted friend, while we were there. She didn't remember the episode with the guns, but she was only a young teen at the time, and Felix may not have told her. (He was much older than she.) We do have Armenta's story of the whole thing (notarized, I believe) written down in the files that Celine has kept.

Tuesday they filmed my interview in the presidential suite of the hotel (classy!). I was able to get across some of the material I had prepared, but the interview was slanted for human interest (Brilliant woman scientist loses career for telling the truth, etc.). They still have all my written material, charts, diagrams, illustrations, slides, and plan to do a lot of voice-overs. What comes across probably will depend on how big a time slot they will allow: They are filming 3-4 other controversial sites. The finished program will be an hour long, and will air late this year. *Mysterious Origins of Man* is the working title.

P.S. Still have not received the book.

The book was The Hidden History of the Human Race, *the abridged popular edition of* Forbidden Archeology.

4.54.18. Letter from Dr. Virginia Steen-McIntyre, May 1, 1995.

Thank you so much for the copy of *Hidden History of the Human Race.* I like the jacket design! And thank you also for the xerox of the newspaper article. I made the *Florida Flambeau!* Woo Woo! [*The* Florida Flambeau *is the student newspaper at Florida State University, Tallahassee.*]

Wanted to tell you that I've sort of located Cynthia Irwin-Williams' research notes. They are in a locked file cabinet at The Desert Research Institute in Reno, her last place of employment. Call it luck? I had written them last November, requesting information on her files. As usual, no reply. Then, in March, an old buddy of Dave's from the USGS stopped by. He lives in Reno, and turns out his lady-friend...works for DRI. I wrote to her. She called me back. Cynthia's co-executor is Dr. Fred Nials, a geomorphologist there. Fred and I were in graduate school together thirty years ago. I've written him requesting information. Fred doesn't have the key to the cabinet.... George Agogino does. George is a retired archaeologist from Eastern New Mexico University in Portales, where Cynthia worked before moving to DRI. He is the one that told me, during a phone conversation, of Cynthia's...[*death*]. No mention of the key at that time, or the filing cabinet, even though he knew I was looking for her Mexico data.

We'll see what develops. I would especially like to photograph the vinyl casts of the artifacts that Cynthia's mother, Kay, made at the site in the early 60's. I'd also like to be able to trace the various artifact-bearing beds across the site plan.

Also enclosed is a first-draft manuscript of an article I'm working on. I hope to see it published in a Christian magazine. My spirit has been very heavy these past few weeks, thinking of what is in store for conservative Christians and how unprepared they are for it. I felt a strong urge (I call it the prompting of the Holy Spirit) to write down much of what the Lord has been showing me since 1982. It concerns just the threat from secular authority. The spiritual battle is something else again.

4.54.19. Letter to Dr. Virginia Steen-McIntyre, May 9, 1995.

Thank you for keeping me up to date on new developments with the Hueyatlaco case. In the copy of your letter to George, you said you enclosed for him a copy of Cynthia's trench diagram. Could you also send me a copy? It will be very interesting to see how George Agogino responds to your request for the key to Cynthia's files. Regarding your article for the Christian magazine, I quite agree with the ideas you express. I hesitate, however, to get into a big discussion about this because it can get frightening (what the government is doing and why). I'll just say that science, with its theory of evolution, is providing the intellectual framework for what the government does. A body is controlled by its brain. The government is like the arms. Big materialistic science is like the brain. So if we can get the brain cured, that will automatically have some results on the government side. The odds against success are tremendous, but if it is God's will, then success may come. I'll repeat my offer to help you get more media coverage. As I said, I think there are a couple of good radio shows that you could go on, if you so desire. You could do the shows by phone, right from your home. Just let me know, and I'll see what I can do. Regarding books, please remember that if you need any copies of *Forbidden Archeology* or *Hidden History* to send to professional colleagues, you can get them free of charge from Alister Taylor in Badger, California. To save postage, you could send your letters to him and he could mail them out with the books from Badger. If we could get some comments in writing from such people, that would be good.

4.54.20. Letter from Dr. Virginia Steen-McIntyre, May 12, 1995.

Here is a copy of Cynthia's southernmost trench diagram, requested in your letter of May 9. Another copy should have been sent to Steve Bernath long ago.

Yet not a word from Nials or Agogino. Perhaps they are consulting?

I won't be interested in radio interviews until the NBC film comes out, sometime late this year. I will be too busy "clearing the decks" for more work on Hueyatlaco—to the point of us building a 3-car garage with work/storage space above where I can spread out the stuff.

4.54.21. Letter to Dr. Virginia Steen-McIntyre, June 21, 1995.

Bill Cote said he sent you a copy of the Sandia Cave article from *The New Yorker*. What do you think? It appears that there were two levels of artifacts. One from just below the stalagmite layer (the Folsom points, with mammal bones) and another from below the layer of yellow ochre (the Sandia artifacts, also with mammal bones). It is the yellow ochre layer that has the uranium series date of 300,000 years, according to this article. Also, according to this article, all the animal bones have carbon dates of less than 14,000 years, but it does not say if the bones were from below the ochre layer. Then there is the testimony about rodent burrows. The ochre, however, had to be laid down during a period of wetness. When could that have been? What is the date for the stalagmite layer itself? There is no mention of that. You had said in a letter that there was a uranium series date for the stalagmite layer of about 250,000 years. Perhaps that was really for the ochre layer? But if it was for the stalagmite layer, then that Folsom point stuck in the stalagmite could be quite old. I guess they want to say all the animal bones and tools were carried in by rodents. But I would like to know how big the bones and tools are and if they were perhaps too big to have been carried by rodents like that, especially little pack rats. Anyway, I would like to hear your thoughts. Maybe you have some contacts that could shed further light on the matter.

4.54.22. Letter from Dr. Virginia Steen-McIntyre, June 30, 1995.

Thanks for your June 21 letter. Now more than ever I wish our garage/storage/office space was built and that I and my files were comfortably ensconced in it! (The contractor is preparing an estimate this week.) As it is, my files and reprints are all packed away in boxes in various parts of the house. Hence, many of my comments here will have to be from memory (not the best).

See Hibben's monograph on Sandia Cave. I don't recall much signs of reworking of the sediment layers shown in his photographs—quite the contrary. If yellow ochre denotes moist conditions, much more dripstone! That would suggest a glacial maximum. I thought the grad student I spoke with referred to the stalagmite layer when he mentioned the 250,000 year date. Perhaps there are two dated units? The dripstone, if that was truly the layer he referred to, was dated by Schwarcz at McMasters. (No, I never received a reply for more information, either from Haynes or Schwarcz.) And weren't Kirk Bryan's charcoal dates from the fire-pits *greater than* dates? (Could as easily be 250,000 years as 17,000 years!)

Look up Haynes' and Agogino's 1986 paper. I believe there wasn't much left to excavate; just skins of sediment, including the pack-rat middens, clinging to the walls. One would expect mixing in such a situation.

Why was the "more promising-looking" Davis Cave sterile? Was it lower in the landscape? If so, there's a good possibility that it had not been exposed by erosion of the local stream at the time Sandia Cave was occupied.

And what about the fluted artifact from *within* the stalagmite layer?

Pull out your *Ancient Man in North America* 4th Ed. by H.M. Wormington. Look up Sandia Cave.

> p. 85-86. (Under the dripstone). "Below this top layer was a crust of calcium carbonate that sealed off the lower levels. This varied from less than half an inch to six inches in thickness. Underneath the crust lay the level known as the Folsom layer because of the occurrence of Folsom artifacts within it...Hibben stated that, in outline, three of these were much like Folsoms, but [the] two that are illustrated exhibit only slightly concave, almost straight bases, and lack the ear-like projections of the true Folsoms. They more nearly resemble Plainview points....Another unfluted lanceolate point narrows markedly toward the base and is leaf-shaped. It is strongly reminiscent of one of the specimens found associated with mammoth in Mexico..." (*And what about those older Mexican sites? No absolute dates. Dated roughly (and I*

> *believe, incorrectly) by correlation over long distances 50 years ago. In well-developed red soils with Hueyatlaco fauna. Mid-Pleistocene would be my guess.)*
>
> p. 253. "...there are extremely striking resemblances between Solutrean points found in western Europe and the single-shouldered Sandias and the laurel-leaf points found in North America."

Look up Ventana Cave (Wormington, p. 178). Leaf-shaped and fluted points. Extinct animals like those found at Hueyatlaco. (With a "volcanic zone" no less! Ash? Pant, pant, pant.)

Angus, Nebraska site (Wormington, p. 43). A fluted point associated with an articulated mammoth skeleton in mid-Pleistocene sediments. "forgery".

I believe what we have here is a bit of intentional muddying of the waters. [*They say*] Don't bother to look in the Americas for the history of Early Man. [*But*] That's where the evidence lies!

If my letter to *The New Yorker* makes it into print, I'll be surprised. For the above reason. The best I can hope to do is to reach Hibben. (Dave just came in the door with a letter from him! Copy enclosed.)

Still no word from Desert Research Institute .... So there sits Cynthia's field notes in that locked file cabinet, with George Agogino holding the key to the lock. Her data might just as well be on the moon.

I've another option possible; a slim one. Copies of all Cynthia's data should, unless they have "disappeared", be on file at the INAH offices in Mexico City. There are new folks in charge there now. Perhaps I'll be given an opportunity to examine and copy them. Neil Steede, a new friend, will send me samples of the Xitli ash at Cuicuilco when he returns from an archeological conference at Quintana Roo in mid-July. When I have a place to work, I'll run it. Want to see if even a small part could be from the TBJ tephra of El Salvador. I'll make a (free) report on what I find and give it to him to pass on to INAH. May open some doors.

4.54.23. Letter to Dr. Frank C. Hibben from Dr. Virginia Steen-McIntyre, June 30, 1995.

Thank you so much for your letter of June 28, just received. I was deeply gratified. Other than kind letters and notes from George Carter and Neil Steede, contact of any kind with archaeologists has been rare indeed!

I've taken liberty to send copies of your letter to George and Neil, and to Michael Cremo and Bill Cote. Michael is one of the authors of

Forbidden Archeology; Bill is doing a documentary on controversial archaeologic sites for NBC. Both men are ferreting-out examples of media hanky-panky when it comes to Early Man data. Your treatment by *The New Yorker* is a classic example.

My husband Dave and I could drive to Albuquerque on short notice to discuss the dating methods used at Hueyatlaco—and Sandia Cave! I could bring my slides and trench diagrams along for an informal presentation. (Necessarily informal! I have only recently returned to the scientific swing of things after caring for elderly relatives for eight years.)

Perhaps you could tell me exactly what the problems are with Hueyatlaco? (Other than it questions pet theories, of course.) In the 22-plus years since we excavated at and dated the site, *not one* archaeologist, including Irwin-Williams, has contacted me to criticize our work. They just say, "It can't be!" As a result, I don't even know what the "charges" are against the site, or we geologists who dated it. All I know is that the only work I can get now is as a flower gardener in a local nursing home!

Again, thank you for writing. I hope to hear from you soon.

4.54.24. Letter to Dr. Vaughn Bryant from Dr. Virginia Steen-McIntyre, August 13, 1995.

George Carter suggested I write to you to learn if it is possible to apply for an adjunct professorship in your department. Mainly so I can "belong" somewhere, now that the Hueyatlaco site is coming into prominence once again. I never thought I'd see the day. For the past several years I've been out of science entirely, caring for family members. Then, the site was mentioned in the books *Forbidden Archeology* and *Hidden History of the Human Race*. The TV program *Sightings* picked up on it from the books, and a group doing a documentary on controversial archaeological sites for NBC learned of it from them. The latter group brought me down to Mexico in March for location and interview shots, my first visit to the site in 22 years. Their film, provisionally titled *Mysterious Origins of Man,* is due to air late this year. Hueyatlaco will be one of several sites featured.

I imagine you are familiar with Hueyatlaco. Dates of a quarter-million years (actually more like 350,000) from butchered bone (uranium series) and overlying ash layers (zircon fission track; tephra hydration). Archaeologists very unhappy because the upper levels (still beneath the dated ash layers and associated with the dated butchered bone) include well made bifacial tools. Cynthia Irwin-Williams was so upset she broke communications with me entirely before she died. To this day, I don't know what "charges" have been brought against the site or our geological work there. No one has had the courtesy to tell me. All I know is that the site killed my career. That's one reason the NBC folk are so interested.

At this date, I don't plan to start my career all over again—I'll be 60 next year—but I would like to finish up loose ends. Especially, I want to describe (or, preferably have someone else describe) the ton or so of stratigraphic monoliths and samples taken from the trench walls at Hueyatlaco in 1973. Right now they are resting, in their original crates, in a rented chicken-coop here in town. Then I'd like to publish the tephra hydration curves for a series of dated Yellowstone ashes I ran for Gerry Richmond at the USGS back in the mid-70's, a beautiful set of curves for tephra layers in the ca 40,000-1,000,000 year range. Great potential for use as seat-of-the-pants dating layers for much of the West. Then there is that ash in glacial Lake Mojave, the 150,000-year-or-so one, the one with the stone flakes beneath...

I would need no lab space and no equipment. I have my own microscopes, sieves, etc., and we plan to build a storage room above our proposed three-car garage where I can work. I do need affiliation in order even to apply for a research grant, at least for most foundations and organizations. I need affiliation to purchase the chemicals I will need, and to publish results. How lovely it would be to be paid once again for my research (the last time was 1978)! However, if I can't find support, I plan to use part of my inheritance to finish up, at least my part of it.

I've enclosed an updated vita along with a copy of the Hueyatlaco article, my latest tephra paper, and some correspondence re the Sandia Cave article in *The New Yorker*. I'd be happy to answer any questions about the vita, and to do an informal show-and-tell talk on Hueyatlaco for your department.

4.54.25. Letter from Dr. Virginia Steen-McIntyre, September 3, 1995.

George Carter has been after me recently to submit my vita to various universities to see if I can "belong" somewhere as an adjunct professor. It would give me a lot more creditability. Don't have much hope, but the exercise has forced me to update my vita and look over my old personnel file. Thought you'd like copies of some of the data, for your files. My stint with the USGS is almost impossible to believe now, but back in the 70's that's how it was for a married woman scientist.

Also included is a copy of my letter to Vaughn Bryant. He's head of the Anthro. department where George is, and was at Washington State University Laboratory of Anthropology shortly after I left. If I have any chance, it may be there, although he has yet to answer my letter.

A copy of my most recent letter to Fred Nials. No word from him, either.

And a copy of a letter to the head of the amateur archaeology bunch here in Colorado, trying to clear up some misinformation circulating about

Hueyatlaco.

Why am I sending you all this? You are a specialist in the history of science. I believe we are at a critical point in that history, and that *someone* should document it!

4.54.26. Letter to Dr. Virginia Steen-McIntyre, October 4, 1995.

Thanks for sending the additional documentation about Hueyatlaco. It will be helpful for future editions of *Forbidden Archeology*. I am waiting to see what happens with those stabilized columns of sediment from the site. Bill Cote called the other day. The NBC show is still on track. He said he shot Charlton Heston introducing my segment recently. There is still no definite date for broadcast though.

4.54.27. Letter to Dr. Virginia Steen-McIntyre, January 10, 1996.

It's great to hear that you will be going on the Laura Lee show. I hope it works out well. I think you will like it. Another good show is the Bob Hieronimus show out of Baltimore. If you like, I can get in touch with the producer and see if they would be interested in having you on the show. Just let me know.

Also, thank you for sending me all the copies of your correspondence related to the reopening of the Hueyatlaco case. I am planning on bringing out a book called *The Impact of Forbidden Archeology.* The book will feature the papers I have given at scientific conferences, reviews from mainstream and alternative science publications (there are a lot), my correspondence with scientists and others, and transcripts of selected radio and television interviews. The book will be useful to scholars interested in scientific controversy, and also may be of interest to people who have read the book. I don't expect big sales. But it will keep the pot boiling and document the responses to the books (and my responses to the responses).

I'm hoping you'll give me permission to use some of our correspondence. I think one of the impacts of *Forbidden Archeology* has been the reopening of the Hueyatlaco case, and it would be good to document it for history. Eventually, I want to bring out a sequel book called *More Forbidden Archeology* or *Forbidden Archeology 2*, with new cases of anomalous evidence, updating of the current orthodox paradigm, and further investigations into cases mentioned in the first book, such as Hueyatlaco.

You'll notice I've moved to Los Angeles. I've split from Richard Thompson due to personal, intellectual, and financial differences. As principal author of *Forbidden Archeology* and *Hidden History,* I've taken control of those books (Richard has given up his interest in them) and placed them with a friendly publisher and distributor. Alister Taylor, who is no longer working for Govardhan Hill, is distributing the books through his company, Torchlight. Alister and I have worked closely together in the past, so I am happy with that.

4.54.28. Letter from Dr. Virginia Steen-McIntyre, March 11, 1996.

Was interviewed by Laura Lee early Sunday morning, March 3. A "B-" job, I'm afraid. My body, try as I would, was in "sleep mode", so I was responding slowly to her questions. I had never heard her program, and the questions were phrased differently than I had expected. To top it all off, we were under an occult attack of an intensity last experienced by me over 10 years ago. First there was a bad static in the phone connection. Then the connection broke completely. Then when I tried to make a point I was "mind zapped" (where the thought left completely with a mental "bang"). Happened *four* times before I could get it out in words! Have no idea which occult group it was—military perhaps? (I wanted to point out that scientists must believe there is an objective reality "out there" and that we have the smarts to at least approximate the objective reality/truths. If scientists don't believe there is objective reality/truth, then why do science at all? Why are they in it? To supply data to their bosses? For what purpose? If we look at history, that purpose would seem to be control of others.) So far not much comment on my segment of "*Mysterious Origins of Man*." Dave checked the sci archeology alt-archeology news group today (we've been away since the third for a family funeral). At least my introduction is on, including the important Hueyatlaco references. We don't have a connection to the Internet ourselves. All calls to connect would be toll calls for us.

4.54.29. Letter to Dr. Virginia Steen-McIntyre, May 23, 1996.

George Carter is certainly correct that from a strictly scientific point of view it would have been better if Bill Cote just did a show about the really solid North American stuff. But such a show might not have made prime time television. In any case, I was satisfied with the show. For one thing, it resulted in sales of about 5,000 copies of *Forbidden Archeology*, which means that 5,000 more people are going to get to read the whole story of Hueyatlaco, among other things. Also, I think the show gives you a platform to do more media work, if you want. You might even be able to go on the university lecture circuit. A booking agency has just agreed to represent me. We'll see what happens.

Regarding the California gold mine discoveries, the hoax stories are connected with the Calaveras skull. Even then, there are several different hoax stories, which means some of the hoax stories are hoaxes, and if some of them are, why not all of them? Whitney himself pointed out that the hoax stories emerged only after the discovery became well publicized and that they emanated from "the religious [i.e. antievolution] press." But leaving aside the Calaveras skull, which I admit is suspect, there are the many other discoveries made in the same region. And their provenance is not as doubtful as George Carter suggests. The strata at Table Mountain have been very well studied, and it is fairly certain that the gold-bearing gravels near the bedrock, where many of the discoveries occurred, are Eocene. So there! I suggest everybody go back and reread those sections of *Forbidden Archeology.*

4.55. Ingo Swann

Ingo Swann is a parapsychologist and author. He wrote a review of Forbidden Archeology *for* FATE *magazine, at the request of Jean Hunt, of the Louisiana Mounds Society.*

4.55.1. Letter from Ingo Swann, August 13, 1995.

[*Parapsychologist*] Alex Imich has been mentioning you to me many times, as had Jean Hunt, and so I've decided to write the note to you I had thought of doing nearly two years ago, but also just to be in touch with you. The book, of course, was wondrous. I already had found mention here and there of very early "anomalous" evidence—and during the 1970s one Joseph Jochmans published some large xeroxed writings recounting even more examples. I've been briefly in touch with him just recently, and it seems he was quite young back then. Perhaps you've run across him. In any event, your book was needed, and you and Richard Thompson did a great job of it. I was shocked to learn someone had actually done what you two did. I was glad to make *Fate* publish the review—not a very high class pub., of course, but perhaps every little bit does help somehow. It goes without saying that I'm drooling for the next one.

There are few to mourn Jean Hunt's passing. I'd not met her personally, and I'm sort of a monk in retreat in my studio and am tired of people in general. But her book was so impressive that I got hold of her over the telephone and we talked about twice a week thereafter. She inspired me.

She was wonderful, and I looked forward to many years of interaction. We were working together on a project. News of her death was awful, still is, and I still grieve. She was the type that can't be matched, and only one issue from that very special mould.

I'd be interested, if you have the time somewhere ahead, to learn how your book has been received in a world which, of course, is neither prepared no really would want it, at least in the alleged scientific sense. I've spent years studying the rejection, bowdlerizing, and hiding of valid psychic evidence by those very strange mind-sets who seem to like to do that. It would seem to me that your archaeological evidence must be somewhat submitted to the same treatment. Yet your book was very impressive, and so it would be interesting what has happened because of it.

Through Alex I've learned something of your interest in psychic matters, a field which is boobytrapped with all sorts of misinformation and mis-concepts, a very limited technology and etc. I'm not the final word on anything, of course, but I've had a lot of unique experience both with the "living" psi factors and with the disgusting superficializing politicking that goes on. So if I might be of some assistance here and there I'd be glad to, at least in honor of Jean Hunt, for your book came to her as an excitement from heaven.

At any rate, blessings on you and your good work.

4.55.2. Letter to Ingo Swann, September 27, 1995.

Your letter of August 13 has just arrived, having been first sent to an old San Diego address of mine, which you might have gotten from the book cover of *Forbidden Archeology*. In any case, I was very happy to hear from you. I have been tracing connections, and Jean Hunt was certainly a center of many good connections for me, including yourself. Your address intrigues me. As soon as I saw the envelope of your letter, with the address 357 Bowery, I was reminded of my *guru* Bhaktivedanta Swami, who lived at 94 Bowery in 1966, in an AIR [*Artist in Residence*] loft. As a matter of fact, your handwritten name "Swann" on your envelope I [*momentarily*] read as "Swami." Of course, when I opened the letter, I was pleasantly surprised to find it was from you. Bhaktivedanta Swami physically left this world in 1977. I wasn't in New York in 1966, but a good friend of mine, Michael Grant, happened to be here in Florida recently. He was one of Bhaktivedanta Swami's first disciples and used to live at 110 Bowery in another loft. So I asked him if he knew where 357 Bowery is, and he said it should be up around Houston Street. That would mean you are just around the corner from 26 Second Avenue, which is where Bhaktivedanta Swami opened his first storefront temple in the summer of 1966. Some of his followers have recently reoccupied the place, and turned it into a living memorial. Anyway, I just found it interesting that you are living in the midst of one of my personal sacred sites.

I've read Jochman's book. We have been in touch. I would like to document some of his cases more thoroughly and perhaps make use of them in another book of my own. I also read Jean's book.

As far as the reception of *Forbidden Archeology* is concerned, I am enclosing a copy of a paper I presented at an academic conference on science and culture. It deals with the impact of the book in various knowledge communities. You'll find it interesting. I am also enclosing a copy of the latest issue of *Ancient Skies,* the newsletter of the Ancient Astronaut Society. It contains the text of an address I gave last month at their word conference in Bern, Switzerland.

I am working away on the follow-up book, which I am calling *Human Devolution*. Much of it is devoted to exploring the implications of psychical phenomena for our understanding of human origins and antiquity. As you might gather from *Forbidden Archeology,* I value the historical approach. Right now I am exploring the work of early researchers in mesmerism, who were reporting a lot of paranormal phenomena in addition to the common hypnotic effects. I've just gone through James Esdaile's *Mesmerism in India* (1846) and *Natural and Mesmeric Clairvoyance* (1852), and William Gregory's *Animal Magnetism.* My main intended audience is those who are not already convinced about psychical phenomena, and I want to show them that for quite a long time a lot of very good evidence has been set aside, as was the archeological evidence mentioned in *Forbidden Archeology.* Of course, I can't deal with everything, but I want to go through all the major categories of paranormal phenomena, from the last century up to the present, and give detailed treatments of the most interesting and convincing research that can be related to human origins. As the book begins to take shape, I will keep you informed of developments. I do very much appreciate your offer to be of help.

4.55.3. Letter from Ingo Swann, October 28, 1995.

Been a little tardy reading what you sent along because I was working to meet a deadline for one of my new books. From the "Reception" paper I see the larger contours of what is involved with your book—and/or, as you state it, your larger or more encompassing purpose.

It was part of my work at SRI to rather extensively investigate the "knowledge filtration" of "accepted wisdom" as it regards ESP, etc., and I prepared several in-house papers for client use. The reason for this effort was that our project at SRI came under extreme attack—largely of the nature typified by Jonathan Marks regarding your book, or rather its implications. The approach and methods of Marks regarding your book are recognizable as the "trademark," so to speak, of a specific, rather small but powerful philosophical group. After the core or inner cell of that group was identified by name it became somewhat possible to disarm them by various methods.

However, who, and why, rules what is considered to be knowledge or not-knowledge is a very interesting thing to consider. I wrote and gave a paper at a Fortean conference about three years ago entitled "The Manufacture and Control of Knowledge," and it summed up the larger picture as I found it.

As I might take it from your paper (correct me if I am wrong), your interest in the "paranormal" would be associated with the Asian background of the Sidhis? Or the Asian concepts of consciousness in general? The English translations are quite dreadful of most of the relevant Asian sources, for example, of the *Yoga Sutras* of Patanjali. If my impression here about your forthcoming book is correct, it will be exciting to see what surely will be a more exact presentation of that whole situation.

In any event, I see you are being very thorough regarding sources, and so you probably don't need my help very much. Regarding Mesmerism and its fallout, I trust you've come across Eric J. Dingwall's *Abnormal Hypnotic Phenomena* (1967, in four volumes). The title is misleading, for the volumes deal with "higher" faculties experimentally revealed by Mesmer's disciples.

Yes, I'm only about a three minute walk from 26 Second Avenue, and I remember when the storefront temple was opened in 1966. I remember the area, even my own building, and what used to go on near 26 Second Avenue, very well from a past life here in NY. Nicola Tesla's laboratory was only two blocks away, too, on Lafayette street.

Well, all in all, I see you've bitten off a large chunk to chew. I very much wish you well and all success. I know your time is valuable, but please keep in touch if you want, and please visit me if you are ever in the City. We can walk through the area.

I did wind up visiting Swann in his Bowery apartment, for a wide-ranging and valuable discussion.

4.56. Dr. Donald E. Tyler, M.D.

Dr. Donald E. Tyler is a physician with an interest in unconventional archeology. His biography appears in Marquis's Who's Who in Medicine and Healthcare 1997-1998.

4.56.1. Letter from Dr. Donald E. Tyler, M.D., May 13, 1996.

After seeing some of the rhubarb on Usenet, I obtained a copy of your book, *Forbidden Archeology,* from the local library.

It is unfortunate that your book was unavailable to me when I wrote about similar subjects. I am assuming that you did not read mine. It would be of interest to me to visit you and learn about your organization; I think you would like to see some of my artifacts.

If you are interested in making a deal, I will send you a copy of my book on *Earliest Man*, etc., for a copy of yours, preferably autographed by both of you. Mine is smaller and in paperback only. But I have good photographs, and an original theory.

I responded positively to Tyler's offer of a book exchange.

4.56.2. Letter from Dr. Donald E. Tyler, October 10, 1996.

Thanks for the copy of your book. I was particularly desirous of having the bibliography. I hope you have received the ones I sent....

My biography was recently selected for publication in *Who's Who in Medicine and Healthcare*. Hopefully, future generations may use my ideas.

I would be interested in the organization reflected in your address. Is it a philosophical society? Has it other publications?

4.56.3. Letter to Dr. Donald E. Tyler, October 1996.

Yes, I did receive your books. Regarding *Earliest Man of America in Oregon, U. S. A.*, the quality of your reporting is good, as is your overall argument. If I had known of your work, I would have cited it in *Forbidden Archeology*. I read through your book rather quickly, and have one comment. For my purposes, it would have been good if you had given some account of the age of the artifacts found at each site, on the basis of the geology of each site, in addition to the general overview you give in the first chapters. In many cases, I find it hard to tell whether you think the artifacts are recent or attributable to some distant era, such as the Pliocene. For example, I find it hard to tell how old you think the Apothecary Creek artifacts are. Also, even in cases where you do give age estimations, it would be good to give explicit arguments why tools found on the surface in such locations were not simply dropped there in recent times. But overall, I find your book a valuable resource.

The Bhaktivedanta Institute is the science studies branch of the

International Society for Krishna Consciousness. Members take inspiration from the ancient Sanskrit writings of India. For further information and lists of publications write to bvi@afn.org and bvi@itsa.ucsf.edu. Be sure to write both. There are different branches of the Institute, with different focuses.

4.57. Dr. A. Bowdoin Van Riper

Dr. A. Bowdoin Van Riper is a historian of science at Kennesaw State College, Georgia. He has a special interest in the history of archeology.

4.57.1. Letter to Dr. A. Bowdoin Van Riper, November 4, 1994.

A publisher's representative...passed on to me a note from you saying you would be interested in receiving a review copy of my book *Forbidden Archeology: The Hidden History of the Human Race.* He mentioned that you sometimes review books for *Isis.* As I still have some personal review copies of the book, I am sending you one. Although you say you are "a firm believer in the conventional account of human origins," I would nevertheless like to hear your frank and honest impression of the book. Thus far it has drawn hot fire and high praise from various quarters.

In your note you also mentioned two other scholars interested in human antiquity—Donald Grayson and David Meltzer. I've read Grayson's book on human antiquity. I found it useful, especially his tracing of the influence of Christian concepts on the developing European ideas on human antiquity. But I think he skipped over a lot of relevant material, especially that related to the Tertiary man debate. In any case, if you can give me the addresses of Grayson and Meltzer, I would be happy to send them review copies of the book as well.

4.57.2. Letter from A. Bowdoin Van Riper, January 5, 1995.

Please pardon my tardiness in thanking you for the review copy of *Forbidden Archeology*. The end of the last quarter was even more chaotic than usual, and I have fallen well behind in my correspondence.

For the same reason, I have not had the opportunity to do more than briefly peruse the book itself. I am looking forward to reading it in more detail, however. The premise is audacious, to say the least, and intriguing to me if only for that reason. I will be delighted to give you my "frank and honest impression" of the book once I've had the chance to look at it more carefully, probably sometime this spring.

I'm afraid I can't help you with Donald Grayson's address, other than "Department of Anthropology, University of Washington" (which you doubtless already know). The last address I have for David J. Meltzer is: Department of Anthropology, Southern Methodist University, Dallas, TX 75275. Meltzer is, as I suggested to the publisher's rep in New Orleans, the leading authority on the history of human antiquity investigations in North America, particularly in the 1877-1927 era. To the best of my knowledge he is, like me, a thoroughgoing supporter of the traditional account of human origins.

Thanks again for the book. I hope to be able to write back to you soon with more detailed reactions. In the meantime, best wishes for the new year.

4.57.3. Letter to Dr. A. Bowdoin Van Riper, February 12, 1995.

I'm happy to hear you received *Forbidden Archeology* and plan on going through it carefully. I'd just as well read your comments in an *Isis* review as in a letter to me personally, but will appreciate your response however you care to deliver it. I recently gave a paper based on the *Forbidden Archeology* material at the World Archeological Congress 3, which was held in New Delhi this past December. There I met an Australian scholar who told me he has reviewed *Forbidden Archeology* for *The British Journal for the History of Science*. Another review, by David Oldroyd and one of his grad students, is forthcoming in *Social Studies of Science.* A rather impolite and scathing review by Jonathan Marks has already appeared in *American Journal of Physical Anthropology*, and a far more civil but likewise negative review by Ken Feder has appeared in *Geoarchaeology.* Other reviews and notices have appeared in *Antiquity, Journal of Field Archeology,* and *Evolution, Ethology, and Ecology*. I've also received some invitations to faculty seminars (history of science). If there are any such opportunities at Kennesaw, please let me know. Kennesaw isn't so far from Alachua, so it would be easy for me to drive up for a day or two's visit.

4.58. Dr. Jenny Wade

Dr. Jenny Wade is a psychologist. I met her at the conference called Toward A Science of Consciousness 96, sponsored by the University of Arizona in Tucson, Arizona. Her book Changes of Mind: A Holonomic Theory of the Evolution of Consciousness *was published by the State University of New York Press in 1996.*

4.58.1. Letter from Dr. Jenny Wade, May 31, 1996.

I just finished *Forbidden Archeology*, and all I can say is, wow! The most useful portions to me were the Introduction and First Chapter where you so clearly, thoroughly, and thoughtfully put forth the case against the scientific prejudice of the prevailing establishment. I'm sure I'll be using and citing that part of your work often. It is brilliantly stated. As for the rest, what a tremendous body of research! I would have been daunted by the scope of your undertaking, not just the volume of the findings but tracing the subsequent reception of the data, the polemics, etc., as well as the many areas of expert knowledge you had to acquire, the translations, locating obscure works,...I can't easily express how impressed I am. What a great job!

4.58.2. Letter to Dr. Jenny Wade, August 1996.

I've just recently returned from Europe. I first went to one of our Krishna centers in Sweden—in the forests outside Stockholm—for the annual meeting of the trustees of the Bhaktivedanta Book Trust. The place used to be some kind of resort, with quaint Swedish wooden buildings. Now it houses a center for translating and publishing our books in thirty European and Asian languages. I had to go there to convince the trustees to keep one of my books in print. At first, it looked like I lost, but then they reconsidered and we came to a compromise, whereby I will make some minor changes in the book (it's about Krishna consciousness and the environment) and they will keep it in print.

Then I went to the Krishna center in Belgium. This one was a chateau on fifty acres of land in the Ardennes Mountains. There was a convention of the European Krishna devotees there, and I gave a seminar on my book *Forbidden Archeology*....

After Belgium, I went to Moscow. I gave some lectures at another devotee convention/festival. This one was held at a Krishna center outside the city, in the forest. The place used to be a rest camp for factory workers. It was built in the late 1940s, so it was a little run down. After that, I went into the city and gave some talks on *Forbidden Archeology* at a conference organized by the Institute for the Study of Theoretical Questions of the Russian Academy of Sciences and at a symposium held at the Institute of Oriental Studies of the Russian Academy of Sciences. The newspaper *Pravda* ran a big and entirely favorable article about *Forbidden Archeology*, and another Moscow paper is running a series of articles based on the book. A Russian friend of mine, a woman who was once imprisoned for her Krishna beliefs by the KGB, brought me to the Russian national television studios and they did a two-hour interview with me, which they will edit into a series of shows that will be broadcast nationwide. I also did a radio talk show, with call-ins from listeners, while I was in Moscow. I then visited St. Petersburg and appeared on a television show there.

Over the next few weeks, I have about 30 radio shows to do on *Forbidden Archeology.* During that time, I will be in Washington State, staying with some friends in Bellingham. Also, on August 6, I am going to be a guest on the Compuserve Literary Forum.

The last week in August, I am going to Australia to meet a friend. We have worked on books together in the past (small ones on reincarnation, meditation, spiritual vegetarian diet, etc.), and we are going to plan out our next one. George Harrison has given us the use of a home he owns on an island in the Great Barrier Reef, off the coast of Australia. I will be back in the USA September 7.

Needless to say, with all this traveling I am not getting much writing done on my *Human Devolution* book. At some point, I am just going to have to stop going to conferences and doing radio and television. Sometime next year, I'll just stay in one place and get down to some serious research and writing.

4.58.3. Letter from Dr. Jenny Wade, August 9, 1996.

I was delighted to get your long and most interesting e-mail. You've had a delightful summer with all your travels from the sound of it, as well as a very busy time professionally. I'm absolutely stunned at the number of broadcast opportunities you've accumulated. Envisioning your stays at the various Krishna centers in different countries—especially the way you described the leadership of the organization—was highly entertaining. You're right—I'd love to have a birds' eye view as a consultant of the dynamics. I can't believe how much work you managed to cram into that time. I hope that, what with the time and cultural changes, you felt like it was restful, as well.

I trust this will get to you in Washington, where I assume you must be now. I've been travelling considerably myself this summer, just back from the Association for Transpersonal Psychology in Monterey, CA and thankfully not having to go to Rhode Island for the rest of the month, as had originally been planned. I've been sustaining a killing pace on a consulting project since some of my colleagues had health emergencies that prevented their doing their part, but we're down to the last stretch now and it looks like we'll make the deadlines with a minimum of fuss. After 18-hour days for the last two months, I did nothing but sleep when I was at the conference in California.

So I have little of interest to report. On the right-livelihood professional front, SUNY said my first printing (2,000 or so) had completely sold out, so they're reprinting already. And the book sold well at two of the conferences I attended recently, much to my delight. I am planning to cut back my consulting obligations drastically this fall, although two of my colleagues are wanting to include me in projects, in order to catch up on all the article-writing I'd planned to do but had to let slip this summer. I also want to start on a popular version of my book, as we had discussed. And I've got the research on gay NDEs to get cracking on some time in the near future, not to mention researching and figuring out whatever my next major writing project will be. I'm with you—when in the world are we going to have time for your *Human Devolution* book and my next big writing project? I'm still looking forward to our working together.

4.58.4. Letter to Dr. Jenny Wade, September 1996.

As I said in my recent note, I will be in Potomac, MD Sept. 23 to Oct. 2 or so, and might be coming to Philadelphia on Sept. 28. It's a busy schedule, as usual, but I am always ready to take advantage of an opportunity to meet with you, even if for a short time. I can be reached by phone at (301) 299-4797 in Potomac. By the way, a devotee friend of mine is a clinical psychologist who specializes in working with AIDS patients. She is prominent in her field, and is always going to international conferences. She lives in San Francisco. I told her about your proposed NDE work, and she said she would be willing to consult with you about it. If you think she could be of help and would like to communicate with her, let me know and I will give you her e-mail address. She might also benefit from your association, as it would help her to see how her psychology might be applied in establishing the existence of a conscious self that exists apart from the brain.

4.59. Dr. Roger Wescott

Dr. Roger Wescott is an expert in linguistics and is president of the International Society for the Comparative Study of Civilizations.

4.59.1. Letter to Dr. Richard L. Thompson from Dr. Roger Wescott, March 8, 1993.

Thanks very much for your letter of 12 Feb. 93 (which arrived late last week) & the two volumes of *Forbidden Archeology*!

I enjoyed your iconoclastic presentation—particularly on Bigfoot & related "wild men"— & was pleased to see that you make use of anomalistic sources like Wm. Corliss's Sourcebook Project.

There is another way to interpret what seem to be human remains from hundreds of millions of years ago. This is to question the regnant chronology & suggest that it is grossly inflated.

For the pre-Holocene age, the most interesting (non-religious) source is, I think, *Solaria Binaria* by Alfred de Grazia & Earl Milton (Metron, Princeton, NJ. 1984). For the protohistoric & historic periods, I would say, it is *Wann Lebten die Pharaonen*? by Gunnar Heinsohn & Heribert Illig (Scarebäus bei Eichborn, Frankfurt am Main, 1990).

I can understand that the Indian Mahayuga tradition would incline you toward the longer chronologies for both man & cosmos. But do at least look at the above books & let me know your reaction.

Best wishes for your bold reinterpretive enterprise!

4.60. Colin Wilson

In 1956, Colin Wilson published The Outsider, *a bestselling exploration of existentialist philosophy. He has since written numerous books on archeology, cosmology, astronomy, and the paranormal.*

4.60.1. Letter from Colin Wilson, March 31, 1995.

I found your *Forbidden Archeology* when I was at the Marion Foundation, near Boston, last weekend—lecturing with Stan Grof, Rhea White and Peter Russell. The Foundation was kind enough to allow me to purchase their library copy (they will re-order), allowing me to read it on the plane coming back to London. The book had already been enthusiastically recommended to me by Paul Roberts and—only two days earlier—by the psychical researcher Alexander Imich. Today, a second copy I ordered from my American bookshop (and had forgotten about) arrived.

It is certainly a most astonishing work, and I shall be quoting it extensively in a book I am writing at present: called *Before the Sphinx*. This deals with the new evidence that the Sphinx dates back to 10,450 BC, which you no doubt know all about. (John Anthony West, who made a TV programme about it, is a friend of mine.)

This is, in a sense, new territory for me. I achieved overnight notoriety in 1956 through a book called *The Outsider*, which (I suppose) was about 'existential philosophy'—the past is not really as irrelevant to my former preoccupations as it might seem.

As you probably know, the maverick Egyptologist Schwaller de Lubicz started all this by remarking that the Sphinx was obviously weathered by water, not wind-blown sand. Yet it seems to me that all this would scarcely matter (what's a mere eight thousand years when your book deals in millions?) if it were not that Schwaller also stated that the Egyptian mentality was, in some profound sense, quite unlike our own. What he seemed to be suggesting was that the earliest civilisations were far less 'left brain' than we are—in fact, that a civilisation might achieve a high level of spiritual, and even technical, accomplishment, without our dependence on language and intellect. What fascinates me about this is the notion that pure intuition might achieve a high level of 'scientific' sophistication without the aid of linguistic symbology.

I had already cited the example of the clay image from Nampa, Idaho, in my typescript—an example I found in Velikovsky's *Earth in Upheaval*—and your book has provided me with a vast amount of new ammunition.

I had also been recommended—by Paul Roberts—to read David Frawley on the Vedas, and was excited to find that he dates the earliest Vedas to before 6000 BC. In New York two weeks ago, I spent a few hours in the NY Public Library reading a 19th-century Mayan archaeologist named Augustus Le Plongeon, who claims that Mayan texts mention a connection between the early peoples of South America and those of India. It is all rather bewildering.

You know as well as I that the problem about writing on this subject is that danger of being identified with the 'lunatic fringe.' The chief opponent of the new discoveries about the Sphinx is Mark Lehner, a highly orthodox Egyptologist. Yet he was not always so. He went to Egypt as a convinced follower of Edgar Cayce, sponsored by the Cayce Foundation, and wrote a book supporting Cayce's assertion that man is millions of years old, and that the Sphinx was built in 10,450 BC. Now he piously insists that he has put these childish things behind him. I don't blame him. In mentioning Cayce early in my book, I acknowledged that his prophecies have occasionally been accurate, but pointed out how often they had been wrong. And, I must admit, I dismissed out of hand his assertion that man might be millions of years old. Now your book has set my head spinning with a whole new lot of speculation.

I don't know yet quite what it all means. My original argument was as follows. Cro-magnon man, originating around 30,000 years ago (perhaps more) was more sophisticated than we believe—even Neanderthal man invented the bow and arrow. There may have been a true civilisation as early as 20,000 BC, possibly in 'Atlantis.' (I hope the word doesn't make you howl in anguish.) Certain evidence suggests that this had spread as far as Lake Titicaca by 10,000 BC, possibly even earlier (the Maya scholar Posnansky estimated 15,000, based on astronomical alignments at Tiahuanaco). By 10,500, seismic convulsions were already sending 'Atlanteans' to Africa and South America, and by the time of the great catastrophe, about 9,500 BC, civilisation was already well established in Egypt and Peru—probably all over Central America.

These people were not primarily 'linguistic', in our sense. (This part of the book will require much argument about how far words 'lead' our thinking processes.) Julian Jaynes even suggests that such people had 'bicameral' brains, hearing 'voices' in the left-brain which they mistook for gods. (I have my reservations about this theory.) Their sense of nature and of reality was certainly less 'alienated' than our own, and for them religion and science were probably identical. (Interesting aside: how did the builders of the Sphinx Temple and Tiahuanaco raise blocks weighing two hundred tons?)

Now if you are correct, and man is millions of years old (on page 813 you even suggest billions), this might at least account for the possibility that the builders of Tiahuanaco and the Sphinx Temple might have achieved a sophisticated, if 'unintellectual', level of civilisation.

Incidentally, it seems to me that Charles Hapgood's arguments for an ancient maritime civilisation around 7000 BC, possibly in Antarctica, are almost irrefutable. And I suspect that my friend Rand Flemath has hit on the right solution when he pinpoints Antarctica as Atlantis, moved south by a slippage of the earth's crust.

I suspect that you could put me on to even more important possibilities, particularly regarding ancient India. Obviously, I'd be deeply grateful for any comments or suggestions you can make.

4.60.2. Letter to Colin Wilson, April 20, 1995.

When I was writing *Forbidden Archeology*, one of my dreams was that it would circulate underground among European intellectuals, not just those in the science field, but in the literary and philosophical fields as well. So you can just imagine my pleasure on receiving your letter. You said the book was recommended to you by Paul Roberts. Who is he? Also, what can you tell me about Alexander Imich? What kind of psychical research is he doing?

I don't see your "venture into the past" as at all irrelevant to your former existential "preoccupations." Rather, the drama of the self has been played out upon a certain temporal stage, and we may well have to reexamine not only the self but the stage, the temporal architecture of our falsely pictured cosmos.

I am familiar with the work of Schwaller de Lubicz, mediated by John West and Robert Schoch. In fact, I am now working with the producers of *The Mystery of the Sphinx* (which featured their work) on a similar television special about the evidence for extreme human antiquity and unexpected cultural accomplishments.

Regarding advanced civilizations with technologies not based on our current scientific methodologies and languages , I quite agree they have existed and continue to exist. In my next book, I will be getting away from the stones and bones and more into that area. In the Vedas, the spiritual realm is described as *a-van-manasa-gochara* (beyond the reach of words and mind). Also, the yogic *siddhis,* which include the ability to manifest material objects (including large buildings, flying machines, etc.) are acquired and executed by meditation, not by getting some engineers together to talk and design and construct. There is also the concept of nonrepresentational language, in which the word is the object, and not a sign.

That is all quite intriguing, more so than stones and bones. But the stones and bones are important in that they provide a temporal framework for human antiquity, and this can be helpful for those who are proposing alternative views of human origins and antiquity (incorporating occult elements).

If you read my Introduction to *Forbidden Archeology*, you should have gained some insight into my purposes in composing such a text. But to give you a more explicit idea, I am enclosing copies of two papers, one delivered at the World Archeological Congress-3 held this past December and the other delivered at an academic conference on science and culture held recently in the United States. From these you might gather some ideas that will be helpful to you. If I can be of any further assistance, please let me know.

You may also want to have a look at Richard Thompson's book *Alien Identities*, which compares modern UFO phenomena to phenomena reported in the ancient Sanskrit writing of India—spacecraft (mechanical and mystical), humanoid entities, etc. The parallels are astonishing. *Alien Identities* is available in the stores in England. If it is not on the shelf, they can easily order it for you.

Richard has also done an interesting study of Vedic star coordinates. In certain Sanskrit astronomical texts, coordinates of stars are given. These vary from the modern star positions. Modern astronomers would attribute the variance to bad measurements carried out by early Indian astronomers. But Richard discovered that the variance of the old star positions is almost always in the direction of the proper motion of the stars. If you look at a constellation, the stars are in the same position from night to night. But if you were to look for hundreds of years, you would see the individual stars gradually shift their positions relative to each other. This is called the proper motion of stars. Each star moves in a different direction. So if the old Vedic star coordinates lie in the direction of the stars' past motions, that is quite significant. It indicates that the variance between the old coordinates and the present coordinates is not the result of mismeasurement. Rather, the old coordinates represent correct measurements made in the past, when the stars were in different positions than today. Since the rate of motion is known, it then becomes possible to calculate the time when the Vedic measurements were made. Richard calculates that a great many of them were made 40,000 years ago. Interesting. Actually, if you go to India, you will find many temples that are said to be millions of years old.

Finally, I am always looking to get more attention for *Forbidden Archeology*. The mainstream press have completely ignored it in England, which is understandable. But if you know any adventurous reviewers, let me know. I'll see that they get copies.

4.60.3. Letter from Colin Wilson, October 5, 1995.

I was very happy to hear from you. I'd been reading your book busily for weeks, and so your ideas were a great deal in my mind. In fact, if you remember, I told you that I had accidentally bought a second copy, which I've passed into good hands—I had a ten-page letter from the friend to whom I sent it only this morning!

I've temporarily had to suspend work on my book *Before the Sphinx*, because I've tended to get overloaded with other work. Since I came back from America a few weeks ago, demands from two separate publishers have completely diverted me—fortunately, I've got a dispensation to deliver the typescript six months later, just before Christmas! (It was due this month.)

As to your questions, Alexander Imich, who recommended your book to me—and showed me a copy in New York—has been a famous psychical researcher for the past forty years—he is now in his eighties, but still as fit as ever. He is just about to publish a book—edited by himself—with an introduction by me. (Unfortunately I've forgotten the title.) This contains some really astonishing material on psychical research, particularly over the next couple of weeks. If you should wish to contact him, his address is [*deleted*]. I'm sure he would be absolutely delighted to hear from you, since he is such an admirer.

I'm also delighted to hear that you are working with the same people as

John West and Robert Schoch. I'm also working with John's close friend Paul Roberts, who has written some very remarkable books on India and Egypt, and with whom I'm hoping to do a television programme. I'm also in contact with Andre Vanderbroeck, who wrote one of the best on Schwaller de Lubicz AL-KEMI, which is still quite easy to obtain in paperback.

Many thanks indeed for the enclosures—I found the article on Puranic time particularly fascinating. I'm also hoping to contact David Frawley, and see if I can get some more solid information indicating that the Vedas date back to the seventh millennium BC.

There has recently been a superb book published in England about this whole subject of an earlier civilisation. It is by Graham Hancock and is called *Fingerprints of the Gods.* The title sounds rather like Von Daniken, but is in fact a very sober and scholarly examination of all the evidence. It has shot immediately to the top of the best seller list in England, and is out in America very shortly.

Please let's keep in touch. In spite of the fact that I've been forced to temporarily abandon the Sphinx book, I'm busily accumulating research for it.

Again, a thousand thanks for your letter.

4.60.4. Letter to Colin Wilson, May 1996.

Since we last communicated, I've moved from Florida to California. *Forbidden Archeology* continues to do well, especially after being featured on a national television program here in the United States called *The Mysterious Origins of Man*. In the fall, I am going on a university speaking tour. I recall that you were going to perhaps be mentioning *Forbidden Archeology* in one of your own forthcoming books. Did you?

4.60.5. Letter from Colin Wilson, May 31, 1996.

Many thanks indeed for your letter and new address.

Yes, indeed I did use your *Forbidden Archeology* in my book, which my publishers insisted on calling *From Atlantis to the Sphinx*. My own title was *Before the Sphinx*, but perhaps they were right—it came out last Monday and already seems to be into a third impression, although God knows what that means by way of sales—perhaps each impression was only 1000 copies...

Now that I have your address I'll airmail you a copy. I'd be extremely interested to hear your reactions to it.

4.60.6. Letter to Colin Wilson, August 4, 1996.

Thank you ever so much for sending me a copy of your book *From Atlantis to the Sphinx*. You have given a masterful summary of what I would call "forbidden archeology." I have gone through the book once, rather quickly. But I am definitely going to read it again quite carefully, as it is an excellent guide to a whole body of literature. In my next book, I am going to be concerning myself in a deep and comprehensive way with the psychic forces and energies connected with human origins and history. Science took a wrong turn about three centuries ago, and I look at our work as a serious effort to bring the world's intelligentsia back on to the right path, which looks more like what is to be found in your book than what is to be found in a standard science text on human nature.

Your treatment of *Forbidden Archeology* was generous, and I am grateful. Some of the details of my personal history, given on p. 165, were not quite correct, which is understandable, considering that you were not able to speak with me directly about this and had to rely on secondhand information. The actual facts are these. In 1973, I joined the International Society for Krishna Consciousness, which teaches a form of Hinduism called Gaudiya Vaishnavism. In 1976, I received initiation from the movement's founder, Swami Prabhupada, who died in 1977. In 1984, I began work with the Bhaktivedanta Institute, which has several branches around the world. At that time, I was in Los Angeles. Swami Prabhupada's suggestion that the question of human origins be treated was given not to me personally, but to the Institute members generally (during Swami Prabhupada's lifetime, of course). But Richard Thompson and I were the ones that took up the suggestion in a serious way. At the time you and I began corresponding, I was residing at the Bhaktivedanta Institute offices in Florida, but I have since returned to Los Angeles, where I have my own small branch of the Institute.

My book *Forbidden Archeology* was recently featured on a major television program in the United States, produced by Bill Cote, and titled *The Mysterious Origins of Man*. Graham Hancock, Roger Bauval, John West, and others were also featured. Since you are friends with them, you have no doubt heard of the show. It was quite good for sales of *Forbidden Archeology* in the United States.

The book has been selling in a limited way through a distributor in England. But I have been considering offering the abridged version of the book to an English publisher. Are you satisfied with Virgin? Do you think they might be interested in the abridged popular edition of *Forbidden Archeology*? It is about 300 pages long (the unabridged version is over 900 pages).

Thanks once more for sending me a copy of your book.

4.61. Eric Wojciechowski

Eric Wojciechowski is a member of the Ancient Astronaut Society.

4.61.1. Letter from Eric Wojciechowski, February 28, 1996.

In August 1993, while attending the Ancient Astronaut Society's 20th world conference, I was browsing the collection of books that was available for sale, courtesy of Arcturus Books Inc. Your book, *Forbidden Archeology*, caught my eye and after flipping through it, I quickly bought it. Unfortunately, it took me until the summer of 1995 to finally decide to tackle its pages.

It would be best, at this time, to explain a little about myself. In 1989, I was introduced to Richard Hoagland's work *The Monuments of Mars.* Since that time, I have been an avid reader of alternate theories relating to the origins of man. I have also been a member of the Ancient Astronauts Society since 1991. It wasn't until reading your book that many of the ideas that I had been contemplating finally began to take shape. Many of the other books and essays that I had read before this were either too speculative to draw any conclusions or the information provided had come from "shaky" sources. What makes your book so different is that your work remains well documented and the data comes from very reliable sources. (In fact, I was impressed by the extensive bibliography which appears at the end of *Forbidden Archeology,* which was one of the main reasons I chose to purchase your book.)

Although I have not finished reading *Forbidden Archeology*, based on the data I have digested so far, I have become convinced of the argument you have tried to make. At this point, I would like to congratulate both of you for the rare material provided within this book which clearly required years of patience to accumulate. I, myself, being only able to read English (as well as having limited resources) could never have found many of the sources you document.

With praises aside, I must get to the point of the letter. I have a few questions that I hope you can answer.

1) Awhile ago, I ordered Mr. Cremo's interview from the Laura Lee show on audio cassette. During the interview, Mr. Cremo stated that the both of you were working on a sequel to *Forbidden Archeology* which was to be called *Human Devolution.* However, as of yet, I have not seen this work on the bookshelves. I am wondering when you plan to release this?

2) Ever since I acknowledged the fact that *Homo sapiens sapiens* have

been around for millions of years, I have been contemplating what we can know about these people. My first step was to examine the ancient writings, descriptions of histories as well as the many legends and myths of the past. So far, in my studies, I have come to conclude that we may never glean any verifiable proof of what life was like by using this method because, although many similarities exist between cultures, there is also a significant amount of data which creates contradiction. And although many of the tales which have come to us from the past may be historical documents (as opposed to "myth"), we will never know for sure what we can accept as authentic representations of past events or what is only folktale and superstition.

So far, I have come to thinking that even if we can accept some of these stories as historical documents, they could not possibly be in reference to events which occurred millions of years ago. This may be my own bias but it is easy to see how historical events become forgotten or corrupted; events which have only occurred in the past few thousand years are currently in dispute. I feel that due to normal human bias, as well as the passage of time, time has probably obliterated anything we can ever know and that the myths and legend which exist today probably have nothing to do with mankind millions of years ago. I hold this opinion because there are many other plausible interpretations of these stories which seem more logical when all the data is examined. And as for those tales which may be referring to actual events, those events may only have occurred in years which could be measured by the thousands, not millions. Unfortunately, time and space prohibit me from becoming specific but I wanted to know what your ideas on the above statement are. In essence, I would like to know if you agree with me or not. And if not, I would appreciate any input either of you may provide.

3) Lastly, I have been attempting to collect any and all critiques of *Forbidden Archeology*. I am interested in what others have to say about your work. I have been jotting down notes of my own for a possible publication of my own in the future. The data in *Forbidden Archeology* has become the center piece of my own studies and although I find the book's argument convincing, I would like to know what the academic community has thought of it. I am enquiring as to whether or not you would be able to provide me with a list of references to this matter or maybe copies of these "book reviews" themselves. As I stated before, my sources are limited and any help from either of you would be most appreciated.

In closing, thank you for addressing this letter as well as my concerns. I look forward to hearing from you soon.

P.S. I am also interested in the Bhaktivedanta Institute. I would also be grateful for further information.

4.61.2. Letter to Eric Wojciechowski, May 23, 1996.

Thank you for your kind words of appreciation for *Forbidden Archeology*. The sequel is still in the writing and research stages, and I do not see it being available until 1998, at least.

Concerning historical records in the writings of various traditions, you are indeed correct that faulty transmission can be a problem. Nevertheless, if one makes a study of all these writings and records, and plots them as data points, and you see a certain pattern emerging, that is certainly suggestive that the different writings are pointing toward a single reality. When examining writings from all over the world and from different times, I do see a general pattern emerging, namely that humans on this earth are part of a cosmic hierarchy of beings in a multilevel universe. Furthermore, when examining different sets of records, one can try to assess how accurate their transmission has been. In the Vedic tradition of India, great emphasis is placed on faithful transmission of sacred texts. Even then the transmission is sometimes broken. The key is that the original transmission comes from a perfect source of knowledge (God). So even if the transmission system breaks down, God will act to restore it. Taking all this into account, it is possible to use sacred texts as a source of evidence, especially if it can be seen that the sacred text is being handed down in a trustworthy transmission system. Otherwise, we have no certain way of what went on in the distant past, and should just admit it.

Regarding critiques of *Forbidden Archeology*, there have been several in both the mainstream and alternative science groups. I am enclosing a paper I have written on this. I am in the process of compiling a book of all the reviews, along with my related correspondence, papers, and interviews. That could be out in 1997.

The Bhaktivedanta Institute is a fairly decentralized small group. For further information I suggest that you write to both of the following addresses: Bhaktivedanta Institute, P. O. Box 1920, Alachua, FL 32615 and Bhaktivedanta Institute, 662 Kenwyn Rd., Oakland, CA 94610. I used to be with the Alachua group, but am now working more independently.

Wojciechowski and I exchanged another set of letters on sacred texts as evidence.

4.61.3. Letter from Eric Wojciechowski, August 12, 1996.

Thank you for your continued correspondence. Your latest letter suggests that I am not the only one who has expressed to you caution in using the ancient writings to support your thesis. Regardless, I am looking forward to your future publications that you have notified me about.

Please keep my name on a list somewhere so that when your books do get published, I can be notified. I am always interested in fresh insights.

Thank you again. Hope to hear from you soon.

P.S. I have been attempting to understand the history and philosophies of India. I already have a few books on the subject but since this is your field of expertise, I thought I might get your opinion. Any recommendations?

4.64.4. Letter to Eric Wojciechowski, September 15, 1996.

Regarding books about Indian philosophy, I am of course partial to the works of my spiritual master Srila Prabhupada, who left this world in 1977. I am enclosing a catalog. I hope you find it of interest.

5
Transcripts of Selected Radio and Television Interviews, with Related Correspondence

In the course of two years, I appeared on over 150 radio and television shows. I have included in this chapter just a few of them. To obtain video and audio cassettes of the interviews in this chapter, see the order form at the end of this book.

5.1. "A Laura Lee Interview with Michael Cremo: Forbidden Archeology." *Townsend Letter for Doctors*, May 1995, pp. 60-69. Reprinted by permission of *Townsend Letter for Doctors.*

Laura Lee hosts a nationally syndicated radio talk show focusing on alternative science and scientific anomalies. I have appeared on several of her shows. The following interview was broadcast from Laura Lee's Seattle studio. It was published in a verbatim transcription in the Townsend Letter for Doctors *(May 1995, pp. 60-70). An edited version of the same was published in* Nexus *(June-July 1995, pp. 11-16). The transcript below is from the* Townsend Letter for Doctors. *For information about the Laura Lee Show, write to P.O. Box 3010, Bellevue, WA 98009, or call (206) 645-1207 or fax (206) 455-1231. Also check out Laura Lee's website at http://www.lauralee.com. Here you will find a chat room, radio bookstore, and show schedule for various radio stations. To call in during showtime dial: 1-800-800-5287, except from Seattle, where you can dial (206) 455-5287.*

Laura Lee: We're back. Let me tell you about some of the most extraordinary finds in archeology. A grooved metal sphere was found in South Africa dating to the Precambrian period. A shoe print was found in Antelope Springs, Utah, dating to the Cambrian period. A metal vase was found in Dorchester, Massachusetts, dating to the Precambrian. An iron nail in a stone was found in Scotland, dating to the Devonian period. A gold thread in a stone and an iron pot were found in Tweed, England, and Wilburton, Oklahoma, dating to the Carboniferous era. What do some of these finds have to say about our early history? Well, they're not allowed to say very much because we happen to be operating under a theory of our prehistory that would rather discard such anomalies which are defined by their being outside the model, than to try to incorporate them or study them fairly. Here to tell the story of forbidden archeology and the co-author of a book by the same name is Michael Cremo (MC). Michael, thanks for joining us.

MC: You're welcome, Laura.

LL: Tell me a bit about *Forbidden Archeology*, and why you decided to study the field and a bit about your background as well, please.

MC: Well, I began studying forbidden archeology in 1984. At that time I was having some discussions with my co-author, Dr. Richard Thompson, about human origins and antiquity, and we had heard a few of the reports about anomalous evidence, and we decided we would do a thorough research, and I was really surprised with what we found.

LL: What did you find? Just give me a general overview, and then we'll get to the details in a little bit.

MC: Well, if you look at what you see in modern textbooks, you'll get an idea that humans like ourselves, *Homo sapiens*, evolved fairly recently, within the past 100,000 years, from more apelike ancestors, and all the evidence that's in these books supports that idea. So it looks like a pretty solid case. But when I started looking into it, I found that over the past 150 years anthropologists have really buried almost as much evidence as they've dug up, and most of that evidence that they re-buried (in a sense of getting suppressed or forgotten or ignored), is evidence that goes against this idea. It's evidence that rather supports the idea that humans like ourselves have been on this planet for hundreds of millions of years, and you named in the beginning a few of the more startling pieces of evidence, such as that exquisite metallic vase that was blasted out of some Precambrian rock in Dorchester, Massachusetts. That would make it over *six hundred million years old.*

LL: Isn't it astounding? Six hundred million years old? Tell me, in your opinion, what would happen if all the evidence were laid out on a level playing field, if there were no discrimination, if there weren't a theory in place, and someone were to come along and say, let's look at the history of the human race on planet Earth and look at the wide range of evidence. What would they conclude?

MC: Well, the first thing you should understand (we point this out in the chart in the back of the book, *Forbidden Archeology*), is that, if you do put all the relevant evidence on the table, it would require several tables, really, instead of just the small portion of the evidence that's currently being studied. So when you take all of that evidence into account, what it looks like is that you have human beings like ourselves co-existing with other sorts of beings on this planet, really as far back as you care to trace—hundreds of millions of years, literally.

LL: The co-existing theory isn't so farfetched because now they think that Neanderthal and Cro-Magnon man co-existed for a hundred thousand years in our history, side by side, contemporaneously. It's the current theory. Isn't that correct?

MC: Right, and there are even people—even scientists today such as Myra Shackley, an English anthropologist—who would say that we still are co-existing with creatures like the Neanderthals. You have these reports of various kinds of wild men in various parts of the world, such as the Yeti or Snowman in the Himalayas, Bigfoot or Sasquatch in North America, and other such creatures in other parts of the world. So yes, I'd say this co-existence idea is quite a valid one.

LL: Well, let's give this a fair look. Let's look at the evidence. We'll start in the beginning and kind of work our way forward, because I was noticing that the farther back you go, you find very, very sophisticated finds, and as you come forward in time towards the present, there's less and less of the highly sophisticated. But you do start dating man, early man, back farther than the general establishment would lead you to believe. It's quite an interesting story, and we welcome your questions. We're talking with Michael Cremo, co-author of *Forbidden Archeology: The Hidden History of the Human Race.* Michael, let's detail a couple of these extraordinary finds that date back millions and millions of years before they're really supposed to, according to the theory in place. Would you like to start with a couple of examples? And what I'd like to know is the circumstances of their discovery and how well documented they are, and where they are today.

MC: Well, I'll give you one very good example from recent history. It's one of my favorites. In 1979 in Laetoli in Tanzania, which is a country in East Africa, Mary Leakey, who is the wife of Louis Leakey, one of the most famous anthropologists of the 20th Century, found in some volcanic ash about three-point-six million years old, some footprints, footprints of three individuals. And many footprint experts looked at these prints, physical anthropologists and others, and this is all documented in *National Geographic* magazine, and various scientific journals which we cite in our book, and you can look it up, chapter and verse, if you like, and see the photographs of these footprints, and they are absolutely indistinguishable from modern human footprints. One researcher said that, if you went out on a beach today and looked at footprints in the sand, they wouldn't be any different than these footprints. Now, what I found very remarkable is that, despite that, despite that, the mind-set of these investigators and researchers was such that they could not draw the obvious conclusion, namely, that these footprints must have been made by creatures very much like ourselves.

LL: Well, on the one hand, they look at the evidence and use it when it supports their case, but they refuse to look at equally valid evidence when it doesn't. I mean, isn't that being a little bit hypocritical?

MC: It is. We call it a knowledge filter, and in one sense we're not talking about some kind of diabolical plot to deceive the public. It's rather a kind of self-deception that these people engage in. For example, with these footprints, they said, well, they must have belonged to *Australopithecus,* even though they know they have footbones of *Australopithecus* which is an ape-man creature that supposedly existed three million years ago in Africa. They have footbones from this creature, but they do not match these footprints.

LL: By a lot or a small degree?

MC: By quite a bit, because the footbones of the very ancient ape-man-like creatures, the Australopithecines, have very long toes, curved toes. They have a big toe that's sort of like our thumb. If you look at a chimpanzee, look at its foot. It has a very large mobile big toe, sort of like our thumb, that it can grasp branches with. So these footprints that were found in Africa and that are dated at three-point-six-million years do not match those *Australopithecus* feet at all.

LL: Why do they choose *Australopithecus* to provide an explanation for them?

MC: Because they think that is the only creature that existed at that time that walked on two legs, and because they ignore all this other evidence, this massive evidence, that shows that human beings like ourselves were around at that time. It just does not enter into their minds to draw the obvious conclusion. You have these human-like footprints. A human must have made them.

LL: What of the Leakeys? How did they interpret the evidence?

MC: Well, Mary Leakey, she wanted to say it must have been some kind of ape-man with human-like feet that made those prints. Now, if that's the only evidence that we had, you could say, well, maybe she was right, but in *Forbidden Archeology* we document many, many categories of evidence. Stone tools, all kinds of artifacts, other human bones, complete human skeletons dating back to the same period. That's what makes us think they actually must have been humans that made those prints.

LL: Because it's corroborated by additional evidence.

MC: Right.

LL: Suggesting that human history has some very interesting chapters which are being ignored simply because we are encumbered by an inadequate theory, other explanations—this is an exciting story, and we'll continue with it.

MC: Well, a very interesting case is a skeleton that was discovered in Africa also, in the early part of this century, in 1913 by Dr. Hans Reck who was from Berlin University in Germany, I believe. He was in what is now the area called Olduvai Gorge...

LL: Very popular for finds, isn't it?

MC: Right. It's where the Leakeys later did much of their work, and in 1913 he found a completely anatomically modern human skeleton fossilized in strata that were over one million—almost two million—years old. Now that's extremely unusual because, according to modern scientific belief, you wouldn't have anatomically modern humans like ourselves until about one hundred thousand years ago or so...

LL: How did they explain it? In 1913 wasn't archeology just getting a feel for man? Why weren't finds like that put into the equation?

MC: Well, at that time the modern ideas were already shaping up with the discovery of Java man in 1894. Now this is very interesting. It gets into what you would call a detective story. Because I noticed when Charles Darwin wrote *The Origin of Species* in 1859 and set off intellectual shock-waves that went around the world, the question most people were interested in was the origin of the human being.

LL: We are rather self-preoccupied, aren't we?

MC: Right. We're not so much interested in the origin of the butterflies or the crab. We're interested in where we came from. So I looked at modern textbooks, and I could see from 1859 when *The Origin of Species* was written until 1894 when the Java man finds were reported, I couldn't find any reports, and I thought, well, that's very mysterious. You would have thought that almost immediately scientists all over the world would have been looking for the missing link and finding all kinds of things. So I asked one of my research assistants to go to a library, get me an anthropology textbook from 1880 or 1885 or so, and let's just look, and I was shocked when he brought in these books, and these were scientists. These were not marginal people. These were genuine scientists of the period and reporting in standard journals all kinds of evidence showing that anatomically modern humans—not ape-man, not missing links—were present ten million years ago, twenty million years ago, thirty million years ago, forty, fifty—as far back as you want to go. I'm not talking about one or two discoveries. I'm talking about hundreds, and we've got a thousand-page book, *Forbidden Archeology*, that describes them all.

LL: It's extraordinary, and I want to know where are these finds? Where do they reside today? What happened to them? Hundreds and hundreds of finds of anatomically modern human skeletons that date back a hundred million years and beyond—what's the earliest, earliest find that you could find documentation of?

MC: Well, the earliest find that we have on record is—of course, now we're talking not just about human skeletons only but human artifacts of all kinds.

LL: Right.

MC: Things that are obviously made by humans and also of skeletons, but the earliest artifact that we have is a grooved metallic sphere which was discovered in South Africa. Many of these metallic spheres were discovered, perfectly round, a metallic object, some of them with three grooves around their equator, and these date back two-point-eight billion years...

LL: Two-point-eight billion with a "b"?

MC: Yes...And the earth is said to be about four-point-three billion years old, according to current scientific estimate, so these are quite old. Now, the oldest human sign that we have is a footprint that was discovered in Antelope Springs, Utah, in 1968, and that dates back to the Cambrian which is about six hundred million years ago.

LL: What about the earliest skeleton remains of anatomically modern humans?

MC: The earliest skeleton that we have in *Forbidden Archeology* is a human skeleton that was discovered in a coal field in Macoupin County, Illinois, in 1862, and this was reported in a magazine called *The Geologist,* a standard journal, and that comes from the Carboniferous period, and it could be about three hundred million years old—so these are quite extraordinary finds. Now, one thing that happens is, if things do not fit the current paradigm, sometimes they don't tend to be preserved. If you have something that fits the current ideas about human origins in antiquity, then it's kept very carefully by the establishment.

LL: It's fussed over, it's published, you read about it...

MC: Right. You hear about it. It's put on display at the museum, you know, they make videos about it and you see them on *National Geographic* specials and things of that sort. If something doesn't fit the current paradigm, then you don't hear about it. It doesn't get preserved, and it's very difficult to track down. I would say that, people like yourself, media people of all kinds, have a duty to look a little bit deeper. It's as if you went to a politician, and they were trying to cover something up, and you just accept their story, and then don't check on it, don't dig for the facts, so I think a lot more of that work has to go on. We can't put as much faith as we do in people just because they occupy a university position or have a Ph.D. behind their name.

LL: I'm in total alignment with you on that point for sure. Michael, you were just stating a very different view of entertaining alternative theories.

You're saying it's one's duty, one's intellectual duty, one's civic duty, that it's really a must to go digging and find out if there's any hidden points of view, any cover-ups, any alternatives, you know, what is the motivation behind the current and reigning paradigm? I've heard so many people express the idea that if you're silly enough to look at ideas that are not endorsed by the establishment, then you're being duped or you're just really being silly and wasting your time. Why this divergent view on participating or at least being willing to entertain alternative ideas, do you suppose?

MC: Well, I think what we have to do is, we have to take the same attitude towards science that we do towards politics. For example, if there is some allegation about some political misdeed, we don't just accept and we don't just go to the politician and accept his statement.

LL: Give me your propaganda, in other words, right?

MC: Right. So I think that sort of attitude is now too prevalent. I think we have to become a little more...

LL: Skeptical?

MC: Skeptical. A little more independent in the way that we look at these things, and sometimes it takes an outsider to really look into a question. People want some independent verification, so you can't always go to the experts and take their opinion in their own field. You need outside people to go in and dig around a little bit and look, and that's basically what we have done with our book. We have gone in there without any preconceptions and dug around, and we've found quite a bit to show that we're not getting an objective look at the past. We're not being told the whole truth. We're not being given all the facts, so in *Forbidden Archeology* we just try to provide all the facts and let people make up their own minds about these things.

LL: What is yours and your co-author Richard Thompson's background in the field?

MC: We don't have backgrounds in this field other than the research that we've done.

LL: How easy was it to do the research? How accessible were some of these reports? How did you know where to go digging if they're not well-publicized or well-known?

MC: Well, it was like a detective story. You go back in time. You look farther and farther back in the records, and then you start finding things, and then you have to trace out obscure little footnotes. It's like a cover-up. It's really like a cover-up. You have to really dig for it. We had to get journals, obscure journals, and reports from all over the world, some dating back hundreds of years, and we had to translate them from German and French and Spanish and Russian and whatever languages they happened to be in, so it took about eight years. It took eight years of very painstaking work to gather this information which we then put into this book, and it's all thoroughly documented so every reference is footnoted. It's got a complete bibliography.

LL: They don't even have to take your word for it. You provided them the paper trail by which they can go back and re-verify the evidence that you provide. Obviously, even in a thousand pages, you couldn't fit in everything. What was your criterion by which you chose what to include in your documentation and your book?

MC: We...we wanted to choose things that were—that we had the most solid evidence for in terms of documentation and scientific testimony. Now, one thing that we did is we learned that this process of knowledge filtration, which we like to call it, is ongoing. It's happening even today, and we were very lucky to run into some modern researchers who had been victims of this suppression, and they were kind enough to give us some information about how the system actually works.

LL: How does the system actually work?

MC: Well, there is money involved. There are positions involved. There's opportunity for publication involved, and there are very small and powerful groups of people that control these positions. You want to become a professor at a university, you need recommendations. If you want your articles to be published in scientific journals, they have to pass what is called anonymous peer review.

LL: People can comment, but you don't even know who they are?

MC: Yeah, you don't know who they are, and in one sense you could say that a dominant group could very easily use that to screen out information that it doesn't want to reach the wider scholarly community, let alone the public.

LL: Is there dissension in academia for this kind of set-up that is so easily abused, or that is so useful to maintaining the status quo, not really expanding the field in terms of its outlook?

MC: Yes, there is. From time to time, I tune in through my computer into the Internet...

LL: So do I.

MC: And there are discussion groups that discuss some of the shortcomings of the current system, and there are people that object to it, but it's still pretty much in place.

LL: What are some of these Internet groups? Are they art groups? Are they libraries? I mean, what are some of these places where you can find some of this information? We have a lot of Internet users among our listener-base.

MC: Well, the one that I found—I can't remember its exact name but it was really odd, because it was a discussion about employment opportunities. That was what the discussion group was technically about, about academic positions and how to apply for them, what jobs were available, but there was all kinds of discussion about the system and how it operated, in that particular discussion group.

LL: Right, no sense talking about the positions unless you figure out how they work.

MC: Right. Because for every academic position that opens up, there are usually hundreds of applicants.

LL: Oh, I see, so you're encouraged just by the competition to conform...Right. you're going to do everything you can if you're really serious about that position.

MC: You have real powerful people in these fields who control position, publication and research money, and if you want to get along you have to go along. That's basically how the system works, and I've had personal discussions with people who have been victims of that system, who have been denied publication, who have been denied position, who have been denied research money.

LL: What kind of heresy were they wanting to report?

MC: Because of their views. Now, one case which we discussed in the book is the case of Virginia Steen-McIntyre, who was a geologist who worked for the U.S. Geological Survey, and she and a couple of other geologists dated a site at Hueyatlaco, Mexico. This happened in the 1970s where some very advanced stone tools were found, stone tools that could only have been made by anatomically modern humans. They dated the site by the uranium series method and other methods, to about 300,000 years. Now, according to current doctrine, you don't have human beings coming into North America until 12,000 years ago, although some people are now willing to extend it to 25,000 or 30,000, but the standard conservative doctrine is 12,000 years.

LL: Right, at the end of the last Ice Age when you had the Siberian land bridge...

MC: Right.

LL: They don't recognize that there could have been other means of crossing an ocean.

MC: So these very, very advanced stone tools at 300,000 years in Mexico are extremely anomalous, and what they speak of—there shouldn't have been any human beings of that type around until about 100,000 years ago. As a matter of fact, stone tools of the type that were found there don't show up in Europe until about 40,000 years ago, so the fact that they are in Mexico where human beings shouldn't have been at all, and at 300,000 years which is 250,000 years before—earlier than any other such stone tools were found that...

LL: That creates a problem, doesn't it?

MC: These researchers, they produced a report but couldn't get it published. Nobody would publish it.

LL: But what happened to the tools? Where are the tools today?

MC: They're stored away in some museum. These were found fairly recently so you can still find them. We tried to get permission to get photographs of those tools to put in our book, *Forbidden Archeology*, but we were told we would be given permission to print those pictures only if we gave the tools a date of less than 25,000 years, and if we were even going to mention that they were 300,000 years old as these geologists reported, that we could not have permission.

LL: You know, it sounds like we're really...all of this information, I mean who's to say how to interpret it. Michael, tell me about another couple of finds, and then I'd like to ask you—isn't all this stuff kind of a world heritage, and who gets to say what about it? What are some of the restrictions, and what are the various methods of dating? How do we know—how are certain dates assigned? What are the various methods of assigning dates? I know a lot of it has to do with the strata in which it is found, but tell me some more of these discoveries, and we'll talk more about the issues surrounding them.

MC: Well, one interesting discovery was the carved shell in the Red Crag formation in England. Now this is a late Pleistocene formation, about two million years old, and Henry Stopes, who is a Fellow of the Geological Society of England, found this shell. It's got a human face carved in it, and, according to modern views, you wouldn't expect art work of this type in Europe until about 40,000 years ago, at the earliest, so at two to three million years it's quite anomalous, and that was discovered in the 19th Century.

LL: Where is it assigned today, for example, this shell?

MC: Now, that should be in a museum in England.

LL: Okay. What's another anomalous find?

MC: Going back to North America and to more recent times, where we have some really good documentation of how evidence can be suppressed, we had a case of a Dr. Lee in Canada, at a place called Sheguiandah on Manitoulin Island in the Great Lakes, who found stone tools there in glacial formation that are about 70,000 years old. Now, as I said, the current idea is that you don't have human beings in North America until about 12,000 years ago. He was working with the National Museum of Canada at the time that he made these discoveries. He had a geologist come in and look at the site and confirm his dating view of it, but he got fired from his job, and they wouldn't publish the report, and he couldn't get another job for years. He was extremely upset by that, and all his stone tools that he found were just taken and stored away somewhere, out of his control, by the museum.

LL: That's how they bury it?

MC: They buried it.

LL: From one burial to another, some of these finds. We're talking about *Forbidden Archeology*, and, in fact, the publisher has a give-away. They'll give away a chart of these anomalies. A 16-page chart for the asking, and you can ask me. I'll be taking care of this, just drop a SASE to Laura Lee at P.O. Box 3010, that's in Bellevue, Washington 98009. And we have Steve in Vista. Hi, Steve.

Steve: Hi, Laura Lee. So good...I had read a long time ago that, and they quoted this and made it sound like this could be science fiction, maybe, but what their idea was, and maybe you've got something to follow up on this, is that actually mankind had risen to quite evolved mechanical and scientific heights seven times in our history, and for one reason or another, like we discovered a bomb or something, for one reason or another, we've forced ourselves back into the Stone Age, just about. Is there any kind of proof to that in what your work turns up, or is there any kind of modern things that show up?

MC: Well, that's a very good question. We do have some modern things that we document in the book. Most of the evidence that we have in the book are discoveries that have been made by scientists and reported in scientific journals over the course of the past hundred and fifty years. Now, at earlier time, in the beginning of this century and the last century, scientists were prepared to find evidence for crude human beings and stone tools and things like that going back ten, fifty, a hundred million years, and that goes against completely modern ideas. They used to have an idea like that, but they weren't prepared to find this evidence of advanced culture, these sorts of things that you're talking about, so we don't find very much in the scientific literature about such things. But we do find that other people, coal miners in particular, have found some very advanced things in very ancient strata, but these aren't so easily documented. Human prehistory is a lot more complex than we have been led to believe, and it could be cyclical, like you say. I think that's one of the interpretations that you could give.

LL: Well, then you must ask, what's the mechanism which causes the rise and the fall. Do you have any speculation on that that you want to share with us and tag it speculation?

MC: Well, this is something I want to go into in detail in my next book, where Richard and I will be giving our view of the alternative to the current ideas.

LL: Okay. Steve, any further comments?

Steve: No, it's just that I find this is another one of your fantastic shows, Laura Lee. You bring us the best shows on radio.

LL: Thanks for your time tonight. We're talking about *Forbidden Archeology*, that is archeology that's ignored by the establishment, by academia, because it lies outside the story they like to tell of human history. This is stuff that dates back much, much earlier than it's supposed to, but I'm asking, why can't this be included? Yes, it would revamp our theories of our early history because this stuff is real. It's well-documented. It's abundant out there, and who owns this stuff? Who has the right to say how it's going to be talked about, photographed, ages are assigned, and such? What are some of the dating procedures? How is it that some of these dates are determined? These are some of our questions. What are the stories of some of these fantastic anomalies?

We're talking with Michael Cremo. He and Richard Thompson have written a stunning 1,000 page compilation of well-documented cases of anomalous artifacts and human remains. They're anomalous simply by their date. They are much, much earlier. They are painting a very different picture of our human history, and it's a fascinating one. Their book is *Forbidden Archeology: The Hidden History of the Human Race*. It's buried and then reburied, in a manner of speaking. Michael, you mostly include in your book those artifacts that can be well documented that you went on a detective hunt for in the scientific literature, but you were mentioning that some of the less well-documented finds come from coal miners, because they're down there digging in very ancient strata. For example, I've even heard of people who are breaking up coal to put into their stoves, and out pops gold chains and metal vases and just incredible things. Tell me some of the stories that are found in some of these very ancient coal strata and things that are found by miners. Where do those finds end up, and what are some of them?

MC: Well, here are some of the ones that we discuss in the book, *Forbidden Archeology*. In 1897, the *Daily News* of Omaha, Nebraska, carried an article titled "Carved Stone Buried in Mine." Now what this was about was a piece of rock that was about two feet by one foot, and the rock was carved into the shape of diamonds. It had marks on it dividing the surface into diamonds, and in the center of each diamond there was an engraving of a human face, a fairly old person, and the question is, how did this get down there? This was a mine dug deep into the ground, a hundred thirty feet deep. The miners said the earth had not been disturbed, and coal miners are very good at this because their life depends on it. If they see that there's—if they're digging somewhere, and they can see that the coal has been disturbed, that maybe there was some shaft there previously that had been filled in—they watch out for that like anything, because they know they could be buried in a landslide or caught underground so they're no dummies. When they say that the coal they were digging in was not disturbed at all, I think we can take their testimony for it. That coal in that part of Omaha—near Omaha, Nebraska—was about three hundred million years old so that's really amazing. Now, where is that object? We tried to track it down, and we could not find it. It was reported in the newspapers. We found a lot of literature about it, but because it's so far out of what modern scholarship would accept, it wasn't put in a museum anywhere. Probably one of the coal miners just kept it, and then when he died it went on to some relative somewhere, and maybe they threw it out...

LL: Or maybe it's sitting in an attic, and nobody knows its true story.

MC: Right, if anybody out there has stories like that...

LL: You want to know about them.

MC: We'd like to hear about it. And we've heard through our little grapevine that coal miners are still finding such things.

LL: What happens to them today? Is there any central clearing house? Do you just—I mean, have you passed out literature to coal miners to say, tell us if you find these things? I mean, where do they end up, just in someone's shoe box?

MC: That's pretty much what happens, and one of the things that we're hoping with the publication of our book, *Forbidden Archeology*—we've got a notice in there asking if people who read the book have any knowledge of any such finds or discoveries to please report them to us, and we are getting a few letters.

LL: Michael, I've heard that additionally there are stone walls buried a hundred and fifty feet. I know there are examples of that in Texas and California, I believe. Are you also—did you come across such huge artifacts, I guess, as buried stone walls?

MC: We have come across reports of such things. We do mention them in our book. There's one such case. It's again from Oklahoma, Heavener, Oklahoma, where we have a report from a coal miner that in the year 1928 he was working in this mine at a depth of two miles, and the way these mines were set up, they had different chambers. Each day they would blast out another chamber so one morning they blasted out some coal, and the miners saw at the end of the room, they call it, a wall of concrete blocks that were very smooth and polished, and it was a wall, built down there. They told the mining officers about it, and they just put the men out of that mine, sent them to another mine, and they filled up that part of the mine.

LL: Does that seem to be standard procedure when such things are found?

MC: Well, there does seem to be a pattern. One thing that might happen is that some of these cases may be reopened and re-investigated, you know, by people with more open minds.

LL: Doesn't it make you wonder that if the earth were transparent for a day and we could actually see all of the dinosaur bones, all of the unusual artifacts, all the human remains, all the footprints, all the things that are buried in the earth from long, long ago—what do you think we would see, Michael? In what abundance?

MC: Well, I think what we would find is a picture of all kinds of creatures, humans and otherwise, going back for long, long periods of time. Now, of course, one problem is that just seeing isn't enough, because a lot of the seeing goes on in our own minds, and you can always explain something away. What we found is that there's a double standard in the treatment of evidence. If something goes along with the current ideas, then...

LL: No problem, right?

MC: No problem. If it goes against the current ideas, then you can immediately pick out all kinds of flaws, because this kind of evidence that you dig up out of the ground, even in the best of circumstances, you can always get some counter-explanation. At the very least, you could just say it was a very elaborate hoax or a fraud...

LL: Right, that someone put it in there or...

MC: Right. But see the problem is, that if you are going to do that and you apply the same standard to the things that you now find in the museums, then you'd have to throw them out, too. For example, if you were to find an anatomically modern human skeleton in some coal near the surface, if it actually belonged in the coal, it would be two hundred million years old, but if someone says, well, it's close to the surface so it must be recent, well...But then most of the archeological discoveries that we have, like Lucy, the most famous specimen of *Australopithecus* found by Donald Johanson in Ethiopia in the 1970s—that was found on the surface, too. As a matter of fact, most of the Java man discoveries in Java were found on the surface. They weren't found buried under the ground.

LL: But they say that's because that strata had eroded away to expose it at that level.

MC: Well, that's what they say, but...

LL: I see, but why the double standard?

MC: I have a letter with me. This is from Dick Muller in Oak Ridge, Tennessee, and he says, I salute you for your rather monumental work, *Forbidden Archeology*. It's satisfying to see this dishonesty so carefully documented. You no doubt know what will happen to your book. Mainstream archeology will try desperately and probably successfully to ignore your book out of existence.

LL: Do you have any where people have told you about certain finds of their own or of things that have been handed down through their family or...?

MC: Well, here's a letter from Ron Calais in Lafayette, Louisiana, that he recently completed reading a copy of *Forbidden Archeology*. "Having conducted nearly 30 years of intensive research work on this very subject myself, I can say without hesitation that this collection of anthropologic discoveries is about the most convincing I've ever digested." He says, "I have a map of similar data on file, some of which the authors have not mentioned in their book." So we've gotten some of his material, and we're looking it over, and we've also had some communications from people who have been investigating human footprints in Utah that are from the Cambrian period, which would be six hundred million years old, so we're evaluating the material that comes in, and we'll try to make use of it in future editions of the book.

LL: I've heard of gold chains dropping out of lumps of coal. I found these documented in certain books on anomalies of every description. Can you tell me some stories on that?

MC: Well, they are documented in several books, including our own. The one very particularly interesting case took place in 1891, and this is recorded in *The Morrisonville Times* newspaper from Illinois. It was actually the wife of the publisher of the paper, you know. Mrs. Culp was breaking coal to put into her stove, and she found a small gold chain, very intricately worked gold chain, in a lump of coal. As a matter of fact, it was so tightly embedded in the coal that two pieces of coal were hanging off each end of it, kind of like a little pendulum after she broke the piece. Now, what we did was actually verify that the newspaper did have a copy of that article describing that find in it, because we'd heard about it, but we wanted to check with the newspaper itself, and they sent us a copy of the article. We also checked with the Geological Survey of the State of Illinois about the age of the coal that that piece of gold chain was found in.

LL: And we're back. Hi, I'm Laura Lee, and we have Ken from Lexington, Kentucky. Hi, Ken.

Ken: Hello. I was wondering if there was any possibility of the correlation between the artifacts that have been supposedly discovered on Mars and the moon and areas of Lemuria and Atlantis?

LL: You mean to connect some of those ancient legends and potential stories altogether? I don't know. Do you have any thoughts on that, Michael?

MC: Yes. I like to research things very thoroughly as we did in *Forbidden Archeology* and in our next book we're going to be looking at those things that you're talking about, but just briefly I'll say that if you look at all the evidence that's available to us at present, I think we would have to conclude that civilization is an interplanetary phenomenon. I don't want to say too much more about that right now, because that's the subject of my next book, which I want to very thoroughly research.

LL: And you want to be able to back up such statements.

MC: Right. We would like to be able to back them up, but that research and writing is in progress now, and we hopefully will be able to come out with a book that will deal with those topics.

LL: Are there artifacts that point directly to such a conclusion, or is it speculation?

MC: Well, you know, if you look at the whole spectrum of evidence that's available to us, when you take into account what people have reported over the past few thousand years, you know, the Egyptians, the Babylonians, going all the way up to the present-day people, reporting strange phenomena from the skies. There is this photographic evidence from Mars. I haven't come to a firm conclusion about it yet, but there is photographic evidence.

LL: It's worth looking at, isn't it?

MC: Yeah, it's worth looking at, and I am looking at it, and I plan to come to some conclusion about it and come out with a very thoroughly documented book that will take all this evidence into account. But before getting into an alternative, to show the necessity for it, and that means looking at the current ideas and the current evidence and just dealing with that, showing that that's not adequate to explain human origins in antiquity, and in *Forbidden Archeology* that's what Richard Thompson and I tried to do, to show the necessity for having an alternative explanation.

LL: Ken, thank you for your call. Michael, what happened to the gold chain? How were you able to document that? It was a story of what someone found in a piece of coal. Other than looking at stories that were written at the time, did you ever find some of these or is somebody wearing it around their neck, thinking it's just a regular antique?

MC: Well, we tried to track down that gold chain...about three hundred million years old, and it was reported in a newspaper, and what we found out was that the owner died in 1959, and all we could find out was that the chain had been passed on to one of her relatives after her death, and we could not track it down any farther than that. Now, that's why we put that case in our appendix of extreme anomalies in that we could not completely document it. Now, the main body of the book is composed of things that are a little less spectacular, but much better documented, and the artifacts are still there, most of them in museums, and you can see them.

LL: What's the story about the eoliths and the early stone tools?

MC: Well, eoliths are very crude stone tools, and they're recognized as a genuine category of archeological evidence. They're found in China at Choukoutien in Olduvai-like ridge, in Tanzania and Kenya, and the problem is that, according to modern scientific ideas, these would be made by *Australopithecus,* and they would not be found any earlier than two to three million years ago. Now, through our research we have found dozens of cases where stone tools of this type have been found in Europe and elsewhere in locations that are, say, up to fifty million years old. Now, that is completely unexpected, according to modern ideas, and you just don't hear about these things. I mean, you can look in your modern textbooks, and you will not hear, for example, that J. Reid Moir, an English scientist, found such tools in the Red Crag formation in England in layers that could be two million, up to fifty million years old, or that a Dr. Rutot in Belgium found them in Belgium in formations that could be up to fifty million years old. Even if you accept the idea that maybe they weren't made by modern humans, anatomically modern humans, that they were made by some sort of ape-man, well, according to modern science, you wouldn't have ape-man that far back. You wouldn't even have monkeys, really, according to their ideas, so they're extremely anomalous. But one thing to keep in mind is that modern tribal people, such as the Australian aboriginals and even modern tribal people in Africa, use the same kinds of tools today. They just pick up a rock, and they knock off a few chips of it, and they use it as a tool, so you can just as well say that an anatomically modern human made them as an ape-man. As a matter of fact, we have the modern—anatomically modern—human bones, as I've said, going back to three hundred million years, according to the discoveries that we document in the book.

LL: We'll take a break, and when we come back I'd really like to hear your explanation about how the current paradigm was formed. What went on, under what agendas or what particular modus operandi, with what governing principles were they putting together this picture of very early man, when at least half the evidence suggests that it's a really very old history of mankind to be seen. I'm Laura Lee. We'll be right back.

LL: We're back, on *Forbidden Archeology*. Tom, in San Luis Obispo, thanks for joining us.

Tom: Howdy. I have some questions on the grooved metallic spheres which were discovered, and I'd like to know how many were discovered, their composition, the method of dating, and was the dating peer reviewed?

MC: Okay. A lot of questions there, but I'll try to answer as many of them as I can. The grooved spheres were found near Ottosdal in South Africa. Now, they have been found for a long time. There have been hundreds of them found. Not all of them have the parallel grooves going around the equator. These have not been reported in scientific journals. That's why we include this particular case in an appendix in the book, but they are being kept in a museum in Klerksdorp, South Africa, and we wrote and corresponded with Roelf Marx, who is the curator of the museum there, and he told us they are a complete mystery to him. He said they look like they're man-made. These are his exact words, that they look man-made, yet at the time in earth's history when they came to rest in this rock, no intelligent life existed. Now, that's his assumption so you see how this works. He said they look like they're man-made, but they *couldn't* be man-made because "I know that no human, not any kind of life existed at that particular time." Now, these were found in a pyrophyllite formation and this is a mineral that has been dated at two-point-eight billion years there. He provided that information from a Professor Bisschoff, the professor of geology at a university there, the University of Potchefstroom. I don't know if I'm pronouncing that exactly right. These spheres are said to be made of limonite—it's a kind of iron ore—but it's a very unusual kind of limonite, because they are extremely hard. They can't even be scratched with a steel point, which means that in terms of mineral hardness they're *extremely* hard. Usually just a limonite ore would be very soft, and these are extremely, extremely hard, so they're very mysterious. Now, as far as peer review, no, because they're not described in any scientific journal as such. I don't think you could find any current scholar who would be willing to consider these as made by a human, although they say they appear to be made by humans.

LL: So who knows? Tom, anything else?

Tom: Yes. Have any of the anomalous archeological discoveries been peer reviewed, and also is Mr. Cremo familiar with Dr. Charles Hapgood's discoveries on certain ancient port-to-port navigation maps which he documented in his book, *Maps of the Ancient Sea Kings: Evidence of Advanced Civilization in the Ice Age*?

MC: Yes, I'm familiar with Dr. Hapgood's work. I think it's very good work. There are maps that are very, very old. Copies of them were made in the Middle Ages, but they, for example, show the coastline of Antarctica under the ice. So those maps must have been made at a time when that coastline was visible, which was a very, very long time ago, and those maps do appear to give some evidence of advanced civilization in early times. Now, that evidence I don't talk about in *Forbidden Archeology*, because I'm mainly interested, in *Forbidden Archeology*, in stones and bones and things of that sort, you know, the hard evidence. But I want to talk about those maps in my next book, and I think they're really worth considering. Now, some of the evidence was peer reviewed. The evidence—most of the evidence in this book, *Forbidden Archeology*, did appear in standard scientific peer-reviewed journals. Now, what happened—this gets back to the question that Laura Lee was raising—what is the history?—is that especially during the 19th Century when they first started looking, seriously, into human origins in antiquity, even people who believed in evolution and ideas like that—they were finding things, but they thought that the process of evolution took place much further back in time than scientists now believe is possible. They didn't have a clear idea yet so they were looking with open minds—they were looking with open minds, and these scientists were finding signs of an advanced human presence going back ten million, twenty million, thirty million years. Now, things changed around the turn of the century when they found the Java man, ape-man in Indonesia, and that was only eight hundred thousand years old...

LL: Oh, so they figured, how could you have something that primitive so late in the record?

MC: Well, what they figured is that if you have something that primitive so late in the record, then what are we going to do with all this other stuff? All these hundreds of other discoveries that have been made of anatomically modern human skeletons...

LL: Well, then why not call that one Java man the anomaly, because you have this whole body of evidence that says otherwise?

MC: Well, that's the case—that's the case that I would make, but what they did instead was—they threw out all the previous evidence, and then anything that ever came up again, and it still does come up—it comes up even today—then they'd say, well, it must be a hoax, you know, it must be...

LL: That's circular reasoning.

MC: They'll explain it away, and there are a thousand different ways you can explain something away if you want to, but if you applied that same standard to Java man, you'd have to throw out Java man, too. And one very interesting thing about that original Java man discovery: it consisted of two bones, a skull fragment and a thigh bone or femur, and in 1973 two British scientists looked at that femur—this was M. H. Day and T. I. Molleson—and they published a report showing that it is not any different—it's not an ape-man femur at all. It's no different than an anatomically modern human femur. It's completely different from *Homo erectus* and every other ape-man that they've found, so that means that that original discovery, which is still talked about and written about in textbooks, wasn't a real discovery at all, and as a matter of fact, you can take it as evidence for anatomically modern human beings in Java eight hundred thousand years ago.

LL: We're back. Maybe a good reason not to believe everything that you're told by those in positions of authority—do we believe the evidence or do we believe their story or do we kind of look at it, and say, there's a dilemma here, a controversy, and it requires further study. A great resource for that is *Forbidden Archeology*, the book. A caller wanted to know off-air about Noah's Ark. A quick comment on that, Michael?

MC: I've seen a lot of reports by different researchers who've gone to Mount Ararat and other places in Turkey and the Middle East, showing evidence of a large ship-like structure, so I don't know too much more about it than that.

LL: Okay, and we have Tom. Do you have any other comments, Tom?

Tom: All the comment I have is that it's disappointing to have this pattern of selfishness and arrogance and fear, covering up corruption and deceit and deception on the part of various professionals, keep re-appearing time and time again. Here it's in archeology. Elsewhere it's been medicine and health and scientific exploration such as planetary exploration which Dr. Hoagland brought up. They're very disappointing. I do thank Mr. Cremo and Thompson for bringing these discoveries to light. Thank you for another good program.

LL: You're very kind, and you put it so well. I want to thank you for the call, and also thank you for a stunning collection, Michael. It's just wonderful to read *Forbidden Archeology*. One last parting shot. A partial shoe sole in Triassic rock from the modern Triassic dating to about two hundred and fifty million years ago. This one had the well-defined thread through which the sole of the shoe was sewn right there imprinted into the rock. How extraordinary. I mean, how can you miss something like that when you can see the stitching around the sole and when the rock is dated to the Triassic? Where is that, by the way, that rock...?

MC: Well, it's another one of these cases of cover-up. Again, that's one more case of a cover-up where you had something—you had a wonderful piece of evidence that was shown to scientists. The person who discovered this, he took it to New York. He took it to Columbia University. He showed it to some of the leading people there at the American Museum of Natural History, he turned it over to them, and we wrote to the American Museum of Natural History, and they just had no information about it. They said that the report is not in their files.

LL: Once again, don't believe everything you hear. Let's check out the alternatives. I think it is a very interesting process and a very important process, and it's certainly one that is going to expand our paradigm. Thank you for joining me in that journey yet once again.

I'm Laura Lee.

5.2. Interview with Michael A. Cremo by Jeffrey Mishlove for the television series *Thinking Allowed*, aired on Public Broadcasting Stations in 1994. Transcript reproduced by permission of Arthur Block, producer.

Thinking Allowed *is produced by Arthur Block. It airs on about 80 Public Broadcasting System stations in the United States.*

Jeffrey Mishlove: Hello and welcome. I'm Jeffrey Mishlove. Today we are going to explore the origins of the human race. With me is Michael

Cremo, who is the co-author with Richard L. Thompson of an unusual book called The *Hidden History of the Human Race.* Michael is a research associate with the Bhaktivedanta Institute in Alachua, Florida, specializing in the history and philosophy of science, and he is the author of a number of books, including *Divine Nature: A Spiritual Perspective on the Environmental Crisis.* Welcome, Michael.

Michael Cremo: Pleased to be here, Jeffrey.

Jeffrey: It's a pleasure to be with you. As a member of the Bhaktivedanta Institute you follow the Vedantic, or Hindu, ancient Hindu traditions of philosophy that offer a very different perspective on the history and origins of the universe, as well as the human race than we find in either Western religious traditions or in the Western scientific tradition. And, your research seems to go quite a long way to suggest that there are some cracks in the construction of our conventional cosmologies that might support more of a Vedantic view of things.

Michael: Well, that's absolutely correct, Jeffrey. Richard Thompson and I do draw our personal inspiration from the ancient Sanskrit writings of India, the Vedic literature. And, in particular, if you look into the Puranas, which are the histories—*Purana* simply means history—you will find that there are descriptions of human civilizations being on this planet for hundreds of millions of years. Now, when we looked at those accounts, we thought if there is any truth to them at all, there should be some, some kind of physical evidence for them. Of course, when we looked in the modern textbooks we didn't find any such evidence. There you will only find evidence supporting the idea that human beings like ourselves have developed fairly recently on this planet from more primitive ape-like ancestors within the past hundred thousand years or so.

Jeffrey: I think the view that I've heard most often is that fully modern human beings have only been around about forty thousand.

Michael: Um, that view has now been superseded on the basis of discoveries in Africa and elsewhere, and most physical anthropologists now working would go for a date of around a hundred thousand.

Jeffrey: They'll concede a hundred thousand. But you're suggesting a hundred million.

Michael: Oh, even...even more than that. Hundreds of millions...

Jeffrey: Hundreds.

Michael: ...there appears to be some evidence for that. So, what we did, Richard Thompson and I, we turned the Vedic account, the Puranic account, into a hypothesis.

Jeffrey: Um huh.

Michael: And we decided to look not just in the current textbooks, but into the entire history of archeology and anthropology over the past one hundred and fifty years. That took us eight years of research. Ah, it meant going to literatures from all over the world, translating them from many different languages. And what we found is that over a hundred and fifty years, archeologists and anthropologists have literally buried almost as much evidence as they dug up out of the ground.

Jeffrey: Uh hmm.

Michael: And what that buried evidence suggests is, indeed, that human beings like you and me have been around for hundreds of millions of years. There's actually a substantial body of such evidence reported by scientists in scientific journals, discussed at scientific congresses, but subsequently, that evidence has disappeared. And, the first idea that entered into our minds was, well, perhaps there was something wrong with this evidence...perhaps...

Jeffrey: Isn't it the case that most of the evidence that you encountered had been presented at scientific conferences prior to the time that the theory of evolution, of Darwin and Wallace, really became accepted?

Michael: Not exactly. Most of the evidence was reported in the years immediately after Darwin brought out the *Origin of Species* in 1859. Now, right after that, everyone was looking for the missing link. All over the world scientists were look—'cause this is the question that most interested them—where did we come from, where did human beings come from? Now, the unusual thing is, they were not finding evidence for a missing link, they were finding human skeletal remains, they were finding all kinds of artifacts, some of a high degree of civilization, showing that human beings of the anatomically modern type had been around for ten, twenty, thirty, forty, fifty million years or more. This went on until around eighteen—the 1890's, early 1890's, when Eugene Dubois, a Dutch physician, went to Java and found Java Man, the first real missing link. The problem there was, the missing link was found in a geological context of about a half a million years old at most. So the problem became what to do with this massive body of evidence that had been accumulated over the previous forty years. And it was systematically suppressed. And I could give you an example of that.

Jeffrey: Well, you use the term knowledge filter to suggest that the anthropological establishment was simply going to filter out all of the data that didn't conform with their theories.

Michael: Yes, that started in the late nineteenth century and has gone on until the very present. Now, one example from that earlier period that I very much like—it has to do with discoveries that were made in California. During the California gold rush days, miners were digging tunnels thousands of feet into the sides of mountains, into solid rock, to get out the gold. And, thousands of feet inside these tunnels, in some cases, they were finding human skeletons, they were finding obsidian spearpoints, they were finding stone mortars and pestles at dozens of different locations. All of these were collected and reports gathered by J. D. Whitney, who was the State Geologist of California at the time, and he wrote a massive book about them which was published by Harvard University. And, we might wonder, well, why don't we read about these today in the textbooks? Why aren't the artifacts displayed with their appropriate ages? The age of these artifacts would have been over ten million years old. We were able to consult with geologists in the state of California today, and learned from them exactly how old those deposits are. They are from ten to fifty million years old. Now, why don't we hear about them today? There was a very powerful anthropologist, William H. Holmes, who was at the Smithsonian Institution in Washington, D.C. He was connected with international science, whereas Whitney was from the provinces, you could say, of California. But, here's what Holmes said about Whitney: he said if Doctor Whitney had understood the theory of evolution as we understand it today on the basis of these Java Man discoveries, then he would have hesitated to announce his conclusions, despite the imposing array of facts with which he was confronted. In other words, if the facts don't go along with the dominant consensus idea that was emerging in the international scientific community, then the facts, even a large number of them, had to be set aside. So, that's an example of this knowledge filtration process.

Jeffrey: Yes. I'm sure most people can understand that. That the theory of evolution is so well accepted at this point that scientists would have to consider facts like that to be completely anomalous and would be set aside until such a time as there's a new theory that could accommodate those facts.

Michael: Yes. But I think people should be aware that that process is going on. I don't think many people are. Which I think is why...

Jeffrey: In fact, what you're suggesting is that data like this could overthrow the Darwinian conception of the origins of man completely.

Michael: Ah yes. This was recognized by William Howells, who is one of the principal architects of the modern concept of human evolution. When I sent him a copy of the book, he was kind enough to reply, and he said—and he noted, he noted the significance of it. Of course it was very obvious to him. He said—he said to me in a letter, ah, that, if the evidence that you present in this book were to be accepted, it would be devastating not only to our idea of human origins, but to the entire conception of evolution, which has been fairly robust up to now. So I think it is very obvious what the impact of the evidence that we document in the book is....regarding that theory.

Jeffrey: This particular instance that you cite, of the California findings....

Michael: Yes.

Jeffrey: ...that would be sufficient in and of itself, but in fact in your book, *The Hidden History of the Human Race*, you bring together about, I think about a hundred instances...

Michael: Yes.

Jeffrey: ...of these kinds of anomalous findings, including objects that go back several billions of years, that appear to be made by some sort of intelligent beings.

Michael: Yes, well this is what is really important to me, and is convincing to me. Even if you had one or two very good cases, the representatives of the dominant consensus could say, alright, well, you have one or two good cases, but how can you set aside the overwhelming mass of evidence that supports our current theory on that basis? What...

Jeffrey: Especially since there might be fraud...

Michael: Right.

Jeffrey: ...there might be some sort of human error...

Michael: Right.

Jeffrey:...involved, there might be shifting of geological...

Michael: Yes.

Jeffrey:...ah, terrain, earthquakes...

Michael: Yes.

Jeffrey:....who knows what might have caused such an unusual fluke.

Michael: That's correct. Now, but when you're able to assemble an equally massive amount of evidence showing the contrary, and also at the same time able to document specific instances of how commitment to a theory had resulted in very good evidence being set aside, then I think we begin to wonder.

Jeffrey: The other thing that you call into question is the evidence that's used to support the conventional theory, and how very, very shaky all of the evidence in anthropology is, that maybe...I think the fundamental point you're making is that it may be premature to arrive at any final theory about the origins of species.

Michael: Yes. This is something that is very well understood within the physical anthropological community itself. For example, I attended the last annual meeting of the American Association of Physical Anthropologists and, of course, Donald Johanson was there announcing some of his new discoveries in Ethiopia, and... But, what people don't realize is that there are other personalities within the scientific community who take a completely different look at the very same evidence. For example, the famous Lucy, the *Australopithecus Afarensis*, who was considered to be one of our earliest human ancestors. Now, according to Donald Johanson...

Jeffrey: ...Found in the Olduvai Gorge, by Richard Leakey as I recall.

Michael: No, ah...Lucy...

Jeffrey: ...I recall wrong...

Michael:...No, Lucy was discovered in the Afar region of Ethiopia by Donald Johanson...

Jeffrey: Ah...

Michael: ...and his coworkers...

Jeffrey: I see...well, excuse me...

Michael: Yes...yes, and I think they, ah...the Leakeys and Johanson are rivals in that sense.

Jeffrey: I see.

Michael: But, when Lucy is presented to the public, she is generally shown to have very humanlike characteristics, being able to walk upright on two legs like a modern human being, for example. But you have other researchers such as Charles Oxnard at the University of Western Australia and some physical anthropologists here in the United States, Randall Susman, and people like that, who look at the same bones and they draw an entirely different conclusion about this creature. They notice, for example, that *Australopithecus afarensis*, Lucy, has an upward pointing shoulder joint which you also find in the gibbon, and other boreal creatures. In other words, she spent a lot of time, perhaps, swinging...ah...from the...

Jeffrey: She's more monkey-like than human, in other words...

Michael: ...branches of a tree. Right. Yes, so this is...it's like you say. This evidence, which is often presented to the public as absolutely factual, may be subject to varying interpretations, widely varying interpretations in fact.

Jeffrey: The conventional view is that the human species is descended from more primitive species that resemble the ape more, and I think what you're suggesting is that modern humans may have existed millions of years ago and been simultaneous with other humanoid species on the planet.

Michael: Yes. What we see is a pattern of co-existence, rather than evolution. Just as today we see that human beings of various levels of civilization and culture co-exist with apes and monkeys. And...now this might get us into a very controversial area, but we may also be co-existing with some very apeman-like creatures if you take very seriously the evidence for the existence of the Sasquatch or the Yeti or Bigfoot.

Jeffrey: Uh hmm.

Michael: So, just as we are co-existing with a variety of primates today, the evidence that we've accumulated suggests that we co-existed with them in the past, very far into the past. One report we read from a scientific journal called *The Geologist* reported that in the 1840's a completely modern skeleton was found in Macoupin County, Illinois, ninety feet down in solid coal. Now, we were not able to track down the actual bones of the skeleton, but the report is there. We were able to determine from the report where it came from, and we contacted the Geological Survey of the State of Illinois and asked them how old are the coal deposits in that area—at that specific location, not just in that area—and they said three hundred million years.

Jeffrey: Uh hmm.

Michael: Now, to have human beings existing three hundred million years ago, that is before the generally accepted age of the dinosaurs, it's before the Jurassic.

Jeffrey: Uh hmm. And they'd be before the existence of mammals in general, for the most part.

Michael: Well, according to the standard ideas, now, before we began this research, Richard Thompson and I, we wouldn't necessarily have suspected that there would be any evidence for human beings existing in those times. So perhaps our whole—we really don't know what we might discover if we were to do a similar investigation of other mammals.

Jeffrey: Uh hmm.

Michael: Ah, I can't really say what would happen, just as I couldn't really say in advance what would happen when we started to look into the history of human fossil remains and those discoveries.

Jeffrey: Well, I think it's time to begin to look at the implications of all of this. I can assure our audience that you have hundreds of similar cases. And, obviously, you are inspired by a religious tradition, as we mentioned earlier...

Michael: Yes, that's correct...

Jeffrey: ...You're a student of the Hindu Vedantic tradition...

Michael: That's right.

Jeffrey: It's very different than the Christian-Jewish-Fundamentalist tradition that insists on a divine creation that occurred some five thousand years ago. The Hindu view of time is not linear, as Western—as we Westerners conceive of time. It doesn't start with a big bang and end with a whimper—it's cyclical; isn't it?

Michael: Right. Yes, the basic pattern is cyclical. And this is actually the subject of a presentation I shall be making in New Delhi to the World Archaeological Congress called "Puranic Time and the Archeological Record." And the concept of time that you do find in the Puranas is in fact cyclical. Now, what might one predict from that? Say if you have civilizations coming and going over vast periods of time, perhaps you might have humans and other apemen-like creatures co-existing. I will mention that the *Puranas* do talk about intelligent races of apelike creatures that use stone tools. This is not an idea that came in with Darwin, it's been there for thousands of years. Now, what might one predict from that? If you were to predict what archeologists might find, you would say, well, they would tend to find a very bewildering mixture of anatomically modern human fossils, apemen-like fossils, crude stone tools, articles indicative of a higher level of culture, all sort of mixed up and going back hundreds and millions of years. I think you might also predict that, given the biases of investigators towards a linear progressive idea of time with things beginning at a very simple state and progressing in a linear fashion to a more advanced state, that they might edit the record to conform to their linear progressive biases. And, indeed, both predictions, we found, in our investigations, do come true. You actually do have that very bewildering mixture of advanced artifacts and bones mixed up with more primitive ones, going back hundreds of millions of years. And you also do find a very systematic editing of this record to conform to a linear progressive—you might call it evolutionary—view of things. Which is quite amazing.

Jeffrey: Uh hmm. Of course, if we have this sort of cyclical picture—circular picture of things—much of conventional science will need to be readjusted, and, naturally, I suppose it's only fair to say that Vedantic thought has many other ideas about the nature of time and space, the nature of spiritual dimensions....I believe that your view of things would be that, rather than say that humanity evolved from simple-celled creatures, that in some sense we rather descended from spiritual planes.

Michael: Ah yes. And this is a matter that Richard Thompson and I intend to go into in great detail in our next book, which we call *Human Devolution*. And this is exactly the idea that we shall be presenting. Now, you're quite correct that the Vedantic, or Vedic, view of reality is somewhat different from that of modern Western science. Of course, it has very much in common with the views of earlier representatives of Western thought, such as the Greeks, for example. You had Aristotle talking of the great achievements, the great accomplishments of human civilization being invented time and time and time again in the course of cyclical time. The Greeks also had that idea.

Jeffrey: They had the legend of Atlantis as well.

Michael: Yes. Of course, it's very interesting in some of the ancient *Puranas* there is some discussion of sunken land masses that were previously inhabited. So, there are all kinds of interesting parallels there. I think another idea that we need to consider when we are looking at the whole question of human origins is are there further dimensions to the human essence, is there more than biochemistry there? Of course, the Greeks had that idea, the Vedantic sages also had that...

Jeffrey: Well, I think every spiritual tradition has that idea.

Michael: Yes. Right. Well, now the question is: is there any empirical justification for adopting such a view? And here again, just as when we were talking about the stones and the bones, there appears to be a pattern of ignoring certain types of evidence because of a commitment to a certain world view. I think when we look into these other areas there are also some very compelling types of evidence that show that, well, yes, perhaps there are more dimensions to our human existence that need to be taken into account.

Jeffrey: There certainly has been a vitalist tradition in biology and philosophy and psychology that has been suppressed over the last hundred years.

Michael: Yes, very systematically as far as I can see. I think we've tended to equate the word science with reductionism and materialism in their most extreme forms. At least this has been true over the past three to four hundred years, but if you were to go back three or four hundred years in European history you would see quite a different picture of reality. You would find a hierarchical multidimensional universe, you would find alchemy and astrology and all types of science that took into account these different, more subtle levels of matter and even spiritual levels of existence. And, if you look today at the academic world, even though there is a reigning orthodoxy which is very much committed to reductionism and materialism, you see around the edges and you see defectors. You see groups of more independent-minded researchers looking more freely into some of these areas that have for so long been considered out of bounds—people looking into all aspects of consciousness, you know, consciousness studies, out-of-body experiences, subtle healing, things of that nature.

Jeffrey: Uh hmm. Well, those are certainly topics that are of great interest to me personally and, I'm sure, of great interest to our audience. I think there is a yearning, a hunger, in the general public for more knowledge, you might say, of the human soul or spirit.

Michael: Yes. And I think that the reason there is a hunger is because we've been starved.

[laughter]

Michael: I think we've denied certain dimensions of our psyche and that has had an effect, not only upon each individual but upon society at large, and how—what values we become committed to.

Jeffrey: Well, Michael Cremo, you've certainly gone a ways in helping to satisfy that starvation. So, thank you very much for being with me, and thank you for your good work.

Michael: You're welcome, Jeffrey.

Jeffrey: [to audience] And thank you for being with us.

5.3. *Mysterious Origins of Man.* An NBC television special, featuring interviews with Michael A. Cremo and Dr. Richard L. Thompson, first broadcast February 25, 1996. Produced by B.C. Video. Excerpts from the narration are reproduced by permission of Bill Cote of B.C. Video.

Mysterious Origins of Man *was produced by Bill and Carol Cote. I met Bill Cote through Jean Hunt, president of the Louisiana Mounds Society. It was a fortunate meeting. I enjoyed working with Bill Cote on this show, which aired twice in 1996 in the United States. It has also been shown in countries around the world. Below are some excerpts from the parts of* Mysterious Origins of Man *focusing on the authors of* Forbidden Archeology *and cases from the book, such as the California gold mine discoveries, the South African grooved spheres, and the work of Dr. Virginia Steen-McIntyre at Hueyatlaco, Mexico.*

Charlton Heston: But sometimes artifacts are found that break all the rules. Archeologists call them "anomalous artifacts." What happens when we find a modern human skull in rock strata far beneath even the oldest of man's ancestors? In their controversial book, *Forbidden Archeology*, Michael Cremo and Dr. Richard Thompson have documented hundreds of these anomalous artifacts which have yet to be explained.

Dr. Richard Thompson: The basic body of evidence that we've uncovered in this book suggests that human beings of modern anatomical type have been existing for many, many millions of years into the past.

Charlton Heston: In 1880 California State Geologist J. D. Whitney was intrigued by an unexpected discovery made 300 feet under Table Mountain. While digging for gold, miners unearthed a variety of stone tools such as mortar and pestles and ladles. Incredibly, the rock strata the tools were reportedly found in was dated as early as 55 million years old. Whitney made a thorough report of these finds and came to an unsettling conclusion. Man could be millions of years older than the current evolutionary model suggests. This bizarre evidence seems to have been well documented, yet the general public and many within the scientific community are unaware of these controversial finds. The question is, why haven't we heard of these discoveries before?

Michael Cremo: Oh, I think we're talking about a massive cover-up. As I said, over the past hundred and fifty years these archeologists and anthropologists have covered up as much evidence as they've dug up, literally.

Dr. Richard Thompson: Basically what you find is something we call a "knowledge filter." This is a fundamental feature of science. It's also a fundamental feature of human nature. People tend to filter out things that don't fit, that don't make sense in terms of their paradigm or their way of thinking. So in science you find that evidence that doesn't fit the accepted paradigm tends to be eliminated. It's not taught, it's not discussed, and people who are educated in scientific teachings generally don't even learn about it.

Charlton Heston: Conventional theory states that modern man originated in Southern Africa around a 100,000 years ago. From there, he migrated north into Europe and Southern Asia. He continued through Asia and crossed the Bering Strait into the New World around 30,000 years ago. He then came down through North America and finally arrived in South America around 15,000 years ago. Yet numerous artifacts have been found across North and South America that are so old they threaten to completely overturn this theory. According to geologist Virginia Steen-McIntyre, she was silenced at the height of her career because of her determination to report the facts. In the summer of 1966 a collection of stone tools including this leaf-shaped spear point was uncovered at Hueyatlaco, Mexico. To find out exactly how old the spear points were, a team of experts from the United States Geological Survey was called in to date them.

Virginia Steen-McIntyre: When we first began to work on the Hueyatlaco site, we thought we had an old site. This was back in '66. And we thought it was perhaps 20,000 years old. And at that time that was considered a very old age for the site. We did what they call radiometric dates, which gives an actual date range. And we used two different methods to do that. One was using uranium atoms; another one was using little zircon crystals. When we finally got the dates in all the different methods we used to date it, it came out to be 250,000 years old. To tell you the truth, I would have been happy with a 20,000-year-old date. It would have made my career. It was very old for the time, but it wasn't so old that it was that controversial. People can take 20,000-year steps. They can't take steps that are over 200,000 years at one time. And I was rather naive—I thought, okay, we've got something big here, but I'm just going to stick with the dates. We've got the information, we've got the facts, let's get the facts out and go on from there. And I didn't realize it was going to ruin my whole career.

Charlton Heston: According to Dr. McIntyre, because she stuck to the facts, all of her professional opportunities were closed off. She's not worked in her chosen field since. The site was closed, and permission for further investigation was denied, forever.

Dr. Richard Thompson: It's not necessarily a deliberate conspiracy in the sense of some people getting together in a smoke-filled room and saying we're going to deceive people. It's something that happens automatically within the scientific community. So when a given piece of evidence disagrees with the predominant theory, then automatically people won't talk about it, they won't report it, and that means science fails to progress in the way that one would hope.

* * * * * *

Charlton Heston: In Klerksdorp, South Africa, hundreds of metallic spheres were found by miners in Precambrian strata said to be a fantastic 2.8 billion years old. The controversy centers around the fine grooves encircling some of the spheres. Lab technicians were at a loss to explain how they could have been formed by any known natural process. According to the curator of the Klerksdorp Museum, Roelf Marx, the spheres are a complete mystery. They look man-made, yet at the time in the earth's history when they came to rest in this rock, no intelligent life existed. 'They're like nothing I've ever seen before.'

* * * * *

Charlton Heston: In the model of the evolutionary tree, man and apes are said to share a common ancestor. However, evidence of that common ancestor is highly contested. That's why it is still called the "missing link." When Darwin's theory of evolution was embraced, it was assumed that in the next century enough fossil evidence would be found to prove that man had evolved from the apes.

Richard Milton: Darwinists have promised us a "missing link" and so they've got to deliver, they've got to come up with one. Any "missing link" will do, it seems. Every so often a skeleton is found in Africa—its discoverers describe it as being the "missing link," the headlines come and go, and then, later on, that skeleton, or those bones, are reclassified, either as human or as ape. And so far, the "missing link" is still missing.

Charlton Heston: One of the most classic examples of this is the story of *Java Man*, discovered by Eugene Dubois in 1892.

Dr. Richard Thompson: Dubois discovered a very primitive-looking ape-like skull cap, and he discovered this thigh bone about 40 feet away. He said, well, obviously they must belong to the same creature. And that creature walked erect like a human being and had an ape-like skull, so that must be a "missing link," the *Pithecanthropus Apeman*. So maybe you had a big ape and a human being living together in Java about a million years ago. The important point to make about the *Java Man* discovery is that it's based on a speculative leap in which two pieces of evidence are put together in a way that's not really warranted.

Charlton Heston: At the end of his life, Dubois realized that the skull cap belonged to a large ape and that the leg bone was from a man. Nevertheless, *Java Man* was prominently displayed at the Museum of Natural History in New York until 1984. Since then, it has been removed. Today, museums all over the world display models of yet another skeleton they call the "missing link," the common ancestor of both man and ape.

Michael Cremo: Lucy, you know, the famous *Australopithecine* discovered by Donald Johanson. He says she was very humanlike. But I was at a conference of anthropologists where many of them were making a case that she was hardly distinguishable from an ape or a monkey.

Richard Milton: These bones have been restored to resemble a "missing link," part human, part ape, and Lucy is now thought of as being our long-lost ancestor. But this is merely an interpretation, an interpretation of one group. Those same bones can be, and they have been taken by scientists and identified as, simply an extinct ape—nothing to do with us at all.

Charlton Heston: Newspapers are constantly reporting new discoveries that add to our understanding of man's origins. But, so far, conclusive evidence of a "missing link" has not been found. So what happens to the evolutionary model if the "missing link" does not exist at all? Without it, there is little support for man's connection with the apes, and the model simply collapses.

Richard Milton: Some people have said to me, how can you criticize a theory if you can't...if you don't have something to replace it with? Well, I don't accept that. If the emperor hasn't got any clothes on, then the emperor hasn't got any clothes on. It's not my fault. It seems to me that if Darwinism is wrong, then somebody has got to point the finger.

5.3.1. Correspondence Related to the Production of *Mysterious Origins of Man.*

5.3.1.1. Letter to Bill Cote, September 27, 1993.

Recently, Jean Hunt of the Louisiana Mounds Society called me and told me you were interested in my book *Forbidden Archeology: The Hidden History of the Human Race.* I happily accept your offer to trade a couple of your videos for a copy.

I will be very interested to see how your program on the Sphinx turns out.

I shall be visiting India from October 11 to November 14. My coauthor, Richard Thompson, will be in the States during that time, but I think he may be traveling.

In any case, I hope you find the book useful.

5.3.1.2. Letter from Bill Cote, October 3, 1993.

Jean Hunt told about your new book and your agreement to barter a copy with me in exchange for two of my videos. I hope you enjoy the enclosed two tapes. They are Hoagland's *Mars: The Terrestrial Connection* and a preview of my work-in-progress, *Marcahuasi and The Mysterious Engraved Stones of Ica.*

I'm sure you will be interested in the section on Dr. Cabrera, who has found a large collection of stones inscribed with images of men with dinosaurs, men with telescopes and men performing advanced surgical procedures. The existence of these stones, if they are verified, could be important in the argument that men did co-exist with the dinosaurs.

Looking forward to reading your book and hearing your reaction to the videos. I also look forward to speaking with you about your own very exciting work.

Jean may have mentioned that I am in the critical final stages of an NBC television special based on the work of John Anthony West about the re-dating of the Great Sphinx. This is based on the observation, by R. A. Schwaller de Lubicz, that the Sphinx was eroded by water and not by wind and sand.

The special is set to air in November (possibly the 10th) at 9PM after *Unsolved Mysteries.* I will probably have more time to catch up on new developments around that time.

Looking forward to hearing from you.

5.3.1.3. Letter to Bill Cote, December 30, 1993.

I've enclosed a copy of Richard Thompson's book *Alien Identities* and a copy of our video *Human Evolution*, which deals with the material in *Forbidden Archeology*.

Congratulations again on the Sphinx special.

5.3.1.4. Letter to Michael Cremo and Richard Thompson from Bill Cote, January 23, 1994.

Thank you so much for Richard's book *Alien Identities*. I am reading now and find it very well done. There must have been a tremendous amount of research to do in this and your *Forbidden Archeology*.

I have enclosed two tapes for you and Richard: *The Mystery of the Sphinx* (95-minute home version) and Hoagland's *Mars: Volume II, The U.N. Briefing.*

I hope you enjoy them and I thank you once again for the books and information. As you know, we are currently creating a proposal for a T.V. Special on the subject of man living at the time of the dinosaurs. I will be in touch with you as we progress, Michael. I think there is a link between your work and what we would like to do.

Thanks, too, for the video *Human Evolution: A Confrontation of Fact and Theory, Part I.* (Is there a Part II?) We found it very informative. What is the future of that project? We would be interested in what is planned.

That's it for now. Good luck and let's stay in touch.

5.3.1.5. Letter from Bill Cote, 1994 (undated).

Since we last spoke, I've been developing [along with my wife Carol and partner John Cheshire] our concept for a TV special. The working title is "Man with Dinosaurs: Fact or Fiction?" We feel that the information you presented in your book *Forbidden Archeology* could be very useful in a segment in our film.

In addition, your video was most helpful in showing some of the work you've compiled, but we still have many questions we'd like to ask you. We will need the best visuals possible to support our arguments. It is our understanding that much of the research for your book was done by Richard Thompson and Steve Bernath. Perhaps some of these questions would best be addressed to them, in which case we would be happy to contact them directly if you could provide us with their addresses.

Whenever possible, we are interested in original newspaper articles and photos, as well as names of persons in articles who found artifacts. Do you know of any living relatives of these people who might be able to recount an anecdote about the find which was passed down in the family?

We are interested in any physical evidence that can be filmed, such as a fossilized welt of a shoe, mortar and pestle, gold chain, or any other artifacts, whether in a museum or private collections, regardless of the country they are located in.

That's it for now. If you have any thoughts on these things, please drop us a line. I am sure we will be in touch with you soon as things progress further. We are looking forward to meeting you and working together on this very exciting project.

5.3.1.6. Letter to Bill Cote, April 25, 1994.

It was good speaking with you on the phone, and I am looking forward to seeing you in May.

Another person you might want to touch base with is Alister Taylor. Alister is a good friend of both Richard and me. For the past couple of years, he has been helping us manage our small publishing company, Govardhan Hill, Inc. He has also been quite helpful in managing our small institute office. Richard and I really appreciate his help, as it keeps us free to concentrate on our research and writing. Over the next year, Richard and I will both be busy promoting our books: *Forbidden Archeology, The Hidden History of the Human Race* (that's the abridged edition), and *Alien Identities* (that's Richard's book on UFOs in light of ancient India's Vedic texts). Alister is organizing that effort, so he will be well-informed about our schedules. Also, if we need to take on someone to do the kind of research we were talking about, he would be the one who would be taking care of that. Furthermore, he is himself very much interested in the topics of our investigations. Alister is originally from New Zealand and is quite a nice person. I think you will like him. I am hoping he can be here when you come in May. I have told him about you and your project, and he would be interested in hearing from you. So if you have time, please give him a call. I think you will find him quite helpful.

5.3.1.7. Letter to Bill Cote, June 28, 1994.

Richard and I have made our move to Florida. We have a small building on a farm in Alachua, which is not far from Gainesville, where the University of Florida is located. The hot and steamy weather is a change from San Diego's dry California beach climate, but I am adjusting nicely. I just wanted to let you know we're here, and give you our new address and phone numbers. If there are any developments with the NBC project, please let us know. The abridged version of *Forbidden Archeology,* titled *The Hidden History of the Human Race*, is now in print. The official release date is in September, at which time I am scheduled to do an author's tour. Hope this finds you in good health and consciousness.

5.3.1.8. Letter from Bill Cote, July 4, 1994.

It was good to hear from you and that you are settled down in your new home in Florida. We hope that your new location will be beneficial for you and your work.

We are fine. The trailer has taken longer than we anticipated: we are forced to confront what the film is about as we decide just what the trailer is about. But that is for the better, since we now have a much better grasp on just what it is we are selling NBC. We expect to have the trailer done in a week, at which point we will show it to a few selected individuals for feedback. Then, after a period of reflecting on what we have and how we like it, make whatever modifications are called for. We hope to be at this point at the end of July.

Of course, we will let you know of our schedules and send you a trailer when it is complete.

Thanks again and let's keep in touch.

5.3.1.9. Letter to Michael Cremo and Richard Thompson from BC Video (Bill Cote, Carol Cote, and John Cheshire), November 12, 1994.

Enclosed is our completed trailer for *The Forbidden History of Man.* Please keep in mind that this will be used as a sales tool. Many of the ideas presented here are simplified for the sake of clarity, but will be more fully developed in the final program.

We wish to thank you for your help in making this all possible. Presently our sales agent is in Los Angeles pitching our show to the networks. We will keep you informed as the project develops.

5.3.1.10. Letter to Bill Cote, January 3, 1995.

The trailer looked good, despite the high percentage of stock footage, which I suppose is necessary at this point. The latest news—that NBC is interested—is quite exciting. I'm sure you must be ready to get this project underway.

I finally got around to writing Robert Schoch. I sent him a copy of *Forbidden Archeology* and will be interested to hear his impressions.

September through November I was doing some touring to promote *Hidden History of the Human Race.* I did about 40 radio talk shows and a dozen television shows, mostly local, but some national. I haven't seen the *Sightings* show yet. Alister Taylor has a tape and is sending it to Florida so Richard and I can look at it.

In December I was in India, where I delivered a paper ("Puranic Time and the Archeological Record") at the World Archeological Congress in New Delhi. The paper is based on our *Forbidden Archeology* material. It created a minor sensation. Our rather primitive *Human Evolution* video was also screened there, and when the address of the Institute came on at the end, I heard lots of notebooks opening and pens scratching.

Regards to Carol and John.

5.3.1.11. Letter from BC Video (Bill Cote, Carol Cote, and John Cheshire), November 17, 1995.

Here is a copy of our show. We hope you like it and want to thank you for all you have done. It was truly a pleasure to meet you and work with you. NBC had initially said they would run it on Jan. 17, 1996, but later they mentioned possible scheduling changes. We will keep you posted when we find out the actual air date.

I will enjoy any comments or reactions you may have.

5.3.1.12. Letter from Bill Cote, January 2, 1996.

In case you have not seen it yet, here is a preview copy of our show, *The Mysterious Origins of Man*. NBC plans to air it on Sunday, Feb. 25th at p.m., but that date could change, so watch the listings.

I want to thank you once again for all the support you, Richard and Alister gave us during the making of the project. As I mentioned to Alister, we will run a tag at the end of the home video telling where to order both your books.

As soon as I catch up a little, I will send you the spheres from Klerksdorp as well as the few additional ones they sent me. I assume they should be sent to you at this address?

Please let me know your reactions to our program and let's keep in touch.

Have a great 96.

5.3.1.13. Letter to Bill Cote, January 10, 1996.

Thanks very much for the preview copy of the show *Mysterious Origins of Man.* It looks just great. Congratulations.

I usually like to have my affiliation with the Bhaktivedanta Institute mentioned, along with a mention of the ancient Sanskrit writings of India as my source of inspiration, but I can understand why that might not have been possible in this case. I'm sure the show will generate a lot of interest in *Forbidden Archeology*, and from the book itself readers will learn more about me and why I am doing what I am doing.

Virginia Steen-McIntyre seems very pleased with the show, and I have been helping her get on some radio talk shows.

There is no great hurry about the grooved spheres. But when you do get around to sending them to me, could you please enclose copies of the metallurgical evaluations you had done on them?

Again, congratulations on the show. I hope it does very well for you and leads to more projects, hopefully some I could be helpful with.

5.3.1.14. Letter from Bill Cote, March 13, 1996.

Here are the Klerksdorp spheres that we borrowed from you and also the new ones that were sent, thanks to Lori, from Africa. These are all the spheres, except for one big one and the block of wonderstone it was found in. This sample is being returned to Africa, as per their request.

It was a pleasure to work with you, Richard and Alister on this project. As you may know, I am carrying both of your books in my inventory, as well as running the ad for Torchlight at the end of the videos as promised.

5.3.1.15. Letter to Bill Cote, May 23, 1996.

I just wanted to let you know I received the South African grooved spheres. Thanks for sending them. The NBC show was good for sales of *Forbidden Archeology*. I am now working on the follow-up book, which I am calling *Human Devolution*. In that book, I will be giving an alternative view of human origins, based on the idea that we are conscious beings who have descended from a higher position. I am enclosing some papers that I have given at academic conferences recently, just to let you know what kinds of things I am working on these days. The material in these papers will form part of the next book. If any of this fits in with ideas you have for future video projects, let me know. I would be happy to work with you again. How are things going with foreign broadcasts of *Mysterious Origins of Man*?

5.3.1.16. Letter to Bill Cote, August 1996.

Based on talks with you and with Alister Taylor, I have been saying the following on some radio shows:

Regarding the California artifacts, these are in a museum controlled by the University of California. During filming of MOM, the producer asked the museum director for permission to film the artifacts. The director denied permission, saying that the museum would have to pay its staff overtime to do this. When the producer came back with an offer to pay the overtime, the director still refused to give permission.

Is this correct? If not, please let me know, and give a correct version.

5.3.1.17. Letter from Bill Cote, August 26, 1996.

Sadly, this is correct. At first we were told they could not make the time. We countered saying we had plenty of time and could wait three or four months. They replied that even their own professors could not get into the museum, and since we were outsiders, we had even less priority. We offered to pay any overtime costs that may be incurred by our special request and they countered that they could not accept payment from any outside source. We patiently went all the way to the head of publicity for the University, but it seems the Museum Director has final say and she said no.

A similar situation occurred when we tried to obtain permission to film the pyramid complex at Teotihuacan, near Mexico City. We approached through proper channels the director of the site. He wanted us to promise that we would make no mention of UFOs or spacemen building the pyramid. I thought this was odd, but since that was not our intention, I replied in all honesty that we would not imply this. Then he demanded a copy of our full script. We had not yet written the scene and told him so, but he insisted. We spent a few hours and drafted a modest version which included the theory of Hugh Harleston Jr., a respected researcher, that the slope angle of one of the facets of the Pyramid of the Sun was aligned to the same degree of latitude as the location of the Pyramid itself in the northern hemisphere. (An easily verifiable fact). In effect, the sunrise over the Pyramid on the vernal equinox would cast a shadow over this facet in an instant, thus making the whole pyramid a sort of giant clock. We were interested in demonstrating the advanced knowledge of the ancients. But we were denied permission to bring our cameras into the site, or even to fly over in a helicopter, despite the fact that we were willing to pay all the appropriate fees, had gone through both archaeological channels and the Mexican film authority. We showed them a letter from NBC asking that they cooperate and told them our show would be seen by 25 million potential tourists, yet we were denied permission to film by one man who thought our proposal was not serious enough.

By the way, we learned later that this was the same man who gave the OK for the dozens of souvenir shops that surround the ancient complex: against the protests of many serious researchers.

The good old boys network continues to hold a powerful control over information that threatens to upset the established view.

5.3.1.18. Letter to Bill Cote, September 1996.

Thanks for the reply about the details of how you were not allowed to film the California gold mine artifacts. I need to ask you one more thing about this. For my files, and for the record, can you give me the exact name of the museum and the name of the museum director—the person who refused permission to film? I do not plan on making much use of that—but in case somebody asks for the name, I would like to have it, or at least know it. Nobody thus far has challenged me on this (and I have spoken about it on a lot of radio shows), but I do want to be ready. It would not look good for me if I could not come up with the exact name of the museum and the name of the director.

I don't talk about the Mexican site of Teotihuacan, but it would be good to have the name of the director of the site, just in case I want to talk or write about it someday.

5.3.1.19. Letter from Bill Cote, September 25, 1996.

Here is the information you requested: Phoebe Hearst Museum of Anthropology in Berkeley, CA: We were rejected by the collections manager, Ms. Leslie Freund. At Teotihuacan, we were denied permission to shoot by the head of Teotihuacan, Mr. Eduardo Matos. I have mailed you copies of some of this correspondence for your files.

By the way, do you know about the John West symposium at the University of Delaware campus this weekend? I mentioned it to Alister Taylor. We are going to shoot it for the Society for Scientific Exploration. It runs from mid-day Friday to mid-day Sunday.

5.3.2. Correspondence Related to Reactions to *Mysterious Origins of Man.*

Mysterious Origins of Man *was shown on the NBC television network in the United States in February 1996. It provoked considerable response, mostly negative, from the scientific community.* Science, *journal of the American Association for the Advancement of Science, reported (March 8, 1996, p. 1357): "The claims of creationists—the young age of Earth, that fossils put the lie to the theory of evolution—routinely send biologists into fits. But those fits pale before the indignation spilling out, mostly over the Internet, since Sunday evening, 25 February, when a major U.S. television network ran a 'special' suggesting that humans coexisted with the dinosaurs, and that the scientific establishment was suppressing the evidence." Furthermore, said* Science, *phones at the National Center for Science Education were ringing constantly, with calls for help from high school science teachers "dealing with fallout from the NBC program." On Usenet discussion groups such as Sci.Archeology, Sci.Anthropology, and Talk.Origins, scientists exchanged hundreds, perhaps thousands, of messages about* Forbidden Archeology *and* Mysterious Origins of Man. *This section includes a sample of this Internet traffic. I regard news releases and individual messages posted and reposted to Usenet discussion groups without copyright notices as public domain material, and I have not therefore sought the permission of the authors to reproduce them. Furthermore, it is appropriate under the "fair use" provisions of U. S. copyright law for me to reproduce material circulated as public criticism of my scholarly work, for the purpose of publicly responding to it.*

5.3.2.1. Internet Press Release from B.C. Video, February 21, 1996. Reproduced by permission of B.C. Video.

From the producers of NBC's Emmy Award winning *The Mystery of the Sphinx* comes another network special.

The Mysterious Origins of Man. This one-hour primetime special is hosted by Charlton Heston and will air on NBC, Sunday, February 25, 1996 at 7pm EST.

"Controversial Evidence That Could Rewrite Man's History"

One of the greatest mysteries of mankind is where did we come from. Although many theories have been put forth, in truth, no one knows. In their search for answers, scientists gather evidence based on what they observe. But sometimes evidence turns up that completely contradicts our accepted theories. In this program we will meet a new breed of scientific investigators, who claim the history of man on this planet may be radically different than what is accepted today.

Documented cases demonstrate that man could be millions of years older than the theory of evolution accepts.

Human footprints, found side-by-side with dinosaur tracks, suggest that man lived at the time of the dinosaurs.

Photographic evidence suggests that the dinosaurs may have survived until the present time.

We will examine the reasons why the scientific community has chosen tosuppress evidence such as this.

Meet the experts who demonstrate that the history of evolution has become so powerful that any evidence against it is ignored.

The ancient city of Tiahuanaco, high in the Bolivian Andes, suggests that technological man could be thousands of years older than history tells us.

New evidence reveals a possible location for the Lost Continent of Atlantis: it could be buried under two miles of ice, at the South Pole.

"Preview Amazing Discoveries!"

Visit the NBC Web Site.
Visit some of our other favorite sites:
The Ancient American magazine
"The Interactive Brainwave Visual Analyzer" Web Site
"The Laura Lee Show"

Copyright 1996 B. C. Video - email "bcvideo@interport.net"

5.3.2.2. Internet Press Release from NBC, February 21, 1996.

Investigate *The Mysterious Origins of Man*

Could it be that man has made the climb from the Stone Age to civilization more than once, and that present-day man is just the latest in this cycle? That's just one of the many compelling theories to be addressed by *The Mysterious Origins of Man*, a one-hour special from the Emmy-winning producers of NBC's *Mystery of the Sphinx*. Airing Sunday, February 25, 7/6 p.m., the special challenges accepted beliefs about prehistoric man.

Other highlights: Did man live with the dinosaurs? Is Darwin's theory of evolution correct? Plus, Author Graham Hancock, author of the best-selling *Fingerprints of the Gods*, investigates the link between Egypt and South America—a link that could rewrite man's history. Narrated by Charlton Heston, this is one mind-expanding special you won't want to miss.

5.3.2.3. Internet Announcement by Alister Taylor, President of Torchlight Publishing, Addressed to Members of the International Society for Krishna Consciousness, February 22, 1996.

Forbidden Archeology, the front-line challenge to Darwinian evolutionists, is receiving national attention on the NBC prime time special The *Mysterious Origins of Man*, which airs this Sunday, February 25th at 7:00 pm. (Check local listings for details.) Following is a portion of a general news release regarding the show.

The producers of NBC's Emmy Award-winning *The Mystery of the Sphinx* present the prime time NBC special *The Mysterious Origins of Man*. Is man the chance descendant of apes, or the descendant of a lost race of advanced humans? Is the current scientific paradigm supported by the evidence?

Headlining the show are authors Richard Thompson and Michael Cremo, whose internationally popular and controversial book *Forbidden Archeology* documents many cases of anomalous evidence, such as bones of modern man found in rock strata dated to over 50 million years old. They explain that we have not heard of many of these finds because a knowledge filter is at work: "People tend to filter out things that don't fit, that don't make sense in terms of their paradigm or their way of thinking. So, in science you find that evidence that doesn't fit the accepted paradigm tends to be eliminated." Is the scientific establishment trying to hide these findings?

Forbidden Archeology and its abridged version, *The Hidden History of the Human Race*, are in the 4th English printing, with 35,000 copies in print in English and German. The Spanish edition will be released by Editorial Diana in Mexico this spring; an Italian edition is due in the fall and negotiations are progressing for Japanese, Russian and Portuguese language editions.

5.3.2.4. Letter to Sci.Skeptic and Talk.Origins Newsgroups, by Anthropologist Dr. John R. Coles, February 21, 1996 (Excerpt).

Feb. 25 (Sun) NBC TV will broadcast a special on the "Mystery" of human origins, apparently with a lot of antievolutionist overtones. Narr[*ated*] by C Heston! ICR [*Institute for Creation Research*] was sent a tape for possible promotion...... NBC PR sounds a lot like Krishna Kreationism as opposed to ICR.

The Institute for Creation Research is Christian and promotes a young-earth geology, limited to ten thousand years. The prebroadcast publicity for Mysterious Origins of Man *mentioned time periods of millions of years for human existence. Coles was familiar with* Forbidden Archeology. *I had once debated him on a radio show. Coles is associated with the National Center for Science Education, a small but vocal anticreationist organization.*

5.3.2.5. Letter to Talk.Origins, Alt.Archeology, and Sci.Archeology Newsgroups from Mike Sullivan, February 26, 1996.

We don't have a TV and my wife teaches high school Earth science. This show will undermine much of what she is trying to teach. I hope you and others will post your critiques ASAP. The kids have short attention spans.

On a more general note about the damage this type of show does: Even if a teacher can debunk it in the eyes of most of her students, they will be left with a great deal of cynicism. How else can a young person react to the knowledge that the leaders of American culture and business will lie outright for a few $$?

It is sad that a young person has to be either cynical or gullible. No wonder they are freaking out.

It is amazing how intensely Darwinian evolutionists desire to maintain total control over how students think about human origins.

5.3.2.6. Letter to Talk.Skeptics Newsgroup from Dr. John R. Coles, February 26, 1996.

Re: NBC & Krishna Kreationism

From preview info, I suspected and predicted that this show was going to hew to the line of the Hare Krsna book, *Forbidden Archeology*, more than the ICR line. Boy, was I right!

5.3.2.7. Letter to B.C. Video from Dr. William Doleman, Archeologist at the University of New Mexico, Albuquerque, February 26, 1996.

Your program *Mysterious Origins of Man* was filled with half-truths, fabricated evidence, and poor logic. In addition, the portrayal of legitimate scientists such as myself as constituting a cabal of evil, evidence-suppressing conspirators is unforgivable. But the worst of your crimes lies in the failure to offer the public a balanced view that compares the overwhelming evidence in favor of evolution theory and conventionally-derived dates for man and dinosaur with the dubious and poorly documented "evidence" the whackos cite. The average citizen knows the difference between fantasy entertainment such as *The X Files* and documentary presentations. To present fantasy and unsubstantiated pseudo-science to the public in documentary format is a pernicious form of anti-science propaganda. The public owns the airwaves, which you rent for pitifully small sums of money. You owe it to them to inform, not bamboozle, them! What really torques me is that you *know* this stuff is garbage, yet you broadcast it as "science" because you know it will sell! Show some backbone and give the pubic the balanced coverage that they deserve. Remember, the technology that you use every day was created by scientists, not kooks!

This is typical of the fundamentalist Darwinian response to Mysterious Origins of Man. *The totalitarian ranting of this true believer is somewhat amusing. Do we really want our tax dollars supporting someone like this at a public university?*

5.3.2.8. Letter to National Association of Science Writers and Various Newsgroups from Jere H. Lipps, Paleontologist at the University of California at Berkeley, February 27, 1996.

Your readers might be interested in the NBC-aired program *The Mysterious Origins of Man,* which generated a good deal of response from scientists in biology and paleontology on various electronic bulletin boards. The problem with the program was its presentation as science, when in fact it was not. I don't care about the content, it's the way science is depicted. The following is an email I posted to these various bulletin boards, NBC, one of its affiliates, and other individuals.

Here's the message posted by Lipps:

I've been asked by a number of emailers what was in the NBC program *The Mysterious Origins of Man.* Here's what the producers, BC Video, have on their WWW site (http://www.bcvideo.com/bcvideo). I don't need to say more. You can order the video off their site for $20, if you really want to. In the meantime, email them to express your views on science and how to do it.

He reproduces the BC Video press release of February 21, then continues:

This is not science. It is a fraudulent attempt to foist a particular view on the viewing public.

It is pseudoscience (photographic evidence that dinosaurs lived to the recent; Atlantis under the South Pole, etc.), lies (carved footprints among dinosaur prints), misrepresentation (scientists have suppressed evidence), and false authority (new breed of scientists, who are unable to get their results published because of the strangle-hold that we—the old breed?—have on the scientific journals).

None of this is revealed to the audience, yet it is obvious to knowledgeable people. Although NBC makes its programming decisions on the basis of numbers of viewers likely and numbers of dollars from sponsors, NBC should be a responsible enough corporate American citizen to properly identify what its programming is and is not. This, like violence on TV, has major impact on American society at large. If the networks and movie studios can label sex and violence, why not science and non-science?

This kind of claptrap needs a response to NBC, its affiliates, and the program's sponsors. I urge anyone with any concern for how our science is done and understood to email NBC and the producers, BC Video, at the emails online@nbc.com and bcvideo@interport.net

We're working on sponsors now. Anyone record the show who can provide the sponsor list?

Here we have a perfect example of fundamentalist science attempting to intimidate the media into following its party line.

5.3.2.9. Letter to Talk.Origins, Sci.Archaeology, and Alt.Archaeology Newsgroups from Mark Isaak, February 28, 1996. "Hollywood 'Moses' Undermines Genesis," by Ken Ham, *Answers in Genesis,* February 1996, p. 5, is reproduced in this book by permission of *Answers in Genesis.*

Folks may be interested in what a prominent Creationist has to say about that program. From *Answers in Genesis,* Feb. 1996, pg. 5:

"Hollywood 'Moses' Undermines Genesis"
by Ken Ham

February 25 at 7 p.m. is the scheduled nationwide showing of the documentary *The Mysterious Origins of Man* on NBC. The program is narrated by famous actor Charlton Heston, who, ironically, played the part of Moses in the Academy Award-winning movie *The Ten Commandments*, yet this documentary actually undermines the writings of Moses concerning man's origins.

At the beginning of the broadcast, Heston asks the question "Where did we come from?" He then (incorrectly) answers by saying, "In truth, no one really knows." (The Creator in His Word, however, tells us.)

While the program does question one particular aspect of evolution, namely the idea that man descended from some ape-like ancestor, the whole thrust of the documentary is to push the idea of millions of years of Earth history. They do this by looking at artifacts and fossils that are "out of place" (i.e., in strata much older than they are supposed to be according to evolutionary theory). They don't question the evolutionary time scale—they place man back in history millions of years before even traditional evolutionists would allow.

The producers also interview a creationist, but then use his material to bolster their own idea that man's history goes back millions and millions of years. (By the way, the evidence from this creationist — such as the supposed man and dinosaur prints in Texas, and the "fossil finger" — is actually highly suspect anyway. According to leading creationist researchers, this evidence is open to much debate and needs much more intensive research. One wonders how much of the information in the program can really be trusted!)

But where are the producers of this program coming from? Well, they interview the authors of the book *Forbidden Archeology* a number of times during the program. This book is dedicated to "His Divine Grace A.C. Bhaktivedanta Swami Prabhupada." It appears the authors are Hare Krishna adherents!

This makes sense of what Heston states at the end, which I believe sums up the philosophy of this program: "It's been said that man has made the climb from Stone Age to civilization more than once, and that our present time is just the latest in this cycle." Everything cycling continuously over millions of years fits well with Krishna philosophy! That seems to be what this program is really all about!

5.3.2.10. Letter to Talk.Origins, Sci.Archaeology, and Alt.Archaeology Newsgroups from Paul Heinrich, February 28, 1996.

While listening to the NBC documentary *Mysterious Origins of Man* (*MOM*), I came across some really mind-boggling distortions of the facts concerning an archaeological controversy. Early in the show, they talked about the Calaveras skull controversy. In this case, a geologist, Dr. D. J. Whitney, was given artifacts and a skull that the miners said were found several hundred feet below, beneath Table Mountain in Tertiary, gold-bearing gravels that they were mining. Because the age of these gravels were about Pliocene age, these finds, as indicated by the references listed below, caused a very public controversy. Later, it came out that the gold miners were doing what miners like to do to stuffed-shirt geologists, as I learned from personal experience as a geologist in the Illinois coal fields. The miners had played a very elaborate prank on him. They had given him artifacts from nearby archaeological sites and a recent skull of a Native American and salted artifacts in gravel before he inspected it. This is all documented in the references that I have listed below.

Anyway, after a very brief and selective summary of the controversy, which strangely omits mention of the skull, the name *Calaveras,* and its resolution, Charlton Heston said: "This bizarre evidence seems to have been well documented. Yet, the general public and many within the scientific community are unaware of these controversial finds."

The last statement is completely incorrect. When I took my beginning, undergraduate course in North America, this case was mentioned and discussed. The archaeologists specializing in Paleo-Indians, who I have as colleagues, know about this case. When the controversy was active, it was even widely reported in the public press as indicated by the below citations. The finds were reported, debated, and when conclusive evidence came forward about the hoax, a judgment was rendered. The whole matter is reported in the scientific literature and public press to read if he had wanted to. Charlton Heston's next statement is: "The question is: why haven't we heard of these discoveries before?"

My answer to this question is: "You have not heard of these discoveries because the writers for *Mysterious Origins of Man* (*MOM*) failed to do their homework." Had they taken the time and trouble to do any sort of research concerning the archaeology of California, anybody would have heard about this controversy. It is quite obvious that the people who did this documentary either refused or failed to do the basic background needed to understand what they were talking about in this case. Otherwise, they would not have written such ill-informed dialogue for poor Mr. Heston....

After that statement, *MOM* had the authors of *Forbidden Archeology*, which is archaeology as seen through a Hindu creationist knowledge filter (Tarzia 1994), complain about how this controversy was a perfect example of the censorship practiced by archaeologists who they claim have buried more sites than they have excavated. The problem is that no censorship occurred in this controversy, which in my opinion exposes the falsehood of their censorship claims. The real problem for it is that modern archaeology fails to conform to certain Hindu creation myths and *Forbidden Archeology* is an explanation through the authors' knowledge filters of why modern archaeology fails to support these beliefs. Evidence for my views are given in (Tarzia 1994). Note: *Forbidden Archeology* was published by the Bhaktivedanta Institute of the International Society for Krishna Consciousness....

Final questions are: If there has been the "massive cover-up" as *MOM* claims, how come I was taught about the Calaveras controversy and instantly recognized what they was talking about, although they had mangled the details of the controversy? Also, if there was such a cover-up, how could I have found the references so quick?

Heinrich's letter is a perfect example of the kind of cover-up Forbidden Archeology *talks about. The statements made by Charlton Heston had nothing to do with the Calaveras Skull, which was discovered at Bald Hill. The statements refer to less well-known discoveries that were made at Table Mountain. When Heinrich says that* Mysterious Origins of Man, *in its discussion of the controversy, "strangely omits mention of the skull, the name* Calaveras, *and its resolution," he is displaying his ignorance of the evidence. When he says he instantly recognized that the discussion was about the Calaveras skull, he was completely wrong. For further clarification, see the following letter by Chris Beetle.*

5.3.2.11. Letter to Talk.Origins, Sci.Archaeology, and Alt.Archaeology Newsgroups from Chris Beetle of Govardhan Hill Publishing and the Bhaktivedanta Institute, March 2, 1996. Copyrighted material reproduced by permission of Bhaktivedanta Book Trust.

Some viewers of *The Mysterious Origins of Man* may think from seeing that show that *Forbidden Archeology* advocates authenticity of the Calaveras skull find. As a reader of *Forbidden Archeology*, I wanted to say that after pages of discussion of evidence on both sides of the Calaveras skull controversy, authors Michael Cremo and Richard Thompson stated: "Can it really be said with certainty that the Calaveras skull was either genuine or a hoax? The evidence is so contradictory and confusing that, although the skull could have come from an Indian burial cave, we might regard with suspicion anyone who comes forward with any kind of definite conclusion."

Other finds in the gold-bearing gravels of California, however, are more convincing. Many such discoveries of ancient artifacts were made under Table Mountain in Tuolumne County, California.

One of the more well-attested cases is that of a mortar and pestle found by superintendent J. H. Neale of the Montezuma Tunnel Company. The stratigraphy at Table Mountain consists of a latite cap dated to 9 million years, andesitic tuffs, breccias, and sediments going back to 21.1 million years, rhyolite tuffs as old as 33.2 million years, and prevolcanic auriferous gravels, some as ancient as 55.0 million years. What follows are some excerpts from *Forbidden Archeology* describing this discovery made in the auriferous gravels of Table Mountain, dated from 33.2 to 55 million years old. For an illustrated version of the same, check out URL: http://nersp.nerdc.ufl.edu/~ghi/ft951022.htm

Excerpts from Forbidden Archeology:

> On August 2, 1890, J. H. Neale signed the following statement about discoveries made by him: "In 1877 Mr. J. H. Neale was superintendent of the Montezuma Tunnel Company, and ran the Montezuma tunnel into the gravel underlying the lava of Table Mountain, Tuolumne County.... At a distance of between 1400 and 1500 feet from the mouth of the tunnel, or between 200 and 300 feet beyond the edge of the solid lava, Mr. Neale saw several spear-heads, of some dark rock and nearly one foot in length. On exploring further, he himself found a small mortar three or four inches in diameter and of irregular shape. This was discovered within a foot or two of the spear-heads. He then found a large well-formed pestle, now the property of Dr. R. I. Bromley, and nearby a large and very regular mortar, also at present the property of Dr. Bromley."
>
> Neale's affidavit continued: "All of these relics were found the same afternoon, and were all within a few feet of one another and close to the bed-rock, perhaps within a foot of it. Mr. Neale declares that it is utterly impossible that these relics can have reached the position in which they were found excepting at the time the gravel was deposited, and before the lava cap formed. There was not the slightest trace of any disturbance of the mass or of any natural fissure into it by which access could have been obtained either there or in the neighborhood" (Sinclair 1908, pp. 117-118). The position of the artifacts in gravel "close to the bed-rock" at Tuolumne Table Mountain indicates they were 33-55 million years old....
>
> In a paper read before the American Geological Society and published in its journal, geologist George F. Becker (1891, pp. 192-193) said: "It would have been more satisfactory to me individually if I had myself dug out these implements, but I am unable to discover any reason why Mr. Neale's statement is not exactly as good evidence to the rest of the world as my own would be. He was

as competent as I to detect any fissure from the surface or any ancient workings, which the miner recognizes instantly and dreads profoundly. Some one may possibly suggest that Mr. Neale's workmen 'planted' the implements, but no one familiar with mining will entertain such a suggestion for a moment.... The auriferous gravel is hard picking, in large part it requires blasting, and even a very incompetent supervisor could not possibly be deceived in this way.... In short, there is, in my opinion, no escape from the conclusion that the implements mentioned in Mr. Neale's statement actually occurred near the bottom of the gravels, and that they were deposited where they were found at the same time with the adjoining pebbles and matrix."

Notes:

Becker, G. F. (1891) Antiquities from under Tuolumne Table Mountain in California. *Bulletin of the Geological Society of America,* 2: 189-200.
Sinclair, W. J. (1908) Recent investigations bearing on the question of the occurrence of Neocene man in the auriferous gravels of the Sierra Nevada. *University of California Publications in American Archaeology and Ethnology, 7(2):* 107-131.

5.3.2.12. Letter to Newsgroups by Chris Beetle, March 2, 1996.

Chris Beetle has contributed this message to a discussion about groove metallic spheres found in mineral deposits 2.8 billion years old in South Africa. The spheres were shown in Mysterious Origins of Man.

Regarding the South African grooved sphere controversy, I am posting an excerpt from *Forbidden Archeology* by Michael Cremo and Richard Thompson. This is copyright 1996 by BBT International. To see the HTML version of this excerpt, which includes the picture of the sphere with three grooves around it, see URL: http://nersp.nerdc.ufl.edu/~ghi/spheres.html

5.3.2.13. Letter to Alt.Archaeology and Sci.Archaeology Newsgroups from Dr. Virginia Steen-McIntyre, March 2, 1996.

Dr. Virginia Steen-McIntyre appeared on the Mysterious Origins of Man *television program in connection with her dating the Hueyatlaco, Mexico, site at 250,000 years.*

Questions about the Hueyatlaco site, Mexico? Here are some important papers published about this site:

Szabo, J.B., H.E. Malde, and C. Irwin-Williams, 1969, Dilemma posed by uranium-series dates on archaeologically significant bones from Valsequillo, Puebla, Mexico, *Earth and Planetary Science Letters, 6*, 237-244.

Steen-McIntyre, V., R. Fryxell, and H.E. Malde, 1981, Geologic evidence for age of deposits at Hueyatlaco Archaeological Site, Valsequillo, Mexico, *Quaternary Research, 16*, 1-17.

Malde, H.E. and V. Steen-McIntyre, 1981, Letters to the Editor, Reply to comments by C. Irwin-Williams: Archaeological Site, Valsequillo, Mexico, *Quaternary Research, 16*, 418-421.

Please reply c/o B.C. Video. They'll forward most by regular mail, and I'll give brief answers to the popular questions here. Or, post your replies here. I may not be able to respond promptly. This posting is via our local library's net link.

For starters, here's a letter [*by Steen-McIntyre*] recently sent to *The New York Daily News:*

> A recent article by Eric Mink (Feb. 23) panned the TV documentary *Mysterious Origins of Man*. In it, he mentions Dr. Ian Tattersall, who asserts that methods used to date stone tools to an age of 250,000 years are unreliable in such a time frame. The scientists who employed these dating methods (zircon fission-track, uranium series) to date materials from the site were leaders in their fields at the time the work was done. Neither saw any problems [with their dating methods.]

I'd be delighted to correspond with scientists about the Hueyatlaco, Mexico site. One of my biggest burdens is isolation. Since we first published the dates in 1981 (*Quaternary Research, v.16*, 1-17), not one anthropologist has contacted me to discuss them.

5.3.2.14. Internet Press Release from B.C. Video, March 4, 1996. Copyrighted material reproduced by permission of B.C. Video.

Producers Respond to Criticism Arising From
NBC's Airing of *The Mysterious Origins of Man*

On February 25, 1996 NBC aired *The Mysterious Origins of Man.* In their search for answers about man's origins, scientists gather evidence based on what they observe. But sometimes evidence turns up that completely contradicts their accepted theories. Here is some of the evidence reviewed in this program.

— Documented cases of human bones and artifacts demonstrate that man could be millions of years older than the theory of evolution accepts.

— Astronomical alignments found in the ancient city of Tiahuanaco, in Bolivia, suggests that technological man could be thousands of years older than history tells us.

— Geological dating methods suggest that modern man was in the New World 250,000 years ago.

— Accurate details in ancient maps suggest the continent of Antarctica was known and mapped before the time of Alexander the Great.

— Human footprints found side-by-side with dinosaur tracks suggest that man lived at the time of the dinosaurs.

Much of this evidence has already been judged false by the scientific community, but many of these judgments may have been based on personal and professional biases, rather than on the evidence itself.

In this show we attempted to re-examine potentially valuable evidence that has been unjustly disqualified. Evidently, we struck a nerve.

Many viewers praised the production "for raising the question in public, even if the scientific community does not believe it..." (R.M., ...alaska.edu)

But the scientific community itself had a completely different reaction.

> "Most of the ideas presented...were so ludicrous as to not even warrant a rebuttal by any honest investigator." (L.W., Mt. Wilson Observatory)

> "I think you should apologize publicly for this show. It was appalling....Frankly, you are either morons or liars." (D.L., ...colorado.edu)

"...the non-scientific public watching this drivel may be inclined to actually believe it and to vote for politicians who also believe it." (J.K., New Mexico State University)

"It's all a bunch of hooey, and my recommendation is to stay away." (B.D., Yale University)

"I recommend people write NBC and protest the presentation of this show as a documentary....Thanks largely to the efforts of people like yourself, the American public is generally not capable of evaluating the 'arguments' and 'evidence' you present." (A.D., University of Texas at Austin)

"You should be banned from the airwaves." (J. J., ALCI)

And so on....

Producers' Response to the Critics

As we expected, the response to our show has been heated. We've been accused of pseudo-science and setting back the course of education in America. But our goal was simply to present the public with evidence which suggests an alternative view to some of our most accepted theories. After all, the theory of evolution is still a theory, not a fact, and therefore alternative views should be welcomed, not banned.

Probably the most common criticism is that the show gave no opposing view from the academic community. The producers' position is that the accepted view has been so frequently presented to the public that only a brief summary by the host was necessary. It was more valuable to focus on the documented anomalous evidence.

For example, if man evolved from the apes around 5 million years ago, then how does the scientific community explain tools of modern man found in rock strata dating to 55 million years old? (J.D. Whitney, California State Geologist, Table Mt. Mine) Those artifacts currently reside in a museum in Berkeley, California. When we applied for permission to film them, we were denied by the museum.

Another criticism is that the information in our show is presented by experts who do not hold degrees in their fields of expertise and therefore their opinions are not endorsed by the scientific community. But Dr. Virginia Steen-McIntyre holds a Ph.D. in Geology and was a fellow with the USGS when she did her field work in Mexico. Her conclusions about the age of the spearpoints she dated (250,000 years BP) were backed by two other USGS members, yet because of their implications, the findings were ignored and her career was ruined.

In the case of the Paluxy River man tracks, to our knowledge, no accredited archaeologist has ever proven the prints to be fake. Furthermore, many scientists have referred us to an article written by Kuban and Hastings who seem to be the experts on this site. They categorically deny that there is any validity to the prints and that the case has been solved.

It is interesting to note that the scientific community refers to this report as if it is definitive proof, when in fact neither gentleman is an accredited archaeologist, anthropologist or paleontologist. If this is to be a fair discussion, let's all play by the same rules.

Many of our critics are using very strong language, calling us morons, liars, and subversive creationists. These are emotional responses, not logical arguments. To set the record straight, we are not creationists or affiliated with any group whatsoever. We are being attacked on a personal level, because we are questioning issues that have been deemed too fundamental to be questioned.

We are fully aware that the information presented is highly controversial. This was re-iterated by Charlton Heston in the show, "We've seen a broad range of evidence, some of it highly speculative. But there are enough well-documented cases to call for a closer look at the conventional explanation of man's origins. "

We never take the stance that we know the answers or in any way suggest that we will provide them. We are merely offering an alternative hypothesis. In this way, we feel that the American public is fully capable of making up its own mind.

Bill Cote, Carol Cote and John Cheshire
Producers of The Mysterious Origins of Man

To follow the controversy on our World Wide Web site:
http://www.bcvideo.com/bcvideo

Copyright 1996: Bill Cote, Carol Cote and John Cheshire.... May reprint with permission. Direct any inquiries to <bcvideo@interport.net.

5.3.2.15. Letter to Sci.Skeptic and Talk.Origins Newsgroups from "Etherman" (R. Cote), March 8, 1996.

The letter from "Etherman" is a typical Internet message composed of responses to pieces of statements from previous messages by Jim Rogers. For ease of reading, I have modified the format somewhat.

Jim Rogers (March 6): Some misrepresentations of fact are so gross it can be difficult to not get emotionally involved and drop to denigrations, in response. I am *appalled* that NBC would air something so shoddy and misleading in a "documentary" format aimed at the general populace. Since the producers acknowledge that the ideas are "controversial" (understatement of the year), they *owe* it to their audience to present some balance, instead of giving a two-minute "summary" of the "other side" from the producer's own biased perspective.

Etherman (March 6): I've noticed that PBS specials on dinosaurs, origins of man, etc. never present opposing views. Perhaps a televised debate would be nice. That way both sides could get their arguments presented.

Jim Rogers (March 7): There have been several.

Etherman (March 8): Funny how I've never seen any. The only thing I've seen that even comes close is presenting both sides of the argument over whether dinosaurs were hot blooded or cold blooded.

Jim Rogers (March 7): How many times must evolution be defended from anti-scientific creationists?

Etherman (March 8): Are you suggesting that evolution is sacred and should not be questioned? My, how unscientific of you!

Jim Rogers (March 7): Do you suggest we air debates challenging whether Apollo missions actually landed on the moon? That, too, is a controversial point with some people.

Etherman (March 8): I'd love it.

Jim Rogers (March 7): Evolution is rather *un*controversial.

Etherman (March 8): Where have you been living?

Jim Rogers (March 7): The Mysterious Origins of Man acknowledged its [*evolution's*] wide acceptance, and labeled their own ideas controversial. Such an acknowledgment would seem to obligate them to present a rebuttal position fairly, which they grossly failed to do.

Etherman (March 8): Why not present your own rebuttal? That way their bias won't affect it.

5.3.2.16. Letter to Sci.Skeptic and Talk.Origins Newsgroups from James J. Lippard, March 12, 1996 (Excerpt).

Etherman: I have a book called *Forbidden Archeology* by Cremo and Thompson that has over 150 examples of what the authors claim is evidence that evolution is wrong. I don't know if they're right or not, but they sure as hell aren't claiming that evolution is wrong because it's contradicted by the Bible.

Lippard: True. Their views are shaped by ISKCON (International Society for Krishna Consciousness), not the Bible.

5.3.2.17. Letter to Sci.Archaeology, Sci.Anthropology, and Sci.Anthropology.Paleo Newsgroups from Dave Oldridge, March 14, 1996.

Restam: The thing is, at least in my opinion, there is no reason to flame NBC for airing their show. Although the opinions they showed are most certainly wrong, people should be allowed say what they think {even if it is, more than likely, only to make money}. Its one thing to disagree with them but to bash them merely for offering an alternative viewpoint is just as silly. Well, just wanted to add my 2 cents....

Oldridge: Let me get this straight now. Is it your claim that NBC should be free to present all kinds of misinformation on their show but that nobody should feel free to criticize them for it? That's just plain crazy!

5.3.2.18. Letter to Sci.Archaeology Newsgroup from Paul Heinrich, March 16, 1996 (Excerpts).

In this letter, Paul Heinrich replies to an earlier exchange between Daniel D. Scripture and Angus Mann.

Daniel D. Scripture: But it [*Mysterious Origins of Man*] ain't science. It is just a sideshow, fun for the ignorant (because it is only plausible to the ignorant).

Angus Mann: Weeeellllll. I have been watching the postings on this newsgroup for a while and if it is a 'sideshow for the ignorant,' then why have we all got so worked up about it? Curious. Because it goes against the beliefs and findings of so many archaeologists?

Paul Heinrich: That is not the reason. An example of why archaeologists and other scientists are so mad about *MOM* is illustrated by a segment starting with Charlton Heston saying:

> "At the end of his life, Dubois realized that the skullcap belong to a large ape and the leg bone was from a man. Nevertheless, the Java Man was prominently displayed at the Museum of Natural History in New York until 1984. Since then it has been removed.
>
> "Today, museums all over the world display models of yet another skeleton they call the missing link, the common ancestor of both man and ape."

Then Mr. Cremo, a co-author of *Forbidden Archeology*, and Mr. Richard Milton indulged in some gratuitous bashing of anthropologists, including Dr. Donald Johanson, by accusing them, in the case of Lucy, of not being able to differentiate ape from human skeletal material and for "restoring" the skeleton of Lucy into the form of a "missing link." In the latter case, Richard Milton clearly was claiming that paleoanthropologists had fabricated a "missing link," transitional human fossil, that does really not exist.

With slander and mud-slinging like that, parts of *The Mysterious Origins of Man*, it was difficult to distinguish from the political attack advertisements shown during the New Hampshire primaries and Iowa caucuses. Just forget the facts, just say nasty things about your opponent. The producers of this documentary obviously have a great future in making political attack advertisements for political campaigns. It is stuff like this that got scientists angry because they were being maliciously called liars, cheats, and frauds and, at best, just plain stupid as a group. Anybody would get upset if someone devoted an hour special to calling that person a fraud and stupid, primarily because their ideas contradict the religious beliefs of creationists like Baugh and Hindu creationists like Cremo and Thompson (authors of *Forbidden Archeology*). *MOM* was based more on assassinating the character of honest scientists than presenting any real evidence. That is why scientists are angry about it.

5.3.2.19. Letter to The Kellogg Company, BC Video and other scholars from Dr. Jere H. Lipps, Paleontologist, University of California at Berkeley, March 15, 1996.

In this letter, Dr. Lipps responds to a letter from a Ms. Pell of the Kellogg Company. After his introductory statement, Dr. Lipps responds point by point to statements made by Ms. Pell in her letter.

Dr. Jere H. Lipps to Ms. Pell of The Kellogg Company: I am quite pleased that you responded for Kellogg Co. Your answer is not particularly satisfactory, as it again suggests that sponsors delegate much of their judgment to the networks. That's a lot of trust!! Do Americans dissociate advertisers from programming? I think not.

Ms. Pell to Dr. Jere H. Lipps: Thank you for your message concerning the ad we placed on an NBC program called *The Mysterious Origins of Man*. Kellogg Company does not sponsor television programs. We do place spot advertising during time frames and on channels where we think we can reach our target audiences. Since we do not sponsor programs, we do not have content, editorial or selection rights. We make our placement selections based on a reviewers synopsis of the program which is then compared to our placement criteria. We make every effort to be sure our commercials appear within television programs that are considered wholesome by the vast majority of the public and we avoid programs which include hard-core violence, explicit sex, and/or obvious anti-social behavior. In addition, we support accurate and fair news coverage and try to avoid programs which are one-sided or put across a singular point of view.

Dr. Lipps: Sounds very good—certainly removes Kellogg from any responsibility. But then everyone associated with this program, from NBC, to affiliates, to those sponsors with the courtesy to respond to inquiries have refused to accept responsibility. Just who does decide what goes onto American TV?

In theory your reviewer system ought to work; in practice it fails because your reviewer cannot properly evaluate what is and what is not science. How good is your reviewer in evaluating violence, sex, anti-social behavior, and fairness? With the boundaries you state for exclusion, a reviewer has a good deal of latitude. Is getting shot hard-core violence? Or only if it shows blood spattering? Or more than one person getting shot with blood all over? I think you need more help in evaluating what Kellogg stands for.

Ms. Pell: In addition, we very much support accurate representation of science and scientific principles.

Dr. Lipps: Thank you. But why then delegate the responsibility to a reviewer who cannot judge?

Ms. Pell: We suggest that you do as we have done in the past, write a letter to the Chairman of the network to let him know of your concern with the distorted information provided by this type of program.

Dr. Lipps: NBC has yet to reply.

Ms. Pell: We have taken strong stands in the past when nutrition and food science issues have been misrepresented (like the ALAR scare), because we have the expertise within our organization to critically evaluate the accuracy of the information and the science behind them. Unfortunately, we are not experts on anthropology or biology and do not feel that we can fairly represent your concerns, but support your efforts in this regard.

Dr. Lipps: My concern is not whether the anthropology or biology was correct or distorted. My concern is that a nonscience program was represented as science. Ask your scientists if they can recognize how science is done. If they cannot, Kellogg is in trouble.

If the public were properly informed about science and nonscience, Kellogg would have fewer problems with ALAR. But then Kellogg refuses to take a stand on what science is or is not, I gather, and thus begets what it eschews.

Perhaps Kellogg's would join with me and other scientists in promoting intelligent science presentations and representations in the media. Sounds like science is as critical to Kellogg as it is to those of us who teach it. How about it? Ask your scientists if they might be interested. They really do know the difference between science and non-science, I assure you, or Kellogg would be in deeper trouble on lots of other issues.

Dr. Lipps seems very determined to make sure that the American television-viewing public never again sees a television program that criticizes the theory of evolution.

5.3.2.20. Letter to Alt.Paranet.Ufo Newsgroup from Orville G. Marti, Jr., Microbiologist, March 18, 1996.

Today I received a letter in the snail mail in response to my letter of several weeks ago complaining about the NBC Special *Mysterious Origins of Man.* My letter was sent to Mr. Todd Schwartz, and the reply appears to be an original, signed, by Mr. Schwartz (Director, Specials, Variety Programs, and Late Night, 3000 West Alameda Avenue, Burbank, CA, 91523 818-840-3009)

The text is as follows:

Thank you for your letter concerning our special, *The Mysterious Origins of Man.* This was one of many entertainment programs NBC has broadcast which speculate about alternative views of our world. Although the show did not contain an opening disclaimer, we feel it was very clear the people interviewed were expressing only their opinions. The program established that their writings and studies were "controversial" and made extensive use of qualifying language such as "claims..." "could be..." "reportedly..." "seems to..." "may be..." "suggests to some..." etc.

The point of the special was not to discredit or discount the results of established scientific research, but to consider a provocative "what if" scenario—a proposition, if you will—which simply asked if modern man could have existed long before what is currently believed. The program was designed only to raise the question, not prove the hypothesis.

Other NBC specials have raised questions about the existence of angels, the fulfillment of prophecies, and the efficacy of alternative medicine, and it was in this context we expected *The Mysterious Origins of Man* to be viewed. We're sorry to hear you were offended by that approach, and we hope to more adequately address your concerns should the same issue be the subject of another NBC program.

5.3.2.21. From a Web Page by Jim Foley for the Talk.Origins Archive, March 18, 1996 (Excerpts).

On Sunday February 25th, NBC broadcast *The Mysterious Origins of Man* (hereafter *MOM*), narrated by Charlton Heston, purporting to be a documentary about scientific evidence that would overturn currently accepted theories of human history.

The film was produced by B. C. Video Inc. (P.O. Box 97, Shelburne, VT 05482, Ph: 800-846-9682), which has set up a web site to publicize it.

Although *MOM* was anti-evolutionary, it was not advocating scientific creationism, even though some of the "experts" and arguments are familiar to readers of scientific creationist literature. Instead, just as scientific creationism is an attempt to use science to support fundamentalist Christianity, *MOM* is apparently an attempt to use science to support Hinduism. Much of the material in the program is based on the contents of two books, *Forbidden Archeology* and *The Hidden History of the Human Race* by Michael Cremo and Richard Thompson, both of whom appeared on the show and are members of the Bhaktivedanta Institute, a branch of the International Society for Krishna Consciousness.

Why are people so upset about *MOM*? Mainly because a pseudo-scientific mishmash of discredited claims and crackpot ideas was presented as cutting edge science, with no attempt at balance. Few, if any, of the experts presented are recognized scientists….

Should you wish to complain about the presentation of pseudoscience as science, you can email NBC at: http://www.nbc.com/mail.html or write to the following NBC employees:

Mr. Robert Wright (President of NBC), 30 Rockefeller Plaza, New York, NY 10019

Mr. Todd Schwartz (acquired the film for NBC), 3000 W. Alameda, Burbank, CA 91523, (818)840-3009

—or to the makers of the film at: bcvideo@interport.net

Keep it polite; Usenet-style flames will not have as much effect as a calm letter explaining why you disliked the film. If you have a letterhead showing your credentials, or university or museum affiliation, make use of it.

The following companies advertised on the show, in case you want to let them know what you thought of it : Coca-Cola, McDonalds, Olive Garden, Toyota, Chevron, Kelloggs, J. C. Penney, Honda, Wendy's, General Motors, LensCrafters, Folger's Coffee, M&M's Candy.

5.3.2.22. Letter to Talk.Origins, Sci.Archaeology, Alt.Archaeology, Sci.Anthropology, and Sci.Anthropology.Paleo by Michael A. Cremo, March 22, 1996.

A friend has forwarded to me copies of messages related to the NBC television show *The Mysterious Origins of Man* that aired February 25. As the principal author of the book *Forbidden Archeology*, which was featured on the show, I have a few general comments in response to the messages that were posted to this group.

First, I do not agree with everything that was presented on that show. For example, I have studied the case of the Paluxy man tracks and decided it is not possible to conclude whether or not they are genuine human tracks. For that reason, I did not include them in the evidence for extreme human antiquity catalogued in *Forbidden Archeology*. Neither do I subscribe to the views that the show presented on Atlantis, massive rapid displacements of the entire crust of the earth, etc.

But I will stand behind the material that came from *Forbidden Archeology* and the conclusions that can be drawn from it. In brief, there is a lot of scientifically reported archeological evidence that puts the existence of anatomically modern humans back tens of millions of years. This evidence was not culled from the *National Enquirer*, but from standard scientific journals of the past 150 years. In examining the treatment of the reports of this anomalous evidence, there appears to be a pattern of unwarranted dismissal, based not so much on the quality of the evidence itself but on its out-of-bounds position relative to orthodox paradigms of human origins.

I have presented academic papers on this topic at the World Archeological Congress 3 in New Delhi, in December 1994, and at the Kentucky State University Institute for Liberal Studies Sixth Annual Interdisciplinary Conference on Science and Culture, April 1995. I am quite pleased that not everyone in the scientific world is reacting to the book with the kind of conditioned negative response that seems so prevalent in the messages posted recently to this group.

For example, Tim Murray, archeologist and historian of archeology at La Trobe University, said in a recent review of *Forbidden Archeology* in *British Journal for the History of Science* (1995, vol. 28, pp. 377-379): "I have no doubt that there will be some who will read this book and profit from it. Certainly it provides historians of archaeology with a useful compendium of case studies in the history and sociology of scientific knowledge, which can be used to foster debate within archaeology about how to describe the epistemology of one's discipline." Tim also guardedly admitted that the religious perspective of *Forbidden Archeology* might have some utility: "The 'dominant paradigm' has changed and is changing, and practitioners openly debate issues which go right to the conceptual core of the discipline. Whether the Vedas have a role to play in this is up to the individual scientists concerned." This is not to say that Tim endorses the conclusions or analytical methodology of *Forbidden Archeology*. He does not, and has personally told me so.

But the point is this—*Forbidden Archeology* is worth reading, and I hope anyone who wants to comment on the elements of it that were communicated in the NBC show would read it before launching into their critiques. The only mention of the book's Vedic perspective is found in a few sentences in the Introduction. Otherwise, the main text of *Forbidden Archeology* is composed of archeological reports and analytical discussion.

5.3.2.23. Letter to Newsgroups from Randy Wadkins, March 22, 1996.

Could you please provide any discipline of science where out-liers and anomalous data are not found?

If I dug through the last 150 years of literature on *any* subject, I could find hundreds, maybe thousands, of cases where experimental data were just plain wrong. Does this mean that all of science is involved in a massive cover-up to protect their pet theories about the nature of the universe (e.g., evolution)?

I eagerly await your next book: "Forbidden Cold Fusion".

5.3.2.24. Letter to Newsgroups by Etherman, March 25, 1996 (Excerpt).

I'm eagerly awaiting evidence that the experimental data presented in *Forbidden Archeology* is "just plain wrong." Just because some experimental data is wrong doesn't mean all anomalous data is wrong.

5.3.2.25. Letter to Newsgroups from Dr. Norman H. Gall, March 25, 1996.

Dr. Norman Gall is from the Philosophy Department of the University of Winnipeg, Canada.

But surely you'll agree that it is the anomalous evidence that requires explanation...by those that would offer it as evidence of some kind. If there is a huge consensus of interpreted data, and some rogue information pops up, we have a prima facie case to treat the new data with some measure of skepticism.

5.3.2.26. Letter to Sci.Archaeology, Sci.Archaeology.Paleo, Sci. Anthropology, Alt.Anthropology, Talk.Origins, and Sci.Skeptic from Etherman, March 25, 1996.

Etherman: Science progresses by falsifying theories. If evidence pops up that contradicts a theory, the proponents of the theory now have to show why this evidence doesn't disprove the theory after all. Otherwise, any evidence that contradicts a given theory can be classified as anomalous and ignored. Then the theory is no longer falsifiable.

Gall: If there is a huge consensus of interpreted data, and some rogue information pops up, we have a prima facie case to treat the new data with some measure of skepticism.

Etherman: Yes, treat it with skepticism, but don't ignore it. Investigate further.

5.3.2.27. Letter to Newsgroups from Dr. Phil Nicholls, March 25, 1996 (Excerpt).

Re: *Mysterious Origin of Man*—A *Forbidden Archeology* Author Comments

Michael Cremo raged: "A friend has forwarded to me copies of messages related to the NBC television show *The Mysterious Origins of Man* that aired February 25. As the principal author of the book *Forbidden Archeology*, which was featured on the show, I have a few general comments in response to the messages that were posted to this group.... "

While I have not read *Forbidden Archeology*, I have encountered people who have on the web before and, if their recitations are in any way accurate, I think this tells me all I need to know about your book.

5.3.2.28. Letter to Newsgroups from Mr. E (jackechs@erols.com), March 27, 1996 (Excerpts).

Re: *Mysterious Origin of Man*—A *Forbidden Archeology* Author Comments

Phil Nicholls (March 25): Michael Cremo raged ...

Mr. E: Do you always modify postings to reflect your propaganda? Mr. Cremo did not rage at all and posted with a significantly higher level of intelligence than your response.

Phil Nicholls (March 25): While I have not read *Forbidden Archeology*, I have encountered people who have on the web before and, if their recitations are in any way accurate, I think this tells me all I need to know about your [*Michael Cremo's*] book.

Mr. E: Well, the book has better references than hearsay.

5.3.2.29. Letter to Newsgroups by Ian Tresman, March 27, 1996 (Excerpt).

Re: *Mysterious Origin of Man*—A *Forbidden Archeology* Author Comments

R. Wadkins: If I had discovered an arrowhead that was dated at 10 million years old, I'd be screaming in the ears of everyone in science. I'd have it dated again and again to prove what I was saying was true.

Ian Tresman: That would only 'prove' that the dating was contaminated. Even the British Museum has been known not to publish dates that fell outside the 'accepted' range, on the grounds of 'contamination'.

5.3.2.30. Letter to Newsgroups from Etherman (rcote@cs.uml.edu), March 27, 1996 (Excerpts).

Re: *Mysterious Origin of Man*—A *Forbidden Archeology* Author Comments

R.M.W.: I am anxiously awaiting confirmation of the data in *FA* by several respectable groups (that's sarcasm, kiddies).

Etherman: It's all appeared in peer reviewed journals.

5.3.2.31. Letter to Newsgroups from Etherman, March 27, 1996.

Re: *Mysterious Origin of Man*—A *Forbidden Archeology* Author Comments

What if the evidence is presented in a peer reviewed journal? What if I had over 150 examples of evidence that contradicts evolution and has appeared in peer reviewed journals? Would you accept that as a good case? That's what the authors of *FA* have claimed.

5.3.2.32. Letter to Newsgroups from Ferret (Duane Brocious), March 29, 1996.

Archaeology is more than digging up neat stuff and doing stat analysis. It is about reconstructing the changes in the course of the planet's history by means of an incomplete temporal projection. Evolution is one of the tools used to explain such transitory changes and is an integral underlying assumption not only to paleoanthropology but to every social science. We have taken the idea of progress and turned it into constant a priori truth. The assertion that evolution is fact only indicates a bias against scientific method, [*or*] truth searching, and simply demonstrates a mindset based on erroneous presuppositions. Something that no doctrine calling itself a science ought to do.

5.3.2.33. Letter to Ferret (Duane Brocious) from Michael A. Cremo, March 31, 1996.

I don't subscribe to any Internet discussion groups as I am just too busy to deal with the information overload. But from time to time, someone forwards me interesting posts. In the midst of a bunch of texts from the discussion that went on about my book *Forbidden Archeology*, I encountered this statement of yours [*reproduced above*], which I very much appreciated.

5.3.2.34. Letter to Sci. Anthropology, Talk.Religion.Newage, Talk.Origins, and Sci.Archaeology Newsgroups from Bill Cote of BC Video, Producer of *Mysterious Origins of Man*, March 31, 1996.

In a recent "review" Frank Steiger offers "to provide the information filtered out by the producers of this video" (*The Mysterious Origins of Man*). Mr. Steiger obviously put a lot of effort into his seven-page report, but he made so many assumptions and got so many facts wrong that we are called upon to set the record straight lest any serious reader be mislead by his imaginative "analysis".

FS writes, "Heston stated that stone tools were 'reportedly' found in Table Mountain in California in 55 million year old strata. This discovery was reported in detail in the fall, 1981 issue of *Creation/Evolution:* conclusive evidence was presented to show that the tools were planted by a local shopkeeper and in fact resembled modern, not ancient, artifacts. Yet the claim was made that the conclusion of an age of 55 million years for these tools 'seems to have been well documented.'"

I have to wonder what evidence could have been found in 1981 to overturn the huge body of first-hand evidence of those who originally reported and investigated the discoveries between 1849 and 1891. This includes reports from miners, scientists and even an investigator from the Smithsonian Institution. Why does the scientific community prefer the claims of a local shopkeeper to the sworn testimony of scientific experts and professional men?

On August 2, 1890, J.H. Neale signed the following statement about discoveries made by him: "In 1877 Mr. J.H. Neale was superintendent of the Montezuma Tunnel Company, and ran the Montezuma tunnel into the gravel underlying the lava of Table Mountain, Tuolumne County... At a distance of between 1400 and 1500 feet from the mouth of the tunnel, or of between 200 and 300 feet beyond the edge of the solid lava, Mr. Neale saw several spearheads, of some dark rock and nearly one foot in length. On exploring further, he himself found a small mortar three or four inches in diameter and of irregular shape. This was discovered within a foot or two of the spear-heads. He then found a large well-formed pestle." Neale's affidavit continued: "All of these relics were found... Close to the bedrock, perhaps within a foot of it. Mr. Neale declares that it is utterly impossible that these relics can have reached the position in which they were found excepting at the time the gravel was deposited, and before the lava cap formed. There was not the slightest trace of any disturbance of the mass or of any natural fissure into it by which access could have been obtained either there or in the neighborhood." (Sinclair 1908, pp. 117-118) The position of the artifacts in gravel close to the bedrock at Tuolumne Table Mountain indicates they were 33-55 million years old.

In 1891, George F. Becker told the American Geological Society that, in the spring of 1869, geologist Clarence King found a stone pestle firmly embedded in a deposit of gold-bearing gravel lying beneath the cap of basalt, or latite. The gravel deposit had only recently been exposed by erosion. Becker stated: "Mr. King is perfectly sure this implement was in place and that it formed an original part of the gravels in which he found it. It is difficult to imagine a more satisfactory evidence than this of the occurrence of implements in the auriferous, pre-glacial, sub-basaltic gravels." Even William H. Holmes, from the Smithsonian Institution had to admit that the King pestle, which was placed in the collection of the Smithsonian, "May not be challenged with impunity." He investigated to see if it could have been embedded more recently, "...but no definite result was reached." (Holmes 1899, p. 454) So this evidence, no matter how unexplainable according to our current theories, is well documented, as we said in the program.

Becker, G. F. (1891) Antiquities from under Tuolumne Table Mountain in California. *Bulletin of the Geological Society of America, 2*: 189-200.

Holmes, W. H. (1899) Review of the evidence relating to auriferous gravel man in California. *Smithsonian Institution Annual Report 1898-1899*, pp. 419-472.

Sinclair, W. J. (1908) Recent investigations bearing on the question of the occurrence of Neocene man in the auriferous gravels of the Sierra Nevada. *University of California Publications in American Archaeology and Ethnology, 7(2)*: 107-131.

5.3.2.35. Letter to Newsgroups from Etherman (rcote@cs.uml.edu), April 1, 1996.

Re: *Mysterious Origin of Man*—A *Forbidden Archeology* Author Comments

Brett J. Vickers (March 28): If all 150 of these pieces of evidence were as strong as the evidence for evolution, then, yes, they would raise serious questions. If, however, you are going to argue that the book *Forbidden Archeology* presents 150 uncontested pieces of evidence contradicting current ideas about human evolution, then you'll probably be met with strong disagreement.

Etherman (April 1): That's what I'm looking for, reasons why the authors have misinterpreted the evidence. As far as I know, the authors don't claim that any of the finds are uncontested. In fact, they go into considerable detail about the issues (the book is 828 pages long).

Brett J. Vickers: While I haven't read the book and can't comment intelligently on the bulk of its contents, I've seen excerpts. The authors apparently feel, for instance, that the Laetoli footprints are incompatible with australopithecine anatomy ...

Etherman: The authors claim that the footprints strongly resemble those of modern humans and indicate bipedalism. Furthermore there are those who don't think the tracks were made by australopithecines (e.g. the Leakeys believed they were made by an ancestor of *Homo habilis*). They also quote an R.H. Tuttle who claims that *Australopithecus* couldn't have made the prints. Apparently it has to do with the toes being too long.

Brett J. Vickers (*cont. from above*): ... and that the Piltdown skull may have been a genuine hominid fossil. If these are 2 of the 150 pieces of evidence you're talking about, then I don't have much hope for the quality of the remaining 148.

Etherman: They agree that forgery was involved but conclude that the skull (not the jaw) was genuine. They base this on iron, chromium, and gypsum content tests. There are marked differences in the contents of these chemicals in the skull and jaw. It would seem that the forger used different techniques of forgery on the skull and jaw or the skull was genuine.

Brett J. Vickers: But since this is all for the sake of argument, let's just pretend that all 150 pieces of archaeological evidence are strong and valid. What would that mean? At most, it would mean that current theories of human evolution are in serious need of revision. We'd have to push the appearance of anatomically modern humans back several tens or hundreds of thousands of years, and we'd have to push their emergence from Africa back. This would certainly be serious, but would we have to discard everything we know about evolution in general? Would we have to toss out common descent? Would we even have to conclude that humans did not evolve from ape-like ancestors? Hardly.

Etherman: If the Laetoli footprints came from modern humans (as the authors seem to contend), then that would make modern humans about 3 million years old. That would be very serious.

Brett J. Vickers: As far as I can tell, the only thing *FA* is contesting is the currently accepted timeline, and perhaps the phylogeny, for hominid evolution. It doesn't do anything to challenge the fact that organisms, human or otherwise, have evolved. So what are these 150 pieces of evidence that falsify evolution?

Etherman: I have neither the time nor the inclination to list all of them. You should read the book because it goes into much more detail than I could. However I'll give a partial listing:

1) Bones marked by flint implements found in St. Prest, France, dating from the Early Pleistocene to Late Pliocene.
2) Intentionally modified bones from the Late Pleistocene found in Old Crow River, Canada.
3) Six mammoth bones with scratches from stone tools dating from the Middle Pleistocene found in the Anza-Borrego Desert, California.
4) Grooved bones from the Early Pleistocene or Late Pliocene found in Val D'Arno, Italy.
5) More grooved bones from the Late Pliocene found in San Giovanni, Italy.
6) Bone implements from the Pliocene to the Eocene found below the base of Red and Coralline Crags of Suffolk, England.
7) Eoliths from the Pliocene found in the Kent Plateau, England.
8) Eoliths from below the Red Crag from the Pliocene to Eocene.
9) Eoliths at Foxhall, England from the Late Pliocene.
10) Eolith at Haritalyangar, Himachal Pradesh, India from Miocene.
11) Paleoliths near Lisbon, Portugal from the Miocene.

12) Paleoliths at Thenay, France from the Miocene.
13) Paleoliths at Aurillac, France from Late Miocene.
14) Paleoliths in Belgium from Oligocene.
15) Sling stone found in Bramford, England from the Pliocene to the Eocene.
16) Bolas from Olduvai Gorge from Early Pleistocene.
17) Anatomically modern human skeleton in Ipswich, East Anglia, England from the Middle Pleistocene.
18) Human skull in Buenos Aires from Early Pleistocene.
19) Human jaw at Foxhall from Late Pliocene.
20) Human skeleton in Savona, Italy from Middle Pliocene.
21) Jaw fragment from Miramar, Argentina from Late Pliocene.
22) Various human skeletal remains from the California Gold Country from Pliocene to Eocene.
23) Human bones at Macoupin County, Illinois from Carboniferous.
24) Human footprints in Rockcastle County, Kentucky from Carboniferous.
25) Human footprint in Central Asia from Jurassic.

5.3.2.36. Letter to Newsgroups from Dean T. Miller, April 1, 1996.

It looks like most of the detractors haven't read the book. Could someone please explain to me how anyone can pose a valid criticism of a book they haven't read? (Yes, I've read the book.)

5.3.2.37. Letter to Newsgroups by Etherman (R. Cote), April 8, 1996.

Etherman (from earlier message on the Laetoli prints): They point out that the Leakeys believed them to be *habilis.* The authors don't claim that they couldn't be from *erectus* or *habilis.* Their argument seems to be more along the lines that we can't rule out modern man.

Phil Nicholls (April 6): Yes, actually we can, because the earliest fossil evidence of anatomically modern *Homo sapiens* is at or about 60,000 years.

Etherman (April 8): The whole point of the book is that your statement is not true: [That there is evidence of modern humans stretching back tens of millions of years ago.]

Phil Nicholls seems to me a typically dense fundamentalist Darwinian. He is slow to comprehend that there is in fact archeological evidence that radi-

cally contradicts conventional Darwinian concepts of human evolution. He seems also not to comprehend that it is no longer cool and daring to be a Darwinian. Darwinism is for those with fossilized minds.

5.3.2.38. Letter to Sci.Archaeology and Alt.Archaeology Newsgroups from Paul Heinrich, May 23, 1996.

Heinrich is responding in typical Internet fashion to a message from Jerry W. Watt, inserting his own comments into the letter from Watt. As usual, I have eliminated the confusing Internet terminology and simplified the formatting for easier reading, but the wording is unchanged.

Watt: We all owe you a great debt for your contributions to debunking the pseudo science which frequents television and certain bbs's today. And I, for one, certainly hope you will continue your informative postings. But, I'm afraid I must respectfully differ with your earlier post concerning the South African grooved spheres ...

Heinrich: I thank you for your compliment. However, after reading the entire post, I feel very much like the honorable men who were similarly so highly spoken about in *Julius Caesar* by Shakespeare for some reason.

Watt: Michael Cremo and Richard Thompson wrote about these spheres in *Forbidden Archeology* (and also in their abridged version of the same book, *The Hidden History of the Human Race*). You correctly point out that Cremo and Thompson cite a weekly tabloid as their source for this information. As you point out, this tabloid, the *Weekly World News*, is known "for its largely fictionalized or completely fictionalized news stories." You are critical of Cremo and Thompson for citing this less than august publication "as a standard scientific journal . . ."

While your debunking of these spheres is persuasive, it is very important to note that Cremo and Thompson did not claim that these spheres were acceptable evidence for a much greater antiquity of man. As they write in the introduction to Chapter 6, "Evidence for Advanced Culture in Distant Ages," in *The Hidden History of the Human Race:*

> The reports of this extraordinary evidence emanate, with some exceptions, from nonscientific sources. And often the artifacts themselves, not having been preserved in standard natural history museums, are impossible to locate. We ourselves are not sure how much importance should be given to this highly anomalous evidence. But we include it for the sake of completeness and to encourage further study.

So, Cremo and Thompson intended the whole of Chapter 6 to be suggestive rather than definitive. They did not quote the *Weekly World News* as a standard scientific publication. They very clearly labeled it a non-scientific source.

Heinrich: In my post of 4/6/96 ...I was responding to both a post by Mr. Chris Beetle and the same material that was posted at http://nersp.nerdc.ufl.edu/~ghi/spheres.html. In both cases, the warning that Mr. Cremo and Thompson had in their book that you mention above ...was left out.

5.3.2.39. Letter to Jerry Watt from Michael A. Cremo, May 26, 1996.

After the above exchange with Paul Heinrich, Jerry Watt wrote to me privately, politely suggesting that the discussion of the grooved spheres be dropped from future editions of Forbidden Archeology *and* The Hidden History of the Human Race. *He felt it distracted attention from the evidence of better quality documented in these books.*

Regarding your suggestion, at present, I do not plan to drop the discussion of the grooved spheres from South Africa from either *Forbidden Archeology* or its abridged version, *The Hidden History of the Human Race*. When it comes time to make a second edition, what I probably will do is extend the discussion to include the kinds of objections raised by debunkers and my response.

The point of *Forbidden Archeology* is that if you want to understand human origins and antiquity by archeological methods, you cannot rely exclusively on the evidence collected in current textbooks and scientific literature. Even the scientifically reported evidence has been heavily filtered for the past 150 years. So to gain some idea of the totality of evidence, you have to recover from the total body of scientific literature a lot of evidence that has been unfairly dismissed. But one has to go even further.

Discovery and reporting of archeological evidence is not confined to scientists and their literature. Others, such as miners and rock hunters, also make discoveries, and these will not normally be reported in scientific literature. But it is still evidence that bears on the question. In any case, a lot of the evidence that is described in scientific literature was originally discovered by farmers, miners, quarry workers, etc., and later brought to the attention of scientists and then reported in scientific literature. But in many cases, the discoveries are not reported to scientists and do not make it into scientific literature. Such discoveries might, however, be reported in nonscientific literature, such as newspapers and sensationalistic journals.

If we are going to consider all the evidence relevant to human origins and antiquity, we cannot turn our backs on this source of potential evidence. It cannot be accepted uncritically. But neither can evidence reported in scientific literature.

So yes, sometime in the 1980s a report about grooved metallic spheres found in South Africa came to my attention, in a clipping from the *Weekly World News.* This is duly acknowledged in Appendix 2 of *Forbidden Archeology*, where the case is discussed, along with others of a similar kind. In this appendix, Richard Thompson and I state about this category of evidence: "The reports of this extraordinary evidence emanate, with some exceptions, from nonscientific sources.... We ourselves are not sure how much importance should be given to this highly anomalous evidence. But we include it for the sake of completeness and to encourage further study." The goal of *Forbidden Archeology* was to explore the full dimensions of the evidence bearing on human origins and antiquity and, therefore, I cannot see the value of excluding potential evidence reported in nonscientific literature. Of course, such reports cannot be accepted blindly; rather they must be carefully evaluated.

So in the case of the South African grooved spheres, yes, the initial information about them did reach us in the form of a clipping from the disreputable *Weekly World News.* But the grooved spheres were not included in the book simply on the basis of uncritical acceptance of the *Weekly World News* report. Well aware of the nature of this publication, we checked the facts. We initiated a correspondence with Roelf Marx, the curator of the Klerksdorp Natural History Museum, where some specimens of the spheres were housed. From Marx, we received confirmation that the objects do in fact exist, and he provided a photograph as well as useful information about the objects and where they were discovered. He also provided correspondence with a South African geologist who attempted a natural explanation of the objects, suggesting they were limonite concretions. The reported hardness of the objects appeared to rule out this explanation (limonite being a fairly soft mineral). So Richard Thompson and I concluded that the objects remain a mystery and that it is possible they are manufactured objects.

Since then, in connection with the NBC television show *Mysterious Origins of Man,* I obtained actual samples of the spheres, some still in their matrix. By arrangement of the producer of the NBC show, some of the spheres were subjected to metallurgical analysis (the ones analyzed are a fairly pure hematite, a form of iron). According to the producer, the analysts were at a loss to explain by natural processes the spherical form of the objects or the rings running around their equators.

During the Internet discussions that took place on sci.archeology, sci.anthropology, etc., after the show, several scientists suggested that such objects were undoubtedly natural concretions. But when I challenged one of them to send me specimens or photographs of perfectly round metallic concretions with grooves around their equators, he said he could not.

So that is where the matter rests at the moment. The objects remain a mystery, and I see no reason for eliminating an artificial origin from the list of possible explanations.

That the objects are now being subjected to scientific discussion is a healthy development. That was one purpose of *Forbidden Archeology*—to draw the attention of the scientific community to such evidence. In this case, that has been achieved. So let us now see how the discussion proceeds. Are we going to see some open-minded consideration of all possibilities, or are we going to see blind, uncritical debunking? All the noise about the *Weekly World News* suggests the latter.

Please feel free to post this message to the appropriate news groups.

The message was posted to appropriate newsgroups.

5.3.2.40. Letter to Sci.Archaeology and Alt.Archaeology Newsgroups from Chris Beetle, May 28, 1996.

This appears to be a response to Paul Heinrich's letter of May 23.

Although in the archaeology newsgroups authors Cremo and Thompson of *Forbidden Archeology* have been subjected to criticism for including a report originally from *Weekly World News* in their book, that is an incomplete understanding of the situation. As they very clearly state in the book, they wrote to the person mentioned in the article, Roelf Marx, curator of the museum of Klerksdorp, South Africa. Marx's response lent credence to the original article and provided further details. Cremo and Thompson went on to include an attempt at a natural explanation advanced to Marx by A. Bisschoff, a geologist, as well as some problems with that explanation. They conclude that the evidence is "somewhat mysterious," not that it is definitive proof of artificial origin.

The whole discussion centered around the personal communication with Roelf Marx, the museum curator. That is the point. The article in the *Weekly World News* is just how they found out about it, but the people criticizing them on the newsgroup don't mention that. Why not? They instead talk about the *Weekly World News*, which the authors themselves would have never considered to be a reliable source, without independent confirmation. This I think is a bit ridiculous.

As far as your criticism of the disclaimers regarding the quality of the evidence in the "Evidence for Advanced Culture in Distant Ages" appendix not being also listed along with the grooved spheres article on the web page, http://nersp.nerdc.ufl.edu/~ghi/spheres.html, I think it is a valid point and I will include it. I put the grooved sphere section on the web because I think people were under a mistaken impression that we put more credence in them than we actually do. I wanted to show what we actually said about them to put the speculation to rest. Actually they are in the final section in an appendix of highly anomalous material, and the people seriously interested in evaluating the veracity of the book would do better to study the earlier chapters, which are filled with arguments by recognized scientists in refereed journals, sometimes supporting and sometimes discrediting hundreds of potential artifacts in very ancient strata.

5.3.2.41. Review of *Hidden History of the Human Race* on Alt.Archeology Newsgroup, by Jerry Watt, May 28, 1996. Reproduced by permission of Jerry Watt.

This is a modified version of the original review, which can be found in Chapter Four of this book.

Mainstream paleoanthropologists will not like *The Hidden History of the Human Race* by Michael Cremo and Richard Thompson. Neither will many "new age" types who delight in alternative theories so they can imagine themselves at the vanguard of a new era. While it is interesting reading, *The Hidden History* is more like an unauthorized biography of a science with more than a few skeletons in its closet (pun intended).

According to Cremo and Thompson, the primary point they make is "...that there exists in the scientific community a knowledge filter that screens out unwelcome evidence." This is not really a very radical view; it has often been noted the it takes a generation or so for new discoveries to be accepted into science. Often it is not until the "old guard" passes away that the ideas of the "new guard" can gain recognition—and, in turn, become ideas of a new old-guard. What Cremo and Thompson do, however, is focus on paleoanthropology, showing that rather than considering all the evidence scientists have been putting skeletons in the closet with the hope they would be forgotten.

An example of the "knowledge filter" Cremo and Thompson write about is the so-called Java Man. This creature, now discredited, was surmised based on a cranium (skull cap) and a femur (thigh bone) found about 45 feet apart. After years of debate and controversy, the Java man became accepted by the scientific community as an authentic record of a "missing link." The cranium is clearly ape-like while the femur is thoroughly modern human. The Java Man has since been discredited for various reasons, but what remains is the human femur found in Java. When the Java Man was accepted, there was no problem with having a human-like femur dating back 800,000 years—and that's 400,000 years before our ancestors were supposed to have wandered into Java. So, is the human femur now accepted as evidence for a greater antiquity of man? Quoting from adverse criticism in the front of the book (yes, they included adverse criticism as well as the favorable), quoting from this criticism: "Your book is pure humbug and does not deserve to be taken seriously by anyone but a fool." And that is from Richard Leakey.

Hidden History makes a good case for the co-existence of anatomically modern humans and the very hominids we are supposed to have evolved from. This leaves paleoanthropologists with the daunting task of trying to explain how we evolved from creatures with whom we shared the plains of Africa (to name just one place). They cite well over a hundred examples of evidence which argue for a greater antiquity of mankind. Even if only a handful withstand the rigors of scientific examination, Cremo and Thompson will have succeeded in showing that paleoanthropologists have been ignoring for the greater antiquity of mankind.

The Hidden History of the Human Race is the abridged version of *Forbidden Archeology*, which was prominently mentioned on the television special, *The Mysterious Origins of Man.* The shortcomings of the television special, it must be noted, are not the shortcomings of either book. Those who are critical of the *Mysterious Origins of Man* should not presume that their criticisms can magically be carried over to *Hidden History* or *Forbidden Archeology*.

Hidden History is an important book, though one that will not be well liked by the scientific establishment. For while it is most instructive to have our shortcomings pointed out to us, it almost always leaves bruised egos by the wayside. Paleoanthropology will get a bruised ego from *Hidden History.*

5.3.2.42. NBC Internet Press Release, May 29, 1996.

In light of the intense campaign of intimidation waged against NBC by the fundamentalist Darwinians in the scientific community, I found the following response from NBC courageous.

Controversy Surrounds *The Mysterious Origins of Man.*

University Profs Want Special Banned from the Airwaves.

Program That Dares To Challenge Accepted Beliefs About Pre-Historic Man Will Be Rebroadcast June 8 on NBC.

NBC's *The Mysterious Origins of Man* sparked heated controversy within the academic community when originally broadcast February 25, 1996, and will be rebroadcast on Saturday, June 8 (8-9 p.m. ET). Professors of science and anthropology from some of the nation's most prestigious colleges and universities voiced strong opinions about some of the theories in the special, which challenged long-accepted beliefs about man's beginnings.

The program presented startling evidence suggesting man may have made the climb from Stone age to civilization more than once; that present-day man is just the latest in this cycle, and that Darwin's Theory of Evolution has serious flaws.

"Our goal was simply to present the public with evidence which suggests an alternative view to some of our most accepted theories," says producer Bill Cote. "We questioned fundamental issues that they (some scientists) felt should not be questioned. The bottom line is, the world is bigger than scientists can explain, and some of them want us to believe they can explain everything.

"We expected some controversy when we produced this show," Cote continues, "but no one was prepared for the enormous cry of outrage from members of the scientific community."

Hundreds of messages jammed Cote's special online website (http:www.bcvideo.com/bcvideo) following the program, and activity continues on several sites dedicated to the program. "While many viewers, including some scientists, praised the production as 'a great accomplishment' and contributing to public education," says Cote, "many scientists expressed outrage and criticism."

Michael H. Gerber (Emmy-winning special *The Mystery of the Sphinx*) and Robert Watts (*Star Wars trilogy, Indiana Jones* trilogy, *The Mystery of the Sphinx*) are the executive producers. John Cheshire, Bill Cote and Carol Cote (*The Mystery of the Sphinx*) are the producers. Bill Cote directed from a script he wrote with Cheshire. Charlton Heston hosts the program from B.C. Video Inc.

Media Contacts:
Dorothy Elery Austin, Entertainment Press & Publicity, (818) 840-3647
Robert Pientrantion, Entertainment Electronic Publicity, (818) 840-3565

5.3.2.43. Letter to Sci.Bio.Paleontology, Sci.Skeptic, Sci. Anthropology. Paleo, and Talk.Origins Newsgroups, from C. Marc Wagner, University of Indiana, May 30, 1996.

This is in response to the NBC press release reproduced above.

Blah, blah, blah...

Let's just chalk it up to more pseudo-science and let this one go! This newsgroup has already wasted too much time pandering to the "lunatic fringe" just because NBC is trying to keep their ratings up. Is this kind of stuff really worth any more effort to address than Jenny Jones (et al.) and their "my mom stole my boyfriend" drivel?

5.3.2.44. Letter to Talk.Origins Newsgroup from Kenneth Fair, University of Chicago Law School, May 30, 1996 (Excerpt).

Re: NBC To Rebroadcast "Mysterious Origins..." Amid Controversy

"Our goal was simply to present the public with evidence which suggests an alternative view to some of our most accepted theories," says producer Bill Cote. "We questioned fundamental issues that they (some scientists) felt should not be questioned. The bottom line is, the world is bigger than scientists can explain, and some of them want us to believe they can explain everything.

"We expected some controversy when we produced this show," Cote continues, "but no one was prepared for the enormous cry of outrage from members of the scientific community."

Quotes like this irritate me no end. There are few things that most scientists feel "should not be questioned." That's not the issue. Nor is the issue that scientists "believe they can explain everything." Any honest scientist knows that's not true.

What they do question are shows like this that present propaganda and questionable research as "scientific"... statements like "Scientists think it shouldn't be questioned" create a sense of paranoia and distrust of authority and appeal to gullibility that plants butts in the seats and sells advertising time for Ford Explorers and Pepsi. It's the same reason that conspiracy theories abound and that the *X-Files* are so popular. People don't bother to find out the facts for themselves and are left looking for scapegoats.

5.3.2.45. Letter to Bill Cote, with Copies to Various Scholars and Newsgroups, from Dr. Jere H. Lipps, Paleontologist, University of California at Berkeley, May 30, 1996.

I appreciate the advance notice of your press release about the reshowing of *The Mysterious Origins of Man.* Can you please provide me a list of the news organizations you sent your release to?

As you expected I am appalled that you and NBC would once again represent that program as the way science in America is done. It does not do you, NBC, or the sponsors any honor whatsoever. It indicates to scientists a large degree of ignorance about how science works.

You seem to think that scientists object to the theories presented. Not in most cases, because everything in the program has been dealt with by legitimate science already. You misrepresent the process of science—that is quite a different and detrimental thing. I can always straighten out bad ideas with my students, but trying to teach them an intelligent way to live their lives in this scientific society is very difficult when TV promotes a fraudulent view of how science works.

I am amazed that NBC [will] show this program again as science when a proper scientific presentation of the same issues would be both beneficial and entertaining to its viewers.

As its writer and director, I can appreciate your desire to use our objections to promote it once again. It is, however, a pathetic way to make a buck, when honesty is so much better and profitable.

I hope no one will be deceived that it was anything other than the anti-Darwinian message of The Mysterious Origins of Man *that provoked such intense reactions from Lipps and others. They did object to the theories presented.*

5.3.2.46. Letter to Sci.Bio.Paleontology, Sci.Skeptic, Sci. Anthropology. Paleo, and Talk.Origins Newsgroups from Dr. Jere H. Lipps, May 30, 1996.

NBC is now proposing to reshow their scientific travesty *The Mysterious Origins of Man,* using the objections of the scientific community as a selling point. This is a major disservice to the general public and misrepresentation of the majority of the scientists' objections.

The problem with the program was less the content, although that was sad, than the representation that this is the way science is done. The program's first showing made science teaching more difficult, and to continue to reinforce the dumbing of America by showing it again as science is irresponsible. Just this week, an Associated Press release published in most major newspapers documented that "only 25% of American adults got passing grades in a survey by the National Science Foundation of what people know about basic science." Few understand the process of science, and that is what this program, its producers and NBC also miss.

If you are worried about science in America, tell your local NBC station, NBC, and its various sponsors that you object to the portrayal of this program as science. America must get smart and we can make a difference!

Mysterious Origins ... and similar programs are a very definite black mark on NBC, its affiliates, the programs' sponsors, and its producers to represent them as science. Entertainment possibly, but not science! Why, I wonder, would a TV network, the makers of a program, and sponsors of the first showing like

Kellogg Co.
Coca-Cola
McDonalds
Olive Garden
Toyota
Kellogg's
J. C. Penney
Wendy's
LensCrafters
Folger's Coffee
M&M's Candy

that are heavily dependent on solid science for their own welfare promote the continuing decline of American prowess in science? Very sad, indeed. As science the program is garbage, but no one I know wanted the program banned from the airways, least of all me, who finds it a wonderful example of pseudoscience. Unfortunately, it represents the failure of the media to understand science or what they are doing to the American public's understanding of it.

Producers of these kinds of so-called science shows demonstrate a lack of understanding of the fundamental process of how science is done, and they pander their ignorance to the public. That is a disgrace!

It is a pity that the producers and NBC are now using the justifiable objections of scientists as a way of promoting their program.

I challenge to NBC and its program producers to have an introduction to the reshowing of this program by a real scientist—one with proper credentials from an established and generally accepted scientific institution—discuss the process of science and tell the viewers that the program may be entertainment, but it is not science!

For further information about this program, NBC, the Producers, sponsors, content, and previous comments, see the *MOM* WWW Site at http://rumba.ics.uci.edu:8080/faqs/mom.html

Ten or twenty years ago the campaign of intimidation waged by Lipps and other fundamentalist·Darwinians in the scientific community would have been sufficient to keep NBC from airing the program again or force NBC to let a fundamentalist Darwinian commentator dictate to the public how they should see the show. That NBC had the courage to stand up to the intimidation and the audacity to use the protests from the fundamentalist Darwinians to promote the rebroadcast of the unchanged original show to the public is a refreshing sign that intellectual freedom is alive and well in America.

The campaign of intimidation waged by Lipps and his cohorts is a real demonstration of how fundamentalist Darwinian science (as opposed to most other science) works. Darwinism is an ideology that fundamentalist Darwinians uphold by unscientific means (after all, what is so scientific about trying to intimidate a television network into taking a show off the air?).

Darwinism is not a concept that can be demonstrated by ordinary scientific means. It is simply an article of faith. And adherents of this faith think that they have a right to impose it upon everyone and silence anyone who speaks against it to the general public. The fundamentalist Darwinians would like a monopoly on access to the thinking of the general public. Fortunately, they do not have it and I hope they never will..

5.3.2.47. Letter to Talk.Origins, Talk.Religion.Newage, Sci.Anthropology, and Sci.Archaeology Newsgroups from August Matthusen, May 31, 1996.

Matthusen cited a sequence of messages by others on the upcoming rebroadcast of Mysterious Origins of Man *by NBC, and then added his own comment.*

Thomas Burgin: The only thing to do is to notice who sponsors the program, let the : sponsors know that you don't appreciate their support for such : nonsense on TV, and informing the sponsors that in response to their : associating their names with deceiving the public, that you will be : purchasing goods from their competitors.

Christopher Wood: Anybody know who the sponsors are? I would like to get an early start : boycotting them. There's always the off chance that some of them will : pull their sponsorship.

Terry Lacy: Might be nice to add something like "Does your sponsorship of the broadcast imply an endorsement of the same intellectual dishonesty? Can the public assume that your firm engages in same?" Maybe someone could set up a 'sponsors of fraud' web page, listing the names.

August Matthusen: You may want to consider going right to the top of the heap: General Electric. They own NBC. The same company founded by Thomas Edison and which makes its money by the implementation of science is behind this. "GE, we bring good things to light." You can write them care of: Chairman and Chief Executive Officer John F. Welch, Jr., General Electric Company, Fairfield, Connecticut 06431

The kind of science represented by Thomas Edison is not the kind of science represented by fundamentalist Darwinism. Electricity is a phenomenon that can be observed and tested in the laboratory. Evolution of life from chemicals and transformation of species are simply articles of faith.

What is really interesting about messages like the above is their depiction of "how science works." The response to unwanted ideas and evidence is not counterevidence and intellectual refutation, but boycotts and intimidation.

5.3.2.48. Letter to NBC and Various Scientists and Newsgroups from Dr. Jere H. Lipps, Paleontologist, University of California at Berkeley, May 1, 1996.

Of special interest is a section at the end of the letter, in which Lipps speculates about the involvement of the "Hare Krishnas" in the production of The Mysterious Origins of Man.

I received numerous replies from scientists, laypersons, and even NBC employees expressing sympathy with my posting about NBC's *Mysterious Origins of Man* (*MOM*) program. I have received no negative comments yet. Science in America and the rest of the world is far too important not to object to this kind of portrayal of it. Nevertheless, I do apologize for the multiple postings and the superlong header (but some people are interested in where this goes—all is now going as blind copies). I understand that some people may not have received a copy of my comments or the producers news release; for a copy, just email me.

I excerpt a few of those more helpful or interesting responses and my replies here:

Response 1: Jere—I just want to congratulate you on your activism on this issue. I know this is cynical, but my sense is that our numbers are so small that NBC can afford to ignore us if it is only the scientists protesting. Is there a way for us to make a louder noise—to enlist an educated public to complain about how they are being treated?

How about the following:

Enlist the support of the press—could science societies sponsor a short course or workshop for journalists—educating science writers about how we do science and how TV misrepresents science?

Enlist major museums with monthly magazines that reach the public—articles that expose misrepresentation of the scientific process with pointed examples.

Material could be put on the WWW to assist teachers and students in watching TV science specials by listing common forms of misrepresentation so that there would be guidelines for flagging bad coverage—perhaps a form for checklisting and rating or ranking scientific content.

One way to combat *Mysterious Origins* specifically would be to show it to students—to critique—kind of "dissection lab." I work with students in all my courses with respect to how science is represented to the lay public, and they pick up very, very fast on "loaded words" and "weasel words," the nature of evidence, the fact that fossils are never ancestors, that hypotheses must be testable, that we never "prove" anything in science, to discount anything that begins with "scientists believe," etc. etc.

Reply by Lipps: Excellent ideas. Some science writers have contacted me already. If you want to use the program as an example of bad science, bad media, public trust, or whatever, you can order it for $19.95 from the producers, if you care to give them any more money. To order, call 1-800-846-9682.

Response 2: I tend to agree with Jere's post and think it has wider relevance than the "dumbing of America", certainly it relates to the numbing of Australia and I suspect too the rest of the electronically connected world.

As Jere points out, it is ironic that counter arguments are likely to be used as publicity fodder without getting much of a hearing in their own right, but this can hardly be an excuse for remaining silent. Again it would be hypocritical of members of the list to object simply to the canvassing of silly ideas in public, since much of our area of interest might be criticized on this ground, some of it no doubt correctly. We should however lend our support to the high school teachers of America and the world as they set out yet again to explain that not all ideas are of equal value and that the best way to distinguish between them is by analysis of the evidence, not by the entertainment value of the presentation.

Response 3: The following posting should be of immediate interest and concern for all educators—especially science teachers [*apparently this respondent had posted to a teachers' newsgroup the original response to* MOM *by Lipps*]. If anyone has suggestions on other listings for teachers where this should be posted, please feel free to pass it on.

Response 4: Yes, I know that we'll lose. But, we cannot stand and watch it happen. Science is a howl. But, too many see us as super-serious pocket protector types. Methodology aside, I'm offended that the story presents scientists as essentially humorless. I got into science because it is fun. I wish that we could convey the enjoyment to the next generation.

Response 5: I'm writing an article for a magazine about how the teaching of evolution is being undermined (mostly by Creationists who are getting their message out due to public misunderstanding of science).

Reply by Lipps: The *Mysterious Origins of Man* program was not a creationist job. Indeed, they wrote a scathing review of it in their Genesis newsletter because it promoted that dinosaurs and man lived together millions of years ago, etc. It seems to be a Hare Krishna job.

From the *MOM* website ...: "*MOM* is apparently an attempt to use science to support Hinduism. Much of the material in the program is based on the contents of two books, *Forbidden Archeology* and *The Hidden History of the Human Race* by Michael Cremo and Richard Thompson, both of whom appeared on the show and are members of the Bhaktivedanta Institute, a branch of the International Society for Krishna Consciousness."

Response 6: Thank you for passing along your press release. Could you pass along Mr. Cote's e-mail address as well? I have already harangued Coke, McDonalds, and Wendy's because I'm a stockholder in these companies.

Reply by Lipps: Yes. Lots of information about that particular program—*MOM* is available on the *MOM* Web Site. The producers have a site that quotes selected scientists (pseudojournalism to match the pseudoscience), as an advertising gimmick probably, at http://www.bcvideo.com/bcvideo/. Mr. Cote, who wrote and produced the program, can be reached at: P.O. Box 97, Shelburne, VT 05482, Ph: 800-846-9682. Email: bcvideo@interport.net (Bill H. Cote) You could also write to: [*gives names and addresses of NBC officers*].

Make your letters reasonable, factual statements rather than outrage or off the cuff remarks that can be used to advertise this program. I think it helps to emphasize that network TV has a public trust to uphold, to quote Ted Koppel (of ABC).

Response 7: Talk about backfires! I don't know what strategy to use; obviously letters from us won't help. This borders on a real news story: Major network deliberately distorts science and the scientific process to hype its ratings. What if a major network were to suddenly revive the Piltdown man as the missing link, rather than as a hoax?

Reply by Lipps: True, some of the more volatile comments are being used as an argument that scientists' own work or theories are threatened by this "expose". This is another example of the producers depth of misunderstanding, ignorance and fear about how science is done, and their greed to make money or promote the Krishna view. I think we should keep writing letters, but to NBC officials, affiliate stations, and the sponsors. We should develop the whole issue of the dumbing of America by the media, politicians, and others that lack an appreciation of how and what science is.

It really matters at all levels: In the *Oakland Tribune* yesterday (p. A9), an article headlined "High Tech Hits Road for More Scholars" noted that semiconductor and biotechnology companies were leaving California and the Bay Area because "the education here is not competitive with...other parts of the country." If we keep dumbing America scientifically, where will we turn?

Response 8: I've forwarded your comments to our station's general manager, program director and news director. Thanks for taking the time to write!

Response 9: How were the Hare Krishnas involved in *The Mysterious Origins of Man*?

Reply by Lipps: The *MOM* homepage goes over some of it. The publisher of the two persons mentioned in that context is, in its own words: "Govardhan Hill Publishing. New Paradigms in Science. Govardhan Hill, Inc. (GH) is a nonprofit educational institution which produces books and videos giving some insight into fundamental questions about human life. How has the universe come into being? How has human life originated? What is the nature of consciousness? Is there other intelligent life in the universe? These are some of the questions we discuss in light of modern scientific evidence and the ancient wisdom of India." (from http://nerdc.ufl.edu/~ghi/index.html)

Response 9 (cont.): Is this program pure, sensationalist pseudoscience?

Reply by Lipps: It is the usual mishmash of pseudoscience techniques: self-proclaimed experts, revisionist history, appeals to authority, self-declarations of scientific inaccuracies without evidence to the contrary, exhumation of old hypotheses that have been proved false long ago, extraordinary scientific claims unsupported by evidence, claims that establishment science is hiding the truth, claims that their scientists can't get their stuff published, etc.

Response 9 (cont.): Do the Hare Krishnas have an agenda?

Reply by Lipps: I dunno. I have generally run from Krishnas at airports, so cannot comment on what it might be. The program was written by Mr. Cote, so he can answer these questions best.

Concluding question [*apparently by Lipps*]: How can NBC be proud of this?

Anticipated response based on comments received [*from NBC*] after last showing: Well, the viewers are entitled to see alternative views.

Correct, Mr. NBC. But get one of your legitimate scientific advisors to help you pass judgment, not the word of Mr. Schwartz, an expert in acquiring programs and not science, or Mr. Cote, a sensationalist writer and producer without known scientific credentials, but who seems to be pushing the views of a particular sect.

Suggestion: Run it if you need the ratings, but label it science fiction. Or introduce it with a legitimate scientific explanation of how to do science. The controversy could help those ratings!

5.3.2.49. Letter to Newsgroups from Keith (littlejo@news.demon.net) June 8, 1996.

This is the same scientific misinformation and falsehoods that Mr. Cremo and Mr. Thompson have repeated on *Mysterious Origins of Man* (which will be repeated tonight on NBC). The claim that Java Man has been discredited is false, as the skullcap alone is quite clearly that of a *Homo erectus* and a significant find by itself....The problem with the femur was that it was a surface find, thus, there is absolutely no evidence that it is 400,000 years old as Cremo and Thompson claim. It is generally agreed upon by scientists that Dubois made a mistake in assuming that the femur, along with some orangutan teeth found nearby, was in any way associated with the skull. This is something that has been long recognized by scientists.

5.3.2.50. Letter to Newsgroups from Jerry Watt, June 10, 1996.

This is a response to the above letter from Keith.

Thanks for your very interesting follow-up. I checked out the websites you listed; however, there was no indication that the "Java Man" femur was a surface find. Cremo and Thompson cite others as saying that the femur was found "in situ," therefore giving us every reason to think it was of the same age as the skull cap.

On page 160 of *Hidden History*, Cremo and Thompson write: "Marcellin Boule, director of the Institute of Human Paleontology in Paris, reported, as had other scientists, that the layer in which the *Pithecanthropus* skullcap and femur were said to have been found contained numerous fossil bones of fish, reptiles, and mammals." In other words, they were not a surface find.

On the same page we have: "The femur, Weidenreich said, was very similar to that of a modern human, and its original position in the strata was not securely established." Again, while its position in the strata was not securely established, it was not a surface find.

On page 175, Cremo and Thompson write: The original *Pithecanthropus* skull (T2) and femur (T3) reported by Dubois were found in situ.... Again, not a surface find.

Is there any documentation that the femur was a surface find?

5.3.2.51. Letter to Newsgroups from Keith, June 10, 1996.

You are right and I was wrong, having gotten this site confused with the practice by Dubois of buying many of his specimens from the local people who gathered them off of the ground. For this mistake, I apologize to you, Cremo, and Thompson.

5.3.2.52. Letter to Newsgroups from Jerry Watt, June 10, 1996.

The letter is addressed to Keith.

There is absolutely no need to apologize. You are an honorable man and, just like me, entitled to honest mistakes. And I'm probably way ahead of you in that department.

For the sake of argument, let's accept as fact that the skull cap should never have been associated with the femur. Further, let's accept as fact that the femur may very well be much younger than the skull cap. Nevertheless, Java Man was accepted (both skull cap and femur) as an authentic *Homo Erectus* by the scientific "establishment" so long as it fit their existing theories. These scientists had no problem with a human femur existing 800,000 years ago—as long as it was part of Java Man. Now that Java Man has been discredited, on what basis do they now "hem and haw" over this unexpectedly ancient human bone? It was okay before, as long as it fit their theories. Now it does not fit their theories and it is discredited. There's a pattern here: theory first, facts second. This is a reversal of the scientific method, which is to ascertain the facts first and devise theories to explain the facts.

Concerning the possibility of the femur being mixed in from a younger stratum, Cremo and Thompson write: "Some have said that the femurs were mixed in from higher levels. Of course, if one insists that the human-like Trinil femurs were mixed in from higher levels, then why not the

Pithecanthropus skull as well? That would eliminate entirely the original Java man find, long advertised as solid proof of human evolution."

On a personal note, sir, I hope that you continue following and contributing to this thread—this in spite of your diabolically fattening cookie recipes.

5.3.2.53. Letter to Alt.Archaeology Newsgroup from Richard L. Thompson and Chris Beetle, June 18, 1996.

Dr. Foley quotes Mr. Charlton Heston as stating [*in* Mysterious Origins of Man] that Dubois later claimed that the Java Man skullcap was actually from a large ape. To this claim, Dr. Foley replied:

> Totally false. Although he did emphasize the ape-like features of the skullcap, Dubois did not say it came from a giant gibbon. He always believed that it was an intermediate between ape and human (correctly), and that the skullcap and thigh bone belonged to the same creature (probably incorrect). Java Man is still recognized as a member of *Homo erectus* by all competent modern scientists (and as an ape by almost all creationists).

The News and Views section of *Nature* magazine, Volume 141, page 362 (February 26, 1938), however, stated just the opposite:

> Dr. E. Dubois, after a prolonged study not only of the material known to anthropological text-books, the skull-cap, thigh bone and tooth from Trinil, but also of other material from the same provenance in his possession, for the most part not previously published, arrived at the conclusion, mainly on the evidence of the long bones, that we are here concerned with a gigantic gibbon.

Regarding whether or not the skull-cap and femur are from different creatures, W. E. Le Gros Clark and G. Campbell in the 3rd edition of *The Fossil Evidence for Human Evolution* (1978) state on page 91:

> Some doubt was naturally expressed (mainly because of their apparent incongruity) whether the femur and calveria belonged to the same individual and whether the femur was really indigenous to the Trinil deposits. However, the accumulation of evidence speaks so strongly for their natural association that this has become generally accepted.

Later, however, on page 110, they say:

> It must, however, be admitted that though the Choukoutien femoral fragments show certain distinctive features (which they share with Olduvai femur, see below) these features are not present in the Trinil femora, which are fully modern in form. This inconsistency, pointed out by Day and Molleson (1973), throws some doubt on the antiquity of the Trinil femora and their supposed association with the famous calotte (p. 91).

Later, in 1984, despite such doubts by scientists, the famous *Ancestors* exhibit in New York City graphically presented casts of both the skullcap and thigh bone as being Homo erectus, for all the world to see. Was this the actual scientific understanding at the time, or was the public presented with outdated information merely for propaganda purposes?

The most charitable view is that there was once a strong consensus among scientists in favor of the idea that the thighbone and calotte belonged to the same individual. Later, the Day and Molleson findings, which we discussed in *Forbidden Archeology* on page 472, caused some scientists to doubt this. At the time of the Ancestors exhibit, the old consensus apparently still ruled. More recently doubt has grown, and the femur now appears to be fully human. But couldn't this be taken as evidence for *H. sapiens* as a contemporary of the Trinil *Pithecanthropus*?

5.3.2.54. Letter to Talk.Origins, Alt.Archaeology, and Sci. Anthropology. Paleo Newsgroups from Dr. Richard L. Thompson and Chris Beetle, June 26, 1996.

This is a typical Internet discussion message, with passages and responses from previous messages included. I have tried to disentangle the sequence and present it as clearly as possible.

Thompson and Beetle (previous message, cited by Jim Foley): The News and Views section of *Nature* magazine, Volume 141, page 362 (February 26, 1938), however, stated just the opposite:

> Dr. E. Dubois, after a prolonged study not only of the material known to anthropological text-books, the skull-cap, thigh bone and tooth from Trinil, but also of other material from the same provenance in his possession, for the most part not previously published, arrived at the conclusion, mainly on the evidence of the

long bones, that we are here concerned with a gigantic gibbon.

Jim Foley: As I previously mentioned, Dubois did emphasize similarities of the skull with apes, especially gibbons. What Dubois said was:

> *Pithecanthropus* was not a man, but a gigantic genus allied to the gibbons, however superior to the gibbons on account of its exceedingly large brain volume and distinguished at the same time by its faculty of assuming an erect attitude and gait. It had the double cephalization [ratio of brain size to body size] of the anthropoid apes in general and half that of man.

Calling it a "gigantic genus allied to the gibbons" is not, to me, the same thing as calling it a giant gibbon.

Thompson and Beetle (this message): We agree here. After all, a gibbon with "exceedingly large brain volume and distinguished at the same time by its faculty of assuming an erect attitude and gait" is not much of a gibbon at all.

Jim Foley (cont.): Nevertheless, one can see that Dubois' statement might, with a bit of a stretch, be claimed to be calling it a giant gibbon, and even many evolutionists have, as shown by the above *Nature* article, said that.

Thompson and Beetle (this message): Yes. We simply accepted what *Nature* said as authoritative when actually it was misrepresenting the facts, thus we got a mistaken idea. Perhaps you might include the *Nature* quote in your gibbon document to show how people could be unknowingly misled.

Jim Foley (cont.): More importantly, Dubois was not claiming that Java Man was just a gibbon, or just an ape. It is clear from the above quote, and others...that Dubois always considered Java Man a human ancestor. *MOM*'s clear implication that the Dubois thought the skullcap had nothing to do with human evolution is false. Dubois said, in 1935 I think: "I still believe, now more firmly than ever, that the *Pithecanthropus* of Trinil is the real 'missing link'."

MOM's statement that Dubois recognized the femur as belonging to a human is also false.

Thompson and Beetle: Yes, that is certainly true. I [*Thompson, apparently*] do not know where the producers of *MOM* got that particular idea. It certainly was not from *Forbidden Archeology*.

Jim Foley: The same quotes make it clear that he always believed the femur and the skullcap belonged to the same creature. You can verify this by looking at *Nature* 136:234 and 136:838 (1935), which are reports of a talk Dubois gave in Holland.

MOM's statement that Java Man had been reclassified as an ape by the scientific community is also false. For over 50 years, Java Man has been considered to be a member of a species intermediate between humans and apes.

Thompson and Beetle: I [*Thompson, apparently*] had difficulty finding the statement in the Java Man section of the *MOM* broadcast that the creature "had been reclassified as an ape by the scientific community." Is that point made somewhere else in the show? Perhaps you could check the transcript you have and let me know.

Richard Milton at one point said that all proposed 'missing links' have turned out to be either apes or humans in the final analysis, but *Forbidden Archeology* does not support that view. It appears that there is real evidence for the existence of beings intermediate between humans and apes. However, because there is evidence of anatomically modern human beings before or during the presence of these ape-men, the supposed evolutionary relationship cannot be supported. For example, the Castenedolo skeletal remains (Ragazzoni, Sergi) provide evidence for anatomically modern humans in the Pliocene of Italy. There are different substantiated reports of bola stones, typically associated with anatomically modern man, from as old as 2.0 million years in Olduvai Gorge (M. Leakey) and 3 million years in Argentina (E. Boman). Who was making these bola stones? There is a great deal of evidence suggesting that *H. sapiens* may have preceded *H. erectus* or was at least a contemporary.

Thompson and Beetle (*previous message, cited by Foley*): Regarding whether or not the skull-cap and femur are from different creatures, W. E. Le Gros Clark and G. Campbell in the 3rd edition of *The Fossil Evidence for Human Evolution* (1978) state on page 91:

> Some doubt was naturally expressed (mainly because of their apparent incongruity) whether the femur and calveria belonged to the same individual and whether the femur was really indigenous to the Trinil deposits. However, the accumulation of evidence speaks so strongly for their natural association that this has become generally accepted.

Later, however, on page 110, they say:

> It must, however, be admitted that though the Choukoutien femoral fragments show certain distinctive features (which they share with Olduvai femur, see below) these features are not present in the Trinil femora, which are fully modern in form. This inconsistency, pointed out by Day and Molleson (1973), throws

some doubt on the antiquity of the Trinil femora and their supposed association with the famous calotte (p. 91).

Jim Foley: Doubt about the Trinil femur persists up to the present day. *It doesn't matter* whether it belongs to the skullcap or not, because the Turkana Boy skeleton (http://earth.ics.uci.edu:8080/faqs/homs/15000.html) has a skullcap which very closely resembles Java Man, and also has an almost totally modern bipedal femur.

Thompson and Beetle (this message): False. The Turkana Boy does not have a totally modern femur (and this makes the modern Trinil femur stand out as significant). In their article "Early Homo erectus Skeleton from West Late Turkana, Kenya" printed in *Nature,* Vol. 316 (August 29, 1985), on page 791, F. Brown, J. Harris, R. Leakey, and A. Walker describe two features of the Turkana femur which are markedly different from the human femur. They cite an article by C. O. Lovejoy, K. G. Heiple, and A. H. Burnstein who, in their article in the *American Journal of Physical Anthropology,* Vol. 38 (1973) on pages 757-780, mention that the biomechanical neck length of the Turkana femur is well over 3 s.d. [*standard deviations*] from the mean of a sample of *H. sapiens.* As well as having a long femoral neck, the neck-shaft angle is very small at 110 degrees, being 5 s.d.s from the mean of the same *H. sapiens* population.

Although the Turkana *Homo erectus* femur has the above differences from *Homo sapiens,* at Koobi Fora, at an older site (least 2.0 million years old), a much more humanlike femur was unearthed. According to studies by anatomist B. A. Wood in 1976, the neck size and neck shape of femurs ER 1472 and 1481 were always within a single deviation from the modern human mean.

Does this mean that evolution went backwards for a few hundred thousand years [*from the time of the Koobi Fora femur to the time of the Turkana femur*] or would a better explanation be that modern humanlike forms existed then [*at the time of the Koobi Fora femur*] as they do now and the ape-men like *Homo erectus* [*of which the Turkana boy is one of the oldest specimens*] have no causal relationship with them.

And since the Turkana *H. erectus* femur differs from a human femur, but the Java Man femur does not, perhaps that Java femur is from a human being after all, and not a *Homo erectus.*

Thompson and Beetle (previous message, cited by Foley): Later, in 1984, despite such doubts by scientists, the famous *Ancestors* exhibit in New York City graphically presented casts of both the skullcap and thigh bone as being *Homo erectus,* for all the world to see. Was this the actual scientific understanding at the time, or was the public presented with outdated information merely for propaganda purposes?

Jim Foley: Depends. If they just said that the femur was found at Trinil, they were right. If they actually said it belonged to *Homo erectus,* then I would consider that, yes, they went beyond what could confidently be asserted. Did they in fact say that? Or could the femur have been shown there because of its historical importance?

Thompson and Beetle (this message): Because of the passage of time, it is difficult to recall the caption on the single case containing the skull cap and thigh bone of the original Java Man find. Still the fact that they were displayed together in the same case in a human ancestors exhibit strongly implies that they were thought to be from a single creature that was thought to be a human ancestor. If the caption spoke to the contrary, that would be similar to a dishonest insurance policy in which the fine print takes away your coverage. [*Thompson was at the* Ancestors *exhibit.*]

Thompson and Beetle (previous message, cited by Foley): The most charitable view is that there was once a strong consensus among scientists in favor of the idea that the thighbone and calotte belonged to the same individual. Later, the Day and Molleson findings, which we discussed in *Forbidden Archeology* on page 472, caused some scientists to doubt this. At the time of the *Ancestors* exhibit, the old consensus apparently still ruled. More recently doubt has grown, and the femur now appears to be fully human. But couldn't this be taken as evidence for *H. sapiens* as a contemporary of the Trinil *Pithecanthropus*?

Jim Foley: The documentation at the Trinil excavation was so sloppy by modern standards that it is impossible to be sure whether the femur was found at the same level as the skullcap.

It's impossible to be sure what the femur was, and it hardly matters. What matters is that the Java Man is not an ape, Dubois didn't say it was an ape, and even if he had said it that wouldn't make it an ape. It's just not an ape, and I doubt any scientist in the last 50 years has said otherwise.

Thompson and Beetle (this message): If the Java Man skullcap is significant as evidence of early man, then the femur is also. If we neglect the femur on the grounds of sloppy excavation, then we should likewise reject the skullcap. Otherwise one is adhering to a double standard. The skullcap, if accepted, implies the existence of a middle Pleistocene ape-man. The femur, which is of modern human form, implies the existence of middle Pleistocene *H. sapiens.*

References:

Boman, E. (1921) Los vestigios de industria humana encontrados en Miramar (Republica Argentina) y atribuidos a la época terciaria. *Revista Chilena de Historia y Geografia, 49(43):* 330-352.

Leakey, M. D. (1971) *Olduvai Gorge.* Vol. 3. *Excavations in Beds I and II, 1960-1963.* Cambridge, Cambridge University.

Ragazzoni, G. (1880) La collina di Castenedolo, solto il rapporto antropologico, geologico ed agronomico. *Commentari dell' Ateneo di Brescia,* April 4, pp. 120-128.

Sergi, G. (1884) L'uomo terziario in Lombardia. *Archivio per L'Antropologia e la Etnologia, 14:* 304-318.

Sergi, G. (1912) Intorno all'uomo pliocenico in Italia. *Rivista Di Antropologia (Roma), 17:* 199-216.

Wood, B. A. (1976) Remains attributable to *Homo* in the East Rudolf succession. In Coppens, Y., Howell, F. C., Isaacs, G. I., and Leakey, R. E., eds. *Earliest Man and Environments in the Lake Rudolf Basin.* Chicago, University of Chicago, pp. 490-506.

Richard Thompson and Chris Beetle: (e-mail: ghi@nerdc.ufl.edu)
Forbidden Archeology Home Page:
(http://nersp.nerdc.ufl.edu/~ghi/fa.html)

5.3.2.55. Letter to Various Individuals and Newsgroups from Dr. Jere H. Lipps, Paleontologist, University of California at Berkeley, June 21, 1996.

I had several responses regarding contacting the FCC about the Krishna's *Mysterious Origins of Man* program aired on NBC. It would seem to be a violation of the public trust to thrust pseudoscientific religious views on the American Public. Two views follow:

1. "I've been away almost 2 weeks and I am just getting caught up on my email. Long before reaching the last letter (I think it's the last) the thought that kept coming to my mind is the violation of public trust. The airways are not the property of NBC. The FCC gives them permission to broadcast on a specific, public frequency, but I plan to view the next one (on the 9th?). After that, I'll be in a position to contact the FCC. If the many scientists who have responded to your wonderful circular letter would do the same, I suspect we could have more impact from NBC."

2. "I have just sent the attached to the addresses listed. If you don't see anything libelous, feel free to share it with whomever. The server just told me the copy to Wright bounced (no such address). I assume the main message got through.

Date: Mon, 17 Jun 1996
To: fccinfo@fcc.gov
Cc: rwright@nbc.com
Subject: NBC violation of public trust

Dear sirs:

The attached document is self-explanatory. If you wish any follow-up, I can be reached at the e-mail address on this message, or by Phone/FAX at 303-443-1375. Thank you for considering this.

Allison R. Palmer,
President, Institute for Cambrian Studies

Re: *Mysterious Origins of Man*, rebroadcast by NBC June 8, 1996.

This e-mail is a request for the FCC to investigate and, I hope, seriously censure the National Broadcasting Company for crassly commercial irresponsible journalism that seriously violates the trust the public should have in materials that are touted as credible by a major network. Last February they produced a program "Mysterious Origins of Man" that purported to be scientifically based, and received massive negative reactions from responsible scientists representing numerous areas of science. Following this response, instead of checking their facts (or factoids) as any responsible journalistic organization is supposed to do, especially before they rebroadcast such a program, they chose to use the reactions of the reputable and responsible science community to generate viewer interest (a la the *National Enquirer*) by distributing PR announcements implying that the content of their show was science that the "establishment" did not want brought before the public. Their show included some Ph.D.'s to establish "credibility".

However, hardly anything they presented could stand even the most rudimentary test of credibility. As only one example: leading up to the suggestion that Antarctica was the lost continent of Atlantis, they cited buildup of ice in the north polar region that would imbalance the Earth's crust, causing it to "slip" taking some previously warmer areas into the polar regions. They never bothered to note the physical impossibility of their proposal and the clear and unequivocal evidence from paleomagnetism that the northern continents have been in much their present latitudinal positions for many millions of years, not to mention that ice cores from the Antarctic continent show continuous ice for well over 100,000 years. The February production could have been viewed as stupid, but to repeat this without checking facts, especially when the criticisms of the show did not come from the lunatic fringe, was the height of irresponsibility. Such behavior only contributes to public confusion about the credibility of almost any information. NBC has done serious public damage that will be extremely hard to undo.

I submit that this sort of behavior should not be acceptable or condoned by the regulators of publicly available network television and that it deserves much more than a slap on the wrist. At the very least NBC should be required to make substantial prime-time apologies to their viewing audience for a sufficient period of time so that the audience clearly gets the message that they were duped. In addition, NBC should perhaps be fined sufficiently so that a major fund for public science education can be established. The public is not being well served if major networks, in the guise of presenting credible information to their audiences, mislead the public and undermine public confidence by tactics more akin to the purveyors of the junk available at supermarket checkout stands. NBC's unprofessionalism damages the credibility of all public television.

Please let me know if there is more specific information you may need to pursue this violation of the public trust."

5.3.2.56. Letter to Dr. Jere H. Lipps, Paleontologist, University of California at Berkeley, from Michael A. Cremo, Late August 1996.

I don't have time to constantly monitor the Internet discussions related to *Mysterious Origins of Man*. But from time to time, unavoidably, it seems, friends send me batches of the traffic. And I see your name has come up quite a number of times. It seems like you have some questions about my book *Forbidden Archeology*, and its connections with the Krishna consciousness movement. You also seem interested in the connection of the book with the show *MOM*. I would be happy to discuss these things with you, and, especially, answer any questions you might have. But first of all, I hope you will get a copy of *Forbidden Archeology* and look it over. If you have any problem obtaining a copy, let me know. I will be leaving tomorrow for a trip to Australia and will be returning to the U.S. on September 7.

Dr. Lipps did not respond to this message.

Bibliography

Conference Papers and Journal Articles by Michael A. Cremo

Cremo, Michael A. (1994) "Puranic Time and the Archeological Record." World Archeological Congress 3. December 4—11, 1994, New Delhi, India.

Cremo, Michael. A. (1995) "The Reception of Forbidden Archeology: An Encounter Between Western Science and a Non-Western Perspective on Human Antiquity." Kentucky State University Institute for Liberal Studies Sixth Annual Interdisciplinary Conference: Science and Culture. Frankfort, Kentucky, March 30 - April 1, 1995.

Cremo, Michael A. (1995) "Forbidden Archeology: Evidence for Extreme Human Antiquity and the Ancient Astronaut Hypothesis." Ancient Astronaut Society World Conference, Bern, Switzerland, August 17-19, 1995.

Cremo, Michael A. (1996) Screams from the Stream: Mainstream Science Reacts to Forbidden Archeology. *The Anomalist*, vol. 4, pp. 94-103.

Cremo, Michael A. (1997) *Forbidden Archeology*: A Three-Body Interaction Among Science, Hinduism, and Christianity (Unpublished). Submitted by invitation to *Hindu-Christian Studies Bulletin.*

Cremo, Michael A. (1997) "The Later Discoveries of Boucher de Perthes at Moulin Quignon and Their Bearing on the Moulin Quignon Jaw Controversy." XXth International Congress of History of Science, Liège, Belgium, July 19-26.

Reviews, Articles and Scholarly Communications about *Forbidden Archeology*

Anderson, Colonel W. R. (1993) *Forbidden Archeology. Vikingship: Bulletin of the Leif Ericson Society,* vol. 29, no. 4, p. 6.

Anonymous (1994) Notice of *Forbidden Archeology*. "Publications," *Journal of Field Archeology,* vol. 21, p. 112.
Anonymous (1994) Notice of *The Hidden History of the Human Race. Sci Tech Book News*, November.
Anonymous (1994) Notice of *The Hidden History of the Human Race. The Bookwatch*, September, p. 3.
Anonymous (1994) Review of *Forbidden Archeology.* "Book news," *Ethology Ecology & Evolution*, vol. 6, p. 461.
Anonymous (1995) *Hidden History* Reveals Major Scientific Cover-Up. *Hare Krishna Report*, No. 5, March, p. 2.
Benedict, W. Ritchie (1995) Review of *Forbidden Archeology. The X Chronicles*, vol. 1, no. 5, November, p. 1.
Belderis, Ina (1995) Will the Real Human Ancestor Please Stand Up! *Sunrise,* April/May, pp. 111-118.
Broodbank, Cyprian (1993) Notice of *Forbidden Archeology.* New books section, *Antiquity,* vol. 67, p. 904.
Conrad, Ed (1993) New book claims man existed on earth long before the apes. *Hazleton Standard-Speaker*, Wednesday, November 17, p. 13.
Corliss, William (1993) Notice of *Forbidden Archeology. Science Frontiers Book Supplement*, no. 89, September-October, p. 1.
Cremo, Michael A. Beyond "Dreck" and "Drivel": A Response to Jonathan Marks's Review of *Forbidden Archeology*, Submitted to Dr. Matt Cartmill for Publication in *American Journal of Physical Anthropology,* April 16, 1994.
Davidson, John (1994) Fascination Over Fossil Finds. *International Journal of Alternative and Complementary Medicine* August, p.28.
Donovan, Diane C. (1993) *Review of Forbidden Archeology. Midwest Book Review*, September.
Earley, George W. (1994) Review of Forbidden Archeology. *The Gate,* July, p. 11.
Erbs, Lori (1993) Unpublished review of F*orbidden Archeology*.
Feder, Kenneth L.(1994) Review of *Forbidden Archeology. Geoarchaeology,* vol. 9, pp. 337-340.
Flynn, Dr. Pierce J. (1993) Foreword to *Forbidden Archeology: The Hidden History of the Human Race*, by Michael A. Cremo and Richard L. Thompson. San Diego: Govardhan Hill Publishing.
Gopiparanadhana Dasa (1993) Book Review: *Forbidden Archeology. ISKCON World Review*, vol. 12, no. 2, August/September, p. 4.
Groves, Colin (1994) Creationism: The Hindu View. A Review of *Forbidden Archeology. The Skeptic* (Australia), vol 14, no 3, pp. 43-45.
Ham, Ken (1996) Hollywood 'Moses' Undermines Genesis. *Answers in Genesis*, February, p. 5. Quoted in Isaak, Mark. <isaak@aurora.com> Letter. 28 February 1996. <talk.origins, sci.archaeology, alt.archaeology>

Heinberg, Richard (1995) The Lost History of Humankind. *The Quest*, Winter, pp 24-31.

Hunt, Jean (1993) Antiquity of Modern Humans: Re-Evaluation. *Louisiana Mounds Society Newsletter*, no. 64, Nov. 1, pp. 2-4.

Johnson, Dr. Phillip E. (1994) Foreword to *The Hidden History of the Human Race*, by Michael A. Cremo and Richard L. Thompson. Badger, CA: Govardhan Hill Publishing.

Kenyon, Douglas J. (1996) Ancient Mysteries: Exposing A Scientific Coverup. *Atlantis Rising*, No. 6, p. 17.

Langbein, Walter J. (1994) Notice of *Forbidden Archeology. PARA*, January.

Langenheim Jr., R.L. (1995) Notice of *Forbidden Archeology*. "Books for Geoscientists," *Journal of Geological Education*, vol. 43(2), p. 193.

Lee, Laura (1995) Notes from the New Edge: OOPS, Our Anomalies are Showing. *Common Ground*, vol. 10, no. 1., January/February.

Lepper, Bradley T. (1996) Hidden History, Hidden Agenda. A Review of *Hidden History of the Human Race. Skeptic*, vol. 4, no. 1, pp. 98-100.

Line, Peter (1995) *The Hidden History of the Human Race. Creation Research Society Quarterly,* June, p. 46.

Lipson, Philip (1995) Michael Cremo and Forbidden Archeology. Unpublished review of *Forbidden Archeology.*

Marks, Jonathan (1994) Review of *Forbidden Archeology. American Journal of Physical Anthropology*, vol. 93(1), pp. 140-141.

Mayer, Andrew M. (1995) Review of *The Hidden History of the Human Race. Small Press,* Winter.

Moore, Steve (1993) Review of *Forbidden Archeology. Fortean Times*, vol. 59, no. 72, p. 59.

Murray, Tim (1995) Review of *Forbidden Archeology. British Journal for the History of Science,* vol. 28, pp. 377-379.

Patou-Mathis, Marylène (1995) Review of F*orbidden Archeology. L'Anthropologie*, vol. 99(1), p. 159.

Rothstein, Mikael (1994) *Forbidden Archeology:* Religious Researchers Shake the Theory of Evolution. *Politiken,* January 1.

Salim-ur-rahman (1994). Spanner in the works. *The Friday Times*, April 21, 1994.

Schwartz, Hillel (1994) Earth Born, Sky Driven: A review of *The Sky in Mayan Literature* (edited by Anthony F. Aveni), *Beyond 1492: Encounters in Colonial North America* (by James Axtell), *Forbidden Archeology: The Hidden History of the Human Race* (by Michael A. Cremo and Richard L. Thompson), *Love and Theft: Blackface Minstrelsy and the American Working Class* (by Eric Lott), *Children of the Earth: Literature, Politics and Nationhood* (by Marc Shell)." *Journal of Unconventional History*, vol. 6(1), pp. 68-76.

Sleeper, Dick (1995) Notice of *The Hidden History of the Human Race.* Dick Sleeper Distribution, Inc. Catalog, Winter/Spring, p.1.
Stanley, Robert (1994) Review of *Forbidden Archeology. UNICUS* vol. 3, no. 4, p. 38.
Stein, Gordon (1994) Review of *Forbidden Archeology: The Hidden History of the Human Race. The American Rationalist*, vol. 39, no. 1, May/June, p. 110.
Stoczkowski, Wiktor (1995) Review of *Forbidden Archeology. L'Homme*, vol. 35, pp.173-174.
Swann, Ingo (1994) Review of *Forbidden Archeology. Fate*, January, p.106.
Tarzia, Wade (1994) *Forbidden Archaeology*: Antievolutionism Outside the Christian Arena. *Creation/Evolution*, 14(1): 13-25.
Tevis, Joni (1995) Krishna scientist debunks archaeology. *Florida Flambeau*, vol. 80, no. 122, Thursday, March 9, p. 3.
Watt, Jerry (1996) The Dawn of Humankind: A Review of *The Hidden History of the Human Race. Life In Action*, vol. 22, no. 5, September/October, pp. 169-171.
Welch, Bill (1995) *Forbidden Archeology. Stonewatch*, vol. 13, no. 2, Winter, p. 10.
Wilson, Katharina (1996) Review of *Forbidden Archeology. The Observer*, vol. 7, no. 3 July-September, pp. 5-6.
Wodak, Jo and David Oldroyd (1996). 'Vedic Creationism': A Further Twist to the Evolution Debate. *Social Studies of Science,* vol. 26, pp. 192-213.

Interview Transcripts

Cremo, Michael A. (1994) Interview by Jeffrey Mishlove. *Thinking Allowed.* Public Broadcasting System.
Cremo, Michael A. (1995) A Laura Lee Interview with Michael Cremo *Townsend Letter for Doctors*, May, pp. 60-70.
Cremo, Michael A.(1995) Interview by Laura Lee. In "*Forbidden Archaeology*" *Nexus* June-July, pp. 11-16.
Cremo, Michael A. (1996) Interview by Christina Zohs. In *Golden Thread*, vol. 1, no. 5, June, pp. 13-16.
Cremo, Michael A. (1996) *Mysterious Origins of Man.* National Broadcasting Company, 25 February [Rebroadcast 8 June].

Correspondence

Alford, Alan F. Letter to Michael Cremo, October 14, 1996.
Ariadne, Patricia A. Letter to Michael Cremo, April 15, 1995.
Armenta, Celine. Letter to Michael Cremo, February 7, 1995.

Barrett, John E. Letter to Michael Cremo, April 6, 1994.
Barron, David P. Letter to Michael Cremo, September 1, 1994.
Bauer, Henry H. Letter to Govardhan Hill Publishing, September 28, 1994.
BC Video (Bill Cote, Carol Cote, and John Cheshire). Letter to Michael Cremo and Richard Thompson, November 12, 1994.
BC Video (Bill Cote, Carol Cote, and John Cheshire). Letter to Michael Cremo, November 17, 1995.
Belderis, Ina. Letter to Michael Cremo, June 24, 1995.
Belderis, Ina. Letter to Michael Cremo, August 1, 1995.
Brown, Geoff. Letter to Bhaktivedanta Books Limited, 1995 (undated).
Burns, Jean. Letter to Michael Cremo, January 29, 1993.
Calais, Ron. Letter to Bhaktivedanta Institute, October 20, 1993.
Calais, Ron. Letter to Michael Cremo, July 10, 1994.
Cartmill, Matt. Letter to Michael Cremo, May 20, 1994.
Cartmill, Matt. Letter to Michael Cremo, August 30, 1994.
Cornet, Bruce. Letter to Dr. Bob and Zoh Hieronimus, August 4, 1994.
Cortner, Laura E. Letter to Michael Cremo, December 18, 1993.
Cortner, Laura E. Letter to Michael Cremo, February 7, 1995.
Cote, Bill. Letter to Michael Cremo, 1994 (undated).
Cote, Bill. Letter to Michael Cremo and Richard Thompson, January 23, 1994.
Cote, Bill. Letter to Michael Cremo, July 4, 1994.
Cote, Bill. Letter to Michael Cremo, January 2, 1996.
Cote, Bill. Letter to Michael Cremo, March 13, 1996.
Cote, Bill. Letter to Michael Cremo, August 26, 1996.
Cote, Bill. Letter to Michael Cremo, September 25, 1996.
Cremo, Michael A. Letter to Dr. Siegfried Scherer, October 20, 1992.
Cremo, Michael A. Letter to Dr. Siegfried Scherer, December 12, 1992.
Cremo, Michael A. Letter to Dr. Siegfried Scherer, December 17, 1992.
Cremo, Michael A. Letter to Dr. Bennetta Jules-Rosette, February 7, 1993.
Cremo, Michael A. Letter to Dr. Charles E. Oxnard, February 7, 1993.
Cremo, Michael A. Letter to Dr. David Bloor, February 7, 1993.
Cremo, Michael A. Letter to Dr. Michael Mulkay, February 7, 1993.
Cremo, Michael A. Letter to Dr. Paul Feyerabend, February 12, 1993.
Cremo, Michael A. Letter to Dr. Siegfried Scherer, March 10, 1993.
Cremo, Michael A. Letter to Dr. David Bloor, May 10, 1993.
Cremo, Michael A. Letter to Dr. Charles E. Oxnard, May 12, 1993.
Cremo, Michael A. Letter to Dr. Michael Mulkay, May 27, 1993.
Cremo, Micheal A. Letter to Dr. Duane T. Gish, July 7, 1993.
Cremo, Michael A. Letter to Jean Hunt, August 3, 1993.
Cremo, Michael A. Letter to Jean Hunt, September 5, 1993.
Cremo, Michael A. Letter to Dr. Charles E. Oxnard, September 18, 1993.

Cremo, Michael A. Letter to Dr. William W. Howells, September 25, 1993.
Cremo, Michael A. Letter to Bill Cote, September 27, 1993.
Cremo, Michael A. Letter to Jean Hunt, September 27, 1993.
Cremo, Michael A. Letter to Leland W. Patterson, October 1, 1993.
Cremo, Michael A. Letter to Dr. George F. Carter, October 9, 1993.
Cremo, Michael A. Letter to Jean Hunt, October 9, 1993.
Cremo, Michael A. Letter(s) to Bill Cote, December 30, 1993.
Cremo. Michael A. Letter to Dr. Thomas A. Dorman, December 30, 1993.
Cremo, Michael A. Letter to Dr. Virginia Steen-McIntyre, December 30, 1993.
Cremo, Michael A. Letter to Mikael Rothstein, December 30, 1993.
Cremo, Michael A. Letter to Dr. George F. Carter, December 31, 1993.
Cremo, Michael A. Letter to Dr. Matt Cartmill, April 16, 1994.
Cremo, Michael A. Letter to Dr. George F. Carter, April 19,1994.
Cremo, Michael A. Letter to Dr. Horst Friedrich, April 21, 1994.
Cremo, Michael A. Letter to Dr. Phillip E. Johnson, April 21, 1994.
Cremo, Michael A. Letter to Dr. Thomas A. Dorman, April 21, 1994.
Cremo, Michael A. Letter to Dr. Virginia Steen-McIntyre, April 21, 1994.
Cremo, Michael A. Letter to Mikael Rothstein, April 24, 1994.
Cremo, Michael A. Letter to Bill Cote, April 25, 1994.
Cremo, Michael A. Letter to John E. Barrett, April 25, 1994.
Cremo, Michael A. Letter to Horst Friedrich, June 27, 1994.
Cremo, Michael A. Letter to Dr. Matt Cartmill, June 27, 1994.
Cremo, Michael A. Letter to Dr. Thomas A. Dorman, June 27, 1994.
Cremo, Michael A. Letter to Bill Cote, June 28, 1994.
Cremo, Michael A. Letter to Dr. Charles E. Oxnard, July 18, 1994.
Cremo, Michael A. Letter to Ron Calais, July 28, 1994.
Cremo, Michael A. Letter to Dr. Jack Donahue, August 16, 1994.
Cremo, Michael A. Letter to Dr. Kenneth L. Feder, August 16, 1994.
Cremo, Michael A. Letter to Dr. Matt Cartmill, August 16, 1994.
Cremo, Michael A. Letter to Dr. Vance Holliday, August 16, 1994.
Cremo, Michael A. Letter to Dr. William W. Howells, August 16, 1994.
Cremo, Michael A. Letter to Dr. Bruce Cornet, August 18, 1994.
Cremo, Michael A. Letter to Dr. Bruce Cornet, August 18, 1994.
Cremo, Michael A. Letter to Dr. Virginia Steen-McIntyre, August 19, 1994.
Cremo, Michael A. Letter to David P. Barron, August 20, 1994.
Cremo, Michael A. Letter to Dr. Phillip E. Johnson, September 6, 1994.
Cremo, Michael A. Letter to Dr. Joseph B. Mahan, October 6, 1994.
Cremo, Michael A. Letter to Dr. Matt Cartmill, October 6, 1994.
Cremo, Michael A. Letter to Dr. Joseph R. Mahan, November 3, 1994.
Cremo, Michael A. Letter to Dr. A. Bowdoin Van Riper, November 4, 1994.
Cremo, Michael A. Letter to Dr. Henry H. Bauer, November 7, 1994.

Cremo, Michael A. Letter to Dr. Robert M. Schoch, November 7, 1994
Cremo, Michael A. Letter to Bill Cote, January 3, 1995.
Cremo, Michael A. Letter to Celine Armenta, January 3, 1995.
Cremo, Michael A. Letter to Dr. Joseph B. Mahan, January 3, 1995.
Cremo, Michael A. Letter to Dr. Robert M. Schoch, January 3, 1995.
Cremo, Michael A. Letter to Dr. Rupert Holms, January 3, 1995.
Cremo, Michael A. Letter to Dr. Virginia Steen-McIntyre, January 5, 1995.
Cremo, Michael A. Letter to René, January 10, 1995.
Cremo, Michael A. Letter to Dr. Rupert Holms, February 10, 1995.
Cremo, Michael A. Letter to Dr. A. Bowdoin Van Riper, February 12, 1995.
Cremo, Michael A. Letter to Dr. P. C. Kashyap, February 14, 1995.
Cremo, Michael A. Letter to Dr. Virginia Steen-McIntyre, February 15, 1995.
Cremo, Michael A. Letter to Dr. Virginia Steen-McIntyre, March 15, 1995.
Cremo, Michael A. Letter to Dr. P. C. Kashyap, March 17, 1995.
Cremo, Michael A. Letter to Celine Armenta, April 11, 1995.
Cremo, Michael A. Letter to Colin Wilson, April 20, 1995.
Cremo, Michael A. Letter to Dr. Virginia Steen-McIntyre, May 9, 1995.
Cremo, Michael A. Letter to Ron Calais, May 10, 1995.
Cremo, Michael A. Letter to Dr. Alexander Imich, May 20, 1995.
Cremo, Michael A. Letter to Geoff Olson, June 1995.
Cremo, Michael A. Letter to Dr. Virginia Steen-McIntyre, June 21, 1995.
Cremo, Michael A. Letter to Gene M. Phillips, June 21, 1995.
Cremo, Michael A. Letter to Dr. Alexander Imich, June 25, 1995.
Cremo, Michael A. Letter to Dr. David R. Oldroyd, June 25, 1995.
Cremo, Michael A. Letter to Dr. Tim Murray, June 25, 1995.
Cremo, Michael A. Letter to Ina Belderis, June 13, 1995.
Cremo, Michael A. Letter to Ina Belderis, July 16, 1995.
Cremo, Michael A. Letter to Ina Belderis, July 16, 1995.
Cremo, Michael A. Letter to Ingo Swann, September 27, 1995.
Cremo, Michael A. Letter to Dr. Virginia Steen-McIntyre, October 4, 1995.
Cremo, Michael A. Letter to Jim Deardorff, October 18, 1995.
Cremo, Michael A. Letter to Bill Cote, January 10, 1996.
Cremo, Michael A. Letter to Dr. Virginia Steen-McIntyre, January 10, 1996.
Cremo, Michael A. Letter to Ivan Mohoric, January 10, 1996.
Cremo, Michael A. Letter to Dr. Tim Murray, January 12, 1996.
Cremo, Michael A. Letter to John L. Cavanagh, March 26, 1996.
Cremo, Michael A. Letter to John L. Cavanagh, May 10, 1996.
Cremo, Michael A. Letter to Duane Franklet, May 15, 1996.
Cremo, Michael A. Letter to Gene M. Phillips, May 15, 1996.
Cremo, Michael A. Letter to Bill Cote, May 23, 1996.
Cremo, Michael A. Letter to Dr. Lonna Johnson, May 23, 1996.

Cremo, Michael A. Letter to Dr. Virginia Steen-McIntyre, May 23, 1996.
Cremo, Michael A. Letter to Eric Wojciechowski, May 23, 1996.
Cremo, Michael A. Letter to Colin Wilson, May 1996.
Cremo, Michael A. Letter to Bill Cote, August 1996.
Cremo, Michael A. Letter to Dr. Beverly Rubik, August 1996.
Cremo, Michael A. Letter to Dr. Beverly Rubik, August 1996.
Cremo, Michael A. Letter to Dr. Jenny Wade, August 1996.
Cremo, Michael A. Letter to Colin Wilson, August 4, 1996.
Cremo, Michael A. Letter to Jim Erjavec, August 4, 1996.
Cremo, Michael A. Letter to P. N. Oak, August 4, 1996.
Cremo, Michael A. Letter to Reverend Father Donald E. Nist, J. D., August 4, 1996.
Cremo, Michael A. Letter to Vash A. Rumph, August 4, 1996.
Cremo, Michael A. Letter to Dr. Wade Tarzia, August 22, 1996.
Cremo, Michael A. Letter to Katharina Wilson, August 24, 1996.
Cremo, Michael A. Letter to Katharina Wilson (undated).
Cremo, Michael A. Letter to Bill Cote, September 1996.
Cremo, Michael A. Letter to Dr. Beverly Rubik, September 1996.
Cremo, Michael A. Letter to Dr. Beverly Rubik, September 1996.
Cremo, Michael A. Letter to Dr. Beverly Rubik, September 1996.
Cremo, Michael A. Letter to Dr. Jenny Wade, September 1996.
Cremo, Michael A. Letter to Katharina Wilson, September 1996.
Cremo, Michael A. Letter to Eric Wojciechowski, September 15, 1996.
Cremo, Michael A. Letter to Dr. Beverly Rubik, October 1996.
Cremo, Michael A. Letter to Dr. Donald E. Tyler, October 1996.
Cremo, Michael A. Letter to Marshall Lee, October 23, 1996.
Cremo, Michael A. Letter to Wiktor Stoczkowski, February 2, 1997.
Cremo, Michael A. Letter to Jo Wodak and David Oldroyd, February 4, 1997.
Cremo, Michael A.. Letter to P. N. Oak, March 15, 1997.
Cremo, Michael A. Letter to Dr. Bradley T. Lepper, June 1, 1997.
Cremo, Michael A. Letter to Dr. Colin Groves, June 1, 1997.
Cremo, Michael A. Letter to Gordon Stein, June 1, 1997.
Deardorff, Jim. Letter to Michael Cremo, October 4, 1995.
Deardorff, Jim. Letter to Michael Cremo, November 3, 1995.
Doleman, William. Letter to B.C. Video, February 26, 1996.
Dorman, Thomas A. Letter to Michael Cremo, August 18, 1993.
Dorman, Thomas A. Letter to Michael Cremo, April 26, 1994.
Emerson, Deanna. Letter to Michael Cremo, June 1, 1996.
Erjavec, Jim. Letter to Dr. Bob Hieronimus, July 16, 1996.
Erjavec, Jim. Letter to Michael Cremo, August 21, 1996.
Feyerabend, Paul. Letter to Michael Cremo, April 26, 1993.

Franklet, Duane. Letter to Michael Cremo, August 9, 1996.
Friedrich, Horst. Letter to Michael Cremo, December 18, 1993.
Friedrich, Horst. Letter to Michael Cremo, April 28, 1994.
Friedrich, Horst. Letter to Michael Cremo, July 5, 1994.
Gish, Duane T. Letter to Michael Cremo, June 28, 1993.
Holms, Rupert. Letter to Michael Cremo, January 22, 1995.
Howells, William W. Letter to Michael Cremo, August 10, 1993.
Hunt, Jean. Letter to Michael Cremo, September 18, 1993.
Hunt, Jean. Letter to Michael Cremo. September 19, 1993.
Hunt, Jean. Letter to Dr. Barry and Reneé Fell, October 1, 1993.
Hunt, Jean. Letter to Michael Cremo, October 11, 1993.
Imich, Alexander. Letter to Michael Cremo, May 29, 1995.
James, R. Wayne. Letter to Michael Cremo, June 3, 1996.
Johnson, Lonna. Letter to Michael Cremo, 1996 (undated).
Johnson, Phillip E. Letter to Michael Cremo, November 30, 1992.
Johnson, Phillip E. Letter to Michael Cremo, May 2, 1994.
Jules-Rosette, Bennetta. Letter to Michael Cremo, April 14, 1993.
Kashyap, P. C. Letter to Michael Cremo, February 13, 1995.
Kashyap, P. C. Letter to Michael Cremo, February 15, 1997.
Leakey, Richard E. Letter to Michael Cremo, November 8, 1993.
Lee, Marshall. Letter to Michael Cremo, August 26, 1996.
Mahan, Joseph B. Letter Michael Cremo, October 11, 1994.
Mahan, Joseph B. Letter to Michael Cremo, November 17, 1994.
Mathrani, Madan M. Letter to Michael Cremo, July 8, 1995.
Meindl, Christopher. Letter to Michael Cremo and Richard Thompson, September 21, 1994.
Mohoric, Ivan. Letter to Michael Cremo, November 20th, 1995.
Mohoric, Ivan. Letter to Michael Cremo, December 1995.
Mooney, Vincent J. Jr. Letter to Michael Cremo, March 20, 1994.
Mulkay, Michael. Letter to Michael Cremo, May 18, 1993.
Nist, Donald E. Letter to Michael Cremo, July 3, 1996.
Nist, Donald E. Letter to Michael Cremo, August 8, 1996.
Oak, P.N. Letter to Michael Cremo, May 31, 1996.
Oak, P. N. Letter to Michael Cremo, February 17, 1997.
Olson, Geoff. Letter to Michael Cremo, January 12, 1995.
Olson, Geoff. Letter to Michael Cremo, June 4, 1995.
Oxnard, Charles E. Letter to Michael Cremo, May 19, 1993.
Patterson, L.W. Letter to Michael Cremo, August 10, 1993.
Phillips, Gene M. Letter to Michael Cremo, June 21, 1996.
Prasad, K. N. Letter to Michael Cremo, November 26, 1993.
René. Letter to Michael Cremo, November 26, 1995.
Rense, Jeffrey. Letter to Michael Cremo, March 1, 1996.

Rothstein, Mikael. Letter to Richard Thompson and Michael Cremo, November 11, 1993.
Rothstein, Mikael. Letter to Michael Cremo, February 2, 1994.
Rothstein, Mikael A. Letter to Michael Cremo, May 16, 1994.
Rubik, Beverly. Letter to Michael Cremo, June 26, 1996.
Rubik, Beverly. Letter to Michael Cremo, August 5, 1996.
Rubik, Beverly. Letter to Michael Cremo, August 25, 1996.
Rubik, Beverly. Letter to Michael Cremo, September 25, 1996.
Rubik, Beverly. Letter to Michael Cremo, September 28, 1996.
Rubik, Beverly. Letter to Michael Cremo, October 3, 1996.
Rubik, Beverly. Letter to Michael Cremo, October 15, 1996.
Rumph, Vash A. Letter to Michael Cremo, July 17, 1996.
Ruparelia, Kishor. Letter to *Back to Godhead* magazine, August 2, 1993.
Sassoon, George. Letter to Michael Cremo, December 31, 1993.
Schoch, Robert M. Letter to Michael Cremo, November 21, 1994.
Sprinkle, R. Leo. Letter to Michael Cremo, February 28, 1996.
Steen-McIntyre, Virginia. Letter to Steve Bernath, October 30, 1993.
Steen-McIntyre, Virginia. Letter to Steve Bernath, November 2, 1993.
Steen-McIntyre, Virginia. Letter to Michael Cremo, January 18, 1994.
Steen-McIntyre, Virginia. Letter to Michael Cremo, September 9, 1994.
Steen-McIntyre, Virginia. Letter to Michael Cremo, October 22, 1994.
Steen-McIntyre, Virginia. Letter to Celine Armenta, November 22, 1994.
Steen-McIntyre, Virginia. Letter(s) to Michael Cremo, November 23, 1994.
Steen-McIntyre, Virginia. Letter to Michael Cremo, January 19, 1995.
Steen-McIntyre, Virginia. Letter to Michael Cremo, March 1, 1995.
Steen-McIntyre, Virginia. Letter to Michael Cremo, April 11, 1995.
Steen-McIntyre, Virginia. Letter to Michael Cremo, May 1, 1995.
Steen-McIntyre, Virginia. Letter to Michael Cremo, May 12, 1995.
Steen-McIntyre, Virginia. Letter to Michael Cremo, June 30, 1995.
Steen-McIntyre, Virginia. Letter to Dr. Frank C. Hibben, June 30, 1995.
Steen-McIntyre, Virginia. Letter to Dr. Vaughn Bryant, August 13, 1995.
Steen-McIntyre, Virginia. Letter to Michael Cremo, September 3, 1995.
Steen-McIntyre, Virginia. Letter to Michael Cremo, March 11, 1996.
Swann, Ingo. Letter to Michael Cremo, August 13, 1995.
Swann, Ingo. Letter to Michael Cremo, October 28, 1995.
Tyler, Donald E. Letter to Michael Cremo, May 13, 1996.
Tyler, Donald E. Letter to Michael Cremo, October 10, 1996.
Van Riper, A. Bowdoin. Letter to Michael Cremo, January 5, 1995.
Wade, Jenny. Letter to Michael Cremo, May 31, 1996.
Wade, Jenny. Letter to Michael Cremo, August 9, 1996.
Wescott, Roger. Letter to Dr. Richard L. Thompson, March 8, 1993.
Wilson, Colin. Letter to Michael Cremo, March 31, 1995.

Wilson, Colin. Letter to Michael Cremo, October 5, 1995.
Wilson, Colin. Letter to Michael Cremo, May 31, 1996.
Wilson, Katharina. Letter to Michael Cremo, August 5, 1996.
Wilson, Katharina. Letter to Michael Cremo, August 29, 1996.
Wojciechowski, Eric. Letter to Michael Cremo. February 28, 1996.
Wojciechowski, Eric. Letter to Michael Cremo, August 12, 1996.

Internet Communication

B.C. Video. "Press Release." 21 February 1996. <http://www.bcvideo.com/bcvideo>.

B.C. Video. "Press Release" 4 March 1996. <http://www.bcvideo.com/bcvideo>.

Beetle, Chris. <afn15317@afn.org> Letters. 2 March 1996. <talk.origins, sci.archaeology, alt.archaeology>.

Beetle, Chris. <afn15317@afn.org> Letter. 28 May 1996. <sci.archaeology, alt.archaeology>.

Coles, John R. <JRC@austen.oit.umass.edu > Letter. 21 February 1996. <sci.skeptic,talk.origins>.

Coles, John R. <JRC@austen.oit.umass.edu > Letter. 26 February. 1997. <talk.skeptics>.

Cote, Bill. <bcvideo@interport.net> Letter. 31 March 1996. <sci.anthropology, talk.religion.newage, talk.origins, sci.archaeology>.

Cremo, Michael A. <105406.257@compuserve.com> Letter. 22 March 1996. <talk.origins, sci.archaeology, alt.archaeology, sci.anthropology, sci.anthropology.paleo>.

Cremo, Michael A. <105406.257@compuserve.com> Letter. 31 March 1996. Personal e-mail to Ferret [Duane Brocious].

Cremo, Michael A. <105406.257@compuserve.com> Letter. May 26, 1996. Personal e-mail to Jerry Watt.

Cremo, Michael A. <105406.257@compuserve.com> Letter. August 1996. Personal e-mail to Dr. Jere H. Lipps.

Etherman [Cote, R.]. <rcote@cs.uml.edu> Letter. 8 March 1996. <sci.skeptic, talk.origins>.

Etherman [Cote, R.]. <rcote@cs.uml.edu> Letter. 25 March 1996. <sci.archaeology, sci.archaeology.paleo, sci.anthropology, alt.anthropology, talk.origins, sci.skeptic>.

Etherman [Cote, R.]. <rcote@cs.uml.edu> Letter. 25 March 1996. <sci.archaeology, sci.archaeology.paleo, sci.anthropology, alt.anthropology, talk.origins, sci.skeptic>.

Etherman [Cote, R.]. <rcote@cs.uml.edu> Letters. 27 March 1996. <sci.archaeology, alt.archaeology, talk.origins, sci.anthropology, sci anthropology>.

Etherman [Cote, R.]. <rcote@cs.uml.edu> Letter. 8 April 1996. <sci.archaeology, alt.archaeology, talk.origins, sci.anthropology, sci.anthropology.paleo>.

Fair, Kenneth. <kjfair@midway.uchicago.edu> Letter. 30 May 1996. <talk.origins>.

Ferret [Duane Brocious]. <dnb105@psu.edu> Letter. 29 March 1996. <sci.archaeology, alt.archaeology, talk.origins, sci.anthropology, sci.anthropology.paleo>.

Foley, Jim. "NBC's *The Mysterious Origins of Man*" The Talk Origins Archive. 18 March 1996. <http://www.talkorigins.org/origins>.

Gall, Norman H. <gall@uwinnipeg.ca> Letter. 25 March 1996. <sci.archaeology, alt.archaeology, talk.origins, sci.anthropology, sci.anthropology.paleo, sci.skeptic>.

Heinrich, Paul <heinrich@intersurf.com> Letter. 28 February. 1996. <talk.origins, sci.archaeology, alt.archaeology>.

Heinrich, Paul. <heinrich@intersurf.com> Letter. 16 March 1996. <sci.archaeology>.

Heinrich, Paul. <heinrich@intersurf.com> Letter. 23 May 1996. <sci.archaeology, alt.archaeology>.

Isaak, Mark. <isaak@aurora.com> Letter. 28 February 1996. <talk.origins, sci.archaeology, alt.archaeology>.

Keith. <littlejo@news.demon.net> Letter. 8 June 1996. <talk.origins, alt.archaeology, sci.anthropology.paleo>.

Keith. <littlejo@news.demon.net> Letter. 10 June 1996. <talk.origins, alt.archaeology, sci.anthropology.paleo>.

Lippard, James J. <lippard@primenet.com > Letter. 12 March 1996. <sci.skeptic, talk.origins>.

Lipps, Jere H. <jlipps@ucmpl.Berkeley.EDU> Letter. 27 February 1996. <alt.archaeology, sci.archeology, talk.origins>.

Lipps, Jere H. <jlipps@ucmpl.Berkeley.EDU> 15 March 1996. Distribution list.

Lipps, Jere H. <jlipps@ucmpl.Berkeley.EDU> 1 May 1996. Distribution list.

Lipps, Jere H. <jlipps@ucmpl.Berkeley.EDU> "Rebroadcast Origins." 30 May 1996. Distribution list.

Lipps, Jere H. <jlipps@ucmpl.Berkeley.EDU> Letter. 30 May 1996. <sci.bio.paleontology, sci.skeptic, sci.anthropology.paleo, talk.origins>.

Lipps, Jere H. <jlipps@ucmpl.Berkeley.EDU> "NBC & the FCC." 21 June 1996. Distribution list.

Marti, Orville G. Jr. <omarti@tiflon.cpes.peachnet.edu> Letter. 18 March 1996. <paranet.ufo>.

Matthusen, August. <matthuse@ix.netcom.com> "Re: NBC To REBroadcast 'Mysterious Origins...' Amid Controvery." 31 May 1996. <talk.origins, talk.religion.newage, sci.anthropology, sci.archaeology>.

Miller, Dean T. <dtmiller@dsmnet.com> Letter. 1 April 1996. <sci.archaeology, alt.archaeology, talk.origins, sci.anthropology, sci.anthropology.paleo>.

Mr. E. <jackechs@erols.com> Letter. 27 March 1996. <talk.origins, alt.archeology, sci.archeology>.

NBC. "Investigate 'The Mysterious Origins of Man.'" NBC Press Release. 21 February 1996. <http://www.nbc.com>.

NBC. "Controversy Surrounds 'The Mysterious Origins of Man.' University Profs Want Special Banned from the Airwaves." NBC Press Release. 29 May 1996. <http://www.nbc.com>.

Nicholls, Phil. <pnich@globalone.net> Letter. 25 March 1996. <sci.archaeology, alt.archaeology, talk.origins, sci.anthropology, sci.anthropology.paleo>

Oldridge, Dave. <doldridg@ra.isisnet.com> Letter. 14 March 1996. <sci.archaeology, sci.anthropology, sci.anthropology.paleo>.

Steen-McIntyre, Virginia. <dub.ent@ix.netcom.com> Letter. 2 March 1996. <alt.archaeology, sci.archaeology>.

Sullivan, Mike. < > Letter. 26 February 1997. <talk.origins, alt.archeology, sci.archeology>.

Taylor, Alister. <105042.3447@compuserve.com> "'Forbidden Archeology' the front line challenge." 22 February 1996. <Free.Forum@com.bbt.se>.

Thompson, Richard L. and Chris Beetle. <ghi@nerdc.ufl.edu> Letter. 18 June 1996. <alt.archaeology>.

Thompson, Richard L. Chris Beetle. <ghi@nerdc.ufl.edu> Letter. 26 June 1996. <talk.origins, alt.archaeology, sci.anthropology.paleo>.

Tresman, Ian. <ianTresman@easynet.co.uk> Letter. 27 March 1996. <sci.archaeology, alt.archaeology, talk.origins, sci.anthropology, sci.anthropology.paleo>.

Wadkins, Randy. <rwadkins@cbmse.nrl.navy.mil> Letter. 22 March 1996. <sci.archaeology, alt.archaeology, talk.origins, sci.anthropology, sci.anthropology.paleo>.

Wagner, C. Marc. <mwagner@indiana.edu> Letter. 30 May 1996. <sci.bio.paleontology, sci.skeptic, sci.anthropology.paleo, talk.origins>.

Watt, Jerry. <jwwatt@pop.erols.com> "'The Hidden History of the Human Race'—A Review." 28 May 1996. <alt.archeology>.

Watt, Jerry. <jwwatt@pop.erols.com> Letter. 10 June 1996. <talk.origins, alt.archaeology, sci.anthropology.paleo>.

Watt, Jerry. <jwwatt@pop.erols.com> Letter. 10 June 1996. <talk.origins, alt.archaeology, sci.anthropology.paleo>.

INDEX

Book Order Form

☎ Telephone orders: Call 1-800-HIDDEN1 (1-800-443-3361).
Have your VISA or MasterCard ready.

FAX orders: 310-837-3363

✉ Postal orders: Torchlight Publishing 3046 Oakhurst Avenue,
Los Angeles, CA 90034 USA

World Wide Web: www.torchlight.com

Please send the following:

○ ***Forbidden Archeology: The Hidden History of the Human Race***
952 pages, 141 illustrations, 24 tables, Hardback $44.95 (Canada $64.95)

○ ***Forbidden Archeology's Impact: How A Controversial New Book Shocked The Scientific Community And Became An Underground Classic***
592 pages, Hardback $35.00 (Canada $50.00)

○ ***The Hidden History of the Human Race***
352 pages, 69 illustrations, abridged version, Hardback $22.95 (Canada $31.95),
Softback $15.95 (Canada $22.95)

○ **Please send me more info on other books published by Torchlight Publishing.**

Company: ______________________________

Name: ______________________________

Address: ______________________________

City:____________________ State________ Zip__________

(I understand that I may return any books for a full refund — for any reason, no questions asked.)

Payment:

○ Check/money order enclosed ○ VISA ○ MasterCard

Card Number: ______________________________

Name on Card: ____________________ Exp. date_______

Shipping and Handling: **Sales Tax: CA Residents add 8.25%**

Forbidden Archeology and ***Forbidden Archeology's Impact***
Book rate USA: $4.00 for first book and $2.50 for each additional book.
Canada: $5.00 for first book and $3.50 for each additional book.
Foreign countries: $6.00 for first book and $4.00 for each additional book.

Hidden History of the Human Race
Book rate: USA $3.00 for the first book and $2.00 for each additional book.
Canada: $3.00 for first book and $2.00 for each additional book.
Foreign countries: $4.00 for first book and $3.00 for each additional book.

Surface shipping may take 3-4 weeks. Foreign orders please allow 6-8 weeks for delivery.